3rd International Conference on Public Relations and Social Sciences (ICPRSS 2023)

AA003595

SHS Web of Conferences Volume 178 (2023)

Xi'an, China
15-17 September 2023

Editors:

I.A. Khan
Y. Chen
C. Shen
J. Rak

ISBN: 978-1-7138-8351-7

Printed from e-media with permission by:

Curran Associates, Inc.
57 Morehouse Lane
Red Hook, NY 12571

Some format issues inherent in the e-media version may also appear in this print version.

For additional information, please contact EDP Sciences – Web of Conferences at the address below.

EDP Sciences – Web of Conferences
17, Avenue du Hoggar
Parc d'Activité de Courtabœuf
BP 112
F-91944 Les Ulis Cedex A
France

Phone: +33 (0) 1 69 18 75 75
Fax: +33 (0) 1 69 28 84 91

contact-edps@webofconferences.org

Additional copies of this publication are available from:

Curran Associates, Inc.
57 Morehouse Lane
Red Hook, NY 12571 USA
Phone: 845-758-0400
Fax: 845-758-2633
Email: curran@proceedings.com
Web: www.proceedings.com

TABLE OF CONTENTS

SOCIAL GOVERNANCE AND PUBLIC RELATIONS PRACTICE

CORPORATE MARKETING STRATEGY AND INNOVATION DEVELOPMENT

Preface

Public relations refer to a series of public activities for an organization to improve the relationship with the public, promote the public's understanding and support of the organization, and achieve the purpose of establishing a good organizational image and promoting commodity sales. Its original intention is that social organizations, collectives or individuals must establish good relations with various internal and external publics around them. However, there are some global problems facing public relations, which could in some way be solved through international communication and dialogue to form a consensus and formulate international standards.

In order to bring in the latest progress in theory, technology and application in the fields of Public Relations and Social Sciences at home and abroad in recent years, and show the latest achievements, 2023 3rd International Conference on Public Relations and Social Sciences (ICPRSS 2023) was held in Xi'an, China from September 15th to 17th, 2023 via virtual form. It was attended by about 80 participants from different countries and regions.

The Conference presented an outstanding program of papers covering the most recent advances in public relations and social sciences, including Public Relations Psychology, Crisis Public Relations, Business Communication, Integrated Marketing Communication, etc. The papers in this Proceedings published by SHS Web of Conferences represent a collection of the invited speeches and other contributions.

At this meeting we had a comprehensive overview of this fascinating field and of future scenarios thanks to the participation of distinguished delegates around the globe. Assoc. Prof. Padma Priya Pillai from Sunway University, Malaysia performed a keynote speech and shared with us her own research experience and professional view with the report Impact of Lecturer's Personal and Impersonal Facebook Account Access on Student-Lecturer Communication Satisfaction. Her study aimed to examine student-lecturer communication satisfaction in personal and impersonal lecturers' Facebook accounts and define the difference between the pre-test and post-test of Student-lecturer Communication Satisfaction (SCS) in personal and impersonal lecturers' Facebook accounts. This study also examined the hypotheses regarding differences in student-lecturer communication satisfaction with feedback on personal versus impersonal Facebook accounts. By clarifying the association of lecturers' self-disclosure on Facebook with students-lecturers' communication satisfaction, this research contributes to the growing body of knowledge on using Facebook in instruction.

We would like to thank all the participants, especially those who contributed speeches, posters and manuscripts, for making ICPRSS 2023 such an exciting and memorable conference. We thank the Technical Program Committee for putting together so outstanding a program. Finally, we acknowledge the members of SHS Web of Conferences for their tireless efforts in the preparation of this volume, and all our colleagues whose friendly and efficient service contributed much to the success of the Conference.

The Committee of ICPRSS 2023

Committee Member

Conference General Chair
Prof. Shakila Yacob, University of Malaya, Malaysia

Technical Program Committee Chair
Prof. Mohd Fauzi bin Sedon, Universiti Pendidikan Sultan Idris, Malaysia
Assoc. Prof. Deepanjali Mishra, KIIT University, India

Publication Chair
Prof. Dr. Intakhab Alam Khan, King Abdulaziz university, Saudi Arabia
Prof. Youbin Chen, Sichuan University of Science and Engineering, China
Prof. Chaoqun Shen, Sanda University, China
Assoc. Prof. Joanna Rak, Adam Mickiewicz University, Poland

Local Organizing Chair
Prof. Yongjun Feng, Shaanxi Normal University, China
Prof. Zhengxi Yang, South China Agricultural University, China

Technical Program Committee
Prof. Madya Dr. Mohd Zakhiri bin Md. Nor, UUM College of Law, Malaysia
Prof. Baoying Zeng, Shaoguan University School of Law, China
Assoc. Prof. Fangping Zhou, University of Malaya (UM), Malaysia
Assoc. Prof. Hanping Zhu, South China Agricultural University, China
Assoc. Prof. Xiaoming Ma, South China Agricultural University, China
Assoc. Prof. Yunhong Zhang, South China Agricultural University, China
Asst. Prof. Gengeswari Krishnapillai, Universiti Tunku Abdul Rahman, Malaysia
Dr. Xia Fan, Wuhan Institute of Technology, China
Dr. Xiaodan Liu, Guangdong University of Foreign Studies, China
Dr. Seriaznita binti Haji Mat Said, Universiti Teknologi Malaysia, Malaysia
Dr. Gengeswari Krishnapillai, Universiti Tunku Abdul Rahman, Malaysia

Statement of Peer review

In submitting conference proceedings to *Web of Conferences*, the editors of the proceedings certify to the Publisher that

1. They adhere to its **Policy on Publishing Integrity** in order to safeguard good scientific practice in publishing.
2. All articles have been subjected to peer review administered by the proceedings editors.
3. Reviews have been conducted by expert referees, who have been requested to provide unbiased and constructive comments aimed, whenever possible, at improving the work.
4. Proceedings editors have taken all reasonable steps to ensure the quality of the materials they publish and their decision to accept or reject a paper for publication has been based only on the merits of the work and the relevance to the journal.

Title, date and place of the conference

2023 3rd International Conference on Public Relations and Social Sciences (ICPRSS 2023)

September 15-17, 2023
Xi'an, China

Proceedings editor(s):

Prof. Dr. Intakhab Alam Khan, King Abdulaziz university, Saudi Arabia
Prof. Youbin Chen, Sichuan University of Science and Engineering, China
Prof. Chaoqun Shen, Sanda University, China
Assoc. Prof. Joanna Rak, Adam Mickiewicz University, Poznań, Poland

Date and editor's signature

Intakhab Alam Khan
Youbin Chen
Chaoqun Shen
Joanna Rak

The Dilemmas of Universal Value and the Possible Solutions

Jiayi Wan*

School of International Relations, Beijing International Studies University, Nanli Community, Dingfu Town, Chaoyang District, Beijing, PRC

Abstract: The concept "universal value", as one of the most fundamental concepts in Western political discourses, occupies an important place in present international politics. These concepts originated from specific historical contexts and can only be applied to these specific historical contexts. These concepts also play a role in the formation of the modern nation state. However, it should be noted that the origins of these concepts are one-sided and derived from some territorial religious thoughts or the cultural movements of certain regions. With the evolution of the international situation, the core concepts are distorted and fixed, carrying intense political purposes. The transformation of ethnic spirits, the increasingly intense international relationships, and the exchanges of world cultures, all of these factors get in the way of the development of these concepts. Present universal values can not satisfy the needs of international society and even bring conflicts and wars to international society in some extreme cases. All of these phenomena raise considerable concerns about the degree of stability in the international relationship structure and the benign development of globalization. Developing universal values is a challenging task, but it is also likely to be the key to unlocking a whole new international situation.

This article will begin with the analysis of the origin of universal value, its development process, and its practical application, following with further research on the actual cases to investigate which dilemma the universal values are stuck in and the concrete reasons behind it. From there it is expected to find the solutions or paths to these problems. The definition of universal value needs to be further clarified, and the content of universal value needs to be carefully considered and set. It is also important to apply all of these discussions to political governance properly. At the same time, it should be mentioned that exploring the concept of universal value should not only pay attention to its own values, but also to the way it is manifested in global orders. Only in this way can the stability of the international relationships be strengthened through the "universal value", which will eventually create good conditions for the realization of the overall development of humankind.

1.Introduction

The concept "universal value" is familiar in present international governance. These universal values played a role in the formation of the modern nation-state, which can be considered a crystallization of human wisdom and a manifestation of social progress. At the same time, universal value is one of the core concepts in Western political discourses. As its name implies, it is considered to be a value that should be accepted by people worldwide, no matter which ethnicity, nationality or race they are from. Ever since the Renaissance and the Enlightenment, the notions contained in universal values have gradually emerged and have been refined along with the development of capitalism. Especially after World War II, its propagation range expanded and its influence increased, which plays an important role in the formation, and evolution of international relationship structure. The practices of universal value triggered considerable political unrest and chaos in the regions of the former Soviet Union and Central Asia and Eastern Europe, which directly led to the troubles of their developments. After several centuries' evolution and transmission, universal value has gained a certain base of influence. At this time, for the better development of international society, it is urgent to explore the dilemmas of universal values and the causes behind them, and to find ways to improve these universal values.

2.The definition of universal value and its development

The concept of universal value is derived from the Western cultural traditions. "Universal" is the literal translation of the Latin word "oecumenicus", which originated from Greek. Since the Middle Ages, it has been used as "ecumenical" by the Christian Church to call people to spread the Gospel of God with an unprecedented sense of mission. Nowadays, it is more and more prevalent to believe in freedom and diversified cultures. The Gospel no longer means converting to God. Rather, it represents the universal values of democracy,

* Corresponding author: jiayiwan0206@163.com

freedom, equality and fraternity. Many Western countries consider these notions to be supra-class, supra-national, and supra-temporal because they are core ideas for promoting the humanization of the world.

In the first place, I think it is necessary to clarify the definition of universal value. The core idea of universal value is the theory of human nature. Human nature refers to the process by which humans detached from animal nature by labour and gradually became mature to create an entirely new self. Therefore, human nature is actually in the constant process of changing and developing along with the evolution of society. The rapid development of universal value is closely related to the advancement of capitalist society. The innovation of science and technology and the need of capitalist commodity economy urged people to exert human self-interest as well as positive individuality, which significantly improved the living standard of human beings worldwide. It can be argued that capitalism is closely in line with human nature at this stage.

However, will human nature stay at this stage forever? From my perspective, human beings will ultimately embark on the path to free and comprehensive development. What human beings chase for will no longer be independent individual development, but the joint development of individual and society, human beings and the environment. Nowadays, with the strengthening of the political cooperation among nations, the economic integration both regionally and globally, and the proposal of environmental governance on the agenda of international conferences, all these phenomena indicate the fact that the international society is ready to move on this path. If we still insist on the human nature as the universal value at present, it will in turn hinder the possibility for human beings to advance to a better stage.

Then I would like to talk about the meanings of the key ideas in universal values.

The idea of "liberty" and "democracy" first embodied in the birth of Athenian democracy and Roman law. With the later establishment and development of the Western capitalist political system, it gradually evolved into modern democratic political thought. The United States is considered to be one of the active advocates of the modern ideological movement of liberty and democracy. Separately, the idea of liberty is based on anarchy, which can be manifested in the loose relationship between the federal government and the states. When the problems of this kind of absolutism were discovered, it gave rise to the emergence of monarchy and strengthened to grant citizens individual liberty. The practical effect of the idea of "democracy" is to restrict the power of government. In summary, the idea of liberty and democracy aims to grant people liberty and restrict government power.

The idea of "human rights" and "contract spirit" belongs to modern Western power politics. Its emergence was relative to the unique religious beliefs and geographical environment of the West. The birth of "human rights" had something to do with the resistance against clerical rights, and perhaps the most representative idea is power politics. By taking the abstract human rights as the basis of national public power, power politics claims that nation or government resulted from the consent or entrustment of people. Likewise, there are numerous small islands, rivers, and seas existing in Europe, which contribute to the prosperity of shipping trades. The constant interactions between these regions gave rise to the establishment of contact spirit, which is indeed similar to the honesty spirit. The contact spirit extending to the political area refers not only to the requirement for individuals to conform to contact but also to the public power. Laws and agreements are also needed to ensure the fulfillment of the contract. It can be argued that the premise of the birth of contact spirit comes from the rule of law. Citizens and governments also need to adhere to the law and conform to contracts.

The Renaissance awakened people's secular morality and the Enlightenment further disrupted the authority of religious discourses, which lays the foundation for the spirits of equality, fraternity, and toleration. At the same time, the establishment of capitalist production relationships and the development of the market economy fueled the bourgeois revolution. The bourgeois thinkers proposed the values of democracy, liberty, equality, the rule of law, and human rights. In certain colonial periods, colonists formed a new set of value systems which is fit for secular society by elevating the morality and ethics in the religion. These values, along with the Christianity, entered into colonies and became the soft power to diverge colonial nations.

Nowadays, the first words that come to people's minds when talking about the values of the Western nations may be liberty, democracy, human rights and contact spirits. The practices of these words help explore and construct new political systems, which accords to the trend to liberate human nature and promote the development of human society.[1] However, these values came to transform into the explicit expressions of capitalist ideology. With the development of history and the change of international situation, these dominant values gradually deviated from the initial cores and the tendency to be hegemonic appeared. The advancement of internet technology provides an efficient tool for Western developed countries to boost the output of these values, from which they realized the permeation of these values and mastered the dominance of value identification. Driven by the purposes of colony and hegemony, some countries impose their values on an improper political environment by adopting forceful practical means, which leads to the unrest of international order.[2] Blindly advocating the socalled western "freedom, democracy, equality and human rights" and other values, ignoring the specificity and uniqueness of their own development,[3] this phenomenon is reflective.

3.Analysis of the practices of universal values

In order to improve the international status and master the dominance of discourses, some Western countries choose to output these values by force without consideration of the diversity and differences of the political environments. Further, they use distorted values to judge other countries, which gives rise to the emergence of double standards. It has been one of the most important reasons which impede the benign development of universal values.

In 2020, the COVID-19 pandemic spread rapidly throughout the world. It was the time the world was supposed to strengthen cooperation and face the threat of viruses together. However, the political judgement under the guise of universal values ruined the cooperation process. The pandemic not only broke out in our lives but also in politics. Many Western countries strongly criticize the lock-down policy of the Chinese government. They claim that the lock-down policy is a manifestation of human rights abuse and restriction of personal liberty. They use so-called universal values to criticize China, which deepens the antagonism of ideologies. The key of this problem arises from the disparity of population between China and Western countries and the different national conditions. The introduction of any policy needs to be balanced among multiple interests. It is hard to guarantee more people's rights to live if the Chinese government only considers guaranteeing the rights for people to move freely. The later herd immunity strategy was taken by the Western countries. There is no need to wear masks and to vaccinate and people gain the "liberty" and "freedom". However, the effects of these measures which aim to prevent and control the disease are not as ideal as originally expected. The number of cases is on the rise within the countries and the mortality rate increases as well. If China adopts the same strategy, the outcome can be terrible. As for the Xinjiang cotton incident, it is considered to be an economic contention under the guise of politics and a discourse contest of ideologies. The Western countries label China concentration camp style forced labor and criticize the so-called oppressive production by using the thoughts of liberty and freedom. It is not only a crackdown on Xinjiang cotton which occupies 20% of the global cotton production, but also aggravates the antagonism of ideologies, which puts China in an awkward position internationally. Some Western countries take this opportunity to propagate the universal values dominated by the West, showing the image of so-called "justice" to the world. In fact, the farmers of Chinese Xinjiang cotton have desirable salaries in picking seasons and their rights are well guaranteed. The production of cotton has already entered a highly mechanized stage. These facts should not be neglected. By contrast, the Western textile industries have done more oppressive things in history. During the enclosure movement period, farmers were expelled and forced to sell their lands at a lower price, and the black slaves were used as tools to pick cotton for the sake of trade development. All of these incidents related the textile industry to the images of violent expelling, cotton industry, slavery, and oppression. In this incident, the neglect of practical evidence and distortion of facts as what McCarthyism did before are presisely the manifestation of double standards.

Compared to public opinion warfare, the influences of real wars tend to be more brutal and apparent. The Russia-Ukraine conflict continues to this day. It can be argued that although the battlefield is located in Ukraine, the battlefront extends to the world as the continuation of Cold War between the East and the West. Many Western countries object to the invasion conduct of Russia, consider Russia as a lack of contact spirit and a violation of the sovereignty of other counties. It is reasonable to make this judgement on its conduct. However, the main cause of the war arises from the expansion of the North Atlantic Treaty Organization to the east. In other words, the Western countries themselves should take the blame for the causes of this war. Despite the North Atlantic Treaty Organization claiming to be an international military organization which aims to realize defensive cooperation, the applications for accession of Russia were rejected four times. They believe the accession of Russia will destroy the cultural and theoretical homogeneity of Europe, which in fact contradicts the universal values dominated by the West. The direction of the North Atlantic Treaty Organization is obvious. Therefore, if we discuss the contact spirit, the fact is actually some Western countries disobey the commitment that the North Atlantic Treaty Organization will never expand to the East at first. In this conflict, some Western countries behind Ukraine started to launch sanctions against Russia under the pretext of maintaining world peace after breaking up the contact spirit. Further, when Russia starts to carry out a series of intensive measures against sanctions, some Western countries use the media to propaganda that Russia disobeys the contact spirit to claim the moral high ground. Undoubtedly, the conduct of Russia violates the peace principle, but the causes of this situation are complicated. Some Western countries also play a role. Some of the America-led Western countries repeatedly launched color revolutions, controlled puppet governments and violated the state sovereignty through thought permeation, which is a manifestation of the destruction of contact spirit. Numerous developing countries are still trying to revive from the "aftershock" of revolutions. Some Western countries use the idea of contact spirit and liberation of human rights as their propaganda argument in self-defense and external expansion. They argue that it is their destiny to do this, but condemn and sanction other countries which hold different ideologies from the West. It is unconvincing.

In summary, the universal values dominated by the Western countries play a great role in the liberation of human rights. However, along with the changes of situation, these values gradually become the objects to be criticized. The main cause of this criticism is not from the subjective prejudice from other countries and regions, but from the measures conducted by the Western countries themselves. They distort the values, output them by force, and establish double standards, regard bourgeois as the inventor, monopoly, exporter and determiner of the universal values such as democracy, freedom, fraternity and human rights,[4] which reduces the influence and recognition of these values.

4.The developing dilemmas of the universal values

4.1 The universal values become the "weapons" of politics

As mentioned above, the emergence of universal values arises from the dissatisfaction of reality oppression and the purpose of it is to liberate human beings and to promote the comprehensive development of human beings. However, it gradually became the explicit expression and implementation means of the capitalist system in recent years.[5]

After World War II, universal values gradually became the tool for the West to promote so-called "democracy" and "liberty", deviating from the principle of "widely accepted and adapted to different cultures, religions and societies". It is not beneficial to improve human welfare or promote harmonious coexistence in the world. With the emergence of cultural relativism and ideological extremism and the conflicts generated in the process of formulating political agendas and gaining economic profits, universal values have become the "weapons" for some countries to reach their own purposes rather than contributing to human welfare. This leads to the indifference or distortion of these values for individual, community and even government, which further destroys the practices or policies of the basic principles.

After the collapse of colonial systems, modern ethnic nation-states started to set up. The Western countries try to establish puppet governments in some developing countries in order to force these supported countries to introduce policies that benefit them. They give full support to the policies based on universal values. At the same time, they favor repressive regimes and oppressive governments that conform to their intentions. The basic human rights of the supported governments are violated instead.

During the process of carrying out policies, they do not conduct strict supervision on the implementation of the values. The conduct of violating democratic principles recur. In spite of the electoral corruptions, restrictions on voting and lack of transparent accountability systems occurring in supported countries, the Western countries keep ignoring selectively. As long as it is in accord with their interest, they do not stop it. Instead, it leads to society unrest and political chaos, and it is the people of the supported countries who suffer all the consequences.[6]

It would be misleading to suggest that the consensus that now exists on the importance of advancing democracy overrides all other priorities in foreign affairs,[7] the consequence of regarding the universal values as "weapons" can be severe. Once the transmission has threatened the survival of countries and nations, it will come back to threaten itself. Some organizations make hate speeches and even conduct terrorism and violence based on the identities, beliefs and backgrounds of individuals or communities, which impedes the stability and advancement of both domestic and international society. In some severe cases, the rights of ordinary people to live are threatened. This conduct gives rise to splits and conflicts worldwide, and it is important to recognize and deal with the deviation in order to maintain and promote the universal values of human welfare.

4.2 The value-oriented diplomacy impedes international communications and human developments

The value-oriented diplomacy is a new diplomatic means based on the recognition of the modern nation-state and the emergence of nationalism. With the change of international situation and national strength, the value-oriented diplomacy between countries becomes gradually active in international politics. After World War II, the United States beaome the center of the world. Economically, they tried to peg the dollar to gold but collapsed at last. Politically, they tried to peg the universal values to political regimes and have gained recognition from many developed countries and developing countries until today. But at the same time, different values such as values from China started to emerge in recent centuries. The Western individualism value and the Eastern collectivism value collide. However, the collision did not generate a better value and regime for the development of human beings. Rather, it destroyed the relationships between countries and aggravated the unrest in international society.

At present, countries still lack consensus on the content of universal values. Different countries have different histories, cultures, and political backgrounds, which influence the comprehension of values and matters of priority.[8] The lack of consensus will disrupt the efficiency of universal value-oriented diplomacy and lead to diplomatic tension. However, the prevalent value-oriented diplomacy not only violates state sovereignty but also manifests the prejudices of the Western countries.

According to the past diplomatic histories between the centered Western countries, it is likely that universal value-oriented diplomacy violates state sovereignty. The values advocated by them are mostly based on the liberty and democracy principles in the contest of capitalist society. But they used to force other countries to accept their values without consideration to the thought and cultural differences in different regions. At the same time, they take tough measures to express their diplomacy and values, which often places their strategic interest on the recognition of gaining values. It can be regarded as the interference in the internal affairs of other countries, which ruins the political regimes and social structures of outputted countries. Therefore, more and more people tend to believe this diplomatic means is actually the tool for the Western countries to advance geopolitics rather than concerning the true universal values.

At the same time, some problems which violate universal values also exist in the West, for example, the remaining problems of human rights. Discrimination and inequality also exist in their societies. The phenomenon of discriminating against particular individuals or communities based on race, sex, religion or social economic status, which leads further to the unequal treatment and chances to gain resources, does not disappear until today. This unconformity destructs the reliability and degrees of recognition, which causes the strong opposition and resistance from target countries where values spread aims for. From the perspective of the development of human civilization, the repulsion and marginalization of non-Western values and views can lead to the perpetuation of cultural hierarchy under the background of prevalent value-oriented diplomacy.

4.3 The universal values expand to various fields and impede the advancement of the developing countries

In 1841, the German thinker Friedrich List proposed a theory called "kicking away the ladder", which means the wisest way for a country is to kick away the ladder by which it climbed up when it has attained the summit of greatness that other countries cannot compete in free competition.[9] The so-called "liberty" value has expanded to economic dimension. When the Western countries were asked for their experiences in developing economies, their answers were free trade and opening markets, and they tried to convince the developing countries by giving practical evidence. However, it was just a way for the Western countries to obtain more economic interests. From the development history of the Western economy, it can be argued that the core factors for the rapid economic development arise from the support for the enterprises of specific fields and the implementation of protectionism policies from the governments at the early stage of market economy. When they reached a certain height, they chose to kick away the ladder. The Western countries weakened the protective effect of the government and proceeded trade exchange with the developing countries strongly, which increased the capital accumulation in transaction after transaction and the consequences were taken by the former.[10] At the same time, it seems like the economic support and product aid of some Western countries towards the developing countries have solved some urgent crises, but in the long term they actually disrupted their fragile industry structures.

Likewise, it happens in the aspect of formulating political agendas. The already favored Western countries initiate the international issues and agendas to realize their political strategies by helping the developing countries develop democracy and liberate human rights. Human rights are critical, though. What people really resist is the violation of sovereignty on the pretext of human rights. The implementation of values is huge obstacles for the developing countries to advance.

In summary, regarding the effects of the universal values it takes to the world, scholars worldwide have different opinions. The advocates of universal values claim that the universal values provide a collective structure for human rights promotion, social justice and global cooperation. Some values, such as the dignity of human beings, equality, liberty, and the right to live, they all deserve universal protection in order to promote human welfare and advancement. It can be argued that these values are based on the collective human nature which is shared by the people from the world, and they can be the basis of maintaining the justice and harmony of global society. On the other hand, the critics of universal values indicate and strengthen the challenges and potential threats of the universal values. The values deeply rooted in the cultural, historical and religious backgrounds and the attempt to implement singular values in the world likely result in cultural imperialism, marginalization and impeding cultural diversity. They stress the importance of cultural relativism, which admits the unique opinions and traditions derived from different social backgrounds and allows them to define their own values in specific backgrounds.

5.The solutions to the universal values in future development

It can be seen that the reason why the universal values are stuck into the dilemmas results from the contradictory to find a balance between promoting the universal values of protecting human rights and ensuring the cultural diversity and autonomy. It is challenging. Furthermore, it is certain to admit and respect cultural differences, but it is also important to deal with conduct which seriously violates human rights and moral principles in specific cultural contexts. In order to break the dilemmas and better apply the universal values to practices, we need to achieve the balance. And here are some possible solutions in the following:

(1) Understand the universal values correctly and abandon the dominant psychology.

It is necessary to take a humble attitude to the development of the universal value itself. Recognition is a dynamic process of constant change, and it should be constantly evaluated and adjusted according to the change of international situations. There are collective values such as equity, justice, and liberty existing, which can transcend ethnicities, races, borders and beliefs.[11] Nonetheless, some values are not accepted or recognized by all the cultures due to the different historical backgrounds. Imposing the values by force can be ineffective and cause rebellion and revenge from dominated countries.

(2) Carry out transcultural exchange and establish an international dialogue mechanism.

It is a critical aspect to admit and respect cultural diversity for better development of universal values. It is better to create an atmosphere where different points of view can be heard and let the collective wisdom of human beings drive the benign development of universal values rather than imposing a set of specific values.[12] It is also important to maintain the constant transcultural dialogues among different countries, regions, communities and interested parties in order to refine the universal values. We need to reach the balance between promoting the universal adaptable values and allowing cultural differences which accord to the basic human rights. The establishment of mechanisms, the promotion of cultural understanding, and the cultivation of sympathy create platforms to share and respect different views for public discussions.

(3) Break up the procedure dilemma and strengthen the legality of the universal values.

According to the definition of universal values, its contents should be cross-class, cross-nation, and cross-ethnic. But the international society has never established reasonable procedures to reach a consensus on this problem. In the context that the Western countries have already had economic and political capital accumulation initially, it is hard to challenge their dominant status. Most developing countries and post-developing countries have already lost the right to speak. At present, the multi-polarization trend tends to be more and more apparent. The effect of regional integration becomes more and more strong. We urgently need to put the issue of the cores of universal values on the agenda of the international meetings and conferences or choose the sound public selection mode to build procedures and work hard for the collective purposes after reaching consensus.

(4) The developing countries should collaborate together.

In recent years, the influence range of universal values has expanded from cultural theories to political and economic fields, which is not beneficial to the developing countries. We have to be alert to the export of universal values and refuse the overstepping collaboration. Under the circumstances where the universal values are still dominant, the developing countries should collaborate together more than ever and strengthen the collaboration for the collective interests. Led by the developing countries with better development trends, it is better to establish the regional coordination mechanism of the developing countries own. By strengthening dialogues, it is possible to express standpoints and needs fairly and impartially. More people means more power. The overlay of national volumes is beneficial to the achievement of the equilibrium in the field of discourse and lets the world hear the voice of the developing countries. Although it can be tough, taking the first step can make a difference.

(5) Pay attention to the development of science and technology and keep up with the time.

The innovation of science and technology brings new chances and challenges to the development of universal values. With the improvement of artificial intelligence, genetic engineering and robot technology, science and technology begin to move into the political field. In this circumstance, it is critical to consider the combination of morality and human rights and integrate the consideration into the development and deployment of new technology. In the context of the dominant Western countries, we need to ensure that the new technology interprets universal values in a proper way and pay attention to science and technology antitrust issues to avoid perpetuating inequality and damage.

The development of universal values needs an inclusive and wise method, which is able to balance the tensions between collective interests and cultural diversity on the basis of ensuring the basic human rights and respecting the unique identities and traditions of different societies. At the same time, the development of universal values is an ongoing process without a certain ending. With the advancement of science and technology and the development of society, both new challenges and chances will emerge and values need to be evaluated, adapted and engaged collectively. By including diversity, engaging in dialogues, and standing on principles of human rights and morality, the development of universal values are able to move in a more inclusive and fair direction and promote the collective advancement of human welfare.

The benign development of universal values need to be carefully considered, protected and strengthened.

References

1. Jimin Chen (2021), The Logic and Dilemma of Western "Universal Values" [J]. Peace and Development, No.180(02):51-66+135-136.
2. Huxue Wang, Wanxin Chen. The common value of mankind and the Western "Universal Value" : Definition, screening and Transcendence [J]. Governance Modernization Studies, 2023, 39(01):72-79.
3. Sang Jiang (2016), Analysis on Western Universal Value Theory[J]. 2nd International Conference on Social Science and Technology Education (ICSSTE 2016):239-242.
4. Yonghong Zhang (2013), Universal Values: Spread in China and the Essence [J]. American Journal of Humanities and Social Sciences, Vol. 1, No. 4, 2013, 258-262.
5. Chen CAI (2022), Paradoxical discourse, essential criticism and value transcendence of "universal values"[J]. Future and Development, 46(07):26-31.
6. Liming Wang (2021), Discussion on Universal Values from the Perspective of Axiology [J]. Journal of Hubei Polytechnic University (Humanities and Social Science), 38(06):58-64.
7. Gershman, C. . (2005). Democracy as policy goal and universal value. Whitehead J.dipl. & Intl Rel.
8. Sen, Amartya. (1999) "Democracy As a Universal Value". Journal of Democracy, vol. 10, no. 3, pp. 3-17.
9. Ha-Joon Chang (2002), Challenge, Vol. 45, No. 5 (SEPTEMBER-OCTOBER 2002), pp. 63-97 (35

pages), https://www.jstor.org/stable/40722165

10. Burton, J. W. (1969) "Universal Values and World Politics." International Journal, vol. 24, no. 4, pp. 673–92. JSTOR, https://doi.org/10.2307/40200285.
11. Tao Xie （2022）, When "Universal Value" meets the common value of all mankind [J]. International Studies, No.211(05):28-31.
12. Tingyou Wang (2020), On Problems to be Clarified in the Criticism of "Universal Value" [J]. Studies on Marxist Theory, 6(01):134-143.

Sentiment Analysis of Yaya Event of Giant Panda in the United States based on LDA Topic Model

Ying Wu*

Aulin College, Northeast Forestry University, Harbin, Heilongjiang Province,150040 ,China

Abstract. With the rapid rise of the Internet and social media, hot topics can often become hotspots of online public opinion. In order to explore the netizens' views on hot topics, this paper constructs an LDA model to analyse the sentiment of the incident of travelling panda Yaya by crawling the Weibo comments on the incident of travelling Giant Panda Yaya in the U.S. The study shows that netizens suspect that Yaya has suffered some abuse in the U.S. Memphis Zoo, and urgently hope that the panda Yaya will come back to China as soon as possible to receive treatment and care in the domestic zoo. The results of the study are helpful for the relevant departments to understand the attitudes and tendencies of netizens towards hot topics in a timely manner, and quickly deduce the development trend of the topics, so as to guide the public opinion accordingly.

1 Introduction

In recent years, Sina Weibo, as a modern large-scale online social media, has been used by a large number of Chinese netizens. As of the first quarter of 2023, the number of monthly active users of Weibo reached 593 million, and the number of daily active users reached 255 million. The number of daily active users reached 255 million. Sina Weibo have gradually become the centre of online public opinion, and many netizens express their views on hot events through Weibo.

This paper will focus on the Weibo comments of netizens on the incident of the Giant Panda Yaya travelling to the United States. Giant Panda Yaya was born on 3 August 2000 in Beijing Zoo, on 7 April 2003, Yaya and male panda Lele took a special flight to the Memphis Zoo in the U.S. to start a 20-year career in the U.S. On 1 February 2023, the sudden death of Lele, who was about to return home, triggered people's attention to the travelling U.S. pandas, which in turn triggered the attention to Giant Pandas Yaya. According to tourists visiting the Memphis Zoo broke the news that Yaya appeared stereotypical behaviour, the physical condition is not optimistic, for this reason, to take Yaya back to China's call is becoming stronger and stronger. 2023 April 27th afternoon, the expiration of the lease period of pandas Yaya by special cargo plane to China [1]. Netizens have a heated discussion on this, and expressed their own views on the incident.

2 Literature review

As one of the largest social media platforms based on user relationships in China, most of the netizens who use Weibo will leave their own comments on some hot topics on the Internet. Literature review Sentiment analysis is a way of analysing the text to determine the sentiment of the sender of the opinion on the event. Sentiment analysis of Weibo comment data can provide an in-depth understanding of netizens' views and emotional tendencies towards the event, and provide valuable reference and analysis basis for the relevant departments. Some researchers have already studied the hot topics of Weibo, such as Yue Yang et al [2], who used a sentiment analysis method containing keyword extraction and machine learning to analyse the sentiment of some Weibo comments related to the crash of China Eastern Airline MU5735, and statistically and analytically analysed the data of Weibo hot topics after the crash. Liu Jingwei et al [3] conducted sentiment analysis on Weibo hot topics by constructing Bi-LSTM neural network, so as to predict the heat evolution of Weibo hot topics.

LDA models are commonly used in social media, image processing, text classification and clustering, community methods and other fields [4]. Xu Heng et al [5] clarified the number of related topics by using confusion and similarity, determined the research focus by topic strength, and studied the topic evolution from the change of topic strength under the time level. Lai Xianjing [6] used LDA model to study the text analysis of MOOC course reviews, which is beneficial for building online education platforms.Overseas scholars such as Inoue Madoka et al [7]. used LDA models to explain the negative impact of the early COVID-19 pandemic on nursing research; Kozlowski Diego et al [8] used LDA models for world trade analysis; Sang-Woon Kim et al [9] investigated a paper classification system based on TF - IDF and LDA.

* Corresponding author: suixin7820@163.com

This study attempts to understand netizens' views on the hotspot event of the Giant Panda Yaya through emotional tendency analysis and LDA theme model analysis, which provides the relevant departments with a more intuitive understanding of the emotional tendency of the netizens, and helps to analyse the development trend of the online public opinion so as to take corresponding measures.

There are the following shortcomings in this study:(1) Since sentiment analysis is a subjective task, it is largely based on the diversity of sentiment expressions and the connections in the context to determine the sentiment tendency. Therefore, in the follow-up study, we should try to construct a sentiment lexicon and use other sentiment analysis models to improve the accuracy of the sentiment tendency in the comment data; Because sentiment analysis is a subjective task, it is largely based on the diversity of sentiment expressions and contextual connections.

(2) LDA is a widely used topic model, but it still has the problem of not accurately identifying Internet comment texts. In this study, we try our best to remove the Internet popular texts that cannot be recognised by the model through word segmentation, lexical annotation, and deletion of deletion words, but there is still subjective uncertainty. In the subsequent study, the LDA model can be improved by further optimising the word segmentation and stopping word removal scheme.

(3) Due to the condition limitation, the experimental samples collected in this experiment are limited, and the text clustering effect shown is not ideal, in the future, we can improve the experimental model by increasing the collection of experimental data samples, and further improve the experimental results.

3 Method

3.1 Research framework

In this study, the LDA topic model is used to study the Weibo comment data of the incident of the Giant Panda Yaya, and to understand the attitude of netizens towards this hot event. Therefore, this study completes the sentiment analysis by crawling the comments of Weibo related to the Yaya incident, data preprocessing, data modelling, analysis and conclusion, etc. The research framework is shown in Figure 1 below.

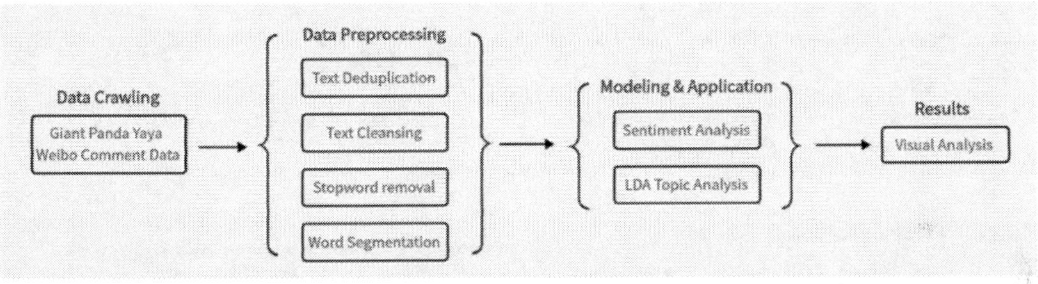

Figure 1. Research framework

3.2 Data acquisition and pre-processing

In this study, we crawled 2467 Weibo comments about the incident of travelling Panda Yaya by python crawler to obtain the netizens' evaluation of the incident of travelling Panda Yaya at different stages, and saved the crawled data into CSV files.

Data preprocessing can remove the insubstantial data in the text and make the data more accurate. The data obtained in this study contains more meaningless letters and numbers, and words such as "Weibo" and "Giant Panda" which appear more frequently are not useful for the experiment, so these words should be deleted before word separation. The phenomenon of multiple meanings of one word and one word with multiple meanings exists in the Chinese language, this study eliminates this phenomenon by means of word segmentation and lexical labelling, so that the computer can identify and process them. In order to reduce the size of the dataset and improve the accuracy of the model, this study also adopts the method of deleting deactivated words to stop word removal.

After removing the deactivated words, this study shows the frequency of keyword occurrence through word cloud as shown in Figure 2 below.

Figure 2. Keyword word cloud

3.3 SnowNLP Sentiment Analysis

3.3.1 Emotional disposition analysis

Sentiment tendency is generally expressed as the subjective tendency of the opinion sender towards the event, which is usually classified into positive, negative and neutral sentiment tendencies. Sentiment analysis is to investigate the attitude tendency with personal subjective emotion revealed in the relevant text data, so as to further judge the sentiment tendency carried in these text data. Sentiment words tend to influence the judgement of sentiment tendency to a great extent, and the connection between the sentiment words and the context can also cause the same result. There is a special class of emotion words in Weibo that can also indicate emotional tendency, the emoticons carried in the comments published by netizens [10], such as "[cry]" and "[angry]", which are more intuitive than the textual data to show the emotional tendency of the netizens, so attention should be paid to the extraction of the emoticons in the study.

3.3.2 Matching Sentiment Words

SnowNLP is a natural language processing library that is widely used in python, but which does not use the Natural Language Processing Toolkit (NLTK), but instead SnowNLP uses the plain Bayesian principle to achieve sentiment analysis, Chinese lexical annotation and other functions, which can better process Chinese text. Therefore, SnowNLP corpus is chosen for the sentiment analysis of the comment data in this study.

3.3.3 View Sentiment Analysis Results

In this study, word clouds are used to present positive comments as well as negative comment data to see the effect of sentiment analysis.

Figure 3. Positive Emotional Commentary Word Cloud

The figure 3 shows that positive sentiment words such as "go home", "flower" and "hug" appear more frequently, and there are no negative sentiment words mixed in the word cloud, so it can be concluded that sentiment analysis can extract positive sentiment comments better.

Figure 4. Negative Emotional Commentary Word Cloud

The figure 4 shows that negative sentiment words such as "abuse", "angry" and "explain" appear more frequently, and there are no positive sentiment words mixed in the word cloud, so it can be concluded that sentiment analysis can extract negative sentiment comments better.

3.4 LDA Modelling

3.4.1 How the LDA model works

LDA model is a three-layer Bayesian model, which can be divided into document, topic and vocabulary three layers, that is, each word of an article is composed by selecting a certain topic with a certain probability, and selecting a certain word with a certain probability from this topic. The final result is: "document - topic" and "theme - word The final result is: "document-topic" and "topic-word". LDA topic model through an iterative way, constantly adjusting the distribution of "document - topic" and "topic - word", in order to get the topic that can explain the text data.

LDA model is a kind of unsupervised learning, that is, without the need to annotate the data, it can be able to filter the high-frequency words in a large number of comment data to extract the text theme, which can help to find the potential theme of the data, so as to analyse the core theme of the dataset.

3.4.2 Finding the optimal number of topics

In this study, the average cosine similarity is used as an evaluation index to determine the optimal number of topics in the LDA model. The smaller the value of average cosine similarity, the better the corresponding number of topics.

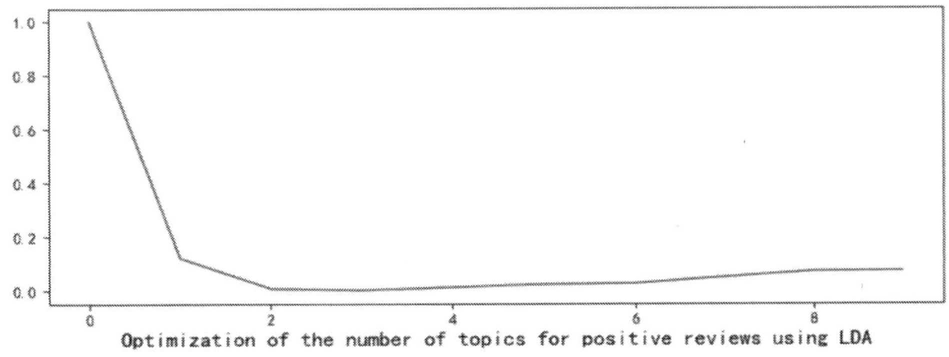

Figure 5. Optimization of the number of topics for positive reviews using LDA

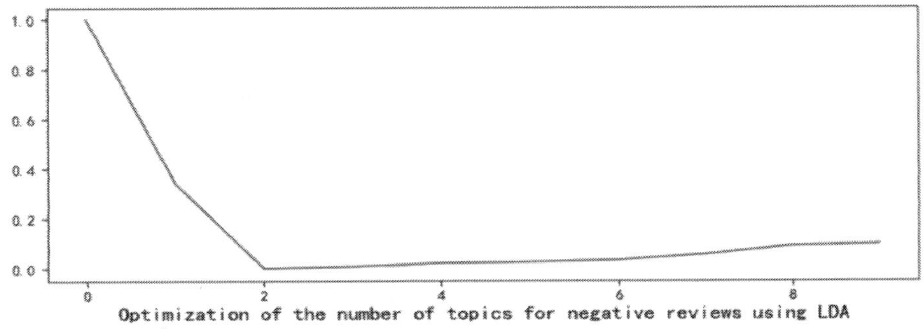

Figure 6. Optimization of the number of topics for negative reviews using LDA

The horizontal axis represents the number of themes and the vertical axis represents the average cosine value, the Figure 5 and Figure 6 represents the change of the average cosine similarity between themes with the increase of the number of themes. According to the results presented in the Figure 5 and Figure 6, it can be seen that for the positive comment text dataset, when the average cosine similarity between the topics reaches the lowest, the corresponding topic degree is 3. Therefore, the optimal number of topics is determined to be 3 by using the LDA model to model the positive comment text dataset, and the optimal number of topics is determined to be 2 for the negative comment text dataset, but considering that the number of topics is too small, which will lead to large errors and the number of positive and negative comment data topics should be the same, the optimal number of topics is determined to be 3 by using the LDA model to model the negative comment text dataset. However, considering that the number of topics is too small, which will lead to a large error in the experiment, and the number of topics should be the same for both positive and negative comment text data, the optimal number of topics is determined to be 3 by applying the LDA model to model the negative comment text data set, taking into account the above factors.

3.4.3 Constructing the LDA model

In this study, the LDA model is constructed for the obtained positive and negative review datasets respectively, and the relevant model parameters are set. After the corresponding LDA theme analysis, ten words with the highest frequency of occurrence will be generated for each theme and the probability of their occurrence will be shown, and the generated themes are shown in the Table 1.

Table 1. Positive Comment Words and Probabilities

order	positive topic 1		positive topic 2		positive topic 3	
	words	probability	words	probability	words	probability
1	hug	0.067	flowers	0.049	video	0.107
2	go home	0.049	think	0.049	love	0.073
3	Thumbs up	0.049	hope	0.038	baby	0.044
4	grateful	0.040	florid	0.027	eat	0.022
5	Ya Bao	0.035	real	0.025	Beijing Zoo	0.021

6	carry	0.032	daddy	0.024	sorrow	0.019
7	take care of	0.024	Thanks	0.024	style	0.018
8	embrace	0.021	happiness	0.021	hard	0.021
9	hee hee	0.019	exhauting	0.020	long	0.014
10	U.S.	0.015	nanny	0.020	more and more	0.013

The above figure can present the latent topics in the positive comment data of the incident of the Giant Panda Yaya travelling to the U.S. The most frequent words in Topic 1 are hug, go home and hope, etc., mainly reflecting that netizens hope that Yaya will be well taken care of and that the Giant Panda Yaya will go home soon after leaving from the U.S., etc. In Topic 2, the frequent words are Thanks , take care of, nanny, etc., which mainly reflect that netizens are grateful to "daddy" and "nanny", who are the keepers of Yaya, for their patience in taking care of Yaya. The high-frequency words in Topic 3 are video, love, eat and baby, which mainly reflect the hope that Yaya will eat more food and become healthy, and that netizens want to have a video session with Yaya so that they can see what Yaya looks like at the moment, and so on.

Table 2. Negative Comment Words and Probabilities

order	negative topic 1		negative topic 2		negative topic 3	
	words	probability	words	probability	words	probability
1	abuse	0.180	skin disease	0.033	Ya	0.045
2	too	0.047	say	0.032	send	0.021
3	open	0.042	catch	0.025	release	0.020
4	point	0.022	add	0.024	please	0.019
5	renew	0.019	Li Daya	0.022	angry	0.017
6	national treasure	0.018	live streaming	0.020	too short	0.016
7	pitiful	0.014	return	0.018	the naked eye	0.016
8	a little	0.013	hairy	0.017	ears	0.014
9	movement	0.011	staff member	0.016	aim	0.013
10	drink	0.009	kiss	0.015	soon	0.013

The Table 2 can present the potential themes in the negative comment data of the incident of the Giant Panda Yaya in the U.S. The most frequent words in Topic 1 are abuse,too, national treasure, pitiful, etc., mainly reflecting that netizens see the appearance of Yaya as a national treasure, and suspect that Yaya has received abuse. The high-frequency words in Topic 2 are skin disease, catch, return, which mainly reflect that netizens suspect that Yaya has contracted a skin disease in the United States and want to take Yaya back to China for treatment. The high-frequency words in Topic 3 are send, angry, etc., which mainly reflect that netizens are angry about the recent situation of Yaya in the United States.

4 Conclusion

In this study, the sentiment analysis of the incident of the Giant Panda Yaya in the United States is based on the LDA theme model, which can accurately reflect the sentiment tendency of netizens for this hot event. It is concluded from the comments of the netizens extracted from the Weibo that the netizens are more concerned about the Giant Panda Yaya and suspect that Yaya has suffered some abuse in the Memphis Zoo in the U.S..They hope that the Giant Panda Yaya will come back to China as soon as possible to accept the treatment and care at home. Most of the words shown in the word cloud are positive emotions, and the word with the highest frequency is "abuse", which shows that the relevant departments were able to find out the emotional tendency of the netizens and intervene in time, reversing the tendency of the netizens to pay attention to the incident.

As the Internet has entered thousands of households, most Internet users like to learn about social problems through the Internet and express their opinions on hot topics. Hot topics on the Internet often have the characteristics of strong dissemination and great influence, so the emotional analysis of the comments under the hot events on the Internet can help the relevant departments to quickly understand the emotional tendency of the netizens, and through the release of information through multiple channels, guide the netizens to correctly express their opinions, and control the development trend of the public opinion in a timely manner.

Reference

1. (2023). Return of the giant panda "Yaya" to the motherland. Chinese General Accountant, 04, 185. doi:CNKI:SUN:ZKJS.0.2023-04-072.
2. Yue Yang, Shan Chenghu, Peng Chunlian, Zheng Xiaolin, Yan Yuxi, & Guqiong. (2023). Evolution of Public Opinion and Sentiment Analysis on the East Airlines MU5735 Crash Based on Sina Weibo Data. Computer Era, 02, 55-60. doi:10.16644/j.cnki.cn33-1094/tp.2023.02.012.
3. Liu Jingwei & Zhang Shuqi. (2022). Evolutionary Analysis of Weibo Hot Topics Based on Sentiment Analysis. China CIO News, 12, 137-140. doi:CNKI:SUN:XXXT.0.2022-12-035.
4. Han Ya'nan, Liu Jianwei, & Luo Xionglin. (2021). A Review of Probabilistic Topic Models. Chinese Journal of Computers, 06, 1095-1139. doi:CNKI:SUN:JSJX.0.2021-06-004.
5. Xu Heng, Zhang Menglu, & Sun Dechang. (2021). Topic Analysis and Evolution Trends in the Field of Domestic Review Mining and Sentiment Analysis Based on LDA Model]. JOURNAL OF HENAN UNIVERSITY OF TECHNOLOGY SOCIAL SCIENCE EDITION, 02, 38-46. doi:10.16433/j.cnki.cn41-1379.2021.02.006.
6. Lai Xianjing. (2023). Text Analysis of MOOC Course Reviews Based on LDA Topic Model. Modern Information Technology, 04, 43-46. doi:10.19850/j.cnki.2096-4706.2023.04.011.
7. Inoue Madoka, Fukahori Hiroki, Matsubara Manami, Yoshinaga Naoki & Tohira Hideo. (2022).Latent Dirichlet allocation topic modeling of free-text responses exploring the negative impact of the early COVID-19 pandemic on research in nursing.. Japan journal of nursing science: JJNS. doi:10.1111/JJNS.12520.
8. Kozlowski Diego, Semeshenko Viktoriya & Molinari Andrea. (2021).Latent Dirichlet allocation model for world trade analysis.. PloS one (2). doi:10.1371/JOURNAL.PONE.0245393.
9. Sang-Woon Kim & Joon-Min Gil.(2019).Research paper classification systems based on TF-IDF and LDA schemes. Human-centric Computing and Information Sciences (1),1-21.
10. Tang Xiaobo & Lan Yuting. (2016). Sentiment Analysis of Weibo Product Reviews Based on Feature Ontology. Library and Information Service, 16, 121-127+136. doi:10.13266/j.issn.0252-3116.2016.16.015.

 https://doi.org/10.1051/shsconf/202317801003

A comparison of healthcare service for international students in United States and United Kingdom: A perspective of institutional complexity

Hangyi Zhou*

Health Science Center, Xi'an Jiaotong University, Xi'an, China 710061

Abstract: As newcomers to the host country, international students face unique challenges when it comes to the healthcare service, such as the language barrier. The challenges lead to a low utilization rate of the healthcare service in the population of international student. This paper aims to find out the main reason behind the phenomenon and explore how to improve the health and welfare of the international students. A nested case study approach is used and the United States and the United Kingdom are the two countries for comparison. We found that benefiting from a simpler system, international students in the UK enjoy superior healthcare access compared to their counterparts in the US. It is recommended that UK and the US should prioritize comprehensive healthcare orientation programs, ensuring students are well-informed about accessing and navigating their respective systems.

1. Introduction

As the two largest destination countries in the world, the US and UK has attracted the largest number of overseas students in the world, and the number is increasing year by year. In 2021, over 0.6 million oversea students are in the UK[1] and over 1.2 million in the US in 2021[2]. And most universities are receiving more international student applications for the 2022/23 academic year[3]. More attention should be paid to this large population. International students usually face a set of challenges that can impact their overall well-being. These challenges include language barriers, cultural differences, financial difficulties, homesickness, academic challenges, and both mental and physical health. Of these challenges, the biggest obstacle for international students is to get used to a variety of unfamiliar institutions, such as the healthcare system. Considering the importance of healthcare system in daily life, the article will investigate how the complexity of healthcare system affects the healthcare service utilization.

Foreign students are not familiar with the new healthcare system so they need time to learn and understand the complex rules and procedures[4]. In particular, international students often struggle to utilize healthcare services in their host country due to the complexity of the healthcare system, language barriers, and the additional time and effort required to obtain information and adapt to the local healthcare system. As a result, some international students may delay seeking medical treatment or even avoid it altogether, which may worse the situation and can finally lead to increased medical costs. To support the health and well-being of international students, it is crucial to pay attention to their health status, willingness to seek medical care, and the convenience of accessing healthcare services.

Nevertheless, the easiness of healthcare services for international students varies across countries. The performances of healthcare systems vary widely depending on factors, such as the access to healthcare, coverage, costs and quality of care. For example, UK (public health service model) is believed to perform better than US (private health service model), in providing universal healthcare coverage for all citizens. Why do them perform differently? Which factor influence the quality of health services for foreign students? It is essential to examine the factors that influence the quality of healthcare services for international students and compare the experiences of international students in different healthcare systems. To address this issue, this article will focus on health service utilization of international students and varied states in different health systems. It aims to describe the challenges confronted by international students and the impact of health system setting on the quality of service.

The content of the article is as follows. First, review the literature on health problems and healthcare utilization of international students, and develop an analytical framework for case study. Second, describe the methodology and data; Third, compare the healthcare service that international students gain in UK and the US, the two countries that we choose for comparative case

* Corresponding author: 1362958009@qq.com

studies. Finally, based on the comparison, draw conclusions and provide policy recommendations to improve the quality of healthcare services for international students.

2. Literature review

2.1 Health-related issues as to international student

There have already been many researches on mental health of international students. Those mental problems are homesickness, loneliness, culture shock, stress about linguistic, academic, interpersonal, and financial concerns, mental exhaustion and burnout, anxiety and depression[5]. For example, Han et al. (2013) illustrated that among 130 Chinese students at Yale University, 45% had symptoms of depression, and 29% had symptoms of anxiety[6]. Among 13,984 full-time students, 35% reported positive for at least one common lifetime disease and 31% reported positive for at least one 12-month disease[7].

There is a growing body of research indicating that international students face a range of physical health problems as they adapt to life in a new country. For instance, they may experience injuries during exercise, contract illnesses like the flu, or even be involved in traffic accidents. Additionally, mental health problems can lead to physical symptoms such as fatigue, headaches, sleep problems, weakened immunity, and heightened susceptibility to other illnesses[8]. International students may also struggle with poor nutrition, lack of appetite, and gastrointestinal issues due to limited access to healthy food and cultural differences in available cuisine[9].

Although there are many health issues, the willingness of international students to seek medical care is not high. Tang Charlotte et al. (2018) investigates challenges that international students face when accessing healthcare in their host countries and the alternative solutions they use to deal with challenges[4]. According to their findings international students encounter several difficulties in understanding healthcare services information, high healthcare treatment costs, unfamiliar and complex health practices, negative beliefs toward western medicine, and communication and language barrier. As a result, many international students must actively seek out information about healthcare services, take preventive measures to stay healthy, and rely on medicine from their home countries to deal with health problems.

2.2 Inclusion of immigrates in different healthcare system

Different countries have different healthcare system. The healthcare system can be broadly categorized into three types on the basis of three aspects of financing, provision and regulation[10]. The types are state healthcare systems, where state actors and institutions assume financing, provision and regulation; societal healthcare systems, where social actors are in charge of financing, provision and regulation; and private healthcare systems, where market actors play the most important role in the three dimension (See table 1). Böhm et al. made a more elaborated classification which included 10 types of healthcare system[11], but the underlying logic of classification was consistent, which was also on the basis of regulation, financing and provision.

Table 1. A typology of healthcare systems

Healthcare system type	Representative country	Regulation	Financing	Provision
State healthcare systems	UK	State	State	State
Societal healthcare systems	Countries of Central and Eastern Europe (CEE)	Societal	Societal	Societal
Private healthcare systems	US	Private	Private	Private

Different healthcare systems provide services in different ways. The NHS in the UK is regarded as a typical state healthcare system, where the government ensures health insurance coverage for all citizens, regardless of their financial status. Immigrants are entitled to receive the same range of services as UK citizens, including primary care, emergency care, and hospital treatment. The NHS does not discriminate based on immigration status, and all individuals are encouraged to access care when needed. Additionally, the NHS offers interpretation and translation services to support communication with patients who may have limited proficiency in English.

Countries of Central and Eastern Europe (CEE) used to be the socialist healthcare system and then changed to social health insurance system[12]. Under that kind of system, the government owns and operates the health system, so the providers are employees in the government.

Unlike many other developed countries that have universal and free healthcare programs, the US healthcare system is a private type. Most Americans are served by a mixture of public and private healthcare systems[13], but access to care can be limited for immigrants. While some immigrants can have access to healthcare through employer-sponsored plans or public programs such as Medicaid, others may be uninsured due to legal status or lack of affordable options[14]. Some states have implemented their own policies to expand coverage for immigrants, like Connecticut, which extended state-funded full Medicaid coverage to pregnant women and children from households with low incomes regardless of immigration status. Nevertheless, the overall landscape remains complex and varies widely depending on the individual's circumstances. Overall, the US healthcare system does not provide universal coverage, and immigrants may face significant challenges in accessing care. Although the health insurance in the United States is more selective, newcomers hold negative views on the insurance system in the United States. They regard the healthcare system in the US as an unpredictable and complicated one which is difficult to use, because in order to utilize the healthcare service, they need to learn a large amount information about the complex system[15].

2.3 Healthcare utilization of international students

International students have a lower utilization rate of healthcare services compared to domestic students. The UK government's data revealed that uptake of the MenACWY vaccination was 45% in EU students and 29% in non-EU international students. Moreover, only 68% of international students knew how to access sexual health services, significantly lower than their domestic peers. Additionally, GP registration was found to be lower among international students, with 48% of unregistered students being unaware of how to access services[16]. Several studies have also indicated international students' mental health service utilization rate is similarly low[17] or even lower when compared to domestic students[18]. It was found that many factors hinder international students from utilizing healthcare services in their host countries. These factors include distrust of western medicine, language barriers, and knowledge gaps, which are commonly encountered by international students in both the UK and the US[4]. These challenges contribute to the lower healthcare utilization rate observed among international students compared to native English speakers.

However, it is noteworthy that the healthcare service utilization rate of international students in the UK is higher than that in the US. According to the CPC-ONS-UUK Survey of Graduating International Students 2017 - Technical report, approximately 67.7% (1312 international students) utilized NHS services in the UK within the last 12 months while studying, including GP consultations, Accident and Emergency visits, and Contraception & sexual health services[19]. In contrast, one study focusing on Asian international graduate students in the US showed that only 6.9% of them utilized any mental health services, significantly lower than the rate of 45.2% among domestic graduate students[20]. The difference in the utilization rate of healthcare services between the two countries may be attributed to the distinctive health systems and their complexities. The health systems in the UK and the US differ significantly. As a result, international students experience varying degrees and types of difficulties when accessing services and seeking medical care. For example, international students in the UK expressed dissatisfaction with the long waiting time of GP appointments and automatic replies from medical centers[21]. In the US, international students often encounter challenges in understanding and utilizing healthcare insurance, along with concerns about the intimidating costs of health services[4].

3. Method and data

The article focuses on a comparative analysis between the US and the UK as case countries. To account for the significant differences in healthcare insurance plans among US universities, this article selects various types of universities for detailed analysis. In contrast, the healthcare service in the UK is provided through a unified system, eliminating the need for a classified discussion.

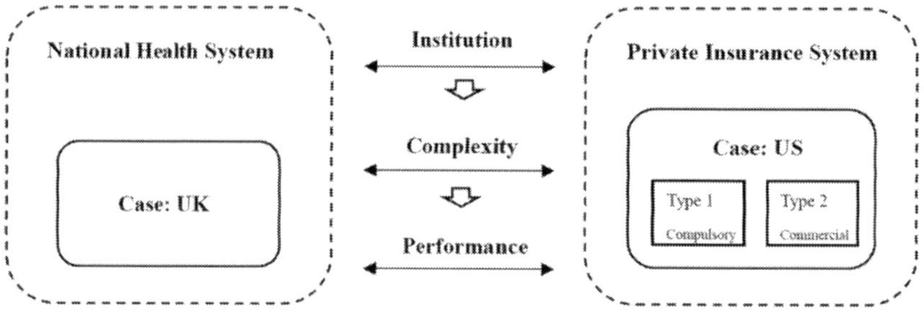

Figure 1. Research design

The article compares the differences between the UK and the US in three key aspects, which are institution, complexity and performance, aiming to analyze how the complexity of healthcare systems (institutional settings) influences the effectiveness of medical services (See figure 1 and table 2). It's crucial to emphasize the importance of studying the "complexity" in healthcare systems. Complexity can significantly influence patients' ability to access and utilize medical services. For example, with varied choices and intricate reimbursement procedures, certain individuals might face challenges in getting the care they need. Moreover, complexity can amplify administrative costs and lead to inefficiencies. Therefore, understanding the complexity of healthcare systems is pivotal to comprehending their impact on the efficacy and equity of medical services.

Table 2. Analytical framework

	Elements	Definition
Institution	Health system	the structure and funding mechanisms.
	Policy for immigrates	fee and coverage for immigrants
Institutional Complexity	Acquisition	ease or difficulty to register
	Utilization	ease of use of the services

	Reimbursement	efficiency of claims and their reimbursement
	Utilization Rate	Utilization rates of the two groups (domestic student and international students)
Performance	Waiting time	average time patients have to wait before receiving medical services.
	Unexpected costs	Unforeseen cost during the course of treatment

The article relies on various data sources, including public statistics from the British and American governments, reports from research institutions, and empirical research conducted by other scholars. These diverse data sources offer valuable information on aspects such as the complexity of healthcare systems and the medical access of international students.

4. Result

4.1 International Students in the UK

4.1.1 UK healthcare system

In UK, residents access both physical and mental healthcare services through the National Health Service (NHS). The NHS is funded through a combination of employee contributions from wages and general government funding raised in the form of taxes. However, the availability of funding may fluctuate based on political and economic factors, necessitating NHS payment system that supports the delivery of improved healthcare within the available resources (see table 3).

Most medical services provided by the NHS are free at the point of use for the patient, but certain procedures may require additional payment. The NHS offers a wide range of services, including primary healthcare at General Practitioners (GPs), dental care, hospital and emergency treatment, family planning and sexual health services, mental health services, and more. People can access these services without any extra cost, except for some specialized NHS services like eye care, dental treatment, and prescriptions for medication.

The British public expresses great satisfaction with their country's health service system and holds it in high regard. A survey commissioned by the King's Fund and conducted by Ipsos reveals that support for the NHS and its founding principles remains robust. The majority of the public, about three-quarters (77%), firmly support maintaining the NHS in its current form, while a smaller proportion (23%) believe it may not be sustainable. This consistent level of support has persisted over nearly two decades[22].

Table 3. A comparison of the structure and financing mechanisms of healthcare system in the UK and the US

Healthcare system		UK	US
Structure	Type	National Health System (NHS)	Private Insurance (PHS)
	Coverage	Universal	Individual
Financing	Collection	Tax-based	Privately financed
	Pooling	Single fund	No pooling
	Providers	Public	Private

4.1.2 The complexity of UK healthcare system

The healthcare system in the UK is relatively straightforward and accessible for international students. Although the NHS may be unfamiliar to many international students, the process of obtaining healthcare through it is simple, and the necessary knowledge is easily within their reach.

As for the acquisition of healthcare service, it's easy for international students in the UK to be qualified for the NHS. As for EU students, are eligible for free NHS treatment with a European Health Insurance Card (EHIC). As for Non-EEA international students, NHS covers foreigners who pay an extra cost called additional immigration healthcare surcharge (IHS) when they apply for visas. Only a small amount of information is required, such as checking their residence time in the UK. Full-time international students studying in the UK for a period longer than six months are eligible to use NHS services. For short stay, the international student should take out private health insurance. International students staying for six months or longer can pay a yearly fee of £470, granting them access to the same NHS services as domestic students.

As for the utilization of the healthcare service, one of the first tasks for international students upon arriving in the UK is to register with a GP practice. They can compare different GP practices based on location, facilities, services, and performance before choosing one that suits them best.

Ensuring that every citizen is served with free medical treatment is the main feature and advantage of NHS over other countries. The NHS protect the health and medical interests of the whole people. The free medical care system in the UK makes medical services accessible and equitable to everyone. Individuals can receive appropriate medical treatment without any financial burden, allowing for timely and comprehensive care. Additionally, the free medical care system in the UK reduces the financial burden of patients, enabling them to focus on their treatment and well-being instead of worrying about medical expenses. Patients can access the medical services they need promptly, leading to better care and improved overall health outcomes.

4.1.3 Medical services encountered by international students in the UK

It is also found that newly arrived migrants tend to utilize healthcare services less frequently than the UK population[23]. Similarly, a study investigating the number of international students' GP visits in the last 12 months within one year revealed that 44% of students had not visited a GP during that period[19]. The specific challenges faced by international students, who predominantly come from non-English speaking countries, include language barriers and cultural differences, which act as significant barriers to healthcare utilization. It is difficult for them to describe their symptoms to the doctor clearly. Not only do doctors fail to correctly obtain their symptoms, but they also can't understand doctors' explanations and medical suggestions[21]. As a result, it increases the lack of sense of security and faith in doctors in the UK and make them feel humiliated. When they get ill, they just use medicines they bring from home such as special Chinese medicines[24] or wait until they return home.

However, the NHS is heavily understaffed and underfunded, which means long wait times for appointments[25]. Some international students may find themselves waiting for extended periods when they need medical services, which makes them painful and uncomfortable[21]. Also, automatic services in medical centers be difficult to operate, leaving students unsure of what steps to take next.

Although the healthcare system still has some deficiencies for international students, which affects their utilization of the NHS service, the overall health condition of international students in the UK appears to be satisfactory. In Jane's survey, the majority of international students rated their general health as very good or good (86.2%), with a smaller proportion considering it fair (12.0%), and only a small minority rating it as bad or very bad (1.8%)[19].

4.2 International Students in the US

4.2.1 US healthcare system

Unlike many other leading countries, the US healthcare system operates on a privately owned and free market system. In short, it means that hospitals and healthcare facilities are driven by profit motives.

The US medical system is not only complicated but also known for its high costs. For instance, a visit to a specialist doctor can cost around $400. A minor surgery at the local hospital may easily amount to thousands of dollars, including charges from the hospital, doctors, surgery, and follow-up care. Given these substantial expenses, having health insurance is crucial for international students during their time in the US. While the monthly health insurance fees can be costly, it provides essential protection against potential financial burdens resulting from medical care. International students in the US typically access healthcare through insurance. They pay a monthly premium for insurance coverage and then make payments when they utilize healthcare services. Without health insurance, seeking medical care could lead to owing thousands of dollars in medical expenses. Health insurance serves as a safeguard against such unexpected costs. Health insurance (Private or Federal insurances) covers the medical expenses of health care treatments of patients, provided by doctors or healthcare providers. Some health insurance plans also covers the prescription drugs.

4.2.2 The complexity of US healthcare system

The United States does not have a national healthcare system, instead there is a distributed, varied private insurance system. For international students, each university give them the flexibility to decide their insurance, although the laws at the national level and state level have some regulations on it. Most American universities provide school health insurance plans, but they are not compulsory. If the student can meet the waiver criteria, which means if student can obtain basic health service from the market, then they can choose other insurance. Universities that don't offer school-sponsored insurance plans require international students to select their insurance from the market.

As a result, international students must invest extra time in acquiring information about their school healthcare plans. They need to know if the university provides health insurance, whether it is compulsory, the waiver criteria, and the process of enrolling in the healthcare system. Additionally, they must familiarize themselves with the details of health insurance, including coverage, prices, and limitations[4]. The diverse information requires international students to prepare well in advance, even before departing their home country, to select the best insurance option for themselves.

In most cases, the healthcare insurance provided by universities is more expensive than that in the market, but both options are cheaper than the average cost of individual health insurance for an American ($6720 in 2023)[26]. For example, the University of California provides UC-ship for international students at $4,410 per year, Columbia University requires all international students enrollment in the Columbia Plan at $4,541 per year, while the Blue Sky 80CA in the market costs only $915 per year. Due to the high price, international students often prefer insurance from the market and need to waive the on-campus insurance. However, although most universities give them the freedom to waive, the fact is that some universities' waive criteria is so restrictive that students can't get through the application, which means the enrollment is mandatory. Some universities don't provide insurance so students have to choose commercial insurance in the market. The following comparison outlines the differences between the two types of insurance.

(1)Students with compulsory insurance

Although the school's compulsory insurance is expensive, it offers superior services. For instance, Duke University is one of the schools mandating students to buy its insurance. It offers a Student Medical Insurance Plan (SMIP) for students. Students may waive the insurance if they does not hold an F1 or J1 visa, but almost all international students are on F1 and J1 visas, so they will be automatically enrolled in the Duke healthcare insurance plan. The annual cost of coverage for all eligible students for the 2023-2024 plan year is $3,282.12. The SMIP grants students access to medical care at the Duke Student Wellness Center and covers a wide range of healthcare services and offers lots of healthcare resources, including primary care, timely care, mental health service ,wellness promotion and nutrition resources. Although dental care is not covered, students can utilize the Basix Dental Savings Program, saving up to 50% on general dental care and 20% on specialty care. Although the annual premium can be a little steep, it will be well worth if international students get injured or ill.

(2) students with commercial insurance

Due to the relatively high premiums of on-campus health insurance or the absence of insurance offerings by some universities, many international students choose to buy off-campus insurance. Ball State University is one of the universities that don't offer an international student health insurance plan, so students there will seek health insurance in the market. While this type of insurance may be cheaper than on-campus alternatives, it does come with certain drawbacks. For instance, Blue Sky 80CA costs only $915 per year, but the deductible and out-of pocket maximum is high, which are $500 and $7000 respectively. Premier 90 costs $1,488, but when you go to the emergency treatment in a non-crisis situation, the loss ratio is 0%. The medical services available to students vary depending on the insurance plan chosen, with generally higher premium plans providing more comprehensive medical services.

In summary, although on-campus health insurance may be more expensive, it often offers better services and coverage. Off-campus insurance can be more cost-effective but may come with limitations and higher out-of-pocket expenses. International students need to carefully evaluate their options and consider their specific healthcare needs when choosing an insurance plan.

4.2.3 Medical services encountered by international students in the US

It is found that international students in the US prefer to go back to their home country for healthcare service during breaks[15]. Their preferences are influenced by various personal concerns, such as distrust of certain health services, fear of privacy breaches, language barriers, cultural differences and discrimination. Consequently, they develop negative beliefs towards Western medicine and place greater trust in doctors and medications from their hometowns[4].

Due to the complexity of private insurance healthcare system in the US, international students encounter difficulties in seeking a desirable health insurance plan. International students find it difficult and frustrating to understand the health-related information. The information available on the university website and pamphlets regarding school-sponsored plans is often insufficient, unclear, or confusing. Additionally, staff at the International Center lack training to provide advice on health insurance, leaving students without practical help or clarification. Moreover, the unfamiliarity of foreign students, particularly Asian students, with the US healthcare system and the intricate insurance terms and terminology poses further difficulty in understanding these aspects at their language level[4].

Accessing medical treatment is also cumbersome for international students. For instance, scheduling appointments can be challenging as doctors often have early off hours, unlike in their home countries where they can visit hospitals without appointments. The US insurance system mandates them to choose from a certain network of doctors only, and requires referrals from their primary care physician before visiting a specialist, but in their home countries, unlike in their home countries where they have more flexibility in choosing doctors[4].

After receiving the medical treatment, international students may experience high out-of-pocket costs depending on different insurances they choose, because the insurance doesn't cover everything and the insurance company pays only a certain amount. It makes them unwilling to see a doctor next time in order to avoid unexpected medical bills. Moreover, the reimbursement process can be cumbersome. International students need to submit the claim for reimbursement in time after the medical treatment ends. Commercial insurance typically allows for direct reimbursement of most medical expenses at the hospital, while insurance provided by the school requires students to submit various reimbursement documents, such as drug receipts, laboratory reports, and doctor's prescriptions, and fill in application forms for insurance company audit. Being unfamiliar with the insurance company's rules can make the reimbursement process difficult, affecting international students' enthusiasm for seeking medical treatment.

4.3 Comparing healthcare service utilization in UK and US

International students often encounter difficulties when living in different countries due to their unfamiliarity with the local medical systems, resulting in a low utilization rate of healthcare services. For instance, they should learn the process of registering with a GP surgery in the UK and understand various insurance models in the US. In addition, communication and language barriers make international students afraid of misunderstanding. It is difficult for foreigner to express their symptoms accurately to doctors, potentially leading to misdiagnosis or inappropriate treatment and fostering distrust in Western medicine. Most international students come from non-western cultures which may hold different views about the causes and medical treatments of illness, so they trust doctors and technology in their home countries instead of the host country. For example, the Chinese medicine uses traditional and unique medicine and therapies like herbal medicine, acupuncture, scraping and cupping. These treatments are not used in western medicine, so when some Chinese students get ill, they tend to use traditional Chinese medicine they brought from home or delay seeking medical care until they return home. Such delays can pose potential risks to public health, particularly in cases of infectious diseases like COVID-19[27].

The complexity of healthcare systems in the UK and the US significantly impacts international students' willingness to seek medical treatment and their utilization of medical resources in these countries (see table 4). In order to fully benefit from the host country's medical services, it is crucial for international students to know the information of service acquisition, medical treatment, and reimbursement. The UK's National Health Service (NHS) offers a simpler and more straightforward system in these respects compared to the complex private insurance system in the US.

In terms of service acquisition, the UK's universal and equal healthcare system requires less information from international students. They only need to pay the Immigration Health Surcharge (IHS) with their visa application according to their visa type, then they can enjoy the same health services as local residents. In contrast, US international students first need to know whether their school provides healthcare insurance plan and whether they can choose off-campus insurance. Comparing coverage, price, and services to select the most suitable insurance option can be challenging, making the US healthcare system less user-friendly for international students. Compared with the UK, the more complex healthcare system is hard to be accepted and used by international students.

In terms of service utilization, most healthcare services in the UK are provided free of charge through the NHS. International students using NHS-covered healthcare services do not have to pay any additional fees. However, certain services like eye care, dental treatment, and prescriptions may incur additional charges. In the US, the cost of healthcare services depends on the insurance coverage held by international students. If international students choose inappropriate healthcare insurance, the cost might be prohibitively high. When they encounter huge medical expenses, they are likely to end up paying out-of-pocket for medical service and face economic difficulties.

In terms of the reimbursement, international students in the UK generally do not need to make payments or seek reimbursements for most NHS services, such as diagnostic tests, hospital treatments and GP consultations . In contrast, after the medical treatment, the reimbursement process in the US is more complex and different from what international students might be accustomed to in their home countries. Students need to submit reimbursement claims after the medical treatment concludes. For commercial insurance, most medical expenses can be reimbursed directly at the hospital. However, insurance provided by the school often requires students to provide detailed medical receipts and reimbursement documents and then undergo an application process for insurance company audit. The unfamiliarity with insurance company rules can make reimbursement challenging. The cumbersome reimbursement affects the enthusiasm of international students for medical treatment.

Table 4. A comparison of of healthcare system and international students' utilization in the UK and the US

Elements		UK	US
Institution	Health system	Regulation, finance, provider all public	All private
	Policy for immigrates	Additional fee in visa application to get the same rights as local residents	No specific regulation, Free choices in Insurance market
Institutional Complexity	Acquisition	Easy	multi sources
	Utilization	Free for most types of services	Services enjoyed depend on insurance plan
	Reimbursement	No need for NHS covered service	Cumbersome
Performance	Utilization Rate	Fairly high rate of utilization, but Asia students use less	Low utilization rate if with poor insurance; Asia students use less
	Waiting time	Fairly Long	Fairly long
	Unexpected costs	No	Sometimes surprise fee

5. Conclusion

The article highlights the challenges international students face in navigating the host country's complex healthcare system. Language barriers, system intricacies, and as a result the preference to seek medical care back home, which often lead to delayed treatments, resulting in worsening health. Addressing these barriers is crucial for both their mental and physical well-being. The article underscores the challenges international students face when navigating the host country's complex healthcare system. Our research indicates that variations of institutional complexity in medical systems significantly influence the healthcare-seeking behaviors of these students.

Benefiting from a simpler system, international students in the UK enjoy superior healthcare access compared to their counterparts in the US. In detail, The UK's National Health Service (NHS) is renowned for providing free healthcare services to its residents, including international students. While this is commendable, one of the persistent challenges the NHS faces is the lengthy waiting times for consultations, largely due to the high demand. Conversely, the US boasts a broad spectrum of healthcare services. However, its complexity, especially regarding acquisition, utilization, and reimbursement, often proves daunting for international students.

To bridge the healthcare gaps in both systems, a few systemic solutions can be considered. Universities in both the UK and the US should prioritize comprehensive healthcare orientation programs, ensuring students are well-informed about accessing and navigating their respective systems. Specifically, in the US, detailed workshops on understanding health insurance can be pivotal. This should encompass aspects like choosing the right plan, understanding reimbursement processes, and more. Furthermore, appointing dedicated healthcare liaison officers can offer students personalized assistance, from booking appointments to understanding medical paperwork. Another avenue worth exploring is telehealth services. Given the wait times in the UK and the intricacies in the US, telehealth can offer prompt consultations and medical advice, eliminating the need for physical appointments. Collaborative health plans, especially for students oscillating between the UK and the US, can be invaluable, utilizing the strengths of both systems. Lastly, training medical staff near universities in cultural sensitivities and multilingual services can further bridge the communication gap, ensuring a holistic healthcare experience for all international students.

Reference

1. International Student Statistics in UK 2023, Studying-in-UK.org, Available at: https://www.studying-in-uk.org/international-student-statistics-in-uk/ [Accessed February, 2023].
2. SEVIS By The Numbers 2021, U.S. Immigration and Customs Enforcement.org, Available at: https://www.ice.gov/doclib/sevis/pdf/sevisBTN2021.pdf
3. Most US colleges report increasing numbers of international student applications for 2022/23, ICEF Monitor.org, Available at: https://monitor.icef.com/2022/06/most-us-colleges-report-increasing-numbers-of-international-student-applications-for-2022-23/ [Accessed 28 June, 2022].
4. Tang, C., Gui, X., Chen, Y., & Magueramane, M. (2018, May). New to a country: Barriers for international students to access health services and opportunities for design. In Proceedings of the 12th EAI International Conference on Pervasive Computing Technologies for Healthcare (pp. 45-54).
5. Mori, S. C. (2000). Addressing the mental health concerns of international students. Journal of counseling & development, 78(2), 137-144.
6. Han, X., Han, X., Luo, Q., Jacobs, S., & Jean-Baptiste, M. (2013). Report of a mental health survey among Chinese international students at Yale University. Journal of American college health, 61(1), 1-8.
7. Auerbach, R. P., Alonso, J., Axinn, W. G., Cuijpers, P., Ebert, D. D., Green, J. G., ... & Bruffaerts, R. (2016). Mental disorders among college students in the World Health Organization world mental health surveys. Psychological medicine, 46(14), 2955-2970.
8. Winkelman, M. (1994). Cultural shock and adaptation. Journal of Counseling & Development, 73(2), 121-126.
9. Thomas, K., & Althen, G. (1989). Counseling foreign students.
10. Wendt, C., Frisina, L., & Rothgang, H. (2009). Healthcare system types: a conceptual framework for comparison. Social Policy & Administration, 43(1), 70-90.
11. Böhm, K., Schmid, A., Götze, R., Landwehr, C., & Rothgang, H. (2013). Five types of OECD healthcare systems: empirical results of a deductive classification. Health policy, 113(3), 258-269.
12. Dubois, C., & McKee, M. (2004). Health and health care in the candidate countries to the European Union: common challenges, different circumstances, diverse policies.
13. Liji Thomas (2023). Healthcare Systems Around the World, News-medical.org, Available at: https://www.news-medical.net/health/Healthcare-Systems-Around-the-World.aspx [Accessed 25 February, 2023].
14. Immigrants and the Affordable Care Act (ACA), National Immigration Law Center.org, Available at: https://www.nilc.org/issues/health-care/immigrantshcr/ [Accessed January, 2014].
15. Jang, S. H. (2021). "I can wait until I return home": healthcare utilization among Korean international students in the US. Journal of American College Health, 1-11.
16. Samuel Honour, Andria Petrou, James Hawkins,& Claire Currie (2018). Student health needs assessment portsmouth, Portsmouth City Council,org, Available at: https://portsmouth.gov.uk/wp-content/uploads/2020/04/portsmouth-student-health-needs-assessment-2018.pdf [Accessed June, 2018].
17. Eisenberg, D., Hunt, J., Speer, N., & Zivin, K. (2011). Mental health service utilization among college students in the United States. The Journal of nervous and mental disease, 199(5), 301-308.

18. Hyun, J., Quinn, B., Madon, T., & Lustig, S. (2007). Mental health need, awareness, and use of counseling services among international graduate students. Journal of American College Health, 56(2), 109-118.
19. Falkingham, J., Giulietti, C., Wahba, J., & Wang, C. (2017). CPC-ONS-UUK survey of graduating international students 2017. Technical report, Southampton: ESRC Centre for Population Change, University of Southampton. Available online at:
20. Yoon, E., & Jepsen, D. A. (2008). Expectations of and attitudes toward counseling: A comparison of Asian international and US graduate students. International Journal for the Advancement of Counselling, 30, 116-127.
21. McMahon, P. (2011). Chinese voices: Chinese learners and their experiences of living and studying in the United Kingdom. Journal of Higher Education Policy and Management, 33(4), 401-414.
22. Kate Duxbury (2017). What do the public think about the NHS? Ipsos.org, Available at:

    https://www.ipsos.com/en-uk/what-do-public-think-about-nhs

    [Accessed 16 September, 2017].
23. Saunders, C. L., Steventon, A., Janta, B., Stafford, M., Sinnott, C., Allen, L., & Deeny, S. R. (2021). Healthcare utilization among migrants to the UK: cross-sectional analysis of two national surveys. Journal of health services research & policy, 26(1), 54-61.
24. Collins, K. F. (2001). International students' perceptions of health care. The Journal of School Nursing, 17(3), 140-147.
25. Anandamayee Singh (2019). How inclusive is healthcare for international students across the world? Braingainmag.org, Available at:

    https://www.braingainmag.com/how-inclusive-is-healthcare-for-international-students-across-the-world.htm

    [Accessed 4 May, 2019].
26. Dan Shepard (2023). Average Cost of Health Insurance (2023), Valuepenguin,org, Available at:

    https://www.valuepenguin.com/average-cost-of-health-insurance

    [Accessed 28 July, 2023].

    http://www. cpc. ac. uk/docs/2017_SoGIS_Technical_Report. pdf.
27. Masai, A. N., Güçiz-Doğan, B., Ouma, P. N., Nyadera, I. N., & Ruto, V. K. (2021). Healthcare services utilization among international students in Ankara, Turkey: a cross-sectional study. BMC Health Services Research, 21(1), 1-8.

Research on the influence of house price to income ratio on female fertility intention and countermeasures – Research proof based on stata

Luo Xiang*

School of Economics, Henan University of Economics and Law, Zhengzhou, Henan Province, 450000, China

Abstract. Low fertility rates are accelerating the aging of populations around the world. Developed countries experienced demographic shifts in the 20th century. Many developed countries are aging because of improved medical care, longer life expectancy and declining fertility rates. In the past decade or so, China has not escaped the impact of an aging population. The increase of the elderly population has brought great challenges to the social security system, so the aging problem in China deserves our attention. As we all know, the factors that affect people's fertility willingness are various, such as economic constraints, social and cultural influences, and traditional family concepts. In China, high housing prices, an important part of economic factors, are increasingly becoming an important constraint for urban population fertility decisions. Based on the data analysis of stata, this paper studies the impact of housing price on women's reproductive intention from 2015 to 2022. Through data analysis, it is concluded that housing price has a negative impact on women's reproductive intention in inverse proportion.

1.Research Background

With the completion of the seventh national census, China has a total of 1.412 billion people, still the world's most populous country. The proportion of middle-aged and elderly people is larger, and the proportion of infants and teenagers is relatively small. In 2021, the average life expectancy of China's population will be 77.3 years, which means that the residents born in the baby-boom background of the last 50 years in China will leave one after another in the near decade, when the death rate of China's population will further increase, and the continuous low birth rate will hardly fill the missing population. In the near future, the situation of negative population growth in our country may appear. The trend of China's aging society is difficult to block, but the stimulus method of increasing the national birth rate is still some. In recent years, China has introduced a series of population and social security policies. Among them, the adjustment of fertility policy aims to alleviate the negative effects of population aging and its low fertility rate. For example, the introduction of the "three-child policy", the provision of education fees, the implementation of maternity and paternity leave and so on. However, the effect of the policy is not as optimistic as expected, according to statistics, in 2021, China's birth rate is 7.52 per thousand, and the total fertility rate is only 1.3, even lower than the fertility rate of Japan. According to this analysis, there are many other reasons that lead to the decrease of birth rate year by year, such as high living cost, urbanization rate, unemployment rate of local population, elderly dependency ratio and young dependency ratio. Our country's housing market carried out the housing commercialization reform in 1998, which changed from welfare housing distribution to market transaction. In the early stage of the real estate industry, on the one hand, there was no reasonable market supervision; On the other hand, the level of housing prices is often accompanied by the high level of local governments' land transaction volume, which makes the real estate industry savage growth.According to data released by the National Bureau of Statistics, housing prices rose from 1,854 yuan per square meter in 1998 to 10,601 yuan per square meter in the first half of 2023. The increase is more than 5.7 times. As an important consumption of the national economy, the soaring housing price has increased people's living cost, and has a huge impact on the development of social economy and people's production and life.As an important consumption of the national economy, the surging house price has improved people's life, and has a huge impact on the development of social economy and people's production and life. American economists Easterlin and Crimmins believe that the cost of living is the key factor that transforms "natural birth" into "conscious birth control" (1985). Exploring the impact of housing price on women's fertility intention is conducive to providing valuable research basis for solving the problem of low fertility rate.

* Corresponding author;1186095358@qq.com

2.Literature review

In the existing studies, scholars have different understandings about the effect of housing price on fertility. Some studies have found that housing prices are directly proportional to the desire to have children. Using survey data from West Germany and the Netherlands, Mulder and wagner[1] found that households that owned property before marriage were more likely to have children, while households that did not own property were more likely to delay childbearing in order to gain happiness (2001); Lovenheim and Mumford[2] showed that for every $100,000 increase in a homeowner's home equity, the likelihood of having a child increased by 16 to 18 percent, based on changes in wealth associated with changes in home prices. Growth in housing wealth increases relative total fertility, and the responsiveness of fertility to housing wealth increases over time (2013). There are also research findings

Housing prices are inversely proportional to the desire to have children. Sato study found that fertility rate is closely related to regional economic development and population concentration degree. Large city size has a positive impact on income, but a negative impact on fertility. Li Yonggang et [3]all believe that the rapid rise of housing prices has a significant inhibitory effect on residents' fertility willingness (2012)

3. Housing price changes and social status quo in Chinese cities

As an important consumption of the national economy, the soaring housing price has increased people's living cost, and has a huge impact on the development of social economy and people's production and life. According to data released by the National Bureau of Statistics, housing prices rose from 1,854 yuan per square meter in 1998 to 10,601 yuan per square meter in the first half of 2023. The increase is more than 5.7 times. With the gradual acceleration of China's reform and opening up, urbanization construction is also closely followed, our living environment and experience continue to improve, but the modern social development environment also means that the cost of birth continues to increase. Raising children involves a lot of costs, both direct and indirect. The direct cost in the economic concept refers to the direct capital expenditure of all costs related to biological survival costs, medical and health costs, education costs, marriage costs, etc. spent in the process of raising a child. The indirect cost refers to the opportunity cost of raising children, including all kinds of transferred income and time lost by each family due to the birth of life, in fact, the most important is the gain and loss of the mother.Because the mother is responsible for the conception of the child, and in traditional family life, the mother is usually responsible for the child care, at the cost of the loss of time value, the sacrifice of labor market income to care for the child, this part can also be seen as the opportunity cost of parenting costs.

Under our established constraints, as the cost of childbirth increases, people's willingness to have children will decline. High housing prices in some areas do affect people's willingness to have children. Starting from the traditional Chinese thought of living and working in peace and contentment, people want to provide a happy environment and place for themselves, their partners and children[4]. Since the 21st century, housing prices continue to rise, rising house prices have been accompanied by high mortgage prices. Housing prices are so high that most buyers have to take out loans to complete the purchase. At the current wage level, if young people want to own their own property, most of the time they need to take out a loan to do so. Housing loans lasting for several decades have doubled the pressure of young people's life and work. Investigation shows that housing loan is the basic constitutive structure of family debt. Among the households with debt, 76.8 percent of them have housing loans, and the average household housing loan is 389,000 yuan, accounting for 75.9 percent of the total household debt. Many people buy homes in their 20s and carry 30-year mortgages; Young people in first - and second-tier cities may even face millions in loans. During this time, when there is economic instability, such as unemployment or health problems, the household's ability to withstand risks is very vulnerable. Especially due to the impact of the global epidemic, many young people are living in fear of unemployment.people's living costs continue to rise, and then because the house in the heart of the Chinese people is the foundation of a family, is just needed. Therefore, in this context, most of the working-age population is unwilling to marry, unwilling to have children, and fully devoted to their own work.Under the influence of cultural and various factors, the enthusiasm of young people to buy a house has not changed significantly against the background of high housing prices. But rising house prices and greater mortgage stress greatly affect People's Daily lives in many other ways. Household consumption in China has trended down since 2015, when a rapid rise in housing prices across the country worsened.The debt burden of the household, which curbs the consumption power of family members. Many young people choose to spend more time at work to pay their mortgage, saving money to cover financial risks.

The high cost of raising children will lead to a decline in the birth rate.

4. Empirical analysis

4.1 Reference model

In order to investigate the impact of the ratio of house price to income on the fertility rate and clarify the causal relationship between them, this paper sets up the following basic model;

$$Ferwill=\beta 0+\beta i lnhp\_in+\beta i\dot{X}i+\varepsilon$$

Among them, I on behalf of individuals, Ferwill, on behalf of the women's childbearing willing hp_in on behalf of the city's housing price to income ratio, beta coefficient after I is prices taking natural logarithms.

https://doi.org/10.1051/shsconf/202317801004

4.2 Variable selection and descriptive analysis

4.2.1. Explained variables

In this paper, the main explained variable is the reproductive intention (ferwill) of the reproductive age group. Drawing on Xiao Guoan and Zhou Muyuan,[5] the survey method is based on the question "If there are no policy restrictions, would you like to have a child?" in the China General Social Survey Questionnaire. Related issues.According to the data table of China Statistical Yearbook, the survey period is generally from June to December of the current year.

4.2.2.Explanatory variable

The main explanatory variable of this paper is the ratio of house price to income (hp_in). According to the Data Table of China Statistical Yearbook[6], the survey period is generally from June to December of the current year, so this paper adopts the data of the housing price to income ratio one period behind as the explanatory variable. Drawing on the practice of Zhang Li et all [7], the ratio of house price to income is measured by the average salary of the respondents in the region and the employees in the same year, which can fully show the difficulty of the urban population of childbearing age to buy a house. The data comes from China Statistical Yearbook, because in this yearbook, only the provincial information of the respondents was disclosed in the comprehensive social survey of China, and the questionnaire did not involve housing price and other related questions, which can effectively avoid the endogenous problem between the control variables at the individual level and the family level.

4.2.3. Variable of control

Based on the existing literature, this paper analyzes from several aspects, and selects urbanization rate, unemployment rate, illiteracy rate, male/female ratio, old-age dependency ratio, juvenile dependency ratio and total dependency ratio as control variables.

The descriptive statistical results of the main variables in this paper are shown in Table 1.

Table 1. Descriptive statistics

VARIABLES	(1) N	(2) mean	(3) sd	(4) min	(5) max
code	7	1	0	1	1
Year	7	2,018	2.160	2,015	2,021
Fertility willingness	7	10.79	2.183	7.520	13.57
Rate of urbanization	7	61.32	2.688	57.33	64.72
House price to income ratio	7	0.304	0.0076 6	0.289	0.313
Unemployment rate	7	3.957	0.245	3.620	4.420
Illiteracy rate	7	4.507	0.911	3.210	5.420
Juvenile dependency ratio	7	24.03	1.361	22.60	26.20
Old age dependency ratio	7	17.19	2.409	14.30	20.82
Total dependency ratio	7	41.35	3.639	37	46.44
Ratio of male to female	7	104.5	0.592	103.3	105.0
New land supply	6	2,781	1,263	301.6	3,676

4.3 Empirical study of the basic model

The empirical analysis of the basic model shows that the impact of housing price to income ratio on fertility intention shows a significant negative effect, and has a strong robustness. As shown in Table 2, from Column (1) to Column (4), after the continuous addition of control variables, the ratio of housing price to income is still significantly negative. Through the regression results, we can conclude that when the control variables of the ratio of men to women, the unemployment rate, the illiteracy rate and the total dependency ratio are successively added, the impact of the housing price-income ratio on the fertility intention of the population is always at the significant level of 1%, which has a strong explanatory power. Income status can comprehensively reflect an individual's education, ability, work experience and so on. On the other hand, it also shows an individual's human capital in work. When human capital is high, the opportunity cost of having children is large, and at this time the income effect of having children is smaller than the substitution effect, leading to a decline in fertility intentions.

Table 2. Results of regression

Fertility willingness	(1) Fertility willingness	(2) Fertility willingness	(3) Fertility willingness	(4) Fertility willingness
House price to income ratio	-192.433***	-190.493**	-69.456	-517.813
	(5.68)	(4.30)	(0.23)	(3.37)
Ratio of male to female	2.120***	2.225**	0.889	3.898
	(6.39)	(3.61)	(0.34)	(3.37)
Unemployment rate		-0.391	-1.377	-15.337
		(0.27)	(0.38)	(-2.39)
Illiteracy rate			1.536	-16.603
			(0.43)	(-2.45)
Old age dependency ratio				-4.680
				(-2.50)
Constant	-269.463***	-281.312**	-115.636	-338.414
	(-8.02)	(-4.49)	(-0.35)	(-2.95)
Observations	7	7	7	7
R-squared	0.861	0.862	0.879	0.989
F test	0.00167	0.0137	0.101	0.0798
r2_a	0.792	0.725	0.636	0.934
F	46.98	23.69	9.099	90.00

Robust t-statistics in parentheses
*** p<0.01, ** p<0.05, * p<0.1

5.Response plan and solution measures

Based on the China General Social Survey data from 2015 to 2022, this paper uses stata model to empirically analyze the impact of housing price to income ratio on the fertility intention of the working-age population. It can be concluded that the ratio of housing price to income significantly inhibits the fertility intention of the working-age population, that is, with the increase of housing price year by year, people's fertility intention decreases year by year. Faced with such a grim situation, the country should vigorously reform to change the status quo.Therefore, this paper suggests to improve the fertility intention of the population of childbearing age from the following aspects.

First, we should make positive macro-control on the real estate market to reduce the housing costs of contemporary young people. In addition, we should vigorously develop the housing rental market and actively improve the housing system with multi-subject supply, multi-channel guarantee and simultaneous rental and purchase.The state should continue to strengthen the control and regulation of commercial housing prices to prevent housing prices from rising too quickly

Second, we should vigorously promote the concept of fertility in the new era and increase the fertility willingness of the right-age population. We should actively construct a traditional value-oriented conception of fertility. We should take the essence of traditional culture and discard the bad ones, and create a good social environment for young people based on the new residual concept of excellent traditional culture, and help the age-appropriate population accelerate the process of marriage and childbearing.

Third,formulate relevant laws and policies to re-plan and build the rental market. We will improve the rental housing market, crack down on dark intermediaries, strengthen monitoring of the housing rental market, and build and maintain a sound rental housing market. While vigorously developing the rental housing market, the country should publicize that "renting is as good as buying", change the housing concept of young people, and improve the related supporting facilities and services of rental housing in terms of policy.

Fourth, promote the fair distribution of educational resources, promote the reform of school choice, curb the culture of school choice, and crack down on malicious speculation of high school district housing.

Fifth, preferential policies for the settlement of talents should be widely carried out, so that excellent labor can be settled even if they do not have houses, and their children can go to school in the region even if they do not have houses.

Sixth, we should curb investment buying and forced buying, and make the price of commercial housing more reasonable. In short, the combination of measures such as controlling the vicious growth of housing prices, creating a good housing atmosphere and improving the concept of renting can make young people feel more at ease to have the next generation, so as to avoid the fertility rate falling off the cliff.

Encourage young people to give birth to good children, and at the same time respect the family key and family values of people of childbearing age. Fertility is not only a private matter, but also related to the strategic goal of national development and sustainable economic development.

Reference

1. Mulder,Wagner The connections between family formation and first-time home ownership in the context of West Germany and the Nether-Lands[J].European Journal of Population,2001 (17) :137-164.
2. Lovenheim and Mumford.Do family wealth shocks affect fertility choices? Evidence from the housing market[J].Review of Economices and Statistic,2013(2) ;464-475
3. Yonggang Li, Xiang Li, Bo Gao. Research on the influence of rising housing price on residents' fertility behavior.
4. Liu Peiwen. Study on the influence of rising housing prices on the fertility intention of Chinese urban population. Journal of Xi 'an University of Finance and Economics 2022. Vol. 35, No. 4.
5. Xiao Guoan, Zhou Muyuan. Effect of housing price to income ratio on fertility intention of child-bearing age population. Journal of Hunan University of Technology, Vol. 28, No.2, 2023.
6. National Bureau of Statistics of China. The Target of Sustainable Recovery and Development of National Economy in 2021 is Well Achieved (2022-01-17) (2022-06-28)
7. Zhang, L., He, J. & Ma, R. How Housing Prices Affect Labor Mobility [J]. Economic Research Journal, 2017,52 (8): 155-170.

Research on the Construction of Digital Government in Heilongjiang Province based on SMARTGOV Maturity Model

Chuan-li Wei [1,*] and Ya-xiu Ding [2]

[1]Harbin, Heilongjiang Province, China, Professor of Finance and Public Management College of Harbin University of Commerce, Master 's supervisor, Doctor of Management. Postdoctoral in Management Science and Engineering, PMP (212202), Member of the American Association of Professional Managers (530902), Research Directions : Public Policy Research, Project Management.
[2]Tongliao, Inner Mongolia, China, is a graduate student of Finance and Public Management College of Harbin University of Commerce. The research direction is public policy research.

Abstract. Digital government aims to provide intelligent and agile personalized public services for government stakeholders. Based on the SMARTGOV maturity model, this paper constructs a digital government maturity model from four dimensions : basic information service, interactive service, integrated service and intelligent service. It is used to evaluate the current level of digital government construction in 12 prefecture-level cities in Heilongjiang Province and the Greater Khingan Mountains and draw conclusions.

1 introduction

Digital government refers to the use of subversive technologies such as artificial intelligence, blockchain, cloud computing, and the Internet of Things by public organizations to provide the public with intelligent, integrated, personalized, and interactive government services. It aims to introduce innovative technologies to transform the supply of public services and improve decision-making procedures, thereby enhancing public service satisfaction. With the rapid development of a new generation of science and technology, the construction of digital government has not only become an important support for promoting the modernization of national governance system and governance capacity[1], but also an important support for innovating government governance model. The State Council 's guidance on ' strengthening the construction of digital government ' also emphasizes the introduction of digital technology into government governance, using the empowerment of technology to improve the efficiency of public services and the satisfaction of the masses, so that the achievements of digital government construction can benefit all the people more and more fairly. Since the party 's 19 th fourth middle school in 2019 first proposed the development of ' digital government construction ', it has achieved fruitful results. However, due to geographical location and economic development, the level of digital government construction in Heilongjiang Province has been lagging behind[2]. Based on the smart government maturity model, this paper evaluates the digital government of Heilongjiang Province and draws conclusions.

2 Model introduction

The model consists of five dimensions, namely, basic information service, interactive service, transactional service, participatory service and intelligent service[3]. Based on the actual situation of Heilongjiang Province, this paper selects three of the five dimensions, namely, basic information service, interactive service and intelligent service dimension, and adds an integrated service dimension on this basis. Combined with the existing literature, 20 indicators are selected to evaluate the construction of each dimension of digital government in Heilongjiang Province, and a maturity model for evaluating the situation of digital government in Heilongjiang Province is constructed[4].

2.1 Basic information services

The first dimension of the digital government maturity model is basic information services. The evaluation criteria of this dimension are mainly to measure the ability to provide contact information ; allow the public to obtain information about laws, regulations, public policies and related documents ; and whether it is possible to obtain relevant information through the search engine 5 indicators[5].

2.2 Interactive service

The second dimension of the digital government maturity model is interactive service, which is the stage of interaction between the government and the public[6]. For example, whether the public opinion collection

* Corresponding author: weichuanli1975@163.com

channel provides email address, and whether the interaction between the government and the public can be carried out online. Four indicators such as cadre supervision and reporting.

2.3 Integrated services

The third dimension of the digital government maturity model is integrated services, that is, the construction of the integrated platform of government services in Heilongjiang Province, including whether the online handling rate reaches 100 %, and the online government service zero running rate, online handling matters That is, the handling rate of five indicators.

2.4 intelligent services

The fourth dimension of the digital government maturity model is intelligent service. This dimension mainly measures the use of intelligent technology in government websites, including whether the government website is on-line data open service, whether it is on-line barrier-free service, whether it can change the size and color of the font, and whether it has intelligent robot / intelligent question answering function.

3 Research methods

Based on the digital government maturity model, this paper visits the government websites of 12 prefecture-level cities in Heilongjiang Province and the Greater Khingan Mountains region. If the target service or function is provided through the federal website, the corresponding indicator is given a ' value 1 '. If not, it is given a ' value 0 '. The digital government maturity value of a government website is the sum of 20 indicators for evaluating the government website. The 20 indicators representing the four dimensions of the standard are evaluated. The 13 prefecture-level city government websites meet each evaluation indicator, that is, the proportion of each indicator in the 13 government websites is shown in the Table 1 :

Table 1. Criteria used to evaluate the digital government of Heilongjiang Province.

Dimensional	Evaluating indicator	Proportion
Basic Information Services	1)Whether the home page of the government website can obtain content through the search engine	100%
	2)Does the home page have specific information on public policy	92%
	3)Does the government website provide links to other local institutions	85%
	4)Whether the government website homepage provides the website mailbox address	46%
	5)Are government websites available in languages other than Chinese	15%
interactive service	6) Whether to provide e-mail address to collect public opinion	92%
	7) Whether the interaction between the government and the people can query the letter information through the online petition	92%
	8) Whether to allow cadres to supervise and report online through government-citizen interaction	23%
	9) Is it possible to file an online personal lawsuit against the public service process through government-citizen interaction	31%
integrated service	10) Does the online service of the government website have a 100 % online service rate %	100%
	11) Whether the online handling of government websites has reached a zero running rate of 90 %	46%
	12) Does the online handling rate of the government website reach 85 %	23%
	13) Whether the compression ratio of online service commitment time limit of government website reaches 95 %	54%
	14) Whether the good and bad evaluation system of government service integration is zero or not	62%
intelligent services	15) Is the government website accessible via mobile devices	100%
	16) Does government website provide open data	15%
	17) Are government websites accessible and able to change font size and color	31%
	18) Does the government website have intelligent robot / intelligent question answering function	23%
	19) Is there a mobile version of the assessed government website	15%
	20) Is the government website on-line aging design / care design	23%

4 Research process

This study evaluates 13 prefecture-level municipal government websites in Heilongjiang Province based on 20 indicators representing four dimensions. The result is shown in the Figure 1, From the graph, it can be found that the Heilongjiang provincial government website has the best performance in the latitude of basic information service, and almost all prefecture-level municipal government websites have reached 50 % ~ 100 % of the relevant indicators in this dimension. That is to say, among the 13 government websites, 12 (92 %) prefecture-level city government websites meet the 50 % -100 % basic information service dimension indicators. In the interactive service dimension, nearly half (46 %) of the 13 government websites meet the 50 % ~ 100 % interactive service dimension index. In the integrated service and intelligent service dimension, the performance is the worst, and few prefecture-level municipal government websites reach 50 % ~ 100 % of the relevant indicators in these two dimensions. That is, among the 13 government websites, only 5 (38 %) and 3 (23 %) websites meet 50 % -100 % of the integrated service and intelligent service dimension indicators.

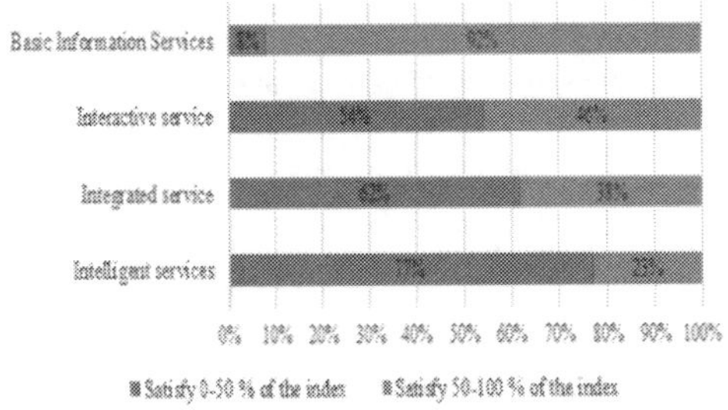

Figure 1 The specific proportion of each dimension of government website

5 Research results

We assign each government website to one of the four digital government maturity levels (i.e., high, medium, medium and low, low) based on its calculated value related to 20 indicators. Government websites with a high level of digital government maturity are at least 15 of the 20 indicators. Government websites that are considered to have a medium level of digital government maturity reach 10 to 15 indicators. The middle and lower groups reached 5 ~ 10 indicators. The government website of the low digital government maturity group has achieved less than 5 indicators.

Table 2 gives the final ranking of government websites based on the total number of indicators they scored in each criterion. The results show that the digital government of all prefecture-level cities in Heilongjiang Province is at a medium and low level as a whole. Among them, there are 6 digital government maturity levels and 7 digital government maturity levels. The top three are Yichun, Jixi and Harbin.

Table 2. Rankings of Government Websites

Government Website	Rank	Score	Level
Yichun,Jixi	1	13	medium
Harbin,Mudanjiang,Daxing 'anling	2	12	medium
Hegang	3	11	medium
Daqing,Shuangyashan	4	10	medium and low
Jiamusi	5	9	medium and low
Qiqihar,Heihe,Suihua	6	8	medium and low
Qitaihe	7	7	medium and low

6 Conclusions

6.1 Basic information services

The government websites of 13 prefecture-level cities in Heilongjiang Province have the best performance in the basic information dimension of the digital government Maturity Model, and 92% of the government websites meet the relevant indicators of 50% to 100%. As shown in Table 1, almost all of the 13 prefecture-level city government websites provide contact information and can quickly obtain relevant content through search engines. In addition, 85%(11) prefecture-level city government websites provide links to other local websites on the home page, which facilitates users to jump to the page and improves the efficiency of public services. However, only 15%(2) of prefecture-level city government websites provide services in languages other than Chinese, indicating that there are serious shortcomings in the globalization of municipal government websites. In the digital age, the integration of governance sinicization and governance globalization Digital governance, as a kind of governance discourse, is the knowledge creation of national governance issues in the digital age[7].

6.2 Interactive service

As shown in Table 1, the construction of digital government in Heilongjiang Province performs well in the dimension of interactive services. Almost all of the 13 prefecture-level city government websites provide email addresses in the public opinion solicitation channels to facilitate the public to speak freely, and people's wishes can be understood through online means. The statistics and analysis of the collected public opinions can provide the maximum value reference for the formulation of policies, so that the formulation of government policies can create greater public value. However, the performance of prefecture-level city government websites in online cadres supervision and reporting and online filing of personal lawsuits is poor. Only 23%(3) prefecture-level city government websites allow citizens to online cadres supervision and reporting through the interaction between the government and the people, and only 31%(4) prefecture-level city government websites can online filing of personal lawsuits on public service processes through the interaction between the government and the people. This indicates that although the websites of various prefecture-level cities in Heilongjiang Province provide some convenient conditions for the masses in the interactive service dimension, there is still a lot of room for progress.

6.3 Integrated services

Heilongjiang Province proposed the construction of an integrated online government service platform for the first time in 2019, and the online handling of entry matters and entry departments in various prefecture-level cities has increased year by year, and the online operation rate of 13 prefecture-level city government websites has reached 100%, which has greatly improved the working efficiency and enhanced the satisfaction of the people. However, among the various prefecture-level city government websites, only half (46%) of the prefecture-level city government websites have reached 90% zero running rate, and only 23%(3) of the prefecture-level city government websites have reached 85% online handling rate, which indicates that although the online service of the government website has reached 100%, However, there are still great deficiencies in the aspect of "immediate application and action". In terms of the good and bad evaluation of the integrated government service system, only 38%(5) prefecture-level city government websites achieved zero error evaluation, which indicates that there is still a lot of room for improvement of the integrated government service system of various prefecture-level city governments.

6.4 intelligent services

Service is the ultimate goal of digital government governance, and digital government is committed to providing personalized services, precise services and intelligent services to enhance the public's service satisfaction and sense of access[8]. As for the intelligent service dimension, the evaluation results show that the performance of prefecture-level city government websites in Heilongjiang Province is generally inferior to other dimensions. This point has been pointed out in the previous 13 government websites, only 23%(3) government websites meet the relevant indicators of intelligent service dimension 50% to 100%. As shown in Table 1, almost all government websites can be accessed through mobile devices, but only 15%(2) have launched the corresponding mobile version of government websites, and most of them have only developed corresponding mobile device apps. On the home page of government websites, only 23%(3) websites have intelligent robot/intelligent question and answer function, and only 31%(4) government websites have barrier-free function, and can change the font size and color and provide voice broadcasting service through this function, which indicates that the construction of digital government in Heilongjiang Province still needs to be strengthened in terms of integration and collaboration. The government of Heilongjiang Province focuses on special groups, and the government service network officially launched the "Service Zone for the Elderly" in 2022. However, in addition to the provincial government websites, only 23%(3) of the prefecture-level city government websites in Heilongjiang Province have age-appropriate design/care design.

Acknowledgment

1.Heilongjiang province education science planning key issues (GJB1422390)

2.Harbin University of Commerce 's 2021 school-level teaching reform and teaching research project : ' Ideological and political brand creation of the core curriculum of administrative management major under the two-line mixed mode ' -Taking ' public policy ' as an example.

References

1. Wang Yimin. The overall architecture and evaluation system of digital government [J].China Leadership Science, 2020 (01) : 65-70.
2. Cai Defa, Li Qing.Research on the Problems and Countermeasures of Digital Government Construction in Heilongjiang Province [J].Commercial Economy, 2020 (12) : 1112 + 19.
3. Hujran O, Alarabiat A, Al-Adwan A S, 等. Digitally Transforming Electronic Governments into Smart Governments: SMARTGOV, an Extended Maturity Model[J]. Information Development, 2021,1-24.
4. Andersen K V, Henriksen H Z. E-government maturity models: Extension of the Layne and Lee model[J]. Government Information Quarterly, 2006, 23(2): 236-248.
5. Chatfield A T, Alhujran O. A cross-country comparative analysis of e-government service delivery among Arab countries[J]. Information Technology for Development, 2009, 15(3): 151-170.
6. Zhao Jinxu, Zhao Juan, Meng Tianguang. Research on the theoretical framework and evaluation system of digital government development-Empirical analysis based on 31 provincial administrative units and 101 large and medium-sized cities [J].China Administration, 2022 (06) : 49-58.
7. Fan Fengchun, Ou Limei. Construction of Chinese discourse system of Digital Government Governance [J]. Journal of Shanghai Administration University,2022,23(04):38-46.
8. Wang Xuejun, Chen Youqian. Digital government governance performance generation path: Qualitative comparative analysis from the perspective of Public value [J]. Electronic Government,2021(08):53-66.

A Study on the Implementation of Civic Teaching in Higher Education Programmes Based on Smith's Model

Xiaoyue Jiang *

Harbin Business University, Administrative Management

Abstract: Education is the foundation of the country, the rise of education is the rise of the country, to achieve the great rejuvenation of the Chinese nation on the road, the curriculum Civic and political teaching to meet the dual requirements of teaching and educating people, is to cultivate talents in the new era is an indispensable and important part. Especially colleges and universities are the main way to export talents for the motherland, so the construction of Civics and Politics in colleges and universities is more important, however, in the process of Civics and Politics programme construction in colleges and universities are faced with many obstructive factors. Therefore, according to Smith's policy implementation process model, we analyse the difficulties in the implementation of Civic Politics teaching from the four levels of policy itself, implementation agency, target group, and policy environment, and point out that the teaching content is not yet perfect, the awareness of the implementation body is weak, and the external environment is not enough to support the implementation of Civic Politics teaching, and put forward the corresponding countermeasures and suggestions to ensure that Civic Politics plays an active and effective role in cultivating China's modern talents. and put forward corresponding countermeasures and suggestions to ensure that the Civics and Politics of the curriculum plays a positive and effective role in cultivating modern talents in China.

1 Preface

Education is a major plan of the country and the party, and the root of educating people lies in educating morality[1]. Since 2004, the state has promulgated policies to strengthen the ideological and political work of college students, and promoted a new round of teaching reform of "subject moral education" in 2005. Shanghai's colleges and universities have taken moral education as a key reform pilot, and gradually transformed from "ideological and political curriculum" to "curriculum ideological and political curriculum" by 2014. Curriculum ideology and politics make college teachers shoulder the burden of moral education, which is the change of educational thought and curriculum concept. The Guiding Outline for the Construction of Curriculum Ideology and politics in Colleges and Universities issued by the Ministry of Education proposed that all colleges and universities, all teachers and all courses should shoulder the responsibility of "educating people", put forward the "good canal" and "planting well", and unify the "explicit" and "dark" education. We will build a comprehensive education pattern for all employees. University is the main battlefield of ideological and political education and ideological and political construction of college students in our country.Curriculum Civics and Politics is to make its position in the curriculum effectively reflected by constructing a whole-staff, whole-process and whole-course moral education,organically combining theoretical knowledge, value concepts and spiritual aspirations, so that all kinds of courses can form a complementary relationship with it, and form the basic educational concept of "establishing morality and educating people"[2]. The basic educational concept of "cultivating moral values".Ideological and political lessons are not incidental to other courses, and being reasonable is not exclusive to ideological and political lessons, and there is no ideological and political truth divorced from reality. Curriculum thinking and politics is through the construction of the whole process, the whole course of moral education, so that its status in the curriculum can be effectively reflected, the theoretical knowledge, value concept, spiritual appeals and other organic combination, so that all kinds of courses complement each other, form the basic education concept of "moral education", so that teachers and students can communicate through the soul. It has a profound effect on thought and behavior without being aware of it.

2 Theoretical framework for the implementation of the teaching of curriculum ideology and politics

According to Graham T. Allison, "the function of programme definition accounts for only 1 per cent of the

* Xiaoyue Jiang E-mail:573797307@qq.com

achievement of policy objectives, while the remaining 99 per cent depends on the effective implementation of policies" [3].Thomas Smith (T.B. Smith) pioneered the basic theoretical analysis framework for policy implementation, and proposed the Smith policy implementation process model (shown in Figure 1, hereinafter referred to as the Smith model)[4].

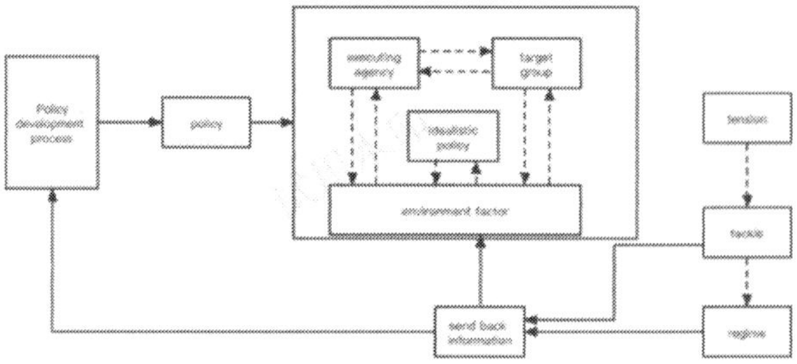

Figure 1 Smith Policy Implementation Process Model

The theoretical framework mainly involves four variables: firstly, the idealised specific teaching content as well as ideas of Curriculum Civics; secondly, the main body of implementation of Curriculum Civics teaching, which is mainly the Central Education Bureau, various local education bureaus, various colleges and universities as well as the lecturers; thirdly, the target group of Curriculum Civics teaching, i.e. the target group of Curriculum Civics teaching, i.e. the students of colleges and universities; and fourthly, the macro-environment of Curriculum Civics teaching, which is mainly concerned with the The overall social environment and atmosphere and economic environment, teaching concepts and so on.

Smith model comprehensively analyzes the various factors that affect the implementation effect in the process of policy implementation and the interaction relationship from multiple angles, and provides a new idea and vision for the follow-up research on public policy implementation. In June 2020, the Ministry of Education issued the Guiding Outline for the Construction of Curriculum Ideology and Politics in Colleges and Universities, which makes a macro-top-level design for the implementation of curriculum ideology and politics. In December 2021, the Ministry of Education issued the "Standards for the Construction of Ideological and Political Theory Courses in Colleges and Universities (2021 Edition)", which provides specific guidance for the orderly development of ideological and political courses . Therefore, the constraints and solutions in the implementation of ideological and political teaching in courses are analyzed with the help of Smith model. It plays a unique role in exerting the sustained effect of curriculum ideology and politics. According to the characteristics of ideological and political teaching of curriculum, this paper constructs the implementation analysis framework of ideological and political teaching of curriculum on the basis of Smith model (as shown in Figure 2).

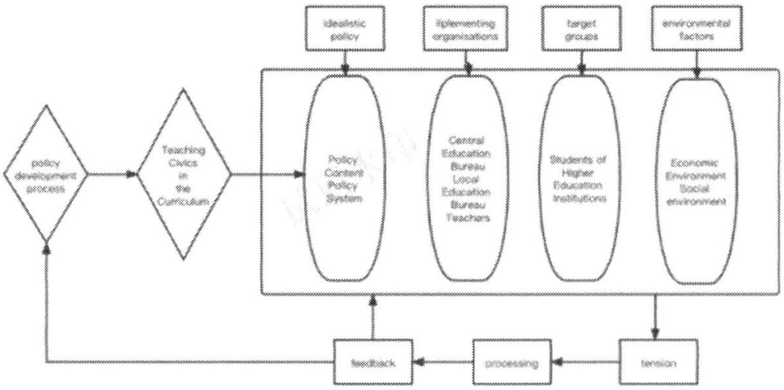

Figure 2 Framework for analysing the implementation of curriculum-based Civics teaching

3 Constraints on the implementation of the teaching of Civics and Politics in colleges and universities

(1) Civic politics teaching content in colleges and universities is not yet perfect.

Thought determines action. On the road to the great rejuvenation of the Chinese nation, we not only need talents with knowledge and ability, but also mouldable talents with correct ideology and harmonious unity with the overall consciousness of the motherland. The concept of curriculum ideology and politics has been put forward since 2014 as a "compass" to guide young people's thoughts and cast correct ideals and beliefs in the new era. It has been successfully implemented in some areas and universities, but there are still some weak links in the teaching process itself.The content of the teaching of Civics and Politics in colleges and universities has not yet been perfected, the teaching design lacks pertinence, and the corresponding curriculum system has not yet been formed, which makes the teaching of Civics and Politics of the curriculum confusing[5].On the one hand, there is a lack of specific teaching content that combines specific disciplines with curriculum ideology and politics, unreasonable arrangement of teaching time allocation in class, review and consolidation form after class, etc., it is difficult for students to form a systematic knowledge system of curriculum ideology and politics, and the external curriculum form is not clear and difficult to internalize in the mind. Secondly, the construction of the curriculum ideological and political system has not yet been perfected. For students of different ages and sections, the curriculum ideological and political system has not yet formed, which makes the curriculum ideological and political teaching chaotic, and the curriculum ideological and political philosophy lacks universality. At present, the curriculum ideological and political concept has been successfully implemented in a few areas in China, and the curriculum ideological and political system has not yet formed in most areas.

(2) Weak awareness of the main body of Civics teaching in colleges and universities

Teachers have not received professional training in Civics and Politics of the curriculum, and their lack of ability and professionalism limit the effect of teaching Civics and Politics of the curriculum.

(3) The target group of Civic Politics teaching in colleges and universities has diversified and complex educational backgrounds.

The big gap between rich and poor and the uneven educational resources cause the implementation effect of Civic Politics in the curriculum to be very different.The object of ideological and political teaching of college curriculum, that is, the target group, seems to be single, but in fact, it is different in region, major, grade and middle school educational background. On the one hand, for students in the eastern region, the educational resources are rich in childhood, while the western region is relatively backward, the wealth gap between the East and the west is large, and the educational resources are unbalanced, which may greatly lead to different implementation effects of ideological and political teaching of curriculum. On the other hand, college students, especially graduate students, have relatively heavy learning and scientific research tasks, strong ideological autonomy, and are easily bored with curriculum ideology and politics. In addition, due to different educational backgrounds in middle schools, liberal arts and science and engineering have different understandings and absorption of curriculum ideology and politics, which seriously affects the implementation effect of curriculum ideology and politics.

(4) Insufficient support of the external environment for the teaching of Civic Politics in higher education programmes.

At present, there is no clear legal regulation on the teaching of ideology and politics in the curriculum, and the documents of the Ministry of Education have not formed a harmonious and unified situation with the curriculum of colleges and universities.First of all, from the point of view of economic environment, due to the large difference of economic development in various regions of our country, there is an obvious gap in the capital investment in education, so the fund support in many regions is insufficient in the construction of curriculum ideological and political construction, which leads to difficulties in the construction of curriculum ideological and political construction. Secondly, from the perspective of social environment, at present, there is no clear legal regulation on curriculum ideological and political teaching, the documents of the Ministry of Education have not formed a harmonious and unified situation with the curriculum setting in colleges and universities, and curriculum ideological and political teaching has not been widely accepted by the public, and a large part of the public have failed to understand the real role and significance of curriculum ideological and political teaching. Curriculum thought and politics is still a very new and unfamiliar teaching mode. Although quality education is strongly advocated, there are still many colleges and universities pay too much attention to exam-oriented education, only the achievement theory, and do not realize the importance of curriculum ideology and politics for the development of students and talent training.

4 Cracking the path of the teaching of Civics and Politics in colleges and universities

(1) Improve the content and system of teaching Civics and Politics in colleges and universities.

Textbooks are teaching books that reflect the specific content of a course comprehensively and systematically according to the curriculum standards and actual needs.Integration of teaching materials to ensure the scientific and integrity of the teaching content at the same time flexibly increase the elements of ideology and politics, to achieve effective articulation, can provide theoretical protection for the reform of the course of ideology and politics teaching.

(2) Enhance the awareness of the main body of Civics teaching in colleges and universities.

Thought is the forerunner of action. Only by strengthening the consciousness of the subject of curriculum ideological and political teaching can curriculum ideological and political teaching be carried out better. The world is undergoing profound changes unseen in a century, and the process of the great rejuvenation of the Chinese nation is unstoppable. To achieve Chinese-style modernization, we not only need high-level and high-tech talents, but also need talents with correct ideological character and values. Therefore, the construction of ideological character and values is particularly important.The Education Bureau should pay attention to the setting and content of the Civics of the curriculum, and change from "supervision" to "guidance" for each institution, giving more support and guidance to the institutions.

(3) Strengthening the awareness of target groups

The acceptance degree of the target group has a significant impact on the effect of policy implementation, and the main body of ideological and political teaching is the students in major universities. Because students have different cognition of curriculum ideology and politics, and there are many contents of curriculum ideology and politics, it is necessary to form a comprehensive and systematic system of collaborative education, so as to optimize the combination of moral education elements, so as to achieve the best education results.Multi-dimensional and all-round investigation should be carried out to explore the paths and methods of breaking through the difficulties of teaching Civics and Politics in the curriculum, closely integrating the curriculum teaching with the curriculum cultivation, and constantly improving the comprehensive quality of students' knowledge, feelings, intentions and behaviours.

(4) Optimising the environment for the implementation of teaching Civics and Politics in the curriculum

In view of the lack of supporting environment for curriculum ideological and political teaching, we should first increase the financial support for educational resources to a certain extent. The economic base determines the superstructure, and financial support is indispensable for the success of curriculum ideological and political construction.Local governments improve the quality of curriculum Civics teaching by establishing corresponding reward systems and providing schools with regular training and assessment of curriculum Civics teaching methods.We will expand education and focus on student development. Ideological and political education itself is extensive. Ideological and political education in colleges and universities should be based on the actual situation and combine with the actual needs of students, such as the employment form of graduates and employment planning and other issues that students are concerned about and need to be solved. Problems can be solved through ideological and political study. Ideological and political education in colleges and universities should play a leading role in the development of students.

5 Conclusions

Curriculum Civics is an indispensable link in the cultivation of modern talents, at this stage, the implementation of curriculum Civics teaching still lacks pertinence, hierarchy and scientificity, and the degree of understanding and implementation of curriculum Civics teaching in various regions are not the same, and the differences are very different, which reflects a great imbalance in curriculum Civics, this paper analyses the constraints on and paths to cracking the implementation of curriculum Civics teaching based on the Smith's model of the implementation process of the policy, mainly divided into two aspects of thought and practice, as well as the government, universities, teachers and students. The path of cracking is mainly divided into two aspects of thought and practice, as well as the four main bodies of the government, colleges and universities, teachers and students, however, perfecting the existing situation of curriculum Civics teaching cannot be achieved overnight, and it requires the common and sustained efforts of all walks of life in the society, to fully explore and use Xi Jinping Socialism with Chinese Characteristics in the new period of the new era of socialism and education, to continuously lead college students to establish the spirit of patriotism, to It is necessary for all sectors of society to make joint and continuous efforts to fully explore and apply education and education in the new socialist era of Xi Jinping Socialism with Chinese characteristics, to lead university students to establish the spirit of patriotism, to cultivate highly sophisticated talents who can shoulder the great rejuvenation of the Chinese nation, and to send usable talents to society.

References

1. Gao Deyi, Zong Aidong. Curriculum Civics:The inevitable choice to effectively play the role of the main channel of classroom education[J]. Thought Theory Education Guide,2017(01):31-34.DOI:10.16580/j.sxlljydk.2017.01.008.
2. Liu Xiaochuan. An analysis of the way forward for the construction of curriculum ideology and politics in colleges and universities in the new era[J/OL]. Contemporary Education Forum:1-10 [2023-08-22]. https://doi.org/10.13694/j.cnki.ddjylt.20230710.001.
3. MatlandR.Synthesizing the Implementation Literature: The Ambiguity — Conflict Model of Poli-cy Implementation [J]Journal of public adminis-tration and research,1995(02) .
4. Smith,T.B.The policy implementation process[J].Policy Sciences,1973(02):197—209.
5. Liang Ping. The Four-Level Goal Theory of "Establishing Virtues and Shaping People" in Curriculum Civics and Politics[J]. Journal of Henan Normal University (Philosophy and Social Science Edition),2023,50(04):151-156.DOI:10.16366/j.cnki.1000-2359.2023.04.22.

A Comparative Study on Child Welfare Policies in China, Britain and Japan

Bo Su *

Harbin University of Commerce, Songbei District, Harbin, Heilongjiang, China

Abstract: Child welfare is an indispensable part of every country's social security system. In the social undertakings of every country, child welfare policies are also extremely important. The quality of child welfare policies not only affects the growth of children, but also has a subtle impact on the future of a country. Most Western countries have conducted research and implementation on child welfare earlier than China, placing greater emphasis on child welfare. China has a large population base and a large number of newborns born each year, making child welfare policies crucial. This article compares and analyzes the child welfare policies of China, the United Kingdom, and Japan, and draws inspiration from the relevant experiences of the UK and Japan for the development of child welfare in China.

1. Introduction

In social welfare, child welfare has always been a key area o f research for experts and scholars from various countries. Providing welfare for children during childhood is a foundation for their growth, protecting them from harm and making their growth healthier and smoother. Being able to adapt to social changes and improve children's survival ability[1].

The main body of the broad child welfare policy is still the relevant policies issued by the state or government to improve the quality of life for all children, benefit their physical and mental health, and promote their future development. Narrowly defined child welfare policies are institutional measures provided by civil affairs departments to improve the living conditions of disadvantaged children[2].

In China, child welfare policies are becoming increasingly important. Child welfare policies not only affect children themselves, but also affect China's fertility rate. Nowadays, child welfare policies are an important factor influencing fertility policies and a crucial part of social security. This article summarizes and analyzes the advantages of child welfare policies in the United States and Japan, and proposes suggestions for China's child welfare policies.

2. Analysis of China's Child Welfare Policy

China started its child welfare policy relatively late, mainly since the founding of the People's Republic of China in 1949, research on the protection of children's rights and interests has gradually become popular. From theoretical research to the introduction of relevant laws and then to practical implementation. Over the past 70 years of development, China's child welfare policies have continuously improved and formed a sound system. From a legislative perspective, we are learning from the relevant policies of other countries on the protection of children's rights and welfare, and establishing a rule of law system for child protection around our country's constitution, gradually making the protection of children's rights more legal and standardized[3]. The main laws include: Compulsory Education Law, Minors Protection Law, Adoption Law, etc. The State Council's "Outline for the Development of Children in China (2011-2020)" clearly states that a portion is related to child welfare. This has further expanded the scope of child welfare. Promoted the transformation of child welfare: from a filling type to an inclusive type. The country has also invested financial resources in child welfare: in 2012, the central government supported 57 million yuan to promote the development of children's public welfare undertakings; In 2013, the central government funded 2.43 billion yuan to ensure the basic livelihood of orphans; In 2023, the National Development and Reform Commission will support 261 livelihood support projects such as child welfare, with an investment of 3 billion yuan from the central budget.

3. Analysis of UK Child Welfare Policy

3.1 Overview of Child Welfare Policies in the UKLearning from the UK

As a typical welfare state in Europe, the UK has achieved high-quality economic growth and social development,

*author: 1641485252@qq.com

 https://doi.org/10.1051/shsconf/202317801007

partly due to its welfare system, including child welfare policies. As early as the 17th century, there have been children's welfare homes in England, with the majority being the private sector. After the 20th century, Britain's child welfare policy entered a period of rapid development[4]. Since the 21st century, there have been over forty pieces of legislation on child welfare in the UK.As shown in Table 1. The scope of coverage includes education, underage protection, child adoption and upbringing, and mental health. So far, the UK has established a comprehensive system of child welfare policies, including Dominican welfare funds, child trust funds, guardianship allowances, child tax credits, and education maintenance allowances[5]. The welfare of children in the UK has the following characteristics: child welfare is based on legislation, with rich content and continuous improvement; Based on the principle of universal enjoyment, taking into account income related benefits; The welfare of children has a government led characteristic; Improve the child welfare service system, etc.

Table 1. Types of child welfare subsidies in the UK

Types of child subsidies	Subsidy age
Child welfare fund	Under 16 years old
Children's savings account	Parents take care of it before the age of 16, and children can use it after the age of 18
Child Tax Credit	
guardian's allowance	Under 16 years old
Welfare policies for children's education	Enjoy different policies at different ages
Early childhood education, childcare, and children's centers	3 to 5 years old

3.2 Overview of Child Welfare Policies in the UK

British welfare system ideology: The progress of thought is the foundation of everything. The UK has gone through a process of "treating children's rights as intangible", "recognizing the legitimacy of children's rights", and then "actively safeguarding children's rights". This also applies to the UK's child welfare legislation, development of welfare programs, and diversification of aid methods. Since modern times, society has not only recognized the importance of protecting the right to survival of impoverished children, but also valued the value of children to society. With the support of this concept, people generally value children's family and social values, and treat their rights and obligations differently or equally with adults. Without losing respect, the government's initial indifference, to actively assuming responsibility for child welfare and intervening in family relationships, all reflect the profound impact of ideological progress[6].

The transformation of the nature of the child welfare system: In the pre industrial era, various aspects of the child welfare system that did not adapt to the development of the times were hindered. In the post industrial era, while reflecting on the development of children's welfare, the government is also actively innovating concepts, creating systems, establishing systems, and improving mechanisms. Every child is important "is the principle of modern British child welfare work.

There are various types of charitable subjects in the UK: as the main bearers of children's welfare, charitable subjects are mainly composed of churches and individual philanthropists. However, with the advancement of ideas and the development of modern society, the subjects of charity have gradually diversified. In the UK, the government has become one of the most important charitable entities, protecting children's rights by formulating welfare laws and implementing welfare programs. At the same time, social groups, non-profit organizations, and commercial enterprises also participate in the cause of child welfare. This transformation of charitable entities not only provides more resources and support, but also brings more professional knowledge and experience to the implementation of child welfare. Government agencies can ensure the fairness and legality of child welfare through legislation and regulation, social groups and non-profit organizations can provide more targeted and innovative projects and services, and commercial enterprises can fulfill their responsibilities through social responsibility projects[7].

4. Analysis of Japan's Child Welfare Policy

4.1 Overview of Japan's Child Welfare Policy

After World War II, Japan experienced economic depression, poverty, displacement of children, and social instability. A large number of children have to engage in illegal activities in order to survive. Due to the prolonged war, Japan's labor force was severely insufficient, resulting in a large number of child laborers being employed.The main contents are shown in Fig. 1.

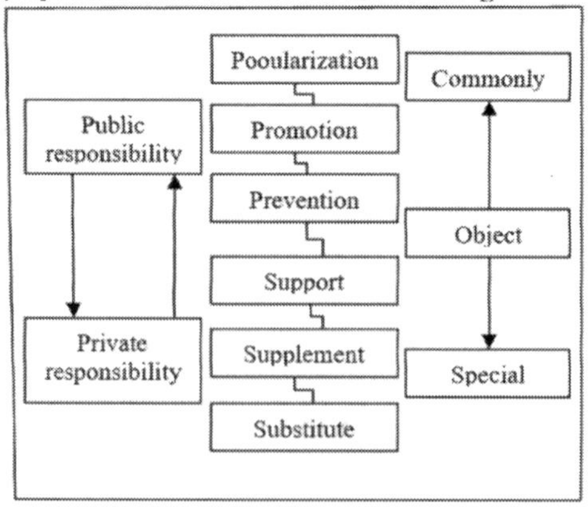

Fig. 1. Japan's Child Welfare Supply Model

The welfare and protection of children in Japan is a serious problem. In 1947, Japan introduced the "Child Welfare Law" to protect the basic rights of children, marking the beginning of child welfare policies. After the

Child Welfare Law, Japan successively introduced the Education Basic Law, School Education Law, Life Protection Law, Child Subsidy Law, and Child Abuse Prevention Law. So far, Japan's child welfare supply model has gradually improved, and child welfare is not only the responsibility of the government but also the responsibility of the private sector. The scope of care recipients is also broader, ranging from general to special[8].

4.2 The main content of Japan's child welfare policy

Child welfare is reflected in family life in the form of allowances: Japanese companies will pay family allowances, medical insurance for their dependents, etc. after establishing a relationship with the employer. Mothers may have different medical subsidies and medical examinations in different regions during pregnancy. Children will also receive cash after birth. Taking 2013 as an example, a newborn will receive 420000 yen. During the process of children's growth, they will also receive government subsidies for children. Japan's child subsidies are roughly divided into: child type subsidies (national subsidies for children after the third child of the family); Child support subsidy (this subsidy applies to a single mother whose father has a serious physical disability or whose father's life or death is unknown); Special child support subsidies (for children with mental and mental developmental delays and physical disabilities); Welfare subsidies for disabled children (with severe mental or physical disabilities that affect daily life); Subsidy for special disabled individuals (who have a serious mental or physical disability and always need to take care of it)

The coverage of child welfare in Japan is broad: Japan's child welfare policies not only apply to all Japanese children, but also to children who are legally staying in Japan (regardless of nationality and race). Japan's child welfare policy involves aspects such as daily life, healthcare, and education.

Japan's child welfare policies are complementary to multiple entities: Japan's policies related to children are operated by the Ministry of Health, Labor, and Welfare, mainly responsible for public services and handling public affairs, while managing child, women, and family welfare in local governments. And responsible for consulting on child welfare administration in the Metropolitan Government and designated municipal child welfare review committees[9].

The policy on the basic direction of future childcare support proposed by various departments in December 1994, which regards the integration of mother and child as the common welfare object, is commonly known as the "Angel Plan". In December 1999, the government further supplemented the Angel Plan and proposed the "Meeting on Measures to Reduce Childhood": the "Specific Implementation Plan for Key Measures to Promote Childhood Reduction" from early 2000 to 2004, commonly known as the "New Angel Plan". Further expansion has been made to childcare services, such as the expansion of childcare centers for young children (0-2 years old),; Promotion of support for homestay parenting; Experience the enrichment of activity intelligence and opportunities; The implementation of children's education environment; Ensuring a safe living environment and gaming environment for childcare. The so-called "Angel Plan" and "New Angel Plan" supplement various aspects of the mother's parenting process, whether in employment or in the home. As shown in Table 2.

Table 2. Types of child welfare subsidies in Japan

Types of child subsidies	Subsidy age
Children's subsidies	6-15 years old
Child support subsidies	Under 15 years old
Special child support subsidy	Under 20 years old
Welfare subsidies for disabled children	Under 20 years old
Special disability allowance	Over 20 years old

5. The Enlightenment of Child Welfare Policies in the UK and Japan on China

5.1 Study the welfare system ideas in the UK

Study the welfare system ideas in the UK.In recent centuries, the welfare system in Britain has gradually developed, and child welfare has benefited first from the transformation of child welfare ideas in the 20th century. Children are the main force for the future development of the country. The equal welfare rights of children should be recognized by the entire society. The concept should also enable the Chinese government to learn in building a child welfare system. Studying the ideas of the British child welfare system to promote the reform and innovation of China's child welfare system. Actively participate in relevant work and promote the development of children's welfare[10].

5.2 Learning from the UK to establish a multi-level child welfare project system

Transition from special welfare to general welfare. Expand the scope of child welfare to enable children to enjoy convenience in life. Make child welfare programs cover more areas such as safety protection, healthcare, education, mental health, and poverty. China can establish a multi-level system to provide comprehensive child welfare programs from multiple aspects. The goal of these projects is to ensure that every child has equal opportunities and care, and receives comprehensive support and protection during their growth process. Learn from the UK to establish a multi-level child welfare program system to ensure the comprehensive development and well-being of children. These measures aim to protect children from abuse, provide a healthy living environment, provide high-quality educational opportunities, and provide child welfare services[11].

5.3 Learning about Japan's welfare object of integrating mother and child

The approach of integrating mother and child as a common welfare object aims to value the relationship between mother and child, ensuring that they receive support and protection together. Incorporate welfare objects that integrate mothers and children, ensuring the common rights and interests of mothers and children. Japan's child welfare is not only aimed at assisting children, but also includes helping mothers of children. The working hours of mothers as workers can be appropriately shortened or flexible. This approach can help improve the well-being of mothers and provide a healthy, safe, and developmental environment for their children. For working mothers, adjustments have been made to their childcare leave and employment environment; Improvement of childcare facilities at work; Support for childcare, retirement, and re employment; Shortening of employment hours or promotion of flexible working hours; Implement diverse childcare services; The expansion of early childhood care, extended care, temporary care, and the adjustment of multifunctional care centers; Increase the implementation of childcare strategies for after-school children and regional maternal and child healthcare systems[12].

5.4 Learning from Japan to Establish an Administrative System for Child Welfare

Learn from Japan's effective and diverse management division of labor model with clear rights and responsibilities for establishing a child welfare administrative system, which complements the central, local, and civilian sectors. China should clarify its responsibilities in children's education, medical care, and other aspects, effectively manage the division of labor. Children's welfare should be managed from grain cultivation to local levels[13]. Continuously ensuring the safety of children's education and living environment, improving the quality of teaching, and effectively protecting children's rights with the assistance of the entire society. In addition, implementing diversified child welfare: through cooperation between government departments, participation of social and non-governmental organizations, professional support from child welfare workers and professionals, as well as the participation of families and communities, we jointly provide comprehensive welfare protection for children. The cooperation of multiple entities ensures the comprehensiveness and diversity of children's welfare, in order to meet the needs of different children and promote their comprehensive development and well-being[14].

References

1. LiWenxiang.(2023) From Diversified Participation to Public Consultation: Orientation Transformation and Mechanism Reconstruction of Child Welfare Construction. Social Science,07: 145-153.
2. HeFang. (2023) Basic Characteristics, Development Logic, and Future Trends of China's Child Welfare Policy in the New Era: An Analysis Based on the "Outline of China's Child Development (2021-2030)". Research on Preschool Education, 05: 10-19.
3. Huang Guan, Qiao Dongping. (2023) Knowledge and Power: A Study on the Discourse System of Child Welfare and Child Protection in China. Hebei Academic Journal,02: 172-179.
4. Chang Jinfeng, Liu Chang. (2023) The Current Situation and Development of Research on Child Welfare in China in the Last 30 Years: A Bibliometric Visualization Analysis Based on CiteSpace. Journal of Taiyuan University, 01: 29-41.
5. Heidi M, E A F, D L M. (2023) Incorporating multiple data sources to assess changes in organizational capacity in child welfare systems . Evaluation and Program Planning,97.
6. Xiaoxi Xiaohe. (2021) Japan's Child Welfare System and Child Welfare Facilities. Social Governance, 09: 21-28.
7. Mayumi K, Daisuke N. (2021) Association between Work Related Trauma Exposure and Postraumatic Stress Symptoms among Child Welfare Workers in Japan: A Cross Sectional Study. International Journal of Environmental Research and Public Health,18 (7).
8. Wei Xiuchun. (2021) An Analysis of the Historical Characteristics of the Development of the Child Welfare System in the UK (1862-1948). Research on Modern World History, 00: 73-95+313.
9. Chen Yanfei. (2019) The Enlightenment of the UK Child Welfare System on China's Child Welfare System. Management Observation, 11: 87-88.
10. Foreclosure. (2015) Directionality study of a Korean child welfare policy through a change process analysis of a Japanese child allowance system. Journal of Japanese Culture, null (66).
11. Qin Baoyu, Yang Hong. (2014) The Child Welfare System in the UK and Its Implications. Human Resources Development, 22: 37-38.
12. Kenji N H K T Y. (2014) Child Welfare in Japan: Children's Care Homes and Mental Health. Adolescent Psychiatry,4 (4).
13. Yoshikawa T, Kaneko H, Nomura K. (2014) Child Welfare in Japan: Children µ s Care Homes and Mental Health. Adolescent Psychiatry, 4 (4).
14. Li Xiaoyun. (2013) Analysis of Child Welfare Policies in the UK and Its Enlightenment for China. World of Labor Security (Theoretical Edition), 09: 177-178.

Research on the Application of Psychometric Assessment in the Recruitment of College Teachers

Yuanwei Li[1] *, Chao Kong[2], Xiuhua Li[3]*

State Grid of China Technology College, Jinan, China

Abstract. To investigate the impact of the use of psychometric assessment in recruitment, the SCL-90 and EPQ personality scales, interviews and surveys were used to test 81 applicants for the selection of talented individuals with sound psychological qualities and personalities. According to the data analysis, recruits were characterized by good mental health, typically low psychoticism and typically low neuroticism, and there was no significant difference in the number of internal and external orientations between recruits and non-recruits. The use of psychometric assessment in talent recruitment can help select the right people more efficiently and reduce recruitment risk, but this work still needs to be improved.

1 Introduction

In the post-epidemic era, traditional on-the-ground teaching models encountered great challenges in the face of massive online learning resources that are efficient and easy to use, not constrained by time and space. To ensure the sustainable development of on-the-ground teaching and the attraction of curriculum resources, competition among teachers has become the core competitiveness of the entire industry in the future. This will lead to more drastic changes in the recruitment process. To adapt to the competitive environment in the future, colleges and universities should make continuous efforts in personnel recruitment, recruit the right talent by improving the reliability of recruitment, and consolidate the human resources foundation for future development. While vocational psychological assessment techniques are not well used judging by their application in recruitment. In view of this, it is important to explore the application of vocational psychometric assessment.

2 Literature review

Yang Jialin et al. [1] believe that the development and growth of a country cannot be separated from talents, and talents better suited for corporate development can be selected fairly and accurately through the assessment of the subject's knowledge level, ability structure, personality traits, career inclination, development potential, and other qualities. Zheng Shengri [2] suggests that there are many ways to select talents. One effective approach is to combine psychology to assess the organizational behavior of the participants, which satisfies the low-cost and high-efficiency characteristics while helping to understand the personality traits and problem-solving abilities of the candidates. Typically, the MBTI personality test, developed by American psychologist Kaylene Briggs and her daughter Isabel Briggs Myers based on Jung's theory of personality, is one of the most widely used tools for testing the professional personality of job candidates. This tool divides human personality into 16 types by analyzing personality from four dimensions, the source of psychological energy, the way of collecting information, the way of making decisions and the way of dealing with the external world, and matches employees with appropriate positions based on personality types, thus providing effective references for enterprises in the process of recruitment [3]. Another commonly used personality test is 16PF, which measures 16 basic personality traits of a person. These 16 characteristics, as the basic factors that affect people's learning and life, constitute a unique personality of a person, reflecting the whole picture of a person's personality [4]. Xue Huifang [5] further proposed that in addition to the written test and psychometric assessment, the interview is also a crucial link in the recruitment process, through which the behavior of candidates can be observed from multiple perspectives, thus providing a more reliable basis for the selection of talents who meet the needs of the post.

However, in the process of market recruitment, many enterprises still have problems such as inadequate recruitment system planning and construction, inadequate job demand analysis, and unscientific recruitment evaluation schemes [6]. The same problem exists, of course, in the recruitment of college teachers. Wu Chunlei [7] found that in the current recruitment process colleges and universities pay more attention to the academic background and professional ability of candidates, but neglect the investigation of their personality characteristics, resulting in some

15216441008@163.com[1] *, kongchao0626@163.com[2], 870674959@qq.com[3]*

corresponding problems and troubles. Li Hui et al. [8] proposed that cost planning and recruitment assessment system should be established, basic management work such as formulating detailed and dynamic recruitment plans for university teachers should be carried out in a timely manner, and scientific talent evaluation system and coping methods should be adopted. Based on the above studies, this study explores the application of vocational psychological assessment techniques in the recruitment of college teachers, with the aim of selecting suitable talent to meet the needs of teachers and better build a high-level faculty teams.

3 Work procedure

3.1 Job analysis and post requirements

Through fieldwork and one-on-one interviews, an in-depth understanding of expectations about the integrated quality of new teachers was achieved, as shown in Table 1, leading to the development of corresponding targeted questions to summarize existing problems, explore the causes of problems, and revise the quality requirements of recruiters.

Table 1. Summary of capacity requirements for teacher posting.

Professional demand	**Comprehensive quality demand**
Scientific in the teaching content	With a high sense of responsibility
Logical in the teaching idea	Being a model of virtue to others
Appropriate in teaching methods	Stable in mood and capable of managing bad moods
Complete and adequate in lesson preparation	Stable in personality
Highlighted in the lecture content	Adaptable to the environment
Articulate, coherent and fluent in speech	Interested in education and teaching
Good at stimulating the students' thirst for knowledge	Skillful in the construction and treatment of human relations
Natural in appearance	Having a personal lifestyle in line with what is expected of a teacher
Able to draw inferences in lectures	Skillful in communicating and dealing with special students
Able to interact with students and create an active classroom atmosphere,	Positive and optimistic in coping style and strong in mental resilience
Good at using new media teaching means	Clear about future career development
Able to deal with teaching emergencies	Having a love for students

3.2 Psychometric assessment scheme

First, it is necessary to revise the recruitment plan after clarifying the job requirements. Clear and specific plans make hiring rules-based, orderly and feasible, thus avoiding random hiring. The main elements of the revision need to cover the list of personnel requirements and the basic criteria for recruitment, such as age, gender, education, work ability and character qualities. The psychometric assessment program includes the location, time, procedures, and the person responsible for the assessment..

3.2.1 Evaluation Tools

Major psychometric assessment scales used include SCL-90 and EPQ for pre-testing to screen out candidates with good mental health and relatively stable personalities. The corresponding results were counted and the work efficiency was qualitatively analyzed. SPSS.20 was used to analyze the correlation between the qualitative study results and the EPQ pre-test results, and the psychometric assessment results were validated twice.

3.2.2 Assessment objects

Target: 81 applicants from one college in 2021 and 2022. Among them were 32 men and 49 women, ranging in age from 23 to 29, all with master's degrees.

4 Data analysis

4.1 SCL-90 scores for applicants

SCL-90 assessments are conducted to screen for individuals with significantly abnormal levels of mental health. Out of the 81 candidates, seven had in their score or a total score greater than 160, as shown in Table 2 below, which required further tests or interviews.

Table 2. Comparison of mental health scores of applicants with national norms

Dimensionality	Applicants (N=81)	National Norm (N=1890)	t
Somatization	1.05	1.42	-2.016*
Obsessive symptom	1.28	1.66	-2.018*
Interpersonal sensitivity	1.06	1.51	-2.559**
Depression	1.1	1.50	-2.037*
Anxiety	1.12	1.34	-1.271
Hostility	1.08	1.50	-2.109*
Horror	1.06	1.27	-1.229

Paranoid ideation	1.11	1.44	-1.799
Psychosis	1.09	1.33	-1.525
Other	1.16	-	
Total point	101	-	

Note: **P<0.01, *P<0.05

As can be seen from Table 2, the scores of applicants in all dimensions of mental health are lower than the national norm, among which interpersonal sensitivity is significantly lower than the national norm at the confidence level of 0.01 and depression, hostility, compulsion, and somatization are significantly different from the national norm at the confidence level of 0.05, indicating that the students participating in the application are generally in good mental health.

4.2 EPQ scores for candidates

Based on the results of previous interviews, the EPQ questionnaire was used to screen for environmental adaptability, emotional stability, and truthfulness in response to questions. Twenty-six of the 81 respondents scored in the typical category, 18 were likely to have a tendency to lie, and eight scored high in neuroticism or psychosis. With reference to later admission results, the EPQ scores of employed and non-employed were compared, as displayed in Table 3:

Table 3. Comparison of scores between the recruited and non-recruited individuals

	Recruited personnel	Unrecruited personnel	t
E(Extroversion-Introversion)	62.03	59.75	2.376
N(Neuroticism)	29.28	32.69	-1.834
P(Psychoticism)	43.17	45.64	-1.564
L(Lie)	55	63.22	-3.099*

Note: *P<0.05

As can be seen from Table 3 above, there is a significant difference between recruited and non-recruited individuals in the masking dimension, where the latter have significantly higher scores than the former, while the difference in the other dimensions is not significant. The recruited achieved higher scores on the dimensions of internal and external orientation and lower scores on the dimensions of psychoticism and neuroticism compared to the non-recruited. At the same time, it is uncertain whether the differences in the other three dimensions are affected by the significantly higher concealment scores of the non-recruited.

4.3 Qualitative interviews in the recruiting department

A semi-structured interview outline was developed in advance, covering overall job evaluation, ability to deal with unexpected issues, relationships with other colleagues, adaptability to the environment, and interaction with students. Five people in charge of an employment department at a university were randomly selected to give job feedback to recruits after half a year on the job. Referring to EPQ, it is found that the internal and external azimuthal dimensions have little influence on the work. Whether the typical extroversion, the typical introversion, or the intermediate types of innerness and outerness have little effect on the main scope of work. But in terms of individual performance, typically extroverted recruits were more proactive and highly cooperative with others, while typically introverted recruits tended to be more grounded and steady in their research. In terms of neuroticism and psychopathy, however, recruits with lower scores performed better in environmental adaptation, stress coping and interpersonal relationships.

5 Discussion

Figures and tables, as originals of good quality and well contrasted, are to be in their final form, ready for reproduction, pasted in the appropriate place in the text. Try to ensure that the size of the text in your figures is approximately the same size as the main text (10 point). Try to ensure that lines are no thinner than 0.25 point.

5.1. Psychometric assessments are not a panacea

Recruiting the right talent is the ultimate goal for HR managers during the interview process. However, due to various psychological effects such as privity effect, recency effect, halo effect and stereotype effect, it is best to conceal the evaluation results before the interview to maximize the objectivity and accuracy of the evaluation results. Therefore, during the interview process, HR managers should try to reduce their psychological bias. That is, psychological evaluations, which are not omnipotent, cannot be used as a vote to reject the candidate in question [9]. As to how to use psychometric assessment, it is important that human resource managers start from the actual needs, pay attention to the content of psychometric assessment itself without going too far in pursuit of form and take a comprehensive evaluation of the interviewee without replacing HR decisions entirely based on psychometric assessments.

5.2 Selection for psychometric assessment

In the course of this study, it was found that individuals recruited by a university were characterized by good mental health, typically low mental quality, typically low neuroticism, and no significant difference in the number of internal and external orientations. At the same time, it can be seen that the relatively small dimensionality of EPQ does not satisfy the requirement of working capacity in the larger range. Therefore, the personality assessment should be changed to a 16PF or MBTI personality assessment with more dimensions at a later stage of the recruitment process, and the stability and security of the assessment environment should be created in order to obtain assessment results with higher reliability and validity.

5.3 Assistive psychological training to increase post -service competency

In order to better improve the post competency of new employees, the relevant person in charge of the human resources department and the full-time teachers of the Mental Health Development Center of the university jointly developed psychological training in addition to the regular new induction training, including group psychological activities and relevant theoretical knowledge on different themes such as teamwork, interpersonal relationship, environmental adaptation and stress coping. At the end of the activity, the psychological quality of the participants was tested and it was found that all aspects of the participants' mental resilience and quality had significantly improved.

6 Conclusion

The use of psychometric assessment in the recruitment can be effective in improving the recruitment efficiency. At the same time, due to the distinct characteristics of job requirements in the university faculty ranks, appropriate psychometric assessments and personality assessment scales can be used to assess the degree of person-to-post match and identify potential candidates at risk in terms of mental health, interpersonal cooperation and other aspects[10] .This will increase the efficiency of the hiring department by reducing the psychological and behavioral risks of employees, enhancing their sense of team cooperation and job loyalty, and reducing employee turnover upon entry.

References

1. Yang Jialin, Lei Meng. Talent assessment：theory and practice [J]. Human Resources Development of China, 2015 (23) : 104-108.
2. Zheng Shengri. Overview of the development of talent assessment theory [J].Transportation Enterprise Management, 2007 (11) : 8-9.
3. Fairfield Donald H. What influences a career: a personal reflection[J]. Perspectives of Earth and Space Scientists,2022,3(1).
4. Zhou Jing, Zhao Yiping. Analysis and discussion of 16PF personality factor test embedded in reexamination of postgraduate students [J].Examination weekly, ly, 2019 (: 193-194.
5. Xue Huifang. Discussion on the norms of public recruitment interview in public institutions [J]. China Health Human Resources, 2014 (08) : 18-20.
6. Wang Hongwei. Research status of effectiveness evaluation of enterprise recruitment [J]. China Collective Economy, 2017 (14) : 104-105.
7. Wu Chunlei. On college teacher recruitment from the perspective of human resource management [J].Academy, 2018, 11 (33): 172-173.
8. Li Hui,Yan Guodong,Ren Xuejing.Cost and risk issue of a university teacher recruitment[J].Economic Research Guide,2023(08):99-101.
9. W. F S ,Sebastian R ,Vesselin P , et al. Multi-trait, multi-method talent assessment through a digital platform: relationship with job performance[J]. Journal of Psychology in Africa,2022,32(4).
10. Yang F ,Chau R . Proactive personality and career success[J]. Journal of Managerial Psychology,2016,31(2).

Democratization Prospects and Challenges in Central and Eastern Europe after the Cold War: A Focus on the Visegrad Group in the Context of European Integration

Longtai Zhang[1]* and Mingqing Zhang[2]

[1]Faculty of Political and Social Sciences, Ghent University, Ghent 9000, Belgium
[2]School of European Studies, Tianjin Foreign Studies University, Tianjin 300204, China

Abstract. This research focuses on democratization and consolidation in Central and Eastern Europe, specifically examining the Visegrad Group (Poland, Hungary, Slovakia, and the Czech Republic) after the Cold War and EU Eastern Enlargement. Using different democratization and European Integration theories, it explores the factors influencing democratization, consolidation, and stability in this region. This study analyzes the transition from communism to democracy, the impact of European Integration on democratization, and the role of civil society organizations in promoting democratic values. It acknowledges the challenges posed by populism and illiberalism and suggests practical policy solutions to strengthen democratic values and stability in Central and Eastern Europe.

1 Introduction

The historical denouement of the Cold War brought forth a momentous inflection point in Central and Eastern Europe, marking the cessation of Soviet hegemony and the transition of this region from communist rule to democratic governance. This transformative phase engendered notable attributes in the democratization trajectory, wherein participatory political systems supplanted erstwhile authoritarian regimes.

During the democratization process within Central and Eastern Europe, the Visegrad Group countries, including Poland, Hungary, Slovakia, and the Czech Republic, successfully navigated various challenging aspects such as developing market-oriented economic systems, ensuring political inclusivity, and establishing the rule of law. Moreover, the democratization process within the Visegrad Group was carried out in the context of European Integration, the EU's significant role in promoting democracy and stability in the region.

However, the process of democratization in Central and Eastern Europe encountered challenges. The rise of populism and the growth of illiberalism presented significant obstacles to the progress made in democratization. As a result, it is crucial to remain watchful and vigilant to protect the achievements and progress attained in this domain.

This study employs an extensive theoretical framework to analyze the challenges and possibilities pertaining to democratization and the consolidation of democratic principles in Central and Eastern Europe, with a particular focus on the Visegrad Group countries. The research explores the historical and political contexts that shaped the democratization process, identifies the catalysts behind democratization, consolidation, and stability within the Visegrad Group, and assesses the influence of European Integration in promoting democratic reforms.

Furthermore, this study puts forth a set of policy recommendations aimed at bolstering democratic institutions and ensuring enduring regional stability.

2 Historical Background

2.1 The Formation of Eastern Bloc after World War II

As World War II concluded, the Soviet Union, the United States, and Great Britain faced mounting tensions regarding the future of Germany. Two crucial conferences in 1945, before and after the war's official end, set the stage for Europe's Cold War division.

In February of that year, US President Franklin D. Roosevelt, British Prime Minister Winston Churchill, and Soviet leader Joseph Stalin convened at Yalta in the Crimea region to discuss the post-World War II European reorganization. Each leader had a unique vision for rebuilding war-torn Europe, with Roosevelt emphasizing Soviet participation in the newly created United Nations and requesting their support to help end the ongoing war in the Pacific against Japan. Churchill, in contrast, called for free and fair elections, specifically in Central and Eastern Europe, with a focus on Poland, to establish democratic regimes. Conversely, Stalin sought a "sphere of influence"

* Corresponding author: vlastimilzhang@gmail.com

for the Soviet Union in Central and Eastern Europe, including Poland, as a geopolitical buffer between the Soviet Union and the Western capitalist world.[1]

During the final weeks of the Second World War, Soviet forces made significant progress on the Western front, pushing the Nazi army to Berlin, and ultimately ending the war with Soviet troops occupying multiple states in Central and Eastern Europe, including the eastern portion of Germany. As the war progressed, the USSR absorbed three previously sovereign Baltic states, Estonia, Latvia, and Lithuania, as well as a section of Romania which was established as the Moldovian Soviet Socialist Republic. In the remaining Central and Eastern European states, they occupied, the USSR extended its assistance to establish hardline communist governments modeled after their own system. While the four major allied powers jointly occupied Germany and Austria, they explored differing futures for the two countries. The victorious Western powers ultimately founded the Federal Republic of Germany in the West, based on market-based systems, whereas the USSR founded the German Democratic Republic in the East, built on hardline socialist principles.

The escalating hostility and fear of further communist expansion prompted the formation of the North Atlantic Treaty Organization (NATO). Conversely, the USSR established the Warsaw Pact in response to the Western allies permitting rearmament in West Germany and the pact's official signing took place in Warsaw. Throughout its existence, the Warsaw Pact's member governments executed repressive policies, including extreme limitations to freedom of speech and opposition party repression. Additionally, extensive spy networks were established to keep track of citizens suspected of opposing the governing regimes. While the states of the Warsaw Pact enjoyed better living standards than the Soviet Union, they still fell significantly behind those of Western Europe.

The political and social control in Central and Eastern European countries was weaker than in the Soviet Union. The region faced added risks due to strong nationalist sentiments against the Soviet Union. After Stalin's death in 1953, the new Soviet leadership made policy adjustments toward Central and Eastern Europe. They granted some economic autonomy to these countries and aimed to improve relations with Yugoslavia to ease tensions within the socialist bloc.[2] In 1955, the Soviet Union and Yugoslavia signed the "Belgrade Declaration," emphasizing non-interference in each other's affairs. The Soviet Union acknowledged the diversity of paths to socialism during the process of normalizing relations. At the 20th Congress of the Communist Party, the cult of personality around Stalin was denounced.[3]

On February 14, 1956, the 20th Congress of the Communist Party of the Soviet Union (CPSU) commenced at the Kremlin. This was the first party congress held after Stalin's death. Khrushchev had already assumed the highest leadership positions within the party and the country. During this congress, Khrushchev addressed three theoretical issues concerning the international situation: peaceful coexistence, peaceful transition, and peaceful competition. On the final late-night session of the congress, Khrushchev delivered a secret report titled "On the Cult of Personality and Its Consequences," primarily focusing on a critique of Stalin. This "secret report" sent shockwaves through the Soviet Union, Central and Eastern Europe, the entire socialist bloc, and indeed, the world at large.

2.2 Resistance of Central and Eastern Europe against the Soviet Model during the Cold War: A Case Study of Poland, Hungary, and Czechoslovakia

Notwithstanding the initial reception of Soviet troops as liberators who expelled Nazi forces from Central and Eastern Europe, the local inhabitants rapidly developed apprehension towards Stalinist doctrine and the newly appointed communist leaders.

2.2.1 Poznan June in Poland

In the early days of June 1956, the city of Poznan in western Poland witnessed a pivotal episode. Workers at the Stalin locomotive factory voiced their demands for improved wages, reduced prices, and lighter tax burdens from the government. Upon facing the refusal of their requests, a delegation of 30 workers ventured to Warsaw, seeking redress for their concerns. Dishearteningly, the delegation encountered an unsatisfactory response, compounded by the unsettling prospect of street protests being met with the deployment of tanks. This unpalatable outcome stirred emotions among the workers at the Stalin factory, leading them to a collective decision: a strike and protest slated for June 28th. On the day designated for the strike, a congregation of more than 10,000 workers coalesced, embarking on a march toward Stalin Square. This demonstration witnessed the organic inclusion of laborers from other factories within the city, alongside a multitude of individuals who spontaneously joined the procession. Among these participants were foreign visitors attending the Poznan International Fair.[4]

A delegation duly elected by the protesters sought an audience with the authorities to ventilate their grievances, but their entreaty was met with rejection. This juncture marked the diffusion of rumors suggesting the apprehension of the representatives dispatched to Warsaw. The propagation of such unsettling tidings further inflamed the sentiments of the assembly, catalyzing a transformation of what had hitherto been a peaceful demonstration into a vortex of violent tumult. Portions of the crowd commenced a fervent assault on government edifices, including the radio station, and audaciously unsealed prison gates, culminating in the release of incarcerated individuals and the conflagration of confidential archives. Further disconcertingly, there emerged reports of individuals seizing control of the prosecutor's office and the courthouse, resulting in the confiscation of firearms and the orchestration of assaults against law enforcement personnel.

In response to the escalating turmoil, the Polish administration invoked the deployment of security contingents and deployed tanks to subdue the escalating discord under cover of nightfall. The events that unfolded in Poznan left a sobering wake: 74 fatalities, more than 575 injuries, the apprehension of more than 658 individuals, and a calculable quantum of material damage approximating 3.5 million zlotys.[5]

2.2.2 Hungarian Revolution of 1956

In the autumn of 1956, a resolute uprising unfurled within Hungary, orchestrated with the aspiration of unseating the rigid communist authorities and ushering in a regime chosen through democratic processes. The crucible of this movement lay in Budapest, where the apprehension of student demonstrators served as the catalyst for a conflagration of revolutionary fervor that consumed the entire nation. The prevailing government structure crumbled, and the shackles imprisoning political dissidents were cast aside as newfound liberty coursed through the social fabric.

During the interregnum, an ad interim administration took center stage, propounding a visionary agenda inclusive of Hungary's disengagement from the Warsaw Pact and the inception of unfettered electoral processes. Regrettably, the arc of destiny took a harrowing turn within a mere span of days: the Soviet Union orchestrated an intervention, an iron-fisted response that ruthlessly suppressed the nascent movement of self-determination. This calculated maneuver entailed a significant cost, with the toll of lives lost stretching beyond mere numbers to encapsulate the shattered dreams of more than 2,500 Hungarians and a conservative estimate of approximately 700 Soviet soldiers, victims of the days of tumultuous violence that ensued.

In the wake of this seismic event, the Soviet Union saw to the installment of a freshly sanctioned draconian regime, one that expediently embarked on a crusade to apprehend and incarcerate all those even tangentially associated with the audacious uprising. The Hungarian Revolution of 1956 served as an unequivocal and indelible testament, laying bare the unwavering resolve of the USSR to quash with unyielding resolve any aspirations harbored by the Central and Eastern European states to extricate themselves from the gravitational pull of Soviet dominion. This pivotal moment underscored the inviolable boundaries of Moscow's influence, effectively delineating the contours of permissible dissent within the Soviet sphere of control.[6]

2.2.3 Prague Spring in Czechoslovakia

During the mid-1960s, the Soviet Union experienced a paradigmatic shift attributable to the tenets of de-Stalinization, a doctrine that precipitated a proliferation of autonomous and clandestine literary and informational expressions, ushering in nascent discourses of transformative change. Amid this evolving landscape, the geographical expanse of Czechoslovakia bore witness to the burgeoning of a novel ideological movement intent on orchestrating a comprehensive overhaul of its ailing economic apparatus and the amelioration of its rigid policy framework.

In the history of Czechoslovakia, a significant turning point was the rise of Alexander Dubcek to leadership in 1968. During his leadership, he introduced a new idea known as "Socialism with a human face." Dubcek took careful steps to make things more relaxed. He allowed more freedom in the media, which meant people could express themselves more openly. He also started to make the controlled economy more open and flexible. These changes led to a new atmosphere where people could talk more openly and discuss topics like communism and their relationship with the Soviet Union in a more detailed way. This was a big change from before.

On the other hand, in the Soviet Union, Leonid Brezhnev rose to power after Nikita Khrushchev was removed from the top position. Brezhnev saw that his predecessor's more lenient policies were causing problems, so he started making the country's rules stricter. This was happening around the same time as changes were happening in Czechoslovakia. Brezhnev thought these changes in Czechoslovakia were a big problem because they went against the main ideas of the Soviet system and the power the Soviet Union had in that area. He strongly told Dubcek, the leader of Czechoslovakia, to stop the reforms he was making. Dubcek, however, stood up against Brezhnev's demands, showing he was determined to continue with the changes he believed in. The situation escalated in August 1968 when Brezhnev used the Warsaw Pact to order a large military intervention in Czechoslovakia. Many troops and armored units were sent in, and they took control of Prague. Despite this show of force, the people of Czechoslovakia didn't give up. They organized peaceful protests in Prague to show their resistance.

The aftermath of this intervention crystallized with the replacement of Dubcek by a more pliant figure, a change that was underscored by his relegation to a subordinate role within the Forestry Service. The events that transpired served as the inaugural volleys of an enduring historical doctrine known as the "Brezhnev Doctrine," a strategic posture that affirmed the prerogative of the Soviet Union to wield force as a potent instrument for the preservation of the rigid communist underpinnings within the domains of the Warsaw Pact. This doctrine laid bare the parameters of permissible dissent within the Eastern Bloc and firmly delineated the contours within which the Soviet dominion could and would wield its influence.[7]

2.3 The Transformation of Eastern Bloc: From Solidarity to Dissolution of the Warsaw Pact

During the 1980s, dissident movements emerging from the Soviet Union and the Eastern Bloc countries aligned with the Warsaw Pact began to capture international attention. Eminent personalities such as Aleksandr Solzhenitsyn, who had experienced incarceration within the Gulag system, nuclear physicist Andrei Sakharov, and Czech playwright Vaclav Havel, undertook the clandestine dissemination of subversive literary works. These clandestine efforts sought to galvanize opposition against the prevailing authoritarian regimes, despite the formidable crackdowns imposed upon such expressions.[8]

Concurrently, the rise of Solidarity, a Polish trade union, gained traction within the nation after a succession of labor strikes that reverberated in 1980. The organization,

 https://doi.org/10.1051/shsconf/202317801009

though confronted with severe state-sanctioned suppression, managed to sustain its activities throughout the decade. The catalytic events of 1980 invigorated anti-communist sentiments and sentiments of dissent. The ideological landscape shifted further with the advent of Mikhail Gorbachev's policy initiatives encompassing glasnost (openness) and perestroika (restructuring), which ushered in a degree of liberalization within the Soviet economic framework and the tolerance for more open discourse. This recalibration breathed new vitality into anti-communist movements across Central and Eastern Europe. However, the entrenched conservative leadership in these Eastern Bloc countries, notably East Germany, remained obstinate in their resistance to Gorbachev's reformist overtures. Strikingly, by 1989, the Central and Eastern European communist states displayed a paradoxical fidelity to the principles of Soviet-style communism that even surpassed the Soviet Union itself.

As 1989 unfolded, burgeoning restiveness precipitated mass mobilizations and street demonstrations across the region. Signaling a departure from the Brezhnev Doctrine, Gorbachev publicly indicated his disavowal of employing Soviet military forces to salvage the beleaguered communist regimes within the Eastern Bloc. A pivotal juncture materialized when, in 1989, Solidarity achieved a historic milestone by becoming the foremost non-communist entity to secure electoral victory within the Warsaw Pact sphere. Simultaneously, communist regimes across the alliance found themselves dislodged from their positions of power during a wave of revolutions. Romania stood as an exception; its transformative trajectory was marred by violent upheaval culminating in the execution of longstanding leader Nicolae Ceausescu on Christmas Day of that same year.

In 1991, the constituent states of the Warsaw Pact collectively opted for the dissolution of the alliance. After this dissolution, several erstwhile member countries have subsequently aligned themselves with NATO, including the trio of Baltic states that achieved sovereign independence after their separation from the Soviet Union.[9]

3 Establishment of the Visegrad Group

The Visegrad Group, an amalgamation encompassing four countries nestled within Central and Eastern Europe—namely Poland, Hungary, the Czech Republic, and Slovakia—materialized as a salient and pivotal entity against the backdrop of the recalibrating geopolitical currents that ensued in the aftermath of the Cold War's culmination. Anchored in the contours of historical junctures, this assemblage, christened in homage to the Hungarian township of Visegrad where its seminal convocation unfolded, occupies a distinct niche within the region's annals. It crystallizes a collective yearning for synergistic endeavors, equilibrium, and a seamless convergence in the swiftly metamorphosing terrain of post-communist Central and Eastern Europe.[10]

3.1 Emergence and Objectives of the Visegrad Group: A Regional Collaborative Endeavor in the Post-Communist Era

The origins of the Visegrad Group can be traced back to a transformative period in Central and Eastern Europe during the late 1980s and early 1990s. This era was characterized by profound changes, notably marked by the fall of the Berlin Wall and the dismantling of Eastern Bloc. These events created a unique opportunity for countries that had previously been under communist rule to chart new courses for their future.

Amid this shifting landscape, Poland, Hungary, and Czechoslovakia recognized the potential advantages of regional cooperation in advancing their political, economic, and security interests. The context of shared historical and cultural ties further solidified the foundation for their collaboration. The core principles of the Visegrad Group were deeply rooted in their historical connections. These principles encompassed a commitment to democratic governance, market-oriented economies, and a strong desire to integrate into Western institutions, such as the European Union and NATO.

The group's early objectives were grounded in the ideals of peaceful cooperation, mutual respect, and a collective ambition to address the challenges posed by the transition from centrally planned to market-driven economies. Formally, the Visegrad Group was established on February 15, 1991, when the leaders of Poland, Hungary, and Czechoslovakia convened in the town of Visegrad and signed a joint declaration outlining their shared goals and aspirations.

Subsequently, the group expanded with the inclusion of Slovakia in 1993, following the peaceful dissolution of Czechoslovakia. In its initial years, the Visegrad Group concentrated on key priorities, with a primary focus on economic transformation and integration. These countries engaged in the exchange of experiences, strategies, and best practices, which proved invaluable in navigating the intricate processes of privatization, deregulation, and structural reforms. This collaborative approach underscored the significance of mutual support and knowledge sharing during the early stages of their cooperation.[11]

3.2 The Visegrad Group's Role in Shaping European Integration and Cooperative Transformation in Central and Eastern Europe

The Visegrad Group achieved a monumental milestone by playing a pivotal role in facilitating the accession of its member states to both the European Union (EU) and NATO. The unwavering commitment of these countries to democratic principles and market-oriented economies proved instrumental during the complex negotiations for membership in these prestigious international organizations. This successful integration not only deepened their political and security ties but also underscored the potency of regional collaboration in achieving shared objectives.

The Visegrad Group's journey is characterized by significant accomplishments, with its members successfully navigating the intricate processes required for EU and NATO membership. Their commitment to democratic governance and market-oriented economic reforms served as a strong foundation for the accession negotiations. These accomplishments were a testament to the group's ability to work cohesively, leveraging their collective strength to overcome challenges and seize opportunities.

However, as the Visegrad Group evolved, it encountered challenges, including disparities in national interests and priorities among its member states. These differences tested the unity of the group and required adept diplomatic navigation to reconcile varying perspectives. Furthermore, the addition of Slovakia to the EU and NATO necessitated adjustments within the group's dynamics. Nevertheless, despite these complexities, the Visegrad Group remained a valuable platform for coordination and cooperation, particularly in advocating for the interests and concerns of its member states within the European Union.

The formation of the Visegrad Group represented a pivotal moment in the history of Central Europe. It symbolized the collective aspirations of Poland, Hungary, the Czech Republic, and Slovakia to transcend their communist past and forge a shared future grounded in democratic governance, economic prosperity, and collaborative security. The group's journey stands as a testament to the power of regional cooperation in shaping the trajectories of countries during times of transformative change. It has left an indelible mark on the history of post-communist Europe, serving as an enduring reminder of the potential for collective action and cooperation in pursuit of common goals.[12]

4 Political Democratization: Tracing the Polity Shift of the Visegrad Group from Communism to the Present Era

After the Cold War, political democratization became an important aspect of Central and Eastern Europe, and the promotion of the establishment of democratic polities was a central element of political democratization. Semi-presidential and parliamentary systems of governance were generally established in Central and Eastern Europe after the Cold War. Overall, these two regimes have played a positive role in promoting democratic transformation in Central and Eastern Europe. However, despite their overall good performance, both systems of governance in Central and Eastern Europe have experienced some problems, such as cabinet instability, conflict between the executive powers of the president and the prime minister, too many parliamentary parties, and monopolization of parliamentary, cabinet, and presidential positions by a single party.[13]

Within the Visegrad Group, both the Czech Republic and Hungary opted for a parliamentary system of governance, while Poland and Slovakia (after January 14, 1999) opted for a semi-presidential system of governance. The establishment of the new form of governance was marked by the introduction of new constitutions.

The timing of the introduction of the constitutions and the establishment of the form of governance in these four countries can be found in **Table. 1.** below.

Table 1. Date of Adoption of New Constitutions and Forms of Governance in the Visegrad Group Countries

Country	Polity	Date of Adoption of the New Constitution
The Czech Republic	parliamentary system	December 16, 1992 (Effective January 1, 1993)
Hungary	parliamentary system	August 24, 1990
Poland	semi-presidential system	April 2, 1997 (adopted by referendum on May 25, 1997, and entered into force on October 17, 1997)
Slovakia	parliamentary system (Before 1999) semi-presidential system (After 1999)	The parliamentary system of governance was established on September 1, 1992, by the adoption of a new Constitution, which entered into force on October 1, 1992, and on January 14, 1999, Slovakia changed its Constitution from a parliamentary to a semi-presidential system of governance.

4.1 Parliamentary Systems of Governance in the Czech Republic and Hungary

In the wake of the Cold War, the Czech Republic and Hungary embraced a parliamentary system of governance driven by historical context and practical considerations.

4.1.1 Parliamentary Legacy and Transformations: Hungary's and Czechia's Political Trajectories in Historical Perspective

Hungary's rich parliamentary heritage traces its origins back to the 13th century, firmly establishing it as one of the earliest examples of legislative assemblies in global history. This enduring tradition has played a pivotal role throughout Hungary's political evolution, with significant parallels to the governance model employed by the United Kingdom. Across the centuries, the parliamentary platform has consistently been at the heart of Hungary's governance, serving as a driving force in shaping the nation's destiny.

As the 20th century dawned, Hungary, like many European countries, experienced a series of seismic shifts in its political landscape. Following World War I, on August 1, 1919, Hungary gained independence from the Austro-Hungarian Empire while still adhering to the idea of a constitutional monarchy. However, the country's course took an abrupt turn with the leadership of Miklós Horthy, whose regency ushered in a period of rapid transformation, ultimately leading to a temporary phase of dictatorial rule.

Similarly, in February 1920, Czechoslovakia established a constitutional framework that solidified a parliamentary system as the cornerstone of its governance. This moves mirrored Hungary's commitment to parliamentary

 https://doi.org/10.1051/shsconf/202317801009

democracy, marking an era where such systems gained prominence across the region.

The end of the Cold War brought about new opportunities for political reconfiguration in Central Europe. Both the Czech Republic and Hungary, having traversed diverse historical chapters, consciously chose to revive their parliamentary systems of governance. This strategic alignment with the pre-World War II governance paradigm represented a deliberate choice, reflecting their commitment to infuse continuity and historical resonance into their contemporary political frameworks.[14]

4.1.2 Practical Considerations behind the Selection of a Parliamentary Form of Governance in the Czech Republic and Hungary

After the Cold War, the countries of Central and Eastern Europe found themselves facing a pressing need for comprehensive legislative reform and the establishment of entirely new state frameworks. This era witnessed a rapid rise in the prominence and influence of parliamentary bodies throughout the region. Particularly, during the early 1990s, a unique opportunity emerged in countries like the Czech Republic and Hungary, where parliamentary institutions began to amass significant power. In contexts where no single dominant political figure held sway over discussions about the shape of institutional architecture, the adoption of parliamentary systems of governance emerged as a pragmatic and logical choice.

Conversely, countries that leaned towards semi-presidential governance structures often did so due to the presence of formidable personalities at the helm of their political landscapes. Figures such as Ion Iliescu in Romania, Zhelyu Zhelev in Bulgaria, and Milan Kučan in Slovenia epitomized this trend. These influential leaders typically favored direct popular electoral endorsement over indirect parliamentary selection, which strongly influenced the inclination towards semi-presidential regimes.

The aftermath of the Cold War ushered in a period of political transformation and experimentation across Central and Eastern Europe. In countries where parliamentary systems gained traction, there was a recognition of the need for inclusive and deliberative decision-making processes. Parliamentary institutions became a symbol of democracy and the will of the people, representing a shift away from autocratic rule.

In contrast, semi-presidential systems often reflected a preference for charismatic leaders who could take swift and decisive action. These leaders, with their direct popular mandates, wielded considerable power and influence over their respective countries' political landscapes. However, this concentration of power could also present challenges to democratic governance, as it relied heavily on the individual leader's judgment and discretion.[15]

4.2 Semi-presidential Systems of Governance in Poland and Slovakia

4.2.1 The Evolution of Poland's Political Landscape: From Presidential Aspirations to Semi-Presidential Governance

In contrast to the countries within the former Soviet Union, Central and Eastern European countries of the early 1990s exhibited a distinct landscape characterized by a relatively balanced political opposition to the former communist parties. For instance, the presence of entities like the Solidarity trade union in Poland set the stage for a noteworthy political dynamic. Solidarity's "your president, our prime minister" configuration laid the groundwork for contemplation regarding a prime ministerial presidential system.[16]

Lech Wałęsa, who assumed the Polish presidency at the close of 1990, aspired to establish a presidential system in Poland that would grant him authority in appointing and dismissing the Prime Minister and the cabinet members. However, this proposition encountered resistance within the Polish parliament. In 1992, the parliament passed a "mini-constitution" that curtailed some of the president's prerogatives. While the president retained the power to nominate a prime minister, the ultimate selection rested with the parliament. Furthermore, the president lacked the authority to oust the prime minister or the cabinet from their positions. Nevertheless, the president retained broader powers, encompassing the initiation, and vetoing of legislation (though vetoes could be overridden by a 3/5 majority of the parliament), the prerogative to dissolve the parliament, and the capacity to appoint the Ministers of Defense, Foreign Affairs, and the Interior.

Subsequent developments in the political landscape continued the trend of constitutional reform aimed at diminishing presidential powers and augmenting those of the prime minister. The center-left camp's victory in the parliamentary elections of 1993, coupled with Lech Wałęsa's defeat in the 1995 presidential race, set the stage for further constitutional adjustments. By 1997, a new constitution was introduced that further delineated the scope of presidential authority. In this constitutional framework, the president's capacity to propose candidates for key ministerial positions was relinquished, and powers not explicitly attributed to a national or local body were vested in the cabinet. The president's veto power remained, albeit subject to override by a 3/5 majority in the parliament.[17]

This trajectory resulted in Poland adopting a Prime Minister-President system characterized by a division of powers between the Prime Minister and the President. However, the presidential powers were notably attenuated, leading to a governance structure often referred to as semi-presidentialism.

4.2.2 The Evolution of Parliamentary Authority and the Semi-Presidential System in Slovakia (1969-1999)

Slovakia, once conforming to the archetype of a nation favoring a parliamentary framework, witnessed the establishment of the Slovak Parliament in 1969, although its substantive influence during the era spanning 1969 to 1989 was marginal. Following the momentous transformations that swept through Czechoslovakia in November 1989, a pivotal juncture that underscored Slovakia's quest for autonomy within the Czechoslovak federation, the paradigm of the Slovak parliament shifted towards active participation. The unfolding of events culminated in the formal separation of the Czech Republic and Slovakia on January 1, 1993. In this partition, the Czech Republic seamlessly assimilated the pre-existing state political apparatus of the erstwhile Czechoslovak federal epoch. In stark contrast, the task of erecting Slovakia's state machinery and political institutions from the ground up was vested in the Slovak parliament. Endowed with the prerogative to draft the foundational constitution and determine the contours of governance, the Slovak parliament ascended to a pivotal role.

This ascent in the authority of the parliament was further cemented by the advent of the new Constitution in 1992, which bestowed exclusive competence upon the parliament to sanction legislative enactments. Additionally, the parliament was vested with the agency to instigate motions of no-confidence targeting the Cabinet or specific ministers, compelling their abdication if successful. It retained the agency to convene premature parliamentary elections and to promulgate legally binding decisions. Furthermore, the parliament assumed the responsibility to nominate and terminate the tenure of the President. The appointment of the Prime Minister and the Cabinet also lay within its purview. Notably, the parliament was entrusted with the selection of the President and Vice-Presidents of the Supreme Court and Constitutional Court, thereby solidifying its influence across the judicial landscape. The parliament's prerogatives extended to the initiation of referendums, encapsulating its multifaceted role in shaping the sociopolitical discourse.[18]

Despite the transition towards direct election of the President of Slovakia in 1999, the nation retained its distinctive semblance as a ceremonial semi-presidential entity. Within this construct, the parliament's centrality within the politico-administrative framework persevered, substantiating its enduring significance in the echelons of political authority.

4.3 Primary Challenges Encountered during the Political Transformation of the Visegrad Group

4.3.1 Dual Executive Challenges in Semi-Presidential Systems: Comparative Insights from Central and Eastern Europe and Western Europe

Prominent scholars of political science, Juan Linz, and Alfred Stepan have highlighted an inherent challenge within semi-presidential systems, the conundrum of a dual executive. This predicament becomes particularly salient when certain conditions are met when the president is not aligned with the majority party in the parliament, when the prime minister lacks parliamentary majority support, or when constitutional provisions remain ambiguous.[19]

The phenomenon of the "dual executive" has also surfaced within Western European semi-presidential countries. However, owing to the more established democratic systems in Western Europe, prolonged practical experience has progressively clarified many of the ambiguities. As a result, these situations generally have not culminated in governmental crises. For instance, Finland has navigated the intricacies of the "dual executive" dilemma by gradually diminishing the president's executive authority. The president's executive powers are primarily circumscribed to foreign affairs and security matters, necessitating consultations with the prime minister, foreign ministers, and defense ministers before pivotal decisions are made. Notably, Finland's constitutional revisions in 1999 further delineated those matters related to foreign affairs within the European Union squarely rested with the prime minister, notably attenuating the president's executive sway.

Contrastingly, in France, the president enjoys a paramount role in the realm of executive authority. This disposition substantially mitigates conflicts between the executive powers vested in the president and those held by the prime minister. When the president and prime minister hail from the same political party, conflicts tend to be minimal. Even in instances of coalition governments, where some degree of contention between the president and prime minister may arise, the situation seldom devolves into political gridlock.[20]

The multitude of political parties in the parliaments of Central and Eastern European countries makes it unlikely, compared to Western Europe, for the President and Prime Minister to come from the same party. This increases the likelihood of having "coalition governments" in Central and Eastern European countries. Additionally, the constitutions of Central and Eastern European countries generally require both the President and the Cabinet to jointly approve important decisions. This rule is meant to encourage collaboration, but it can often be exploited as a tool by one party to exert pressure on the other. Furthermore, these Central and Eastern European countries are experiencing significant historical events like nation-building, political transformations, and accession to the EU and NATO. This often leads to the emergence of highly esteemed political figures. When these figures become President, they sometimes exceed their constitutional powers and engage in power struggles with the Prime Minister.

When Donald Tusk assumed the role of Prime Minister in November 2007, heading Poland's Civic Platform party, he found himself in a strained and conflict-ridden relationship with the then-President, Lech Kaczynski, who led the Law and Justice party. Their discord reached a boiling point in October 2008 during a contentious dispute regarding who should represent Poland at an EU summit. This dispute escalated to such an extent that it had to be adjudicated by the Constitutional Court. The Constitutional Court's ruling clarified that the President retained the right to maintain a symbolic "presence" at international events while emphasizing that the nation's diplomatic authority rested with the government. In essence, while acknowledging the President's symbolic role, the Constitutional Court limited his capacity to "represent" Poland on the international stage without the consent of the Cabinet. Nevertheless, the conflict between the President and the Prime Minister remained unresolved, leading to a persistent standoff on nearly all crucial political matters. Kaczynski wielded his veto power over a multitude of bills passed by the Parliament, effectively stalling the functioning of the Polish Cabinet.

4.3.2 Shifting Paradigms in Parliamentary Systems and the Rise of Conservatism

In the context of Central and Eastern Europe, the parliamentary system exhibits certain potential flaws. Under this system, once a political party secures a parliamentary majority, it often gains considerable influence over both the prime minister and the president. Conversely, in semi-presidential countries, when a party attains parliamentary control, its authority primarily extends to the prime minister, as the president is not chosen by the parliament. Moreover, in semi-presidential systems, the president is elected by the entire populace, necessitating primary accountability to the citizenry rather than exclusive allegiance to the ruling parliamentary faction.

The concept of accountability in this regard is underscored by the observations of Hungarian scholar Ágh Attila. In the post-Cold War era, Central and Eastern European parliaments evolved by modernizing pre-World War II democratic traditions and drawing inspiration from the successful Western parliamentary model. This transformation elevated parliaments to prominent roles in the region's democratic consolidation, consolidating their dominance in political affairs, thus culminating in a phenomenon described as "supra-parliamentarization".[21]

An illustrative example of this transition is evident in the case of Slovakia, which adhered to a parliamentary system of governance from 1994 to 1998. During this period, the cabinet led by Prime Minister Vladimír Mečiar wielded significant influence, manipulating the parliamentary majority while concurrently maintaining control over the presidency. However, in 1999, the newly formed ruling coalition opted to amend the constitution, effecting a transformation from a parliamentary to a semi-presidential system. This pivotal change mandated the direct election of the president, thereby reshaping Slovakia's system of government into a semi-presidential framework.

Following the global financial crisis of 2008, a prevailing trend in Central and Eastern European countries has been the enduring dominance of right-wing political parties across parliamentary, executive, and even presidential offices. In the April 2010 parliamentary elections, the "Young Democrats' Union-Hungarian Civic Union" party in Hungary secured an overwhelming victory, capturing 52.73% of the popular vote and securing more than two-thirds of the parliamentary seats. In the subsequent 2014 parliamentary elections, while the party's vote share experienced a slight decline, it still retained control over 133 out of 199 seats, comprising over two-thirds of the total, thereby enabling it to establish an independent government. Upon assuming parliamentary authority, the party promptly assumed control of the prime ministerial position and the cabinet. Simultaneously, it clinched the presidency in that year's election. Furthermore, this same party exercised authority over both the legislative and executive branches. In 2012, despite facing limited opposition, it succeeded in pushing through a revised constitution, albeit one that exhibited certain vulnerabilities. In a swift maneuver in 2010, the President of the Supreme Court, whose term had not yet concluded, was replaced by an individual aligned with the Alliance of Young Democrats-Hungarian Civic Union, cementing their influence over the judiciary.[22]

A 2014 report from the Economist Intelligence Unit's Democracy Index revealed a concerning decline in democracy across Central and Eastern Europe in recent years. Hungary occupied the lowest position among new EU member states, experiencing a notable deterioration in both its "Democracy Score Ranking" and "Democracy Index Score Ranking." Consequently, Hungary found itself at the bottom of the list among the new EU member states in these key democratic assessments.[23]

Since the onset of the 2008 financial crisis, Czech politics and political parties have exhibited a notable trend towards conservatism. In the 2010 elections, despite the leftist Social Democratic Party (SDP) securing the highest vote share at 22.09%, it faced the challenge of forming a cabinet due to the predominantly right-leaning nature of parliamentary parties. Consequently, the Civic Democratic Party (CDP), the second-largest party, assumed a dominant role in the government. The Civic Democratic Party (CDP), holding the second-largest vote count, took the helm of the cabinet. Subsequently, the conservative right-wing forces gained control over both the parliament and the executive branch. Furthermore, Vaclav Klaus, who has held the presidency since 2003, also hails from the Civic Democrats, resulting in a unique scenario where a single party exercises authority over the parliament, the cabinet, and the presidency of the Czech Republic.[24]

5 European Union's Contribution to the Political Democratization of the Visegrad Group in the Context of European Integration

During the process of democratic transition in Central and Eastern European countries, the European Union (EU) applies pressure on candidate countries to implement reforms that align with EU standards. This pressure is primarily exerted through the establishment of accession criteria, the formulation of normative treaties and specific requirements in various domains, and the periodic assessment of progress. Simultaneously, the EU actively supports and guides the democratic transition in Central and Eastern European countries by offering financial assistance, technical expertise, and sharing advanced experiences.

5.1 The Evolution of Legal and Constitutional Reforms in Central and Eastern Europe: A Case Study of Poland

During the 1990s, Central and Eastern European countries faced a shared set of structural and socio-economic challenges as they transitioned from their communist pasts to embrace democratic principles and market-oriented economies. To effectively adapt their national legal systems to the evolving social and economic landscape and to maintain the stability of their legal orders, many of these countries turned their gaze toward the European Union (EU). The EU's accession standards became a guiding light for these countries, offering a framework for necessary reforms and improvements in governance. These reforms were further incentivized and catalyzed by the EU's periodic evaluation rounds, which scrutinized the progress of candidate countries.

In this context, the framework and trajectory of reforms within the judicial domain of Central and Eastern European countries evolved through their engagement with the EU. Through this interaction centered around EU norms, these countries cultivated a legal system endowed with legitimacy. The constitutions of these countries delineate the mechanisms for the exercise of state power, encompassing both restrictive and empowerment aspects known as "authorization functions." The legitimacy of political authority is intricately linked to the multifaceted functions of their respective constitutions. Prior to EU membership, the principle of sovereign protection embedded in the constitutions of Central and Eastern European countries accorded them exclusive rights over domestic governance without sharing authority with external entities. However, EU accession necessitated the voluntary relinquishment of a portion of their sovereignty to EU institutions. Acknowledging EU norms became imperative for legitimacy, and consequently, the legislative function had to align with and be constrained by EU norms. These countries embarked on a path of reform under the scrutiny of EU accession criteria.[25]

For instance, Poland embarked on a series of constitutional revisions in 1997, a crucial step in establishing and enhancing legal procedures for the transfer of state power. These revisions provided a robust framework for sovereignty transfer, aligning Polish law with European standards. In 2000, Poland took another significant step by establishing the EU Legislative Committee of the National Assembly. This committee was instrumental in further harmonizing legislative efforts, facilitating the absorption and internalization of EU laws and regulations, and ensuring that Poland diligently worked to meet EU membership standards. These proactive initiatives by Poland garnered attention and commendation from the EU, signifying the country's commitment to aligning its legal and political systems with the broader European community.

5.2 Advancing Judicial Reform and Legal Development in Central and Eastern Europe: A Case Study of Hungary and the Czech Republic

In the EU Strategy Document and Assessment Report, the European Commission lauded the remarkable progress achieved by candidate countries in Central and Eastern Europe in their efforts to adopt fundamental legislation. The Commission recognized that well-defined, forward-looking laws are essential for providing clear guidance for effective action. However, the Commission also emphasized the critical importance of having an independent and impartial judiciary to support the pursuit of justice. This judiciary must adhere unwaveringly to the principles of natural justice and remain accessible to individuals seeking redress for perceived violations of their rights. The process of judicial reform embarked upon by these countries was viewed as a transformational journey, one that aimed to reshape the legal and political legacies of Central and Eastern European countries. This transformation, in turn, contributed significantly to legitimizing new regimes and establishing self-regulatory mechanisms that would serve as the bedrock of democratic governance.

Furthermore, the European Commission outlined specific action details in its proposal for Central and Eastern European countries. It underscored the need to bolster human resource management within the judicial sector, emphasizing that judges in these candidate countries should be well-trained, efficient, highly regarded, and compensated fairly. Efforts should be directed towards enhancing the working conditions of judicial personnel, refining the mechanisms for executing judgments, and improving citizens' access to the judicial system. This multifaceted approach aims to build and enhance public confidence in the judicial systems of Central and Eastern European countries.[26]

For instance, in the 2001 Hungarian report, the EU highlighted the crucial importance of continuous professional development for judges, with a particular emphasis on clarifying their powers and responsibilities. This was seen as a means to ensure the expeditious and effective processing of cases, thus preventing case backlogs and ensuring that justice was delivered in a timely manner. In the Czech reports of 2001 and 2002, it was noted that judge training should encompass diverse areas of expertise, including human rights, market economics, and EU law. By broadening their knowledge in these crucial areas, judges were better equipped to handle complex cases and make informed decisions, ultimately enhancing their competence and understanding.

5.3 Promoting Human Rights and Minority Protection in Central and Eastern Europe: A Case Study of Slovakia

Central and Eastern Europe boasts a rich tapestry of ethnic diversity, with various minority groups contributing to the cultural, social, and historical fabric of the region. This diversity includes Hungarian communities in Slovakia, Roma populations scattered across the Czech Republic, Slovakia, and Hungary, as well as foreign nationals who arrived during the communist era for educational or work purposes. Regrettably, these minority groups often face discrimination and unfair treatment within their host countries, prompting Central and Eastern European countries to make human rights reforms a central focus of their agendas.

The European Union (EU) has emerged as a pivotal player in advancing human rights protection in candidate countries within Central and Eastern Europe. This endeavor is not only aimed at harmonizing standards but also at minimizing the potential impact of their accession on existing EU member states and fostering peaceful coexistence among new member states, with a particular emphasis on reforming minority protection policies.

In 1993, during the Copenhagen Conference, the EU formally incorporated "respect and protection of national minorities" into its standards. Subsequently, the EU introduced various measures to facilitate the reform of minority protection policies in Central and Eastern European countries. Given that EU member states are predominantly members of the Council of Europe and the Organization for Security and Cooperation in Europe, EU policies regarding ethnic issues either directly align with the policies of these two organizations or incorporate their fundamental principles. Notable examples include the European Framework Convention for the Protection of National Minorities, the European Convention for the Protection of Human Rights and Fundamental Freedoms, and the European Charter for Regional and Minority Languages. In 2000, the European Union enacted key legislative instruments such as the "Charter of Fundamental Rights of the European Union," the "Racial Equality Act," and the "Employment Equality Act." These measures served as benchmarks for Central and Eastern European candidate countries, prompting them to adopt similar laws and align their domestic legal frameworks with EU standards.[27]

For instance, in 2001, Slovakia introduced a "Comprehensive Anti-Discrimination Law," a significant step towards addressing discrimination and promoting equal rights for minority populations. This was followed by the adoption of Slovakia's New Labor Code in 2002, effectively incorporating EU anti-discrimination legislation into Slovak legal norms. Furthermore, international organizations like the OSCE have issued valuable recommendations, including the "Hague Recommendations," "Lund Recommendations," and "Oslo Recommendations." These recommendations have played instrumental roles in promoting initiatives related to education, self-governance, political participation, and resistance to ethnic-based discrimination targeting minority populations in Central and Eastern Europe.

6 Illiberalism and Populism: Impeding Factors to Political Democratization of the Visegrad Group

The emergence of illiberalism and populism has become a significant political trend in recent years within Central and Eastern Europe. This rise in populism is rooted in the ongoing challenges of democratic transition that the region has been facing since 1989, particularly within the framework of liberalism. Illiberalism, fundamentally characterized as national populism, has shifted the political landscape, diminished the traditional left-right party divisions, and highlighted a schism between nationalist conservatives and pro-European liberals.

6.1 Populism and Shifting Sentiments: Central and Eastern Europe's EU Journey Amidst Crises

Following the dissolution of the Soviet Union, Central and Eastern European countries aspired to "return to Europe" and reconstruct what they considered a "normal society." During the 1990s, the European Union (EU) generally enjoyed a positive perception in Central and Eastern Europe. Riding the wave of public sentiment, the EU played a pivotal role in overhauling the economies of these countries. It mandated applicant countries to modernize their economic structures and establish viable market-oriented systems.[28]

At its core, the EU's strategy for expanding eastward aimed to align Central and Eastern European countries with the Western political bloc, solidifying their geopolitical advantages. As part of this endeavor, the EU imposed stringent political prerequisites on applicant countries, demanding the preservation of "democracy and freedom" and the genuine implementation of "political pluralism." In 2004, the European Union incorporated the former Central and Eastern European Group into its strategic sphere, giving rise to the term "New Europe" for the relevant countries. While extending economic support to these countries, the EU continued to instigate changes in their systems and disseminate its values.

Initially, during the early stages of transformation in Central and Eastern European countries, most citizens harbored positive expectations of EU accession, primarily driven by the anticipation of economic advantages. However, they did not fully comprehend the EU's political system and value framework. Their preparedness for potential adverse consequences of EU membership was limited. Consequently, when crises emerged, there was a notable shift in public sentiment. Membership in the EU was akin to a double-edged sword. While it opened greater opportunities within the Western European market, the high degree of reliance on foreign trade rendered these countries acutely sensitive to fluctuations in the EU market. Even minor shifts had swift repercussions in these countries. Furthermore, the process exacerbated social disparities, leading to increased polarization between the affluent and disadvantaged, along with the consolidation of social strata, due to institutional changes. A substantial portion of the middle and lower-income populations still failed to reap the economic benefits promised by EU accession.[29]

Starting from 2010, the European Union has faced a series of significant challenges, including the European debt crisis, the refugee crisis, and Brexit. Consequently, the outlook of people in Central and Eastern European countries towards the EU has shifted, and doubts among the middle and lower classes regarding EU membership have resurfaced. The European debt crisis marked the first major economic shock that "New Europe" encountered after joining the EU. Since 2011, the EU economy has struggled with stagnation and recession. This economic downturn initiated a detrimental cycle in the economies of Central and Eastern European countries. Investors grew increasingly skeptical about their economic prospects, resulting in higher interest rates, increased unemployment, lower wages, and reduced incomes and living standards for many households. People in these regions came to the realization that the stability of the Eurozone's economy was not as secure as they had envisioned.

However, the pivotal moment that intensified the centrifugal tendency of Central and Eastern European countries within the EU was the European refugee crisis. In 2015, over a million refugees from the Middle East and North Africa arrived in Europe, catching the European Union off guard. Central and Eastern European countries were allocated a significant number of refugees, sparking widespread panic and discontent in the region. Between May 2015 and April 2016, due to ongoing terrorist attacks and the EU's imposition of refugee quotas, the proportion of Polish citizens strongly opposing refugee acceptance surged from 21% to 43%. Notably, younger generations in Poland displayed a more negative stance toward refugees, with a decreasing likelihood of acceptance among Poles under 45.[30] In 2015, Hungary became the second-largest destination for asylum applications after Germany, with the highest per capita rate of refugee acceptance in Europe. The country also expressed dissatisfaction with the handling of the refugee crisis by Austria, Germany, and the EU. According to a European Commission poll in the autumn of 2015, 82% of Hungarians and 81% of Czechs expressed resistance to immigrants from outside the EU.[31]

These successive crises disrupted the previously harmonious relationship between the EU and Central and Eastern European countries, bringing to light potential conflicts between "old" and "new" Europe. Despite more than a decade passing since their accession to the EU, these countries maintain distinct religious, ethnic, and cultural traditions that have contributed to a profound divide between them and the traditional "Western" societies. Tensions have escalated between upper-class elites embracing Western European culture and lower-class individuals nostalgic for the previous system. These circumstances have provided fertile ground for the emergence of populism as a significant political force, with the refugee crisis serving as a catalyst for the transformation of deep-seated dissatisfaction in Central and Eastern Europe into populist movements.

6.2 Illiberalism in the Visegrad Group: Viktor Orb án's Ideological Shift and Its Regional Impacts

On July 26, 2014, Hungarian Prime Minister Viktor Orbá n delivered a speech at the "Bálványos Summer Free University and Student Camp," where he expounded on his illiberal ideology. During his address, Orbán identified 2008 as a significant turning point. He believed that the era before 2008 was characterized by the dominance of the "liberal worldview," while the period following 2008 saw a growing interest in understanding the successes of non-Western liberal democracies. Orbán argued that it was imperative to recognize that democracy does not inherently align with liberalism. He further asserted that societies organized around the principles of liberal democracy may face challenges in maintaining their global competitiveness unless substantial changes are enacted. This necessitated a departure from liberal principles and societal organization, marking a general shift away from a liberal understanding of society.

In the realm of interpersonal relationships, Viktor Orb án, the Prime Minister of Hungary, offered a critical analysis of the prevailing liberal social structures. He pointed out that these structures were firmly grounded in the principle that individuals have the right to act as long as their actions do not infringe upon the freedoms of others. This principle, often associated with Western liberal democracies, had been embraced by Hungarian society in the two decades leading up to 2010. During this period, Hungary had adopted many of the fundamental principles of Western Europe, aligning itself with the prevailing liberal ideology. However, Orbán argued that in practice, the determination of what constituted mutual freedom was not guided solely by abstract principles of justice. Instead, it was heavily influenced by the power dynamics at play within society. The most powerful individuals and groups, Orbán contended, had a disproportionate say in defining the boundaries of freedom, potentially leading to unequal access to opportunities and resources for the broader population.

Orbán also raised concerns about how liberal democracy, as it was practiced in Hungary, posed challenges to national interests. He argued that both the liberal democratic ideology and the liberal-oriented Hungarian government had failed to adequately protect the collective assets of the community and to shield Hungarian families from the impacts of foreign debt crises. This perceived failure to safeguard national interests led Orbán to emphasize the importance of prioritizing the community as the cornerstone of national reconstruction. In Orbán's view, there was a profound interconnectedness between individual labor and interests with the well-being of the community and the nation-state. He believed that strengthening the community and ensuring its prosperity were essential for the overall health and stability of the nation.[32]

Drawing from these ideas and concerns, Orbán concluded by characterizing the Hungarian state that he and his administration were constructing as an "illiberal state." This label signaled a deliberate departure from the liberal principles that had previously guided Hungary's political landscape. It marked a shift towards a more centralized and controlled form of governance, where the state played a more active role in shaping society and protecting what Orbán saw as the nation's core values and interests.

In line with his illiberal ideas, Orbán bolstered the governing party's administrative authority through constitutional and legislative amendments. Additionally, he increased control over media outlets, cultural institutions, and similar entities. In terms of economic policy, there was a concerted effort to cultivate a favorable investment climate and strengthen economic partnerships with Eastern countries, including China, Russia, and countries in Central Asia. This strategy aimed to reduce Hungary's economic reliance on Western countries. Measures were also taken to generate employment opportunities domestically, especially through public works projects aimed at assisting the underprivileged. Initiatives like providing free meals to school and kindergarten students were implemented to address social welfare concerns, resulting in a reduction in Hungary's poverty rate.

Analysts have noted that the trend of illiberalism is not limited to Hungary but has also manifested itself in Poland. Illiberal tendencies have surfaced in Slovakia and the Czech Republic as well, with notable features being anti-immigration stances, skepticism towards Western values, efforts to consolidate authoritarian governance, and a focus on preserving economic and cultural distinctiveness while catering to national interests and voter preferences. Some studies have even suggested that while Hungary and Poland are at the forefront of this trend, similar patterns are emerging in countries like Slovakia and Croatia. In recent years, doubts have arisen regarding the liberal model that emerged after 1989. This skepticism primarily challenges the legal framework of liberal democracies, emphasizes nationalist narratives, and underscores the importance of what is often referred to as the "culture war."[33]

6.3 Strengthening Democracy in the Visegrad Group: A Comprehensive Approach

To fortify and enhance democracy within the Visegrad Group, a comprehensive and collaborative approach is imperative, involving both governmental bodies and civil society organizations (CSOs). This multifaceted strategy is vital for safeguarding democratic values and ensuring long-term stability in this region.

A primary focus should be on nurturing the growth and independence of CSOs. This entails providing financial support, legal safeguards, and essential resources to bolster their capabilities. CSOs serve as critical pillars of democracy by actively promoting democratic principles, scrutinizing government actions, and advocating for the rule of law.

Media freedom is an essential cornerstone of a vibrant democracy. Governments must play a role in ensuring a level playing field in media ownership by enacting and enforcing regulations that prevent undue concentration of media ownership. Encouraging investigative journalism and ensuring media outlets have access to diverse sources of information is crucial. Furthermore, supporting independent media watchdog organizations can help maintain fairness and integrity in reporting. [34]

Civic education programs within educational institutions should be fortified to impart a deep understanding of the significance of democracy, human rights, and the rule of law. These programs should also cultivate critical thinking skills and media literacy among citizens, enabling them to discern accurate information from misinformation. An informed and engaged citizenry is pivotal for sustaining democracy.

The fight against corruption necessitates the establishment and enforcement of stringent anti-corruption measures and transparency regulations. It is essential to create an independent body dedicated to investigating and prosecuting corruption cases. Additionally, governments should adopt practices that promote transparency, such as disclosing government contracts and expenditures, thereby ensuring that public resources are used ethically and efficiently. [35]

Active collaboration with the EU and other international organizations committed to democratic values is of paramount importance. The EU can play a significant role in holding the Visegrad Group accountable for adhering to democratic standards. Establishing partnerships and fostering dialogue with other countries and international groups can facilitate mutual learning and joint efforts to bolster democracy's strength and stability.

In summation, the Visegrad Group can secure the longevity of its democracy by translating these practical proposals into action. By doing so, the region can uphold the fundamental principles of democracy, including fairness, human rights, and transparency, ensuring that its democratic institutions remain robust and resilient in the face of challenges.

7 Conclusion

In post-communist Central and Eastern Europe, the transformation towards democracy represented a significant historical turning point. The Visegrad Group played a critical role in this shift from communism to democracy and free markets, bringing about substantial political, economic, and social changes. During the democratization process, the Visegrad Group sailed through several challenges such as developing market-oriented economic systems, ensuring political inclusivity, and establishing the rule of law. The transition was, however, not without obstacles, including weak democratic institutions, corruption, and political instability. Nonetheless, their commitment to democratization paid off, resulting in some of the region's most successful and stable countries.

The democratization process in Central and Eastern Europe was inextricably linked with the region's alignment with the European Union. The EU played a significant role in promoting democracy and stability in the area by promoting democracy as a prerequisite for membership and offering several incentives and support programs that accelerated the democratization process in the territory. Furthermore, by providing a framework for democratic consolidation and institutional strengthening, European Integration contributed to the Visegrad countries' successful transition to democracy.

However, democratization in Central and Eastern Europe faces ongoing challenges despite the progress achieved. The erosion of civil society, the rise of populism, and the emergence of illiberalism pose significant threats to the stability and long-term sustainability of democracy in the region. Hence, it is crucial to remain watchful and continue to strengthen democratic institutions, promote civic engagement, and support sustainable governance practices to safeguard the region's democratic gains. To this end, this study has suggested various policy strategies and interventions that aim to address the democratization process's challenges and opportunities. Such recommendations include strengthening democratic institutions, promoting media freedom, enhancing civic education, and improving transparency and accountability. Implementing these measures will enable the Visegrad countries, and other countries in the region, to establish resilient and sustainable democratic systems that enhance stability, prosperity, and security.

Overall, this research contributes to a better understanding of democratization processes in Central and Eastern Europe, mainly the Visegrad Group countries, and the EU's role in promoting democracy and stability. The study draws from a theoretical framework and provides practical recommendations to address the region's ongoing challenges and opportunities. This research is anticipated to facilitate efforts to sustain democracy and stability in Central and Eastern Europe and beyond.

References

1. "The End of WWII and the Division of Europe." *The Center for European Studies (CES) at the University of North Carolina*, 2018, europe.unc.edu/the-end-of-wwii-and-the-division-of-europe/.
2. Hunter, Brian. Soviet-Yugoslav Relations, 1948-1972. *Scholarly Title*, 1976.
3. Brown, Michael Barratt. "In the Straggle for Freedom, Yugoslavia: The New Class in Crisis and Economic Development of Communist Yugoslavia, 1947–1964." *International Affairs*, vol. 47, no. 2, Apr. 1971, pp. 426–429, https://doi.org/10.2307/2613993.
4. Horowitz, Donald L. "Democracy in Divided Societies." *Journal of Democracy*, vol. 4, no. 4, 1993, pp. 18–38, https://doi.org/10.1353/jod.1993.0054.
5. Kenney, Padraic. "The Gender of Resistance in Communist Poland." *The American Historical Review*, vol. 104, no. 2, Apr. 1999, p. 399, https://doi.org/10.2307/2650371.
6. Lieblich, Eliav. "The Soviet Intervention in Hungary (1956)." *SSRN Electronic Journal*, 2017, https://doi.org/10.2139/ssrn.2960865.
7. Matlochová, Irena. "Comparison of Czechoslovak and Hungarian reform of socialism in the 60th years" *Masaryk University*, 2011, is.muni.cz/th/yuqhe/?lang=en.
8. Kuran, Timur. "Now out of Never: The Element of Surprise in the East European Revolution of 1989." *World Politics*, vol. 44, no. 1, Oct. 1991, pp. 7–48, https://doi.org/10.2307/2010422.
9. Roeder, Philip G. "The Revolution of 1989: Postcommunism and the Social Sciences." *Slavic Review*, vol. 58, no. 4, 1999, pp. 743–755, https://doi.org/10.2307/2697197.
10. Dangerfield, Martin. "The Contribution of the Visegrad Group to the European Union's "Eastern" Policy: Rhetoric or Reality?" *Europe-Asia Studies*, vol. 61, no. 10, Dec. 2009, pp. 1735–1755, https://doi.org/10.1080/09668130903278934.
11. Schmidt, Andrea. "Friends Forever? The Role of the Visegrad Group and European Integration." *Politics in Central Europe*, vol. 12, no. 3, 1 Dec. 2016, pp. 113–140, https://doi.org/10.1515/pce-2016-0019.
12. Dangerfield, Martin. "The Contribution of the Visegrad Group to the European Union's "Eastern" Policy: Rhetoric or Reality?" *Europe-Asia Studies*, vol. 61, no. 10, Dec. 2009, pp. 1735–1755, https://doi.org/10.1080/09668130903278934.
13. Elgie, Robert. "Varieties of Semi-Presidentialism and Their Impact on Nascent Democracies." *Taiwan Journal of Democracy*, vol. 3, no. 2, 1 Jan. 2008, pp. 53–71, https://doi.org/10.29654/tjd.200801.0004.
14. Scully, Roger M. "MEPs and the Building of a "Parliamentary Europe." *The Journal of Legislative Studies*, vol. 4, no. 3, Sept. 1998, pp. 92–108, https://doi.org/10.1080/13572339808420566.
15. Russo, Federico, and Matti Wiberg. "Parliamentary Questioning in 17 European Parliaments: Some Steps towards Comparison." *The Journal of Legislative Studies*, vol. 16, no. 2, June 2010, pp. 215–232, https://doi.org/10.1080/13572331003740115.
16. Sedelius, Thomas, and Olga Mashtaler. "Two Decades of Semi-Presidentialism: Issues of Intra-Executive Conflict in Central and Eastern Europe 1991–2011." *East European Politics*, vol. 29, no. 2, June 2013, pp. 109–134, https://doi.org/10.1080/21599165.2012.748662.
17. Legvold, Robert, and Herman Schwartz. "The Struggle for Constitutional Justice in Post-Communist Europe." *Foreign Affairs*, vol. 79, no. 5, 2000, p. 146, https://doi.org/10.2307/20049937.
18. Schimmelfennig, Frank, et al. "Costs, Commitment and Compliance: The Impact of EU Democratic Conditionality on Latvia, Slovakia and Turkey." *JCMS: Journal of Common Market Studies*, vol. 41, no. 3, June 2003, pp. 495–518, https://doi.org/10.1111/1468-5965.00432.
19. Coller, Xavier, et al. "Problems of Democratic Transition and Consolidation. Southern Europe, South America and Post-Communist Europe." *Johns Hopkins University Press*, no. 76, 1996, p. 291, https://doi.org/10.2307/40184001.
20. Poulard, Jean V. "The French Double Executive and the Experience of Cohabitation." *Political Science Quarterly*, vol. 105, no. 2, 1990, p. 243, https://doi.org/10.2307/2151025.
21. Luciano, Bruno Theodoro. "Inter-Parliamentary European Union–Latin American Caribbean (EU-LAC) Relations and the Increasing Political Convergence among Latin American Regional Parliaments (2006–15)." *Parliaments, Estates and Representation*, vol. 37, no. 3, 15 June 2017, pp. 318–334, https://doi.org/10.1080/02606755.2017.1336326.
22. European Election Database: https://www.politico.eu/europe-poll-of-polls/european-elections-data-visualized/
23. "Democracy and its discontents, A report from The Economist Intelligence Unit," http://www.sudestada.com.uy/Content/Articles/421a313a-d58f-462e-9b24-2504a37f6b56/Democracy-index-2014.pdf
24. Hloušek, Vít, and Lubomír Kopeček. "Caretaker Governments in Czech Politics: What to Do about a Government Crisis." *Europe-Asia Studies*, vol. 66,

no. 8, 14 Sept. 2014, pp. 1323–1349, https://doi.org/10.1080/09668136.2014.941700.

25. Schimmelfennig, Frank, and Ulrich Sedelmeier. "Governance by Conditionality: EU Rule Transfer to the Candidate Countries of Central and Eastern Europe." *Journal of European Public Policy*, vol. 11, no. 4, Jan. 2004, pp. 661–679, https://doi.org/10.1080/1350176042000248089.
26. Dimitry Kochenov. "Behind the Copenhagen Facade. The Meaning and Structure of the Copenhagen Political Criterion of Democracy and the Rule of Law." *Social Science Research Network*, 13 July 2004. https://papers.ssrn.com/sol3/papers.cfm?abstract_id=563743.
27. Guliyeva, Gulara. "Joining Forces or Reinventing the Wheel? The EU and the Protection of National Minorities." *International Journal on Minority and Group Rights*, vol. 17, no. 2, 2010, pp. 287–305, https://doi.org/10.1163/157181110x495908.
28. Balcerowicz, Leszek. "How Capitalism Was Built. The Transformation of Central and Eastern Europe, Russia, and Central Asia. Anders Åslund." *Eurasian Geography and Economics*, vol. 49, no. 2, Mar. 2008, pp. 228–230, https://doi.org/10.2747/1539-7216.49.2.228.
29. Podobnik, Boris, et al. "Predicting the Rise of EU Right-Wing Populism in Response to Unbalanced Immigration." *Complexity*, vol. 2017, 2017, pp. 1–12, https://doi.org/10.1155/2017/1580526.
30. Public Opinion Research Center of Poland, 2016, *CBOS*: http://www.cbos.pl/EN/home/home.php.
31. BBC, Migrant crisis: Migration to Europe explained in seven charts. *BBC news*. 2016,4 March. http://www.bbc.com/news/world-europe-34131911.
32. "Prime Minister Viktor Orbán's Speech at the 25th Bálványos Summer Free University and Student Camp" July 30, 2014, *Website of the Hungarian Government*, https://2015-2019.kormany.hu/en/the-prime-minister/the-prime-minister-s-speeches/prime-minister-viktor-Orb á n-s-speech-at-the-25th-balvanyos-summer-free-university-and-student-camp.
33. Hooghe, Liesbet, and Gary Marks. "Grand Theories of European Integration in the Twenty-First Century." *Journal of European Public Policy*, vol. 26, no. 8, 17 Jan. 2019, pp. 1113–1133, https://doi.org/10.1080/13501763.2019.1569711.
34. Ž ižek, Slavoj. "Against the Populist Temptation." *Critical Inquiry*, vol. 32, no. 3, Mar. 2006, pp. 551–574, https://doi.org/10.1086/505378.
35. Greskovits, Béla. "The Hollowing and Backsliding of Democracy in East Central Europe." *Global Policy*, vol. 6, no. 1, June 2015, pp. 28–37, https://doi.org/10.1111/1758-5899.12225.

Metaphor-constructed Terrorism Discourse

Leiming Wang[1]*, Shanti C Sandaran [2]

[1]Universiti Teknologi Malaysia, Johor Bahru, Malaysia
[2]Universiti Teknologi Malaysia, Johor Bahru, Malaysia

Abstract. The US launched "war on terror" has stirred hot debate globally. However, it seems that nobody has stopped to think about the nature of terrorism, and it appears that very few people including quite a few scholars have not realized the discourse nature of terrorism. Therefore, this paper will study the production mechanism of terrorism discourse. The discourse analysis thoey and metaphor theory will be used to make a detailed analysis of the former US president's speeches and relevant data to disclose the discourse nature of terrorism and anti-terrorism with an aim to finding out the production mechanism of terrorism discourse through varied discourse strategies and tactics.

1. Introduction

Political rhetoric is actually the use of rhetoric in the political process. What distinguishes it from general rhetoric mainly lies in the rhetorical activities that political subjects take political affairs as the content and revolve around political power, whose purpose is to achieve political persuasion and achieve a certain political purpose.The reason why this paper puts forward and forms such problems as political rhetoric is that rhetoric is indispensable, universal and plays an important role in the political process, especially in the operation of political power, and therefore,let social members recognize the legitimacy of political power, its ways and methods, but one of the important ways is to convince them that the political power is legitimate, this way of specific rhetoric to express political views is the legitimacy process of political rhetoric, namely political subject using political language of appropriate wording, myth, metaphor, tradition, ideology, etc., or adopt various rhetoric means of political persuasion.[1] Political rhetoric rules out all the possible terrorism discourses to legitimize the so-called anti-terrorism war together with all the concomitant operations.

2. Literature Review

The US "War on Terror" began in the War in Afghanistan in 2001. The study of the political rhetoric of the US "war on terrorism" has certainly followed. In a few years, there were not many academic works specializing in the political rhetoric of the American "war on terror", but several academic works did study political language and political rhetoric from the perspective of language, symbols and politics, and war relations. These works include: 《Collateral Language: Guide to America New War, edited by John Collins and Ross Gloff, and Joseph S. Tuman's 《Communicating Terror: the Rhetorical Dimensions of Terrorism》, Sandra Silberstein:《War of Words: Language, Politics and 911》, Richard Jackson: 《Writing the War on Terrorism: Language, Politics and Counter-terrorism》 .John Collins and Ross Gloff argue that language is a terrorist constitution; that the legitimacy of the "new American War" stems from linguistic tyranny.Sandra Silberstein believes that in the war on terror, all the familiar images and arguments serve the Bush administration and the support for the war in Afghanistan. While the American people are not just blind followers of government and media propaganda, the media are increasingly focused on the elaboration of rhetoric and image, and, through multicultural interactions, generate consensus among the public around the war. In short, political rhetoric is essentially a rhetorical activity of political subjects around political power with political affairs as the content. Its purpose is to realize political persuasion and give full play to political influence.The US "war on terror" was the first protracted and massive war of the 21st century. The "war on terror" launched by the United States, like any war in world history, deserves its legitimacy. In order to justify the "war on terror", the United States must prove the necessity of war that the United States needs war and has to conduct war. To prove the necessity of war began with the American government constructing an "age of terror," where the US was in a dangerous era of terrorism surrounded by terrorism. At the heart of this "age of terror" is the "threat", a threat to terrorism. The American government constructs the terrorist "threat" through political rhetoric, making the American society in the general extreme fear, so as to persuade the American people to support its war policy, and thus obtain the legitimacy of the "war on terror". Why should the US government construct a "threat" to put the American society in widespread

* wlmsufe@126.com

extreme fear to gain the legitimacy of the "war on terror"? This should start with the relationship between fear and legitimacy.

In order to launch a "war on terrorism", the United States first through the construction of terrorist political discourse, to create the upcoming, will lead to great pain or destruction of the disaster, to create a general fear in the American society, to prove the legitimacy of its "war on terror", to obtain the people's recognition to launch a "war on terror". Popular recognition is an important pillar of the legitimacy of modern countries. The higher the public identity, the lower the cost of ruling the country, and the lower the public identity, the higher the cost of ruling the country. Therefore, modern countries want to get the best of their people based on the cost of ruling. If the tool of violence is an explicit rule, then obtaining the psychological recognition of the explicit rule is an inevitable trend of modern politics; to realize the implicit rule of public psychological identity, it is necessary to persuade the people through political rhetoric.[3]Especially for the United States the strong sense of civil rights, has the precedent of the Vietnam war, the Bush administration launched the "war on terrorism" especially need to win the general recognition, especially need to construct through the political discourse extremely frightening state of emergency, to persuade people generally agree that the "war on terrorism" is justified.

3. Research methods and theoretical analysis

It should be noted that when the US government constructs the legal legitimacy of the war reasons for attacking Afghanistan, it is achieved through the defining function of political rhetoric. The United States officially invoked Chapter VII, Article 51 of the UN Charter, and Security Council Resolution 1368. The problem is that the official US political rhetoric constructs the 9 / 11 events as an "act of war". The entire composition of American official discourse is based on building terrorist attacks into "acts of war."It is the "act of war" political rhetoric construction of the "911" incident into the category of international law, so that the United States can properly quoted the relevant provisions of the UN charter, and according to the relevant provisions of the UN charter, make the United States can freely choose what time to attack, who attack, how to attack, this is the exercise of the right to self-defense.[2]On the one hand, the US official use the moral justification of post-punishment to illustrate the justify the "war on terror". US officials have repeatedly stressed that the purpose of "anti-war-terrorism" is to give terrorists their due punishment and achieve justice. Mr.Bush's most famous expression of justice is: " Whether we punish our enemies with justice, or whether we punish our enemies with justice, justice will be done."While then Deputy Secretary of State Mark Grossman affirmed that the war was consistent with international law and had legal legitimacy, he did not forget to construct the legitimacy of morality: I think what we do met the legal requirements, and what we do met the moral requirements of...... The right to self-defense covers all of this. On the other hand, US officials use the moral justification of punishment to the fore to justify the "war on terror". If the reason for the war in Afghanistan is the legitimacy of the punishment after the event-because the perpetrator of 9.11, al Qaeda is in Afghanistan, then the reason for the war in Iraq is the legitimacy of the punishment in advance. Because "evil regimes" like Iraq have the potential to provide weapons of mass destruction to terrorists, which could do far more harm to the American people than the 9.11 incident. This kind of moral and legitimate political discourse of prior punishment has often appeared in the speeches of the US president, Secretary of State and other US officials at different times and on different occasions.

Citing international support for war further illustrates the legitimacy of war. On the one hand, the political discourse that terrorist attacks are an "act of war" persuaded not only the majority of the United States, but also the majority of Congress, thus gaining domestic legal support and public opinion support, and making the Bush administration as a legitimate authority the subject of the declaration of war. At the same time, the Bush administration also showed through the political term "coalition forces" that the legitimacy of the war reasons can also be justified by the joint actions of the international community. It shows that the United States is not fighting for the "war on terrorism". It has won the support of international public opinion, and it has real international political and military support.

In the following text, the U. S. government arouses legal support around the world. A strong coalition of nations is "said" to "support American action," anti-war fear " is not unilateral action, but the world solidarity:

We have support around the world, not just verbal support, but real support. Whatever happens ahead, this support is firm. It is a full league, it is a trustworthy league.

More than forty countries from the Middle East, Africa, Europe and Asia have granted us the right to use their airspace and territories. More countries are sharing their intelligence. We are supported by the collective will of the whole world.

In this cause, the world is united... and in this cause, our partners represent countries and peoples of all cultures, of all religions, and of all races.

This is a powerful political discourse that clearly marks the "war on terror" as a just war. Here, it is no longer an American war, nor a simple conflict between a single civilization, but an expression of the common will of all cultures, all religions, and all races, including all civilizations, and a symbol of human civilization and progress. The enemy of this common will is the "extremely dangerous enemy of terrorism", the "axis of evil". Through this construction of political discourse, the "war on terror" has a strong moral power. It is no longer a unilateral military action, but a "joint military action" formed by the international support and good cooperation of many countries in the world.

From "right to self-defense", "the legitimacy of

punishment" to all cultures, all religions and all races of the expression of the "common will", the American government's political discourse has a clear logical level, the legitimacy of the war reason gradually revealed, and constantly strengthen, finally complete the legitimacy of "famous" construction.

3.1 The political discourse of the US "war on terror" constructs the legal legitimacy

"Internal enemies" is one of the major battlegrounds of the US "war on terror".In the U. S. government's view, "the new era of terror", "the state of extreme emergency", "extremely dangerous terrorist enemy" and other political words to describe the threat of the terrorist enemy seems to be not huge enough, fear of the last scene in the imagination of American fear: the threat is not only foreign enemies, it also hidden in the United States. As Bush said, it is necessary to "get law enforcement the more means needed to trace the terrorist forces at home."This sentence reinforces the idea that "at home", a comfortable and safe place, has been infiltrated by terrorism. John Ashcroft's construction is even more convincing.This construction of internal enemies has a considerable rhetorical power, which further shrinks the already terrifying psychology of the American people. The success of this political rhetoric built the legitimacy for the United States to launch a "war on terror".In order to prove the legitimacy of the "war on terror", the United States not only proved that the war is necessary against threats, but also constructed the "just war on terrorism" from the perspective of the just war theory.After the "911" incident, on the one hand, the US government constructed a very frightening "era of terror" and justified the "war on terror"; on the other hand, it positively constructed a "just war" in line with the humanitarian principles to prove the legitimacy of the "war on terror". announced in the national speech that the nature of the incident is the "act of war" against the United States, because the terrorists attack is the American way of life and American freedom, not just two civil aircraft crashed into the twin towers, therefore, "our troops at home and abroad have been on high alert". Clearly, Bush sees terrorist attacks as an act of war and put the troops at war. Now that "acts of war" against the United States, the United States has the right to fight a war in self-defense. Because, legally speaking, Article 51 of Chapter VII of the Charter of the United Nations stipulates that: " If Member States of the United Nations are armed attacked, this Charter shall not prohibit Member States from exercising their natural rights of defense alone or collectively until the Security Council takes measures to maintain international peace and security."In this way, the US government is" the most important result of the construction of terrorist attacks is to allow the United States to respond within the framework of correct acts of war in accordance with international law." Therefore, a military strike against Afghanistan is a "self-defense" operation, fully consistent with international law. So, then U. S. Defense Secretary Donald Rumsfeld said: " What the United States of America did is what I said, and it is fighting against terrorists to defend the United States."

3.2 The political discourse of the "war on terror" in the United States constructs the moral legitimacy. Morally speaking, the United States defines the "war on terror" as a legitimate punitive measure. In western culture, there are two kinds of punishment legitimacy: one is the legitimacy of later (backward-looking) punishment; the other is the legitimacy of prior (forward-looking) punishment. The legitimacy of later punishment comes from the intentional injury to people has occurred, from the moral (deontology, the Greek "necessary"), it is a legitimate revenge, revenge, tit for tat. Those who intentionally harm others deserve the punishment, and revenge is justified, and it is a necessary act to "correct the wrong behavior" to restore the moral balance. So, the terrorists must be punished. The legitimacy of prior punishment lies in the legitimacy of its results, which is to prevent the possible intentional harm against people, eliminate the risks that people face, and show to the potential criminals that such harm is intolerable.[4]

3.3 The US government cited the international support of the war to further illustrate the legitimacy of the war.The "war on terror" was announced by Bush in the days following the terrorist attacks. His political discourse that terrorist attacks are an "act of war" persuaded not only the majority of the United States, but also the majority of Congress, thus gaining domestic legal support and public opinion support, and making the Bush administration as a legitimate authority become the subject of the declaration of war. At the same time, the Bush administration also showed through the political term "coalition forces" that the legitimacy of the war reasons can also be justified by the joint actions of the international community. It shows that the United States is not fighting the "war on terrorism". It has the support of international public opinion, and it has real international political and military support.The army invested in the war on terror is legally organized, a just division, and the US is not doing a solo performance. So the focus of the official US discourse is on the strong alliances of all the countries collectively involved in the "war on terror."In the daily official discourse, the troops fighting in Afghanistan and Iraq always say "coalition forces," the official words that rarely fail.

In one speech, Bush said the victory is certain, and the only question is when it will happen. Through hard work, he thinks, the day will soon come: " By fighting side by side, we will soon see the day of victory." This authoritative statement of the president of the United States is the most influential symbol of American political power. It gives full play to the character appeal in the political rhetoric, and is full of persuasion, making the victory of the war become a definite understanding of the American government and the people.

The fact that winning the "war on terror" is impossible is just a false metaphor. Conceptually, winning a "war on terror" is similar to winning a victory in suppressing an uprising. One kind of terrorist political violence is always favorable to certain protagonists (states, dissidents, guerrillas, revolutionaries), and in some circumstances (a nation is under the brutal occupation of a supermilitary power). In fact, the US "war on terror" has never survived by the time when the Bush administration stepped down in 2008. Instead, predictions about a repeat of the Vietnam War are coming true. Today, the security situation in Afghanistan and Iraq shows that the "war on terror" is impossible. However, no matter what the result is, the US government has established a firm belief in the justice of the war by the US government and the American people through the construction of the political rhetoric, which has obtained the political guarantee for the smooth implementation of the "war on terror".[5]

4. Research results and conclusions

4.1.Character appealing

The character argument in the political rhetoric of the war on terrorism is mainly the credibility of the US government. The subject of political rhetoric, the United States government includes the president of the United States and officials at all levels of the United States government. Therefore, the character of the US government actually contains the credibility of the US president and US government officials at all levels. The government of any country is credible as long as it enjoys authority among its people. As a national institution of the United States, the United States government enjoys the authority needed for its rule among the American people. Therefore, the United States government has the character needed for political rhetoric.

4.2Rational demands

There is no doubt that reason is also an important argument for the political rhetoric of the US "war on terror". In fact, the US government has been using rational arguments in the process of constructing the "war on terror".

The rational arguments of political rhetoric are generally accomplished through syllogisms. Bush tried to explain the American attacks by appealing for reason."America is being targeted because our beacon of freedom and opportunity is the brightest and brightest in the world."If the political rhetoric syllogism including premise and conclusions is reconstructed, it can be analyzed as follows:

premise: Terrorists target those who promote freedom and opportunity.

Small premise: America's beacon of freedom and opportunity is the brightest and brightest in the world.

Conclusion: The American beacon of freedom and opportunity is the brightest and brightest in the world, so the terrorists target the United States.

4.3 Emotional rendering

Emotional argument is an important means for political rhetoric to maintain, strengthen or change a certain political attitude of the audience. When creating and developing political discourse, the political rhetorical subject realizes political persuasion on the basis of his character and rationality, but arousing the emotions of the audience is the final embodiment of realizing political persuasion. Whether the political persuasion can be realized, the main criterion is to see whether the political discourse maintains, strengthens or changes the emotions of the audience, so that the audience is in a certain state of mind that the political rhetoric subject wants, so that they can support the political views of the political rhetoric subject.The political rhetoric resources brought about by the 9 / 11 incident are only the first step in the political rhetoric of the war on terrorism. More importantly, the US's "war on terror" also needs to find the target of its military strike, that is, where the enemy is. Through the intersection of rationality and emotion, the US government has not only logically proved that Bin Laden and the rival regimes in Afghanistan, Iraq and Iran are the enemies of the United States, but also demonized and dehumanized the enemy emotionally. In this way, from the 9 / 11 attacks to the war on terror.

4.4Dual-opposed political rhetoric

The political language structure of binary opposition constitutes the political rhetoric of binary opposition. Binary opposition of political rhetoric is the most commonly used and persuasive political persuasion strategy, which can integrate the power of political rhetoric into the system of political language. In the political discourse of the "war on Terror" in the United States, the political rhetoric of binary opposition includes the political cultural framework and the political language framework of binary opposition, which is embodied in the subject opposition, objectives, means and moral appeal in the process of political rhetoric.Like other typical American "war on terror" political rhetoric (for example, the political rhetoric of the Cold War), it adopts a binary and antagonistic political and cultural framework. It opposes and excludes other political cultures different from American political culture, and uses this binary opposing political cultural framework to achieve the purpose of political persuasion, proving the legitimacy and legitimacy of the "war on terror".

The political rhetoric of the "war on terrorism" uses the binary political and cultural framework, but also uses the framework of binary political language.[6] On the one hand, the political expression framework of the binary opposition is the rational expression of the political language, and on the other hand, it is also to construct the political culture that arouses people's hatred and opposition. Political terms of binary opposition in the political rhetoric of the US "war on terror" abound. Here summarizes the opposing political terms used in the political rhetoric of the "war on terror" as grouped as possible. They include: Civilization / barbarism,

Freedom, Despotism, Democratic / authoritarian, Hero / Cowards, Peace / War, Justice / evil, just / evil, Wealth / poor, Advanced / backward, internal / external, domestic / Foreign, self / foreign, Self / Other, we / them, fighters / terrorists, native / foreigners, etc. Although the binary political terms of "war on terrorism" can only be summarized here, they are typical political terms commonly used, which are enough to explain the actual content and political function of the political rhetoric of "war on terrorism". These binary and opposing political terms in political rhetoric are only intended to persuade the audience that "we" are "good" and "kind", while the "enemy" is "bad" and "evil", so that the subject of political rhetoric can obtain the political recognition and political support of the people. When it comes to the political rhetoric of the US "war on Terror", the US government intends to convince the American people that the American government and the American people have the excellent qualities of generosity, generosity, gentleness, wisdom, courage, sacrifice, friendship and unity, while the enemy is barbaric, narrow, evil and cowardly terrorists.In short, through the functional analysis of the political rhetoric of the US "war on terrorism", it can be seen that the US government has constructed the "era of terrorism" and the "just war on terrorism" from two aspects when constructing the "war on terrorism".

References

1. Sandra Silberstein : War of Words : Language, Politics, and 9/11, Routledge, 2004, p.19.
2. Richard Jackson, Writing the War on Terrorism: Language, Politics and Counter-terrorism, Manchester University Press, 2005 , p.125.
3. Rumsfeld and Myers Briefing on Enduring Freedom 2:45 p.m.(EDT) ; October 7, 2001, http ://www.yale.edu/lawweb/avalon/sept_1 1/dod_brief 22. htm.
4. Verna V.Gehring (ed) , War after September 11, Rowman & Littlefield Publishers, Inc., 2003, pp.ll-12.
5. Remarks by the President to the People of Poland, http://www.whitehouse, gov/ news/releases/2003/05/20030531 -3. html.
6. Joseph S.Tuman, Communicating Terror : the Rhetorical Dimensions of Terrorism, Sage Publications, Inc., 2003, p.90.

Research on the Improvement Path of College Online Teaching Quality based on TQM Theory

Pan Di[1,*], Jin Ju[2]

[1]Harbin University of Commerce, Harbin Postcode, Heilongjiang Province, China
[2]Harbin University of Commerce, Harbin Postcode, Heilongjiang Province, China

Abstract. The gradual development of online teaching has become an important field of personnel training in universities, strengthening the quality of online teaching has become a new task for the development of higher education in our country. In view of the challenges faced by online teaching in universities in quality assurance, quality control and quality management, based on the requirements of applying TQM theory to online teaching in universities in an all-round, whole process and full participation, starting from establishing online teaching quality assurance mechanism, optimizing online teaching quality control system and improving online teaching quality supervision measures, the quality of online teaching in universities is effectively improved.[1] Help our country higher education cause reform and development.

1 Feasibility analysis of applying TQM theory to online teaching quality management in universities

1.1 Overview of TQM theory

Total Quality Management (TQM) is a quality management model that focuses on quality, takes full participation as the basis, and achieves long-term success through customer satisfaction and benefits for all members of the organization and society. The core idea of TQM theory is to take quality as the center, to meet customer needs as the basic starting point, to emphasize the systematic point of view as the guidance for all-round comprehensive management, pay attention to the whole process of affecting quality, highlight the important role of all staff participation in quality management, and continuously improve product quality. The conceptual framework of total quality management includes quality assurance, quality control, quality management and so on.[2] Its basic requirements can be summarized as all-round guarantee, whole-process control and whole-staff participation in management, that is, quality improvement requires all-round hardware equipment and objective elements to provide guarantee, quality control should run through the whole process from beginning to end and all aspects of the entire system, and quality management needs the joint participation of every employee at all levels of the organization. And continuous and endless improvement of quality. Therefore, the total quality management theory, as a complete quality management system and scientific management technology, is conducive to the establishment of a systematic, dynamic, personalized and characteristic quality management mechanism, so that the entire production process is in a controllable state, to achieve sustainable management and development.

1.2 Significance of TQM theory applied to the improvement of online teaching quality

TQM theory is of great significance to the improvement of online teaching quality in universities. First, TQM theory is beneficial to improving the quality of online teaching. At present, many universities lack perfect online teaching quality assurance mechanism, the main reason is that they fail to take quality assurance measures for the multi-directional factors that affect the quality of online teaching. TQM theory is an important means to ensure and improve the quality of products or services. A set of practical and effective quality assurance mechanism is constructed by using the all-round guarantee thought of TQM theory. On the one hand, the quality of online teaching is fully guaranteed. On the other hand, it is helpful to build a complete online teaching infrastructure to ensure the improvement and development of online teaching quality. Second, TQM theory can help control the quality of online teaching. Although the traditional online teaching quality control mode focuses on process control and sets up many key links or measures, it lacks the assistance of system theory and is more a collection application of conventional practices. The whole process quality control system of TQM theory emphasizes the idea of people-oriented and continuous improvement, so that the whole online teaching process is in a controllable state. 3Universities are required to continuously improve the quality according to the changing needs of students,

* Corresponding author: 535450014@qq.com

emphasize continuous innovation, improvement and change, play a long-term role, and help effectively improve the quality of online teaching.[3] Third, TQM theory helps to supervise the quality of online teaching. TQM theory pays more attention to the teacher-student orientation, and emphasizes the construction of quality supervision mechanism involving universities, teachers, parents and other subjects. All staff, teachers and students have important responsibility for the improvement of online teaching quality. Relying on TQM theory, full participation management model can provide new ideas for guiding the improvement of online teaching quality.

2 The new challenges to improve the quality of online teaching in universities

2.1. Quality assurance dilemma

For the quality assurance of college teaching, the development of online teaching is both an opportunity and a challenge.[4] One is the virtualization challenge of online teaching environment. On the one hand, the classroom environment has changed from face-to-face teaching between teachers and students in the same classroom to online video communication.[5] On the other hand, the institutional environment is becoming virtual with the change of teaching mode. All system and management activities will be carried out on the virtual network platform, based on the digital teaching feedback generated by the online teaching background. Second, the continuous upgrading of teachers' network literacy challenges. The introduction of online teaching mode brings a new upgrade of college teaching from form to method, and many teachers will face the test of network literacy. Network literacy includes not only network technology, but also network consciousness and network thinking. In the new era, only by keeping up with the pace of The Times and improving network literacy can improve the quality and build a new model of higher education teaching. Third, students lack the challenge of independent learning ability. Due to the secret characteristics of the network, students can listen to lectures in an environment that is too familiar and relaxed, and the learning effect is poor, the learning interest is not high, and the learning attitude is not correct.[6]

2.2 Quality control dilemma

Quality control of online teaching in universities refers to that relevant teaching management departments find the deficiencies in the process of online teaching in time, find out the reasons, and take corresponding measures by collecting, analyzing and sorting out all aspects of online teaching. At present, the quality control of online teaching is faced with many problems. First, the main position of students in teaching is weakened.[7] More online teaching activities are carried out in the form of "lectures", with theoretical explanation as the main teaching approach and relatively fixed teaching content. Teachers only pay attention to their own teaching rhythm, ignore students' learning needs and future development needs, and lack the mechanism of online teaching mode controlled by motivation dimension. At the same time, the participation rate of students in teaching activities is reduced, and the laws of student development and education are ignored, which is not conducive to the quality control of online teaching mode. Second, the lack of reflective mechanism in online teaching. Excellent teaching quality depends on continuous reflection and exploration, teaching reflection is to give full attention and feedback to the past and ongoing teaching activities, so that it can be continuously improved, perfected and developed. Online teaching mode should continuously improve the teaching level, meet the diversified needs of students, and continuously improve the teaching quality.

2.3 Quality management dilemma

College online teaching quality management mainly includes online teaching evaluation, online teaching supervision and so on. At present, online teaching quality management has the following difficulties. First, the main body of online teaching evaluation is single. At present, the online teaching evaluation of universities mainly consists of four parts: leadership evaluation, peer evaluation, teacher self-evaluation and student evaluation. Due to the difficulty, complicated procedure and limited resources in the implementation of leadership evaluation and peer evaluation, college online teaching evaluation is mainly based on student evaluation. However, the professional degree of students' evaluation is not enough, the evaluation results are too subjective, lack of science, objectivity and comprehensiveness, and can not truly reflect the teaching quality. Second, the difficulty of online teaching supervision is upgraded. The complexity of supervision work has increased, the lack of unified and reasonable supervision standards, and the judgment of teaching effect only based on online teaching time, teaching materials submission and other standards is too one-sided. At the same time, the supervision of online teaching is not enough, the boundaries of rights and responsibilities of personnel in various departments are not clear, and there is no clear division of labor and cooperation system, which makes the management of online teaching loose, unable to achieve effective quality management and meet quality requirements.

3 New requirements for improving the quality of online teaching in universities based on TQM theory

3.1. The quality of online teaching should be fully guaranteed

All-round guarantee of online teaching quality refers to the comprehensive management and improvement of various factors that directly or indirectly affect online teaching quality, so as to ensure the improvement of online teaching quality in universities. The quality assurance of online teaching in universities mainly includes three aspects: the first aspect is the online teaching environment, which mainly includes the teaching equipment environment and system environment. The second aspect is the preparation of online teaching, which mainly includes the network literacy composed of network technology, network consciousness and network thinking. The third dimension is students' online learning, which mainly includes learning motivation, learning attitude and learning interest. TQM theory requires universities to guarantee the quality of online teaching in an all-round way, to construct, manage and supervise the factors that affect online teaching.[8]

3.2 The quality of online teaching should be controlled throughout the whole process

From the perspective of the whole process, the formation and development of any product or service quality is a complete process, and the whole process control system has two core concepts: one is customer-centric. Customer satisfaction theory emphasizes that quality improvement is embodied in the satisfaction of customer needs. By meeting the explicit needs of customers and exploring the hidden needs of the future, we can achieve the guidance of customer needs and ensure the durability of benefits. The second is continuous improvement. In order to make all aspects of quality, including technical, management and human factors, always under control, it is necessary to establish error correction and early warning mechanisms, so as to achieve the purpose of continuous quality improvement. Therefore, the continuous improvement of online teaching quality requires universities in two aspects. First, student-centered, universities are required to understand and master the needs of students in the process of online learning, and make clear the subjective status of students in the process of online teaching. Second, insist on continuous improvement and enhancement of online teaching quality. Universities should take the initiative to adapt to the changes in students' needs, constantly adjust their goals and quality standards, drive development with innovation.

3.3 All staff are required to participate in the management of online teaching quality

TQM theory holds that all employees are responsible for quality, should establish quality awareness, and actively participate in quality management. The realization of the quality goal depends on the cooperation of the management and front-line employees. Full participation in management reflects the humanism of enterprise management, and requires teams at all levels to achieve mutual cooperation and supervision, so as to promote the formation of team spirit and common management consciousness. All the staff and students in the university are the main body of online teaching and management activities. Teaching quality improvement is not only the responsibility of principals and leaders at all levels, staff of all departments, all teachers and students have the responsibility to participate in the monitoring of teaching quality and ensure the improvement of teaching quality. All teachers, students and staff should participate in, care for and support the teaching quality work. Full participation in online teaching management requires two aspects of work. First, online teaching evaluation needs the participation of all staff. Second, the supervision and management of online teaching is the common responsibility of all teachers and students, and all staff need to participate in online teaching supervision activities. The university must make all members and departments aware of the goals and objectives of the university, and involve every faculty member in the management of the university by empowering them to take responsibility for the quality of their work.

4 Based on TQM theory of college online teaching quality improvement path

4.1. Establish online teaching quality assurance mechanism

First, improve the online teaching platform. Under the background of education information construction in our country, we should make use of emerging intelligent information technology to build a unified and intelligent online teaching platform. Universities can make online teaching equipment more adjustable and applicable by establishing a unified online teaching platform consistent with their teaching objectives, campus culture and school-running characteristics. In addition, the teaching management departments of universities can use the platform to get involved in teaching procedures and extract teaching data, which is more conducive to the development of teaching management.[9] Universities also should make full use of the convenience of the Internet era and the development of information technology to build an intelligent platform. It senses the online teaching situation and teaching interaction process, and can meet the individual needs of college students.

Second, strengthen the construction of teachers' network literacy. Teachers should constantly improve the awareness of online teaching, update the knowledge structure on the basis of improving the application ability of information technology, and complete the teaching objectives. Universities need to strengthen the construction of network literacy and provide sufficient conditions for all teachers. Colleges should regularly carry out online teaching technology and other training courses, establish a teacher training system, and regularly assess the online teaching technology of teachers to help the continuous improvement of teachers' teaching technology, which is conducive to fully exploring the compatibility of online teaching mode.

4.2 Optimize the online teaching quality control system

The first is to build students' subjectivity model. The online teaching mode should highlight the students as the center.First of all, teachers should take students' needs as teaching guidance. Secondly, as the guide and supervisor of online teaching, teachers should play the role of online guidance and supervision, pay attention to students' learning rules and learning characteristics, and take effective teaching measures to mobilize students' learning enthusiasm, so as to improve the quality of online teaching. Finally, online teaching should give students more opportunities to communicate and show themselves online. Creates more opportunities and exhibition space for students, which can enable students to fully tap their own potential and help create a student-oriented online teaching model.

The second is to carry out regular online teaching reflection activities. Teaching reflection is to give feedback to the current situation of teaching activities and think about improvement plans. The once-and-for-all teaching method can no longer meet the learning and development needs of college students in the new era. Universities should regularly organize teachers to carry out teaching reflection activities, timely identify problems and adjustments. Finally, universities must pay attention to the effectiveness of the online teaching reflection program. The regular online teaching reflection mechanism will constantly raise and discover new problems, and teachers need to implement improvement strategies in a timely manner, so that the reflection mechanism can become a powerful help to control the quality of online teaching.[10]

4.3 Improve online teaching quality supervision measures

First, improve the online teaching quality evaluation system. Universities should establish an implementation mechanism to promote online teaching quality evaluation with full participation. First of all, university leaders must attach great importance to online teaching evaluation, actively participate in teaching quality evaluation and mobilize all departments to work together to improve the online teaching evaluation system. Secondly, the teaching management department should track the teaching dynamics and progress in time, evaluate the quality of teachers' lesson preparation materials and the interaction between online teaching time and process, and understand the completion of teachers' teaching work through the professional degree and teaching attitude of the submitted materials, so as to reflect the quality of online teaching.

The second is to innovate the collaborative supervision mode. Universities should clarify the rights and responsibilities of families and schools, achieve collaborative supervision. University and relevant administrative departments should perform the duties and play a leading and supervising role. We should not only capture the highlights of online teaching, but also explore potential problems, communicate with teachers and students regularly, make online teaching activities orderly. At the same time, online teaching activities should be supervised by parents. By introducing home-school collaborative supervision into online teaching, parents can enter online classes as spectators, which is an invisible constraint for universities. It effectively supplements the management system and accepts opinions from society, which promotes the development and progress of online teaching in universities.

5 Conclusions

Online teaching mode plays an important role in the construction of high quality education system. Through establishing quality assurance mechanism, optimizing quality control system and improving quality supervision measures, high-quality development of online teaching can be achieved.

References

1. Zhang J. Design of College Teaching Quality Evaluation System Based on Cloud Platform[C]. ETMHS 2018,2018:309-312.
2. Zhao Jian, Guo Shaoqing. Research on Quality Assurance Mechanism of “Internet+” Volunteer Teaching Service in Normal Universities[J]. China Educational Technology,2021(10):79-85+104.
3. Musundire A,Dreyer M J. Effectiveness of the Clinical Supervision Strategy as a Tool for Improving Teaching Quality: Perceptions of South African School-Based Managers and Educators[J]. Africa Education Review,2019,16(2).
4. Moffett P,Eaton P. The impact of the Promoting Early Number Talk project on classroom mathematics talk[J]. Early Child Development and Care,2019,189(11).
5. Yu G. A Study on the Application of the Flipped Classroom Model in Contemporary Teaching of Operatic Ensemble Courses[C].ICMEEM 2019.Francis Academic Press,2019:359-363.
6. Yang S. Construction research on index system of teaching quality of distance education[J]. Journal of Discrete Mathematical Sciences and Cryptography,2018,21(6).
7. B. J A. Institutional governance and management systems in Sub-Saharan Africa higher education: developments and challenges in a Ghanaian Research University[J]. Higher Education,2018,75(2).
8. Shujian L,Ya G. WITHDRAWN: Study of Cognitive Education Cultivation for Students of Science and Engineering College[J]. Cognitive Systems Research,2018.
9. Xifeng L. The satisfaction of entrepreneur instructors and higher vocational students majoring in economic and business administration with the

cooperative instruction in the work post practice[J]. Journal of Physics: Conference Series,2019,1423.

10. Lei C. Analysis of the Work of Hidden Ideological and Political Education in the Media Era[C].ACAELT 2019.Francis Academic Press,2019:1585-1589.

A Research on the Current Situation and Countermeasures of English Writing Teaching for Junior English Majors

Yuxin Jin, You Chen*, Sixin Yang, Yijin Chen, Shiyang Liu, and Wenjie Wu

College of Foreign Studies, Guangdong University of Science and Technology, 523083 Dongguan, China

Abstract. At present, the effectiveness of English writing teaching for junior English majors is not satisfactory, with a lack of interactivity in the classroom, inflexible teaching methods, a deficiency of targeted teaching for students' problems, and insufficient emphasis on writing training. The main reasons for these problems are that teachers have failed to conduct the student-centered English writing teaching model, emphasizing one-way output, ignoring students' writing problems, and resulting in the shortage of writing training. In this regard, teachers need to transform the teacher-centered English writing teaching model, consider students' actual problems, establish common writing goals, construct online teaching resources for English writing course, and apply multi-modal writing teaching methods to boost the effectiveness of English writing teaching.

1 Introduction

Writing teaching is an important aspect of foreign language teaching. For English learners, English writing ability is one of their essential core competencies and an important means of language output to the outside world [1]. However, English writing is a soft spot for most junior English majors [2]. In the Test for English Majors-Band 4 (TEM-4) , students generally lose more points in the writing section [3], which indirectly reflects the urgent need for improvement in our English writing teaching. Therefore, how to enhance the effectiveness of English writing teaching and promote students' English writing level has become a focus of increasing attention for teachers and students.

2 Research design

2.1 Research object

This study involved a total of 211 participants, including 32 males, 155 females, and 24 teachers. The participants are 157 freshmen and 30 sophomores majoring in English, as well as 24 English writing teaching teachers from College of Foreign Studies of Guangdong University of Science and Technology.

2.2 Research methods

2.2.1 Questionnaire survey

In order to better grasp the current situation of English writing learning and teaching among young undergraduate students majoring in English, this study carried out a questionnaire survey on the current situation of English writing learning and teaching among 211 participants (students and teachers) from College of Foreign Studies of Guangdong University of Science and Technology in early June of 2022. Students participated in a survey on the current situation of English writing learning among junior English majors, while teachers were engaged in a survey on the current situation of English writing teaching among junior English majors.

2.2.2 Interview

In mid-June of 2022, the research team randomly selected 21 junior English majors (11 freshmen and 10 sophomores) and 5 English writing teachers from Guangdong University of Science and Technology for in-depth interviews. This study first conducted interviews with randomly-selected English majors to understand the situation of English writing learning and satisfaction with writing teaching. Then, interviewing the English writing teachers to perceive their understanding of English writing course design, classroom teaching, and student learning status.

3 Research results

3.1 Lack of interactivity in English writing class

In terms of the lack of interactivity in the writing class, the results of students' survey show that 66.67% of the participants believe that there is a shortage of teacher-student interaction in the English writing class, while only 33.33% of them state that the interaction in the English writing class is rich. For the classroom

* Corresponding author: *chenyou@gdust.edu.cn*

atmosphere, 59.62% of students feel that the English writing classroom is dull, which is confirmed in interviews with 21 students. 66.67% of the interviewees claim that there is insufficient interaction and communication between teachers and students in English writing classes; 61.90% of respondents feel that teachers lack teaching enthusiasm and the content they teach is difficult to understand; and also 30% of students emphasize that teachers should pay more attention to the feedback or opinions provided by students, rather than just relying solely on teachers' unilateral output. In addition, in the teachers' questionnaire survey, 58.33% of the tested teachers proclaim that English writing class is a relatively boring course, and the content taught tends to be theoretical, which is not conducive to teacher-student Q & A (Question & Answer), interaction and communication. At the same time, out of 5 teacher interviews, only one of them (taking up 20% of total sample size) designs writing teaching based on students' characteristics and levels, and focuses on interacting with students. The other four teachers (amounting to 80% of total sample size) indicate that they are more inclined to explain the knowledge and exercises of textbook in writing classes, and then assign relevant writing assignments after class, neglecting classroom interaction and communication.

3.2 Inflexible methods for teaching English writing

Firstly, regarding the teaching methods of writing in the classroom, 62.82% of the students surveyed deem that the current teaching mode of English writing in the classroom is traditional and emphasize a single form; 31.01% of participants reckon the teaching mode in writing class belongs to indoctrination teaching; 41.77% of subjects assume that there is insufficiency for supplementing extracurricular contents in English writing teaching; and also 37.97% of them opine that teachers are wanting in flexible teaching methods. Secondly, in interviews with 21 students, 47.61% of them hold that the form of English writing teaching in classroom is singular. They believe that teachers' classroom teaching methods are relatively ambiguous, mostly depending on the teaching resources provided by textbooks, and lacking extracurricular expansion. Meanwhile, the results of the teachers' questionnaire manifest that 50% of teachers consider their teaching methods in English writing classes are not flexible enough, and they fail to design teaching based on students' actual writing level and existing problems. They mostly complete teaching content explanations according to the course teaching plan. Furthermore, among the interviews with 5 English writing teachers, 3 teachers (accounting for 60% of total sample size) express their dissatisfaction with the current situation of English writing teaching. They assert that many writing teachers apply teaching methods improperly (such as teacher-centered, indoctrinated teaching, lack of interaction, neglecting writing training, etc.), and ignore flexibility and diversity in the forms of English writing teaching.

3.3 Deficiency of teaching tailored to students' problems

According to the results of the students' questionnaire survey, 55.32% of students acknowledge that the teaching content of English writing in the classroom does not meet their needs. In terms of writing training, 65% of the participants report that the content taught by the teacher lacked writing knowledge related to tests (such as TEM-4), and even 72.5% of testees believe the current English writing class excludes targeted and planned teaching and training to work out their writing weaknesses. Simultaneously, in students' interviews, 42.86% of respondents relate that in English writing class the teacher cannot consider students' actual writing problems and situations, resulting in the absence of designing teaching to address students' existing problems; and 57.14% of students declare the knowledge taught by teachers in the classroom is only focused on textbook contents, and teaching activities are carried out in a copycat manner, ignoring the fundamental needs of students. This is coincidental in the interviews with five teachers. Four interviewed teachers (making up 80% of total sample size) allege that they strictly follow the content of the teaching plan and do not adjust the teaching contents in accordance with the actual needs of students, which leads to content deviation in English writing classroom teaching.

3.4 Insufficient emphasis on writing training

First, when it comes to English writing training, the students' questionnaire results imply that 89.39% of students feel that teachers do not attach enough importance to writing training in teaching, lack practicality in the class, and have less relevant English writing training, mostly explaining concepts and theoretical knowledge. Then, in students' interviews, 71.43% of the interviewees indicate that most of their writing training is limited to the writing class, and teachers rarely assign writing training after class (perhaps only 4-5 times in a semester). Finally, in the teachers' questionnaire survey and interviews, the data results also confirm this point. That is, 83.33% of the teachers in the questionnaire mention they assign about 4 times of writing training during the whole semester, while the rest of the assignments mainly involve previewing textbook content, group presentations in class, and completing textbook exercises. Besides, in the interview, most teachers explain they assign writing tasks based on the school requirements (e.g. at least how many times per semester) for teachers to score students' assignments, without any mandatory requirements or regulations for other assignments relevant to writing training.

4 Discussion

In the first place, there is a lack of interactivity in the English writing class for junior English majors, which is due to teachers' failure to achieve a student-centered approach to writing teaching. In the teacher-centered teaching model, there is insufficient overlap and cohesion between the theoretical input of students' English writing and the output of English writing practice, which makes it difficult for students to solve difficulties in English writing in a timely manner, let alone improving their English writing proficiency [4]. However, the student-centered teaching model requires teachers to complete the transformation from teacher-centered to student-centered, adhere to student-centered teaching, deeply explore students' interests, and use them to stimulate students' interest in classroom learning, cultivate students' ability to discover, analyze, and solve problems, thereby achieving the role of improving teaching effectiveness.

In the second place, as to teaching methods in the writing class, teachers often have a single form of teaching in the writing class, mostly in the shape of cramming or indoctrination. This is because teachers blindly emphasize one-way output and neglect the combination of theory and teaching events. Under this one-way output teaching mode, student feedback shows that the knowledge taught by the teacher is easily forgotten, and learning effect is not satisfactory [5]. Thus, in the teaching of English writing courses, teachers should maximize students' participation in the class, increase interaction and communication between teachers and students, guide students to indulge into the classroom learning, and utilize flexible and varied teaching methods (such as task-based teaching, communicative teaching, comprehensive teaching, etc.) to elevate teaching quality [6].

In the third place, for English writing classes, teachers lack targeted teaching to address students' problems, which is because they have not taken into account the current writing level of students and have neglected the problems of students' writing. In the students' questionnaire, 57.38% of the participants state they are not able to receive teacher feedback and comment during writing training, and 14.29% of students express that they often encounter difficulties in writing and cannot seek help from teachers. In this way, what they can only do is choose to use translation software to assist in completing writing tasks, which to some extent reflects the teacher's neglection of students' writing problems and failure to teach students according to their aptitude.

In the last place, with regard to English writing training, teachers do not attach enough importance to writing training. The key reason is that many English writing teachers currently adopt the traditional teaching method of "teacher proposition-student writing-teacher correction", which fails to effectively utilize today's online automatic correction writing software (such as Microsoft Aim Writing, Youdao Dictionary, Grammarly, etc.) or websites (such as Composition Correction Network, Youdao Writing, Aim Writing, etc.) to conduct a large number of automatic correction of students' compositions. In writing correction, manually correcting students' compositions by teachers is not only a challenge for teachers, requiring a lot of time and effort to modify students' writing exercises and assignments, but also brings work pressure to teachers. This is also one of the significant reasons why many teachers do not pay enough attention to after-school writing training.

5 Countermeasures

5.1 Changing the model of teacher-centered writing teaching

In English writing teaching, teachers should transform the current model of teacher-centered writing teaching into a student-centered teaching model. The traditional teacher-centered teaching model, where teachers basically carry out teaching according to pre-designed contents. However, the student-centered teaching model requires teachers to have a clear understanding of students' English writing proficiency, writing needs, course objectives, and output results [7]. In the meantime, teachers need to pay attention to emotional communication with students, encourage students to actively participate in classroom interaction, and help students solve writing problems. Writing is actually a communicative activity, with participants including the author, peer-reviewers, and teacher-reviewers [8]. In order to make peer evaluation become scientific and meaningful, teachers should provide related training and demonstration for peer reviewers. In addition, the student-centered writing teaching needs to be combined with networked teaching tools (such as essay correction websites, Chaoxing Learning Platform, etc.), and process writing ought to be integrated into English writing course teaching to improve the effectiveness of writing teaching. For example, teachers can make full use of the Composition Automatic Correction Network to supervise the entire process of students' writing, including conception, discussion, drafting, revision and feedback, gain a deep understanding of the main problems in students' writing, and provide writing feedback through various evaluation methods such as self evaluation, student evaluation, teacher evaluation, and teacher-student interaction through the correction network system. This not only effectively reduces the burden of review and helps students solve common errors in writing, but also makes students get involved in writing activities and enhances the interaction and communication in the process of writing teaching.

5.2 Emphasizing students' writing problems and establishing common writing goals

In writing teaching, teachers should center on students' actual writing problems, clarify whether students have problems with vocabulary scarcity, weak grammar, discourse structure and writing strategies, and then establish common writing goals with students together. First, teachers have the responsibility and obligation to inform students of the learning objectives of the writing course, so that students can have more planning and motivation in their after-school learning. Second, teachers need to adhere to the teaching philosophy of "student-centered and teaching according to students' aptitude", constantly update their teaching concepts, enhance their sense of belief as the "guide" role, and cannot simply ignore students' actual needs and problems only for the sake of completing teaching tasks. Third, teachers should have clear teaching objectives, sufficient teaching preparation, establish a harmonious teacher-student relationship, and construct a writing evaluation and feedback mechanism. Clarifying teaching objectives and fully preparing for teaching require teachers to inform students of areas for improvement after writing classes, and to assess students' English writing abilities, in order to adjust and better suit students' writing classroom teaching, thereby improving teaching effectiveness and students' English writing level.

5.3 Constructing online teaching resources for English writing course

Nowadays, the application of online teaching platforms can effectively solve the two major problems of limited writing hours and difficulty in correcting essays. The Internet has become the main carrier of massive amounts of information, and the convenience brought by the Internet has led to a high degree of acceptance of online resources by students. The construction of online teaching resources for English writing course is also an unstoppable trend. Taking "Chao Xing Superstar Learning Platform" as an example, the teaching functions of the platform include: message chat, live streaming, question answering, classroom voting, course creation, attendance, grades, academic situation, mobile library, mobile lesson plans, personal notes, and assignments, etc. [9]. Teachers can directly create online teaching resources for English writing course in this platform. For instance, teachers develop online course syllabuses, exam outlines, teaching plans, teaching courseware, and lesson plans based on the contents of English writing textbooks, accompanied by videos explaining the knowledge points in each chapter, corresponding exercises, and tests. Concurrently, teachers can also upload relevant knowledge points and video explanations based on students' actual needs to satisfy their diverse and differentiated learning needs. What is more, teachers are in a position to regularly distribute questionnaires to collect students' problems in English writing, and create relevant online learning resources based on their feedback, and provide students with a high-quality English writing learning platform.

5.4 Applying a multi-modal writing teaching model

Multi-modal teaching refers to the use of language, images, sounds, actions, and other modes by teachers in teaching to coordinate them into effective ways of expression, providing more authentic contexts for English learners, making students willing to participate in learning inwardly, and thereby improving teaching effectiveness [10]. In other words, English writing teachers can use a multi-modal teaching model, based on their own teaching objectives, combined with the textbooks set by the school curriculum, and select appropriate multi-modal teaching materials to carry out appropriate teaching activities, in order to increase the fun of English writing class. Meanwhile, in English writing teaching, teachers can use multi-modal teaching materials to set up group cooperative activities such as "connecting multi-modal materials to tell stories" [11]. In this kind of activity, teachers should first of all set activity requirements on the basis of learning contents to achieve the purpose of reviewing what students have learned. Then, group cooperation can promote communication and mutual learning of creative ideas among students. Next, after completing the storytelling, the teacher can build up a session of student evaluation to allow students to provide feedbacks on any inappropriate connections in the story. Lastly, the teacher makes some supplementations and summaries. After completing this activity, students are able to receive writing thinking training, as well as how to make the transition and development, and how to add up details. These similar writing learning processes are obviously more interesting than single-text teaching, not only increasing teacher-student interaction, but also making it easier to capture students' attention, which is beneficial for improving the effectiveness of English writing teaching.

6 Conclusion

In summary, English writing teaching is important for improving the English writing ability of junior English majors [12-13]. However, improving English writing ability is a relatively lengthy process that requires the joint efforts of students and teachers. On the one hand, students have to determine their own learning goals for English writing, accumulate knowledge of vocabulary, grammar, discourse structure, writing strategies, and practice more to achieve the level of proficiency. On the other hand, teachers need to first establish a student-centered teaching model in English writing teaching, attach importance to students' actual writing problems, establish common writing goals with students, actively guide students to conduct writing training, construct online learning resources, and use multi-modal writing teaching models to promote students' English writing proficiency.

Funding

This paper is funded by Innovation and Entrepreneurship Training Program for College Students of Guangdong University of Science and Technology in 2022 (Project Name: A Research on the Current Situation and Optimization Path of Junior English Majors' English Writing Level; Project No.: School-level 2022005)

 https://doi.org/10.1051/shsconf/202317801012

References

1. Cen, Q. F. An Empirical Study on the Application of Discourse Coherence Theory in English Writing Teaching for Non English Majors [D]. Guangxi Normal University, 2022.
2. Adamson, B. China's English: A history of English in Chinese education[M]. Hong Kong University Press, 2004.
3. Wang, C. & Ran, X. X. The Cultivation of Students' Critical Thinking Ability in Writing Teaching of TEM-4 [J]. Journal of Sichuan Tourism University, 2017 (02): 97-100.
4. Liu, J. Problems and Countermeasures in English Writing Teaching in Application-oriented University [J]. Journal of Liaoning Economic and Management Cadre College, 2018, (2): 107-110.
5. Nicol, D. J. & Macfarlan-Dick, D. Formative Assessment and Self-regulated Learning: A Model and Seven Principles of Good Feedback Practice[J]. Studies in Higher Education, 2006, 31(2): 199-218.
6. Meng, P. A Research on the Application of Multi-modal Theory in College English Writing Teaching [J]. Journal of Liaoning Normal University (Social Science Edition), 2021 (01): 67-69.
7. Yang, L. The Application of the Teaching Model of "Reading-based Writing" in English Writing Teaching in Universities [J]. Journal of Jilin Agricultural Science and Technology College, 2022,31 (03): 118-121
8. Yuan, J. A Research on the Application of Automatic Composition Evaluation System in College English Writing Teaching [J]. Journal of Jiamusi Vocational College, 2022, 38 (06): 81-83.
9. Zhang, F. A Research on the Application of the "DingTalk, Superstar, QQ" Platform in Online Teaching [D]. Hebei Normal University, 2021.
10. Wei, Y. M. A Research on the Impact of Multi-modal Writing Teaching on the English Writing Level of Non English Majors [D]. Jiangxi Normal University, 2020.
11. Zhang, D. L. Design and Modal Invocation of Multi-modal Foreign Language Teaching [J]. China Foreign Language Journal, 2010, 7 (3): 48-53, 75.
12. Zhang, C. Necessity and Rationality of a Ladder Model for Teaching College English Online Writing: Teaching Students Based on Their aptitude and Being Student-centered [J]. Journal of Chengdu University of Traditional Chinese Medicine (Education Science Edition), 2011, 13 (3): 38-40.
13. Zhang, J. A Research on AI-Based English Writing Teaching Practice [J]. Journal of Hebei Energy Vocational and Technical College, 2022,22 (03): 85-88.

Research on National Ideological Security:Guiding Contemporary Chinese Social Thought with Marxism

Qi Li [1,a], Zhuanxiong Hu [2,b], Chen Li [3,c]

[1] Macau University of Science and Technology (Macau, China):Master,
[2] University of the Chinese Academy of Social Sciences (Beijing, China):Doctor,
[3] Lanzhou University (Lanzhou City, Gansu Province,China):Master,

Abstract: National ideological security is one of the important contents of the overall national security concept. As a tendency and development trend of thought, social ideological trends reflect people's certain interests and demands in specific environments, and have a broad impact on people's thoughts in social life. However, the current world is complex and ever-changing, and contemporary China is also in a "unprecedented great transformation", which determines the long-term coexistence of various ideological trends, especially the mixed social ideological trends of neoliberalism, historical nihilism, democratic socialism, and universal values, challenging the mainstream ideological status of Marxism. In this context, it is particularly important to play the guiding role of Marxism as the mainstream ideology. Adhering to the mainstream ideology of Marxism is conducive to strengthening the Party's leadership in ideological work, resisting the negative effects of other social trends, and better maintaining national ideological security.

1.Introduction

The current ideological struggle at home and abroad is complex and sharp. From the mainstream perspective, the 18th National Congress of the Communist Party of China has strengthened ideological work and achieved a fundamental improvement in ideology. However, it should also be noted that the leadership of non Marxist opposition parties and the ideological movements of the socialist system are still rampant. In the report of the 20th National Congress of the Communist Party of China, it was pointed out that "ideological work is a work that establishes the heart and soul of the country and the nation. We firmly grasp the leadership of the Party in ideological work, fully implement the responsibility system for ideological work, and consolidate and strengthen the mainstream ideological and public opinion in the new era." [1] As a result, the construction of national ideological security has entered a new era. As the world's largest socialist country, it is incumbent on China to use Marxism to guide its ideology. Our party has accumulated experience and lessons in exploring different periods suitable for China's revolution, construction, and reform, providing reference for strengthening Marxist ideology at present. But since the reform and opening up, various social ideologies have been constantly emerging, impacting the guiding position of Marxism. We need to carefully study the manifestations and impacts of contemporary Chinese social trends, in order to recognize the important role of Marxist theory in the construction and development of Chinese society. Only by adhering to and maintaining the guiding position of Marxism in the ideological field can we promote the great rejuvenation of the Chinese people and maintain the long-term stability of the country.

2.The Manifestation and Influence of Contemporary Social Thought

Since the reform and opening up, the mainstream ideology of Marxism has encountered counter current impacts from Western capitalist countries, such as "new liberalism, democratic socialism, historical nihilism, and universal values". This has brought serious challenges to China's ideological construction. Only by applying Marxism can we see through the phenomenon, see the essence and harm of these ideological trends, and then deeply analyze them, To better consolidate the strong position of the mainstream ideology of Marxism.

2.1 Neoliberalism

Neoliberalism is a popular social trend in the economic and ideological fields of Western capitalist countries. Neoliberalism advocates economic diversification and marketization, while politically it advocates political diversification and multi-party system. Since the beginning of the reform and opening up in the last century, it has been introduced to China. The introduction of neoliberalism has sounded an alarm for China's ideological construction. Although the mainstream ideology guided by Marxism is the mainstream, the harm brought by neoliberalism cannot

[a]1477317642@qq.com* (corresponding author)
[b]3243401859@qq.com, [c]1760012032@qq.com

be ignored. The secret of neoliberalism is to disguise it in a "legitimate" way, thereby deceiving society and the public, disrupting their thinking. For neoliberalism, we must be vigilant, truly understand its values and concepts, as well as its potential harm to the development of Chinese society, make contingency plans and precautions, and even engage in necessary counterattacks. Neoliberalism is based on criticizing Marxist, socialist, and communist ideas, which is the main line of neoliberalism that never changes without its roots. In fact, it is to challenge the guiding position of Marxism in ideology, and its core is to compete with us for ideological leadership.

Neoliberalism often emphasizes individual 'freedom', which is essentially opposed to Marxist 'freedom' and has evil political intentions, smearing Marxist ideology as backward, conservative, and rigid; Equate socialism with communism, totalitarianism, and authoritarianism, while promoting the essence of socialism such as public ownership, distribution according to work, social equity, and common prosperity as utopia. Neoliberalism also has an undisguised theoretical flaw: it "confuses the 'freedom' of basic human rights with the 'freedom' of the rich to accumulate as much wealth as possible,"[2] leading to a negative understanding of freedom.Neoliberalism often likes to use words such as "individual, property, freedom, human rights, rights, equality, democracy, and diversity" [3]to spread its confusing views. For neoliberalism, we need to strengthen the propaganda and ideological work to promote the harm of neoliberalism. It can be seen that if the trend of neoliberalism is allowed to spread unchecked in China, it will inevitably gradually corrode the thoughts and minds of the masses, and the dissemination of socialist core values among the masses will be affected, thereby weakening the guiding position of Marxism in ideology and bringing catastrophic consequences to the future of the party and the country. It is necessary to understand the differences in the "freedom" advocated by neoliberalism, strengthen the research on the discourse, interpretation, and practice rights of defining "freedom" in socialist discourse power, apply Marxist positions, viewpoints, and methods, deepen and develop Marxist research on freedom, and combine freedom, equality, justice, and the rule of law through socialist core values, and combine the interests of the state, collective, and individuals, Let the people realize that capitalism cannot fundamentally achieve human freedom and comprehensive development.

2.2 Historical nihilism

Historical nihilism "stands in opposition to Marxist historical materialism in its historical perspective, promoting the viewpoint of historical idealism. In terms of content, historical nihilism often lacks a concrete, dialectical, and logical essence for the history. It often negates the materiality and objectivity of human historical development through some superficial phenomena, arbitrary cutting and piecing together history. When denying a country's material possessions, it also negates its erroneous views and ideological tendencies in terms of spirit, history, and culture. In recent years, historical nihilism has been constantly changing its guise, relying on online self media to shift from an open and systematic "overall defense" to a covert and fragmented "flexible attack", spreading some so-called "historical truth" and "historical restoration" to deny the revolution led by our party and the objective laws of history, excessively exaggerating the temporary dark side in the historical process and the mistakes made by our party, Separating the objective connections of historical development, attributing all the achievements of our party to "accidents" and "individuals", denying the scientific viewpoint that the people create history, and denying the historical inevitability of China's socialist path.

Specifically, historical nihilism is based on the wrong idealistic methodology of history, distorting the history of modern China since the Opium War, attacking progressive classes, advanced figures and national heroes, and belittling the groundbreaking role of the birth of the CPC by generalizing, fabricating and doubting history. To dismiss the heroic figures who emerged in the process of national independence, people's liberation, national prosperity, and national rejuvenation as useless, and to negate the historical inevitability of the socialist path, as well as the historic achievements of socialism with Chinese characteristics in the new era.Historical nihilism also negates China's traditional culture and intends to deconstruct the spirit of Chinese culture, stating that the Chinese nation is a "inferior nation" and denigrating China's significant contributions to human civilization over 5000 years ago, as well as the historical mainstream of China's peaceful development. Historical nihilism also negates the guiding position of Marxist theory and regards the Western modernization model as the norm, dispelling the development of China's modernization under the guidance of Marxism. The ancients said, "To destroy a country, one must first go through its history." Historical nihilism distorts history to exaggerate, and its fundamental purpose and sinister intentions are to attempt to subvert the leadership of the Party and the socialist system. This has sounded an alarm for us to adhere to the mainstream ideology of Marxism.

2.3 Democratic Socialism

Democratic socialism, also known as social democracy, is an ideology that is opposite to scientific socialism and is a political theory believed in by the Social Democratic Party, Socialist Party, and Labour Party in some contemporary Western countries. Although democratic socialism literally carries the word "socialism", it is seriously opposed to the scientific socialism of Marx and Engels. Democratic socialism denies the dictatorship of the proletariat and the leadership of proletarian political parties, and denies the lofty ideals and goals of communism. Essentially, it is the reformism of the Western bourgeoisie. Regarding the fallacy of democratic socialism, Marx once referred to the "Red Party" in the mid-19th century's "Class Struggle in

France from 1848 to 1850", which was formed by the combination of the Mountain Party and Socialists of the petty bourgeoisie democratic faction in France, as the "Democratic Socialist Party" or the "Social Democratic Party" [4]. At the end of the 19th century, Bernstein revisionism emerged within the Second International, endowing social democracy with revisionist connotations.[5] Democratic socialism used the new changes in capitalist society in the 1880s and 1990s as an excuse to deny the theory of class struggle and proletarian revolution, advocating for "democratic reform and economic reform as means to achieve socialist transformation of society." By the mid-20th century, Democratic socialism has officially become the guiding ideology and goal of social democratic parties in various countries. After World War II, democratic socialism is essentially a continuation of Bernstein's revisionism, emphasizing that democratic socialism is a "third path" different from capitalism and communism, advocating for freedom of belief and diversified guiding ideology, and taking freedom, equality, and fraternity as its own values, Believing that socialism should be redefined as' democracy ', believing that democracy' determines everything ', and realizing the socialist system can only be achieved through a' democratic 'approach and parliamentary reform.

In the mid to late 1980s, democratic socialism held a dominant position within the ruling Communist Party of the former Soviet Union and socialist countries in Eastern Europe. Historical experience and lessons have shown that allowing democratic socialism to spread will inevitably have serious consequences. One important reason for the upheaval in the Soviet Union and the East was the comprehensive denial of the history of the Soviet Communist Party, the denial of Soviet leaders, and the promotion of "humane and democratic socialism" within the Soviet Union. These countries replaced the original Marxism Leninism with "democratic socialism", leading to the overthrow of the dictatorship of the proletariat, the overthrow of the socialist system, and the tragedy of the upheaval in Eastern Europe and the disintegration of the Soviet Union. History once again shows that "democratic socialism" has dispelled the truth of scientific socialism and easily made the masses lose confidence in communism. Since China's reform and opening up, the trend of democratic socialism has also begun to permeate our country. Democratic socialism advocates distorting the scientific truth of Marxism, shouting the fallacy that "democratic socialism is the orthodoxy of Marxism" and "only democratic socialism can save China". From this, it can be seen that democratic socialism is fundamentally opposed to the mainstream ideology of Marxism, and it is necessary to constantly strengthen the construction of the mainstream ideology of Marxism.

2.4 Universal Value Theory

The "universal value" was originally derived from the anti feudal revolution of the Western bourgeoisie, but with the development of history, today capitalism is increasingly moving towards the late stage of monopoly capitalism and has lost its historical progressiveness. The "universal value" is essentially a concentrated manifestation of capitalist ideology, especially in today's world where socialism and capitalism coexist. The "universal value" has been re adapted by Western countries as a tool for promoting socialist countries. On the surface, the "universal values" promote "freedom", "equality", and "fraternity", believing that capitalism is eternal and denying that the socialist system is the direction of human development. In international relations after World War II, the "universal values" advocated by Western imperialist countries such as the United States often paved the way for hegemonism, power politics, and neointerventionism. In the past decade or so, the "color revolution" in some countries and regions around the world has caused serious harm to these countries and regions. The "universal value" is completely a pretext for Western capitalist countries to infiltrate our ideology. They use the so-called banner of freedom to actually serve the aggressive policies of imperialism such as the United States. In the era of anti globalization, it is also an excuse and tool for exercising long-term jurisdiction over developing countries.

3.Leading Contemporary Chinese Social Thought with Marxism

3.1 Adhering to the Guiding Position of Marxism in the Field of Ideology

Marxism is a scientific method for understanding and transforming the world, an action guide for China's socialist modernization construction, and a driving force for our socialist spiritual civilization. Adhering to the guiding position of Marxism in the field of ideology helps to correctly guide the erroneous ideological trends that exist in current Chinese society, resist the invasion of various erroneous ideological trends, and open up a good beginning and take a good step for the development of our country in the new era. It is precisely because of the guidance of the mainstream ideology of Marxism that we can embark on a new journey towards the second centenary goal. It is very important to vigorously promote the scientific spirit of Marxism in the process of criticizing neoliberalism, which is beneficial for us to better understand the current environment, make scientific judgments, and comprehensively grasp the historical mission and position of our party and country. We must apply Marxism and adhere to its guiding position in the field of ideology in China, To better study new situations and solve new problems.

3.2 Guided by Xi Jinping Thought on Socialism with Chinese Characteristics for a New Era

In contemporary China, adhering to the mainstream ideology of Marxism means arming the entire party with Xi Jinping Thought on Socialism with Chinese Characteristics for a New Era, educating the people, promoting the sinicization, modernization, and popularization of Marxism with vivid examples of China's development, and enhancing the discourse and initiative of Marxism in leading contemporary Chinese social trends. Xi Jinping Thought on Socialism with Chinese Characteristics for a New Era,It is formed under the historical conditions that socialism with Chinese characteristics has entered a new era and the CPC is facing the new test of governance. It is another new leap in the party's theory. In the face of the current complex social trends, it is necessary to constantly invade various non Marxist social trends, firmly defend the mainstream ideology of Marxism, and better maintain the security of national ideology. At the same time, we must adhere to the Party's leadership over ideology and successfully fight against historical nihilism, new liberalism, and other ideologies. We must adhere to the Party's leadership and be guided by theXi Jinping Thought on Socialism with Chinese Characteristics for a New Era.

3.3 Enhancing the Inspiration of Socialist Core Values

In contemporary China, adhering to the dominant position of Marxist ideology means cultivating and promoting excellent traditional Chinese culture with socialist core values, and resisting the ideological corrosion of new liberalism, democratic socialism, historical nihilism, and "universal values" on young Chinese people in the new era. Tell the vast majority of young people about the inherent conflict between the universal values advocated by Western society and China's socialist core value system. The essence of universal values is to promote the so-called economic, political system, and spiritual culture of capitalism to non Western countries, but the socialist core values serve the socialist cause with Chinese characteristics, The purpose of socialist core values is to provide spiritual guidance for us to build a modern socialist country and achieve the great rejuvenation of the Chinese nation.Consolidating the core socialist values into a spiritual force for achieving the great rejuvenation of the Chinese nation, widely promoting the common values of humanity, promoting the awareness of a community with a shared future for mankind, and resolutely resisting the infiltration of universal values in Western society. Advocate for an international system with the United Nations at its core, resist the invasion of hegemonism, unilateralism, and power politics, work together with countries around the world to help each other and sail in the same boat, and promote the construction of a fair and reasonable new international political and economic order.

3.4 Always adhere to the path of socialism with Chinese characteristics

Adhering to the dominant position of Marxist ideology is to promote the historical inevitability of China's socialist path. We should do a good job in the work of socialist ideology from the perspectives of academic theory, principles and methods, and propaganda models. Tell the story of China well, spread the voice of China, and strengthen the dissemination of contemporary Chinese Marxism in a language that the people enjoy. Strengthen the propaganda of the superiority of the socialist system, especially in the context of the current COVID-19, China's success in combating the epidemic compared with the "epidemic crisis" in western countries, to further illustrate the superiority of China's socialist system in safeguarding human rights, protecting human rights, and achieving equality. We will resolutely respond to the epidemic rumors spread by Western countries. Since the reform and opening up, we have always adhered to the comprehensive deepening of reform, continuously achieved new results, and promoted the rapid development of China's socio-economic level. China's development and progress not only provide a material basis for today's resistance to the epidemic crisis, but also demonstrate that through the inspection of the "epidemic crisis", only the socialist system is truly a system for protecting human rights, Only by taking the path of socialism with Chinese characteristics can we better reflect the superiority of the socialist system with Chinese characteristics.

4.Conclusion

In the new situation, maintaining national ideological security means using Marxism to lead contemporary Chinese social trends, arming people with scientific theories, shaping people with excellent spirit, and guiding people with correct consciousness. Whether we can do a good job in ideological work is related to the future and destiny of the Party, the long-term stability of the country, national cohesion and centripetal force. It can be said that adhering to Marxism as the fundamental guiding ideology for building the Party and the country is the common theoretical foundation for the unity and struggle of all ethnic groups in the country. At the same time, using Marxism to lead social trends can not only effectively consolidate the spiritual strength of the entire society and maximize the consensus of the people, but also fully stimulate the patriotic enthusiasm of the masses, thus forming a strong cohesion of 1.4 billion people in the entire society.Therefore, with the continuous development of the cause of socialism with Chinese characteristics, only by holding high the banner of Marxism and persisting in the guidance of the mainstream ideology of Marxism for a long time can we stably promote Chinese society to embark on a new historical journey under the Xi Jinping Thought on Socialism with Chinese Characteristics for a New Era, and thus achieve the Chinese Dream of the great rejuvenation of the Chinese nation as soon as possible.

References

1. Xi Jingping.Report of the 20th National Congress of the CPC on holding high the great banner of socialism with Chinese characteristics and working together to build a socialist modern country in an all-round way.People's Daily, (2022)

2. Clayton, Heinzek. Organic Marxism: Ecological Disaster and Alternative Choice of Capitalism. Translated by Meng Xianli, Yu Guifeng, Zhang Lixia.People's Publishing House, (2015)
3. Chen Peiyong. The Potential Harm of Neoliberalism to China's Future Development, Red Flag Manuscripts, 24, (2017)
4. Marx, Engels. Selected Works of Marx and Engels, People's Publishing House, 448,(1972)
5. Bernstein. The Premises of Socialism and the Tasks of the Social Democratic Party, Sanlian Bookstore, 168,(1958)

The Evolution of Domestic Waste Classification Policies and Their Outcome Analysis: The Case of Shanghai Municipality and Beijing Municipality

Wei Wang[1], Yunqin Wang[2]*

[1]Harbin University of Commerce Harbin,Heilongjiang,China
[2]Harbin University of Commerce,Harbin,Heilongjiang,China

Abstract: With the development of the times, the phenomenon of garbage surrounded by cities in China is becoming more and more serious, in 2016, China's Development and Reform Commission issued the "13th Five-Year Plan" for the construction of harmless treatment facilities for urban household garbage in China, China officially began to carry out garbage classification, Shanghai and Beijing as a pilot city in the practice of garbage classification policy is more representative and Shanghai and Beijing, as pilot cities, are more representative and typical in the practice of garbage classification policy. This paper adopts the method of content analysis to analyze the changes in the nature of China's domestic garbage classification policy and the results of Shanghai and Beijing . Suggestions are made to develop simple and easy-to-understand garbage classification standards, improve garbage classification facilities, increase publicity, develop online publicity channels, and reduce the operability of regulations.

1. Introduction

As China's economic strength continues to grow, bringing not only the improvement of the national happiness index of life, urbanization produces a large number of various kinds of living garbage caused by the phenomenon of garbage surrounded by the city, the water source and the land is polluted resulting in many residents body has a variety of health problems.Recycling is becoming ever more important as waste generation rates increase globally. Policy-makers must decide which recycling practices to implement from the host of options at their disposal to best divert waste from landfill [1]. The government is consulting on updated guidance on waste planning policy. In order to solve this problem, China's Development and Reform Commission (NDRC) released the "Thirteenth Five-Year Plan for the Construction of Harmless Treatment Facilities for Urban Living Waste" in December 2016, and China officially began to categorize garbage.In March 2017, the NDRC released the "Implementation Plan for the Living Waste Categorization System", which began the trial work of garbage categorization and selected 46 cities for the trial implementation of mandatory living waste. cities to implement mandatory classification of domestic garbage on a trial basis.In January 2018, the Ministry of Housing and Urban-Rural Development (MOHURD) issued the Notice on Accelerating the Work of Classification of Domestic Garbage in Some Key Cities, which explicitly required the 46 trial cities to plan specific work on garbage classification, and to delineate the objectives and specific deadlines for the work.In June 2019, the MOHURD and other departments issued the Notice on Comprehensively Carrying out the Work of Classification of Domestic Garbage in Prefectural-level and Above-level Cities in the Country. Notice on Comprehensively Carrying Out the Work of Classifying Domestic Waste in Cities at the Prefectural Level and Above Across the Country", requiring 46 pilot cities to formally implement the requirements for garbage classification.In July 2019 Shanghai formally implemented the "Shanghai Municipal Regulations on the Management of Domestic Waste".The "Decision on the Revision of the "Beijing Municipal Regulations on the Management of Domestic Waste", adopted by the Standing Committee of the Fifteenth People's Congress of the Beijing Municipal Government in the sixteenth meeting of Beijing Municipal Government in November 2019, was amended. Shanghai and Beijing officially started the transition from an advocacy-based waste classification policy to a mandatory waste classification policy. The new garbage classification policy has achieved very obvious results after its implementation. According to Xinhua News Agency's report on April 28, 2021, Beijing's average daily volume of 20,600 tons in the first four months of 2021 was 6.36% lower than the average daily volume of 2020, and more than 7,100 tons lower than the average daily volume of 27,700 tons in 2019, which is equivalent to the construction of two fewer 3,000-ton per day waste incineration plants. Shanghai City, the implementation of the "Shanghai Municipal Domestic Waste Management Regulations" for two years, the city's domestic waste volume "three increases

* Corresponding author: 15145453508@163.com

and one reduction" of the effectiveness of the gradual stabilization of the first half of 2021 than before the implementation of the "Regulations" on July 1, 2019, the city's dry garbage control volume of 14,847 tons / day (28% year-on-year decline), the amount of wet garbage sorting 10,311 tons/day (an increase of about 89% compared to 2019), recyclables recovery volume of 7,104 tons/day (an increase of about 1.65 times compared to 2019), and hazardous waste classification volume of 2.02 tons/day (an increase of 14 times compared to 2019) (Source: Party Building Leads a New Fashion Garbage Classification Refueling - The Shanghai . Based on the above understanding, according to the existing literature and information summary and the need for optimization of garbage classification policy, the process of change in the nature of garbage classification policy is taken as the object of study to analyze the necessity of changing from the advocacy-type garbage classification policy to the mandatory garbage classification policy, so as to provide a basis for the subsequent development of garbage classification policy.

2. Literature review

In the recent discussion on waste separation, a controversial issue is whether it is more effective for the government to implement mandatory waste separation or for social organizations to guide waste separation[2]. On the one hand, some people point out that a mandatory waste separation policy is more effective than an advocacy policy. From this point of view, the government should not overdo the policy by advocating people to separate waste in order to make it work well[3]. However, on the other hand, others argue that waste separation itself is a complex human behavior, and that the policy can only be well implemented if people are subjectively willing to separate their waste. In the words of Chen Lu, the main proponent of this view, "the public should take the main responsibility and obligation of waste separation[4]. All in all, the question is whether the government should enforce the policy of garbage separation or let the public fully understand the significance of garbage separation and then make them take the initiative to separate garbage.

3. History of the development of waste separation policy tools

3.1. 1992-2010: publicizing and educating, encouraging categorization and piloting in selected cities

China's garbage classification policy can be traced back to the "Regulations on the Management of Urban Amenities and Environmental Sanitation" in 1992, but at that time, it only put forward the concept of urban garbage classification and advocated the classification of urban garbage, but there were no specific measures, no clear classification standards and no responsible departments. 2000's "Notice on the Announcement of Pilot Cities for Separate Collection of Domestic Garbage" classified eight cities, including Beijing, as pilot cities for the classification of domestic garbage. In 2000, the "Notice on the Announcement of Pilot Cities for Separate Collection of Domestic Waste" classified eight cities, including Beijing, as pilot cities for the classification of domestic waste, but the effect of the policy was not satisfactory because there were no clear classification standards and the time set aside for the formulation of the policy and implementation of the policy was too short. Therefore, although publicity and education were carried out at that stage, because the government only played the role of advocating and did not actually promote, so few cities responded and few residents volunteered, and the garbage classification policy did not play a real role in that period.

3.2. 2010-2017 Formalization of pilot projects and development of incentive-based policies for domestic waste segregation

As the phenomenon of garbage encircling the city worsened, the State put forward the "Technical Guidelines for Domestic Waste Disposal" in 2010, which delineated in detail the categories of domestic waste and put forward specific procedures and requirements for the collection, transportation, treatment and disposal of domestic waste, and continued to promote and advocate the publicity of garbage classification, plan the responsibilities of the government, and formulate the policy objectives in the subsequent policies. For the previous phase of the waste classification policy, the second phase of the waste classification policy is significantly more detailed. In terms of pilot cities the Notice of the Ministry of Housing and Urban-Rural Development and Other Departments on Carrying Out the Work of Model Cities (Districts) for Classifying Domestic Waste, issued in 2014, clearly indicates that a pilot city can only be declared after it is in a position to formulate the relevant local laws and regulations. To a certain extent, this policy has strengthened the enthusiasm of cities to implement waste classification policies, making it possible to implement waste classification policies in some cities, but because the policies formulated by cities are still mainly incentives, they have achieved some results but have not yet met the objectives of the policy.

3.3. 2017-present domestic waste classification policy tends to improve, from advocacy policy to mandatory policy and nationwide large-scale promotion

In March 2017 the state issued the "implementation plan for the domestic garbage classification system", selected 46 cities as key cities to try out the garbage classification policy, and in January 2018 issued the "notice on accelerating the work of domestic garbage classification in some key cities", which explicitly requires 46 pilot cities to plan the specific work of garbage classification, and to divide the work objectives and specific deadlines. After the introduction of mandatory living garbage classification policy, the key cities have corresponding national requirements, developed a series of living garbage classification policy, Shanghai and Beijing have achieved remarkable results, Beijing is for the garbage classification policy introduced relevant laws, which shows that the mandatory type of garbage classification policy is more powerful, the implementation of the effect is better, the simple advocacy type of policy can not achieve the policy objectives, mandatory living The mandatory waste separation policy is more effective in restraining the residents' behavior of waste separation.

4. Problems in the implementation of the policy on the categorization and treatment of urban domestic waste in China

According to a questionnaire survey jointly conducted by the Office of the Shanghai Municipal Commission for the Construction of Spiritual Civilization, the Shanghai Volunteer Association, and a group from the Institute of Sociology of the Shanghai Academy of Social Sciences for the city's registered volunteers and citizens at the end of 2019 released by the Surge News, 50.7% of the 1,477-point questionnaire believed that the standard of garbage classification was too complicated, 50.5% believed that the garbage classification facilities were not perfect enough In the questionnaire, 48.5% thought that the residents' awareness of garbage classification was weak, 46.9% thought that the subsequent disposal of classified garbage was difficult, 46.2% thought that the residents lacked the relevant knowledge, 35.6% thought that the relevant regulations were not operable, 12.4% thought that the government did not publicize the issue enough, and 1.2% thought that there were also other difficulties. According to the Beijing Municipal Bureau of Statistics issued by the urban and rural residents of Beijing garbage classification awareness and the status of the survey report, only 16.5% of the people chose to have the installation of intelligent information technology means to assist in the classification of the put, for the publicity means of the policy is mostly barrel station duty guidance, hanging banners or posters, household publicity and guidance, the main body of the organization of the publicity and activities, points rewards, gift redemption, and so on. According to the questionnaire can be seen in the two cities although the implementation of mandatory waste classification policy but there are still the following problems.

4.1. Waste classification standards are too complicated

Shanghai's garbage classification standard is divided into dry garbage, wet garbage, hazardous garbage and recyclables. Dry garbage and wet garbage but from the literal understanding of the more difficult, and dry and wet garbage in the variety and complexity, although there is a proverb explaining that wet garbage is the pig can eat the garbage, dry garbage is the pig can not eat the garbage, but some residents may have a cognitive problem with the classification, can not be very well remembered all the categories of garbage, resulting in every time before the classification of the type of garbage before inquiring about the type of things for the classification of the residents to bring the inconvenience. Beijing's waste classification standards for recyclable waste, food waste, hazardous waste and other waste, these names are easier to understand, but also more convenient for residents to classify, and will not cause residents to throw garbage every time before they have to go online to find the specific classification and other trouble.

4.2. Inadequate waste separation facilities

During the first few months of the implementation of the garbage classification policy in Shanghai, there were problems such as insufficient funds and poor procurement processes, which resulted in the failure of materials and supplies to be delivered in many areas, and residents were unable to start classifying garbage immediately. Subsequently, there is also because of the wind and sun caused by the original garbage cans above the classification of words blurred, so that the public is difficult to really fulfill the responsibility of garbage classification. Some areas of Shanghai also exist in the residents of the mixed throw, the community specifically set up staff to follow up on the problem of garbage classification, not only is a waste of manpower, but also did not develop the habit of residents to carry out garbage classification.

4.3. Residents do not have enough knowledge related to garbage classification and publicity efforts are insufficient

Residents only know that they should separate their garbage, but they are not clear about how to do so and the need to separate their garbage.Policy implementation had no effect on the reported influences of social and moral norms on the intention to separate household waste[5]. Because residents do not have the habit of separating garbage, although incentives have always existed in the past, there is no mandatory means to really separate garbage. The sudden implementation of a mandatory waste separation policy will cause residents' incomprehension and rebelliousness. However, because the government's publicity measures are mostly offline or use incentives such as giving gifts and accumulating points, the effect is not obvious to young people or residents who need to work. Moreover, the content of offline campaigns is too boring, which not only fails to reflect the necessity of waste separation, but also fails to attract the attention of the target group. The government used more interesting and attractive ways to attract the attention of the residents during the implementation of the New Crown Seal City policy, which is also a mandatory policy, so that the residents understand and agree with the implementation of the policy.

4.4. Low operationalization of relevant provisions

Shanghai garbage classification not only has requirements for classification, but also for putting, for example: waste paper needs to be laid flat, there must not be anything inside the bag, and three-dimensional packages should be flattened after cleaning before putting and so on. There are also garbage needs to be placed at regular intervals and other requirements. These requirements are somewhat stringent and have had an impact on the lives of Shanghai residents. Beijing's garbage classification policy is more lenient than Shanghai's, and is easier to implement. Not only is there no requirement to put garbage out at regular intervals, but it is also stated that if the public is unable to accurately distinguish between food waste and recyclables, they can place them in the other garbage.

5. Policy recommendations on waste classification

5.1. Develop simple and easy-to-understand standards for waste separation

Simplify the criteria for garbage classification, and set up a category that is tolerant of errors, such as other garbage in Beijing, so that if residents are unable to recognize the category of certain garbage, they can put it in other garbage. This method is not only easy for residents to understand, but also convenient for residents to classify and put into the garbage, greatly reducing the difficulty of residents' garbage classification, and avoiding the phenomenon of residents putting garbage into the garbage casually because they can't distinguish the type of garbage.

5.2. Improvement of waste separation facilities

In the implementation of the process of waste classification should be well prepared materials and materials, in the official implementation of the policy before the placement of equipment, in the implementation of the policy in the early stage of the completion of the placement of equipment for the residents to prepare for waste classification. Avoid residents want to respond to the policy but found that there is no trash can be classified downstairs dumpster. Do a good job of follow-up maintenance of garbage bins, to avoid the emergence of garbage bins above the specific categories of words blurred, residents do not know where to put garbage things. Also do a good job of monitoring the residents of garbage classification, rather than part of the community to complete the garbage classification for them. This is not only not conducive to the subsequent implementation of garbage classification, but also a waste of human and material resources.Residents should develop the habit of sorting their own garbage, so that they can continue to do so if they move away from the area. Garbage separation should be universal rather than regionalized.

5.3. Increase publicity and develop online publicity channels

Nowadays, the popularity of the Internet is extremely high, and most of the residents will learn about current affairs from the Internet, especially young people who will take the Internet as the best tool to learn about the world. The Government can make use of software with a high number of visitors to publicize the necessity of garbage classification, so as to make the residents understand that garbage classification is imperative, or it will lead to serious consequences. Some hotspots can be used to publicize the policy of waste separation, and some Internet celebrities can be used to popularize waste separation in an interesting way, so as to increase the residents' understanding of the standard of waste separation and increase the attractiveness of waste separation to the residents. Publicizing the waste separation policy online can increase residents' initiative in waste separation, so that they will participate in waste separation voluntarily and be willing to supervise people around them to participate in waste separation together.

5.4. Reduced operationalization of provisions

Although the current waste separation policy is a mandatory policy, it should also show its humane degree, and should be both rigid and soft, not a one-step tough at the beginning, which will reduce the acceptance of the residents. For example, the Shanghai Municipality requires waste paper to be kept flat and wet garbage to be drained before putting it out, which is troublesome to operate. In the implementation of mandatory policy in the early residents have a certain degree of incomprehension and rebelliousness, the operation of the increased difficulty will only make it more reluctant to carry out the classification of garbage, be mandatory implementation of waste classification policy is only a means of implementation of the policy, so that the residents take the initiative to voluntarily carry out the classification of garbage is the implementation of the policy objectives. It should be carried out in a gradual and orderly manner, so that the residents can voluntarily accept the waste classification before more requirements are made.

6. Conclusion

Adequate material preparation, simple classification standards, appropriate publicity policies and gradual policy implementation are all essential in promoting waste separation activities. The government should take the lead so that residents are willing to actively participate in waste separation, and it should also provide simple and clear methods and perfect facilities to help residents adapt to waste separation.

References

1. Mueller W, Waste Management.**3**, 33(2013)
2. J. Li, C.Y. Li, Journal of Shanghai University (Social Science Edition). **38(05)**, 68(2021)
3. S. H. Guo, Y. Li, Research on Socialism with Chinese Characteristics. **1**, 60(2022)
4. J. H. Zhou, Y. Jiang, Industrial Engineering and Management. **1**, 20(2022)
5. Issock I B P,Roberts-Lombard M,Mpinganjira M, Social Marketin Quarterly, **26**, 99(2020)

The Comparative Analysis of Fair Use of Works in Machine Learning

Yunlan Zhu[1,2,*]

[1] Shenyang University of Technology, Shenyang, China
[2] School of Law, Sapienza University of Rome, Rome, Italy.

Abstract. Before generative AI outputs the content, it copies a large amount of text content. This process is machine learning. For the development of artificial intelligence technology and cultural prosperity, many countries have included machine learning within the scope of fair use. However, China's copyright law currently does not legislate the fair use of machine learning works. This paper will construct a Chinese model of fair use of machine learning works through comparative analysis of the legislation of other countries. This is a fair use model that balances the flexibility of the United States with the rigor of the European Union.

1 Using works in Machine Learning: principles and legal logic

1.1 How Machine Learning works

Machine Learning is a subfield of artificial intelligence that involves automatically detecting meaningful patterns in data and using the detected patterns for certain tasks. Briefly, the learning process involves an algorithm that takes training data (representing past knowledge or experience) as input and output information that other algorithms can use to perform tasks such as prediction or decision making [1].

For example, ChatGPT, the world's popular chat robot, is a large language model among generative machine learning models. In ChatGPT, a large amount of text is first pre-trained to generate a general model [2]. When the user inputs a question or information, ChatGPT will generate a corresponding answer or expression based on the previously learned language model.

1.2 Copyright involved in Machine Learning works

The development and application of machine learning relies on the "feeding" of massive information including copyrighted works, which specifically includes multiple behaviors such as extracting, copying, and format transcoding of works [3]. The process is likely to fall into the regulatory scope of the right of reproduction.

In order to carry out machine learning, staff need to translate, label, organize, and summarize the data. These behaviors may infringe the derivate right. In order to conduct machine learning or to achieve the verifiability of research results, relevant personnel need to transmit data or text through the Internet, or upload it to the "cloud", etc. From the perspective of copyright law, it may infringe the right of communication to the public [4].

There are three copyright rights involved in the use of machine learning works. So, as artificial intelligence creation using machine learning technology becomes more and more common, it not only brings prosperity to cultural creation, but also brings challenges to the traditional copyright system [5].

2 Fair use of works is necessary for machine learning in China

2.1 The current fair use rule lacks usability

The fair use clause in China's Copyright Law has listed twelve specific situations that fall under fair use, but it is not clear which situation the use of works by machine learning falls under.

On this basis, we can seek legal interpretation to determine whether the use of works by machine learning can be covered by the circumstances clearly enumerated in Article 24 [6]. However, in fact, these situations all have applicability issues to a greater or lesser extent and cannot be applied to the utilization of machine learning works involved in the development of artificial intelligence applications. The first paragraph limits the subject and can only be used by individuals. Paragraph 6 limits the quantity of use, but the use of machine learning in works is large and extensive. Paragraph 8 limits the usage, which can only be displayed or preserved.

Although China's new "Copyright Law" in 2020 has added the general provisions of "other circumstances

* Corresponding author: elainejoo@126.com

stipulated by laws and administrative regulations" in the terms of Fair Use, the specific circumstances of my country's copyright Fair Use are stipulated in the "Copyright Law" and "Information Network Communication". In addition, there are no other laws or administrative regulations that stipulate the Fair Use system, and the behavior of using works by Machine Learning is difficult to be included in the scope of these situations. Therefore, when laws and administrative regulations do not provide for other circumstances, the catch-all clause cannot play a catch-all role, nor can it solve the dilemma of applying the Fair Use system to Machine Learning.

2.2 Rationality of Machine Learning fair use

After the authors create the works, they are copyright owner who can get interests from works. As a kind of private right, copyright law mainly focuses on the protection of private right. Others, such as the algorithm company and platform users, represent the public interest in public domain. However, Machine Learning may break the balance between the intellectual property and public interest [7]. Under the guidance of the public interest, copyright law delineates the scope of the public domain, so that works within this scope can be used freely and without compensation.

For example, the fair use system in copyright law. The purpose of fair use is to limit the property rights of copyright and reduce the scope of economic rights of the copyright owner. In this case, the copyright owner loses the ability to restrict the behavior of certain users and reasonably transfers some of its rights and interests [8]. Because of the dynamic balance between exclusive rights and the public domain in IP law, the fair use system has become a very important legal system for economic and social development and technological innovation in contemporary world.

3 Representative legislative models of machine learning fair use of works

3.1 United States: Application of the transformative use rule

Although machine learning is technically further and more complex than text data mining, the two are closely related. If text data mining is a copyright exception, then this copyright exception applies equally to machine learning [9]. Currently, statutory laws and case law in many countries have established copyright exceptions for text data mining.

U.S. jurisprudence has established an "unconditional exception" model that does not set any conditions for use of copyright exceptions for text mining, and has established a "transformative use" rule.

The U.S. copyright law adopted open legislation with "four elements" as its core in 1976. U.S. courts flexibly weigh the proportion of each factor in the judgment of a specific case, thereby expanding the application space of the fair use system in practice.

Google launched the Google Library project in 2004, which unites the world's major research libraries and collects a large number of book resources, converting these books into digital texts. However, the Writers Guild of America took Google to court in 2005, arguing that the book resources scanned and uploaded by Google contained many works that were used without the authorization of the original copyright holders, which was a copyright infringement [10].

In 2013, the U.S. court ruled that Google's actions did not constitute copyright infringement. According to the "four elements of fair use" stipulated in Article 107 of the U.S. Copyright Act, the Google Library Project's actions had a "transformative purpose." In 2016, the Writers Association appealed to the Federal Supreme Court, which ultimately upheld the original verdict [11].

However, the recent "TV Eyes case" made a judgment different from the Google case, which to a certain extent reflects the court's rigorous attitude towards judging the Fair Use of Machine Learning. The defendant in this case, TV Eye, is a media company. The company records the programs broadcast by various TV channels, digitally classifies the recorded content, and provides clips of no more than 10 minutes long to customers, so that customers can locate the programs they want according to their preferences.

The completely different judgment results of the Google Books case and the TV Eyes case show the attitude of the US courts on the Fair Use of Machine Learning. That is, the recognition of the transformational use of Machine Learning works constitutes Fair Use, but the presentation of the output results needs to be considered independently. If the effect of public communication on copyright content has the possibility of replacing the original work and occupying the market of the copyright owner, it will be deemed as an infringing use.

3.2 Japan, the United Kingdom: independent and clear exceptions

3.2.1 Japan

In 2009, Japan's "Copyright Law" added a new exception to copyright law - "Copying for the purpose of analyzing information", which is considered to be the earliest legislative example involving the provision of copyright exceptions for the use of artificial intelligence works. However, this computer exception is limited to the scope of "information analysis", and it is difficult to cover new ways of using works under the background of Machine Learning and deep learning technology.

The Japanese government has also pointed out that the provisions of the old law are not enough to meet the current development status and long-term needs of the artificial intelligence industry. Although this provision only applies to works and does not apply to databases, in general, compared with other countries. Japan's copyright legislation has a system of restrictions on the exercise of copyright that gives considerable freedom to text and data mining [12].

Japan's "Copyright Law" revised in 2009 added new restrictions to the right of reproduction, that is, when using a computer for information analysis, the user has the right to cache or adapt the work within the necessary limits (including Cache of Derived Works). Japan's Intellectual Property Protection Organization (AIPPI) pointed out in a report on "Exceptions to Copyright Protection and Licensing of Works in a High-Tech Digital Environment" that the type of work does not affect the application of this clause. However, at the same time, this article stipulates that the use of databases does not apply to this rule, which is also consistent with the "three-step test method" stipulated in the "Berne Convention", that is, the laws of member states have the right to allow the reproduction of the above-mentioned works under certain special circumstances. As long as this copying does not damage the normal use of the work or unreasonably harm the legitimate interests of the author.

3.2.2 UK

The United Kingdom revised its copyright law in 2014, and made changes to copyright law based on the recommendations in the famous "Hargreaves Report" to adapt to the development of the times, including the copyright system for the Fair Use of text and data mining [13]. The "Hargreaves Report" believes that it is necessary to establish a Fair Use copyright system for text and data mining, because this technology is very practical and representative in the new era, and cannot be restricted by copyright laws and cannot play its role.

Unlike the "unconditional exception" in the US, the text and data mining exception in the UK is a "conditional exception". In the main element, purpose element and form element, there is a certain degree of restriction. In terms of the main requirements, the subject must obtain the work legally. In terms of purpose, it must be for non-commercial purposes. In terms of formal requirements, it is required to show respect to the copyright owner by indicating the source in the mining results. If there are orphan works, the database information of the source of the work should also be indicated [14]. While the United Kingdom stipulates that text and data mining can be used reasonably, it also stipulates transfer restrictions, purpose restrictions, and transaction restrictions to better protect the legitimate interests of copyright holders.

The copyright exception system for text and data mining in the UK, on the one hand, conforms to the basic content of the "three-step test method" embodied in the TRIPs agreement. Abuse also helps society rationally utilize the existing scientific and cultural achievements and realize the unification of personal interests and social interests.

3.3 European Union: relatively prudent exceptions

The European Parliament and the Council of the European Union approved the "Digital Single Market Copyright Directive" on March 26 and April 15, 2019, respectively. One of the goals of this directive is to adapt the exceptions and limitations of copyright law to the development of the digital market. Among them, Articles 3 and 4 stipulate copyright exceptions to TDM, and the domestic laws of EU member states will establish TDM exceptions that meet the requirements of the "2019 Copyright Directive" within two years.

First of all, with regard to the applicable subjects, there are only two types of scientific research institutions and cultural heritage institutions. Then, regarding the applicable elements, the first is "legal source", that is, TDM users must have legal access to the mined text or data. Users who obtain texts or data illegally or illegally cannot benefit from the EU TDM exception [15]. The second is "safety protection measures". One of the applicable conditions of the EU TDM exception is that TDM users take appropriate security measures for the mined text and data. The lack of security protection measures will lead to the risk of copyright infringement of the mined content, and the TDM exception still applies at this time will result in the loss of rights holders' interests. The requirement of safe preservation measures can prevent the abuse of TDM exceptions and protect the legitimate interests of right holders [16].

4 The comparative law perspective of fair use of works in Machine Learning

4.1 The features of various countries' legislation

Generally speaking, the countries that stipulate copyright exceptions for TDM are basically developed countries in Europe and the United States, and generally have the following two characteristics: (1) Existing legislation basically limits TDM exceptions to the field of scientific research, (2) Text analysis and research activities that are exempt from liability under the TDM exception should be for non-commercial purposes. Among them, Japan is quite special. Its legislation does not explicitly limit the TDM exception to the use of works for non-commercial purposes, nor does it use the term "scientific research" with non-commercial implications.

A major development in the legislative practice of copyright exceptions for TDM is the Directive on Copyright in the Digital Single Market passed by the European Parliament in March 2019. Article 3 of the Directive stipulates a mandatory exception for TDM applicable to research institutions, which means that the copyright exceptions for TDM will be extended from countries such as Britain, France, and Germany to other EU countries, because EU members The country is obliged to transform the substantive content of the new EU Copyright Directive into domestic law within the statutory period. After the legal practice of many countries and the theoretical recognition of academic research, it can be considered that in non-commercial scientific research activities, TDM has gradually become a legal exception to copyright at the international level.

The United States is at the most lenient end in comparison, and its case law has basically agreed that the act of copying works by artificial intelligence does not constitute copyright infringement if there is no public communication [17].

4.2 The enlightenment on the establishment of the fair use rule in China

The newly added "three-step test method" in Article 24 of the Chinese Copyright Law is semi-closed and is limited to "use in the following circumstances", which essentially constitutes a restriction on the interpretation of 12 specific situations, rather than an open authorization regulation. Courts cannot "interpret" new limitations or exceptions under this article in specific cases. In comparison, Article 107 of the U.S. Copyright Law adopts a flexible legislative model, which provides a broad and flexible space for interpretation by U.S. courts.

For text and data mining technology, the United States has a broad market and prospects, and has relatively mature judicial precedents [18]. The "conditional exception" represented by the UK and the EU is to limit some necessary content in the Fair Use system to better adapt to social development and respond to social voices [19].

So, there are two options for my country to optimize the legislative model of Fair Use: one is to adopt the flexible and open "Fair Use + specific enumeration" model of the United States, which stipulates four considerations for Fair Use. And then lists common Fair Use methods. The other is to follow the mainland. The "three-step test method + specific enumeration" model of the legal system clearly stipulates that "use for AI learning and creation" is Fair Use when amending China's "Copyright Law Implementation Regulations".

5 Conclusion

The Fair Use system has played an important role in coping with the development of information technology and maintaining the balance of interests many times throughout the history of copyright development. In the current situation which artificial intelligence has become a new engine of economic development and a new opportunity for social development, which is rapidly changing human life and becoming the focus of international competition. It will once again assume this important responsibility in China.

To sum up, this paper proposes adding a new type of Fair Use of machine learning after Article 24, Paragraph 1, Item 12 of China's current "Copyright Law", namely: "copying, adapting other people's works for machine learning, and using creation achievements are made available to the public in the form of broadcasting and information network dissemination, but the metadata of the obtained work shall not be publicly disseminated in its original form."

References

1. M. Sag, Journal of the Copyright Society of USA. The New Legal Landscape for Text Ming and Machine Learning. *66*(**2**), 346–36 (2019)
2. L. Lessig, *Penguin Books*. Remix: Making Art and Commerce Thrive in the Hybrid Economy (2008)
3. H. Richter, Max Planck Institute for Innovation and Competition Research Paper. Position Statement of the Max Planck Institute for Innovation and Competition on the Mode of European Copyright Rules. (**2**), 7–10 (2017)
4. M. McJohn, L. McJohn, Northeastern University Law Review. Fair Use and Machine Learning. **12** (2020)
5. A. Li, Electronics Intellectual Property. Copyright Analysis of Machine Learning Works -- Non-Work Use, Fair Use, and Infringing Use. (**06**), 60–70 (2022)
6. Z. Xuan, Publishing Research. On Fair Use of Copyright in AI Creation from the Perspective of Classification Protection. (**03**), 81–87 (2022)
7. A. Foer, A. Durst, The Antitrust Bulletin. The Multiple Goals of Antitrust. 508 (2018)
8. HD. Wu, *A Study on the Fair Use System of Copyright*. China Renmin University Press (2020)
9. E. Lee, Southern California Law Review. Technology Fair Use. *83*(**4**), 808 (2010)
10. H. Piwowar, J. Priem, V. Larivière, PeerJ. The State of OA: A Large-Scale Analysis of the Prevalence and Impact of Open Access Articles. *6*(**4**), e4375 (2018)
11. G. Spindler, International Review of Intellectual Property and Competition Law. Copyright Law and Artificial Intelligence. *50*(**9**) (2019)
12. T. Pihlajarinne, International Review of Intellectual Property and Competition Law. Should We Bury the Concept of Reproduction–Towards Principle-Based Assessment in Copyright Law? *48*(**8**) (2017)
13. Y. Bathae, Berkeley Technology Law Journal. Data privacy and user exception. 35(2020)
14. B. Sobel, Columbia Journal of Law & the Art. Artificial Intelligence Fair Use Crisis. 41(**1**), 49–79 (2017)
15. M. McJohn, L. McJohn, Northeastern University Law Review. Fair Use and Machine Learning. 12 (2020)
16. J. Boyle, *The Public Domain: Enclosing the Commons of the Mind*. Yale University Press. 29 (2008)
17. M. Lemley, 13Southwestern Journal of Law and Trade in the Americas. New Balance betwen IP and Antitrust. 2(**3**) 237-248 (2007)
18. D. Schönberger, Social Science Electronic Publishing. Deep Copyright: Up-and Downstream-Questions Related to Artificial Intelligence and Machine Learning. *10*(**1**), 163-167 (2018)
19. C. Geiger, J. Griffiths, R. Hilty, European Intellectual Property Review. Toward a Balanced Interpretation of the “Three-Step Test” in Copyright Law. 30(**12**), 489–496 (2008)

Research on Ideological Discourse in Foreign Language Education

Jinyu Dong[1,a*] , lu Zhao[2,b]

[1]School ofForeign Languages, Northeast Petroleum University, Daqing, Hei Longjing 163318,China
[2]School ofForeign Languages, Northeast Petroleum University, Daqing, Hei Longjing 163318,China

Abstract: The discourse system is an important carrier of ideology, and ideology must rely on the discourse system to express and disseminate. The discourse system of ideological and political education carries the content and expression of ideological and political education, and the degree of its improvement and the role it plays directly determine the effectiveness of curriculum ideological and political education. The research on discourse in ideological and political education in foreign language education mainly focuses on what the connotation of discourse in ideological and political education is, which elements it includes, how it operates and plays a role in teaching practice, how educators complete discourse conversion and carry out effective discourse dissemination, and so on. Through these researches, the discourse system of ideological and political education in foreign language education is established. Combined with Marxist theory, discourse criticism theory, and communication theory, the basic concepts and content of the discourse system of ideological and political education are clarified, and the basic elements and operation process of the discourse system of ideological and political education are analyzed.

1. INTRODUCTION

The discourse issue has always been a hot topic in Western social science. With the linguistic shift in Western philosophy, the discourse quandary has gradually broadened from the field of philosophy to many fields such as sociology, political science, and education, receiving wide attention from scholars. Numerous translations and scholarly papers related to Western discourse theory have been incorporated into the academic landscape of domestic social science, and a large number of academic achievements on Western discourse theory have emerged. Concurrently, a growing number of domestic scholars with an intensified sense of theoretical and academic cognizance are incorporating Western discourse theory into their scholarship, centering on issues relevant to China. This endeavor aims to formulate a discourse theory imbued with distinctly Chinese attributes, which may serve as a guiding force for China's actuality. Importantly, during the Symposium on Philosophy and Social Sciences Work, General Secretary Xi Jinping has repeatedly mentioned "discourse" and "discourse system", emphasizing the coordinated development of the discourse system, discipline system, and academic system as an important task to accelerate the construction of Chinese characteristics in philosophy and social sciences. "Building a discourse system of philosophy and social sciences with Chinese characteristics, Chinese style, and Chinese temperament"[1] has become an important goal of academic research, and the issue of discourse and the discourse system has gradually become an academic focal point. In addition, since the 18th National Congress of the Communist Party of China, the Party and the state have attached great importance to ideological and political education in universities. The National Symposium on Ideological and Political Work in Higher Education and the National Symposium on Ideological and Political Education for Teachers have both made important instructions for ideological and political education, and have issued a series of important documents to provide institutional guarantees for the reform and innovation of curriculum ideological and political education, which has objectively promoted the research on hot issues in ideological and political education by Marxist theoretical workers. In this research trend, there have been continuous achievements in the "discourse system of ideological and political education."[1]

2. THREE STAGES OF RESEARCH ON THE POLITICAL AND IDEOLOGICAL DISCOURSE

During the initial phase, spanning from 2007 to 2011, the research was in its nascent, based on the rise of research on ideological and political education discourse. In 2007, discourse research began to enter the research scope of ideological and political

*Corresponding author:[a] eilleen_dong@163.com
[b]3394579644@qq.com

education disciplines, and gradually developed. In this period, a significant amount of research was devoted to exploring the fundamental theoretical aspects concerning the discourse of ideological and political education. This included inquiries into topics such as composition, reform and innovation, and the construction of an effective system. Notably, Dr. Qiu Renfu from Shanghai University was a prominent figure in this area of scholarship, providing a theoretical basis for subsequent research endeavors. While there exist studies on the discourse of ideological and political education, political discourse has received insufficient attention in academic circles, resulting in scarce findings with just a few articles being published each year. This shows that the academic community has not given enough attention to discourse in ideological and political courses, not to mention a dearth of corresponding theoretical investigation. Research was more focused on the teaching practices of ideological and political courses, with little involvement in fundamental theoretical issues, or direct replication of the theoretical results of ideological and political education discourse. At this time, the term "political and ideological discourse system" had not yet been proposed.

During the second stage, spanning from 2012 to 2015, the emergence of the concept of the "ideological and political discourse system", related to Party's theoretical innovation and General Secretary Xi Jinping's series of important speeches, resulted in a surge of research activity on the topic. Throughout this period, the academic community increasingly turned their attention towards exploring the "ideological and political discourse system" as a prominent research object, ultimately fueling a noteworthy expansion in research outputs. In August 2013, General Secretary Xi Jinping proposed at the National Propaganda and Ideology Work Conference: "Carefully construct a discourse system for external communication, innovate external discourse expression, create new concepts, new fields, and new expressions that integrate Chinese and foreign cultures... Strive for international discourse power and increase the affinity of cultural communication." [2]In November 2013, at the Third Plenary Session of the 18th Central Committee of the Communist Party of China, it was proposed to "expand cultural exchanges with foreign countries, strengthen international communication capabilities and build a discourse system for external communication, and promote Chinese culture to the world."[2] For a while, discourse systems have garnered significant attention. While the contemporary notion of 'discourse systems ' is predominantly discussed in the context of international communication with an emphasis on Chinese characteristics, academic circles have already noted and reflected upon the issue of discourse systems. This has particularly manifested in research on ideological and political education, specifically within the realm of ideological and political courses, giving rise to the concept of an 'ideological and political discourse system'. There has been a noticeable trend in recent academic research towards exploring the evolution of the discourse surrounding political and ideological matters. This has led to an increase in scholarly focus on this topic, with researchers devoting substantial efforts towards delving deeper into the intricacies of discourse. It is worth noting, however, that current research has largely concentrated on practical application in teaching political and ideological courses, with only limited exploration of fundamental theory, such as the various connotations, types, and elements of the political and ideological discourse system.

In the third stage, spanning from 2016 to the present, investigations pertaining to the 'ideology and political theory discourse system' have undergone a decrease in activity, subsequently followed by an upward progression. In May 2016, Xi Jinping mentioned "discourse system" multiple times at the symposium on philosophy and social sciences. A scholarly surge towards the analysis of discourse systems has been ignited within the philosophical and social sciences academic community. Not only is this a pragmatic necessity for the advancement of the Party and state, but it also entails a crucial duty and accountability for philosophers and social scientists to cultivate the Party's theoretical innovation. In light of the significance of ideological and political education, the Ministry of Education has declared 2017 as the "Year of Teaching Quality of Ideological and Political Courses in Higher Education."[2] Furthermore, in April 2018, the Ministry of Education has released the "Basic Requirements for Teaching Work of Ideological and Political Theory Courses in Higher Education in the New Era" [3] in order to facilitate the advancement of instructional quality in ideological and political courses. In March of the year 2019, General Secretary Xi organized a national seminar targeted towards teachers of ideological and political education with a primary focus on highlighting the integral nature of the course in fulfilling the fundamental objective of instilling proper morals in students. The organizer put forward a set of 'eight cooperative requisites ' necessary for restructuring and innovating ideological and political education while concurrently outlining 'six requirements' aimed at the individuals charged with facilitating such teachings, which were formulated with the intent of optimizing the effectiveness of ideological and political education schemes. In August 2019, the General Office of the CPC Central Committee and the General Office of the State Council jointly issued "Several Opinions on Deepening the Reform and Innovation of Ideological and Political Theory Courses in the New Era", [3] fully implementing the spirit of the speech of Xi Jinping at the symposium of ideological and political theory teachers in schools. Therefore, there has been a resurgence in the investigation of the discourse system pertaining to courses on ideological and political theory within the academic realm. This phenomenon not only signifies the implementation of the spirit of reform and innovation in colleges and universities promoted by the Party and the state, but also serves as an imperative for elevating the caliber and standard of ideological and political theory courses in higher education institutions. In this particular stage, scholars have adopted a greater spectrum of viewpoints and attained an amplified degree of profundity and scope in their investigations, thereby culminating in a multitude of theoretical progressions.

3.RESEARCH ON IDEOLOGICAL AND POLITICAL DISOURSE IN FOREIGN LANGUAGE EDUCATION

3.1 Research on the Subject of Ideological and Political Discourse

This article posits that discourse is derived from the spoken word, which serves as a means of expression, dissemination, and communication. The individuals responsible for these actions become the subjects of discourse. Consequently, the subject of a discourse system must encompass those who produce, disseminate, and receive speech, thereby forming a cohesive "subject group." Utilizing the "theory of subject-interaction" as a specific analytical framework, it can be postulated that discourse producers within the foreign language education domain's ideological and political language system encompass not only educators and learners, but also the textbook authors and subject researchers who are responsible for the development of educational materials. The rationale underpinning such a view is predicated upon the notion that both educators and learners are regarded as disseminators of discourse; a reality that effectively epitomizes the subject-interaction mode of "subject-subject" relationship wherein educators are deemed to occupy a stronger role in the guidance, propagation, and refinement of discourse. Educational practitioners and learners are identified as among the recipients of discourse, with the latter assuming a more prominent role in the processes.

The discourse objectives and functions vary across different disciplines. From the perspective of language producers, textbooks are the basic basis and guidelines for teaching, fundamentally carrying the integration of the ideological and political curriculum. However, most foreign language education majors use English original textbooks, which naturally contain content about Western culture. American linguist Kramsch believes that "learning and understanding the target language culture should be combined with contrasting it with the background knowledge of the native language culture. Cultural teaching in language education includes both target language and native language cultural teaching."[4] Hence, it becomes imperative to extensively absorb political and popular discussions by analyzing the content of textbooks. This is crucial for developing textbooks that are not only based on scientific principles and standardized guidelines but also easily comprehensible. Therefore, textbook discourse is formulated.

During the actual teaching process, the textbook content is adapted to suit the students ' specific circumstances. The discourse is then refined and reproduced before being conveyed to the students, thereby ensuring effective dissemination. When students are presented with language, they have an exchange with their teachers, generate novel discourse, and provide feedback. Such interactions carry the linguistic production and dissemination, with students serving as ambassadors of language and teachers fulfilling the role of recipient. This interactive dynamic highlights the subjective nature of the relationship between teachers and students, both of whom are the subject of the discourse surrounding ideological and political education.

3.2 Research on the Discourse Context of Ideological and Political Education

The term 'context ' denotes the surroundings or circumstances in which language is employed. The renowned philosopher, Frege, introduced the celebrated 'principle of context', positing that "words should be analyzed in relation to sentences, not in isolation." While expounding on the topic, Robbins delineated the context as a 'primary structure that enables the elucidation of information regarding the function and connotation of a discourse'. [5] The theory of pragmatics dictates that the utilization of any language must be contingent upon discrete contextual parameters, with the connotation of speech being inherently tied to its contextual surroundings. As such, context is a crucial component in the articulation, transmission, comprehension, and recognition of speech. In the suitable context of oral communication, individuals exhibit greater comprehension and concurrence with the ongoing conversation. With varying contexts, the semantics of words undergo substantial alterations.

In communication studies, context is called communication context. In a narrow sense, it refers to the scene where specific communication activities (such as a dialogue between two people) take place, including time, place, and the people present. In a broad sense, communication context also includes the larger environment of the community, organization, system, language, culture, and other subjects engaged in communication behaviors. The transmission of context frequently gives rise to fresh interpretations that extend beyond the symbolic text itself and affect the significance of the symbolic text, thus creating situational meaning. "The meaning conveyed by the situation will have a repercussion on the symbolic text system's meaning and exert limitations on the dissemination outcome." [6]Consequently, it is evident that context also encompasses the notion of a 'field'.

The pragmatics and communication studies offer valuable insights for contextual analysis of ideological discourse systems. In particular, a contextual examination of the elements that comprise ideological discourse systems in foreign language education holds significant importance for promoting effective communication in this field. Drawing upon principles from both pragmatics and communication theory, as well as insights derived from systemic functional linguistics, with particular attention paid to the division method employed by linguist Hu Zhuanglin, the study of ideological discourse in foreign language education is segmented into three distinct contextual components. These components are comprised of textual context, situational context, and cultural context.

3.3 Research on ideological and political discourse content

The essential factor determining the ideological and political education discourse system and setting it apart from other discourse systems lies in its core content. This content is vast and varied, and can be classified according to different criteria. By examining its distinctive features, the content of ideological and political education discourse can be divided into three categories: political discourse, theoretical discourse, and popular discourse. This study aims to analyze the content of ideological and political education discourse in foreign language education through the lens of these three aspects.

3.4 Research on the Discourse Forms of Ideological and Political Education

Form refers to the structural and expressive methods by which the different components constituting an entity are consolidated. Concerning the composition and presentation of discourse, ideological and political education discourse has two forms: textual form and spoken form.

Textual form is commonly defined as written discourse. In the context of foreign language education, ideological and political discourse is primarily presented in the form of textbooks and teachers' instructional materials. Textbooks serve as the central reference for teaching discourse, providing standards and guidelines for teaching activities. On the other hand, teachers' instructional materials encompass various forms such as handwritten notes on blackboards, multimedia presentations, videos, audios, and pictures. These materials are the outcome of teachers' creative adaptation of textbook texts, incorporating their interpretation of the textbook and curriculum, and considering the practicality of teaching and the needs of students.

Spoken form: As implied by its name, it refers to a mode of expressing information or ideas through spoken words. It involves both the form of verbal expression and the manner in which thoughts are conveyed. This constitutes the fundamental and primary mode of instructional delivery within the classroom setting.

3.5 Research on the generation of ideological discourse

The production of conversation in foreign language education and ideological education is not singular, static, or isolated. The discourse of ideological education in foreign language education should draw from the assets of traditional Chinese culture and moral education, demonstrate the significant historical accomplishments of revolution, construction, and reform under the guidance of the Party, exhibit deep historical reasoning, and also encompass profound theoretical reasoning. It should act as the primary arena for moral education and embody lively practical reasoning.

3.6 Research on the Conversion of Ideological Discourse

The production of discourse provides resources for foreign language education and the ideological discourse system. However, in order to effectively educate and guide students' thoughts and behaviors, it is necessary to convert the generated discourse into spoken form. The ideological discourse system, which contains ideas and values, uses the production of discourse as a starting point and aims for the effective dissemination of discourse. The conversion of discourse serves as a vital intermediate link between production and dissemination. Teaching is a dynamic interaction between teachers and students, and the challenge now lies in activating static text-based discourse in order to meet the needs of teaching and students - a complex yet crucial problem.

3.7 Research on the Dissemination of Ideological Discourse

Communication scholars argue that information dissemination within human society is characterized by evident processivity and systematicity. Taking a macro perspective, the discourse system of ideological and political education within foreign language education is seen as a communication system itself. The generation and transformation of discourse are viewed as integral components of the communication process. However, in the present study, the focus is on the micro perspective, specifically on the specialized field of ideological and political education within the foreign language education curriculum. The aim is to investigate discourse dissemination as a fundamental element within this context.

4 CONCLUSION

The research is to be conducted from various disciplinary perspectives, including linguistics, Marxist theory, and education. This thorough and systematic approach intended to enable a comprehensive analysis and understanding of the subject, content, context, form, generation, transformation, and dissemination of ideological language in foreign language education. The study aims to clarify the fundamental concepts and associations of the ideological language system, while also analyzing the basic components and operational processes of such a system. By examining the practical issues concerning ideological language systems in foreign language education, with reference to related theory in linguistics and education, an innovative system for teaching ideological language in foreign language education can be constructed. Research into ideological discourse in foreign language education has the potential to benefit educators in their efforts to guide students in the development of proper perspectives on life, values, and the world. Additionally, this research can facilitate the conscious adoption of correct socialist core values among students while strengthening their sense of collectivism and patriotism.

ACKNOWLEDGMENTS

This research was financially supported by Guiding innovation fund of Northeast Petroleum University—Research on Ideological Discourse in Foreign Language Education (Grant NO. 2022YDW- 18)

Northeast Petroleum University Higher Education Teaching Reform Research Project— Reform and Practice of the Teaching Model of Integrating Ideological and Political Elements into English Majors under the Background of Journalism: Taking "Introduction to Linguistics" as an Example (Grant NO. DYJG2021067)

REFERENCES

1. Minfang Xie. A Study on the Innovation of Teaching Discourse of Professional Courses in Universities from the Perspective of Curriculum Ideology and Politics[J]. Journal of Hefei University, 2021,6:122-127.
2. Wenzhong Lu.The Structural Form and Optimization of the Discourse System of Ideological and Political Theory Courses in Universities[J]. China Higher Education, 2017, (11): 29-31.
3. Wu Chunlei. The Construction of Marxist Ideological Discourse in China in the New Era[J]. Chongqing Social Science, 2017, 000(010): 11-17.
4. Claire Kramsch. Context and Culture in Language Teaching[M]. Shanghai: Shanghai Foreign Language Education Press, 2004.
5. R.H.Robins. Concise History of Linguistics[M]. China Social Science Press, 1997:42.
6. Qingguang Guo. Communication Studies Tutorial[M]. China Renmin University Press, 2011.

SHS Web of Conferences **178**, 01017 (2023)
ICPRSS 2023
https://doi.org/10.1051/shsconf/202317801017

The Impact of Parental Pension Insurance on The Children's Marriage Market

Jiayi Kang*

International College Beijing, China Agricultural University, Beijing, 10083, China

Abstract: In recent years, China's marriage rate and fertility rate have continued to decline, and the divorce rate has been rising, accompanied by the increasing aging problem in China. This paper empirically studies the impact of parental pension insurance on children's marriage market, and classifies the views of young respondents on the impact of parental social insurance on children's marriage market from their perspective and the perspective of the next generation. The conclusion shows that the positive effect of parental pension insurance on the choice of children's marriage market is more significant, but there are still some differences between age, gender, urban and rural areas and family types, which provides a scientific basis for improving China's marriage rate and fertility rate.

1. Introduction

China has traditionally valued the importance of marital happiness and family harmony. However, in recent years, the marriage rate in the country has declined significantly, leading to a greater impact on China's aging process. As it shown in Table 1, based on data from the National Bureau of Statistics, the number of marriage registrations in China has decreased from 13.4693 million in 2013 to 7.643 million in 2021, marking a nearly 50% decrease over the past decade. The widespread concern from various sectors arises due to the potential impact of this trend on China's birth rate, which could further intensify the problem of aging.

Table 1. Number of marriages in mainland China in the past decade

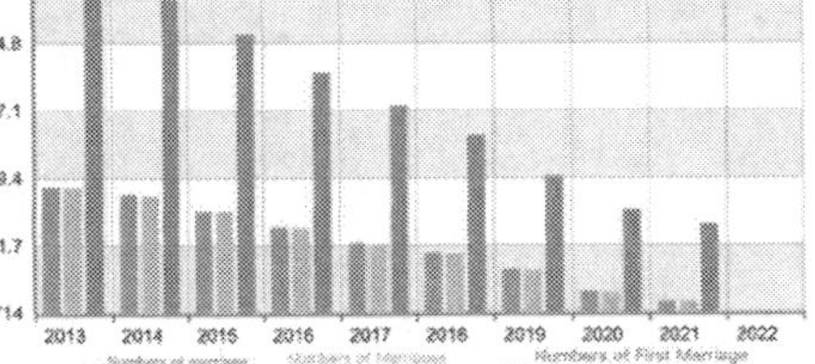

Source: http://www.stats.gov.cn[1]

There are a number of reasons for the decline in first marriages, including increased female awareness, better education and increased stress in life. For example, one study found that for every 1% increase in house prices, the rate of first marriage fell by 0.31%.[2] In addition, it is widely believed that parental stress also has an impact on marriage rates, and this includes not only the stress of raising the next generation, but also the stress of caring for the generation of parents. However, empirical evidence shows that pension insurance can replace the financial support and medical supply of children to a certain extent, which can improve the medical participation rate and reduce their medical worries. When the individual's security ability is low, the new rural insurance and new rural cooperative medical care will play a more obvious role in the replacement of their children's pension. [3]

Old-age insurance in China is classified into four types: urban employee old-age insurance (for state-owned, private, and foreign enterprise employees), government employee old-age insurance (for civil servants and public institution employees), urban and rural resident old-age insurance, and commercial old-age insurance. The importance of pension insurance is growing among individuals, especially as China's population ages. China's New Rural Pension Scheme is replacing the traditional family pension model with a "social pension" model, according to research. Pension schemes lead to increased household income and food expenditure, reduced agricultural Labour, improved health and lower mortality. [4]

Over the past decade, the number of people participating in endowment insurance in China has increased annually, according to data from the National Bureau of Statistics. As of the end of 2022,503.49 million people participated in the basic endowment insurance for urban employees, which is up by 22.75 million from the previous year. Additionally, 549.52 million people participated in the basic endowment insurance for urban and rural residents, which is up by1.55 million. [1] These figures indicate a gradual increase in the proportion of elderly individuals in China,

* Corresponding author: kangjiayi227@163.com

as well as a growing emphasis on the concept of 'social pension'.

2. Literature review

At present, a large number of academic studies have been carried out in the field of pension insurance. It is understood that the endowment insurance is considered as a social welfare measure that can improve the well-being of the elderly.

However, since pension insurance still requires payment, it is closely linked to the defensive savings of the insured and their families. Research shows that in high-income plans, the increase in insurance-induced consumption is caused by a lower incentive to save defensively.[5] Despite this, there are relatively few studies on the impact of parental pension insurance on the marriage market, with most studies focusing on its impact on physical health, income level, life satisfaction, labor supply time, and female fertility. For example, Some economists reckon that the public pension model has the effect of lowering the fertility rate, which indirectly leads to a decline in the population and a decline in the marriage rate of the next generation.[6] Another study have found that basic old-age insurance for urban and rural residents can improve the health of the elderly in rural areas by increasing medical spending, increasing leisure activities and maintaining a balanced diet.[7] This result may make children more willing to choose elderly people with pension insurance as marriage partners, thereby reducing the family's medical costs. Pensions can also reduce the income gap among rural elderly, thereby reducing welfare inequality, which helps to improve household income and economic resilience,[8] while also encouraging children to pay more attention to their parents' pension status. Moreover, although different pension systems have different effects on fertility, the cost of dependency remains a constraint on marriage and fertility.[9] Therefore this study aims to fill the research gap on the relationship between parents' participation in pension insurance and children's choice of spouse, and provide scientific basis for promoting marriage and fertility in China through questionnaire survey and Pearson correlation analysis.

3.Empirical analysis

3.1 Data source and statistical expression of variables

The data of this study are from random questionnaires. 513 questionnaires were collected and the questionnaire mainly includes two parts: the basic characteristics of the respondents (gender, age, whether there is old-age insurance, the type of old-age insurance, whether there are children, the type of household registration, whether it is a single-parent family, etc.) and the importance of old-age insurance in romantic relationships, which provides data support for this article.

Table 2 shows the statistical analysis of variable interpretation and description of the results of 513 statistical questionnaires, including variable name, variable interpretation, mean value and standard deviation

The gender distribution of the respondents was relatively uniform, with 42.11% men and 55.95% women. The respondents had an average age of37.13, and81.2% had pension insurance while 18.71% did not. This indicates that most respondents were under40 and had a good understanding of pension insurance. Of those surveyed,86.1% had children and 13.9% did not. The respondents' household registration types were also relatively evenly distributed, with44.59% having rural household registration and 54.05% having urban household registration. Among the respondents, 12.16% were from single-parent families, while 87.84% were from non-single parent families. Of the single-parent family respondents, 54.88% had a father/mother with pension insurance. In terms of the parents' pension insurance status, 61.54% of the respondents' parents had pension insurance, while 18.46% did not. The survey on types of pension insurance found that the father's pension insurance type was distributed fairly evenly among the four types, with urban employee pension insurance at 26.67%, institutional pension insurance at 16%, urban and rural resident pension insurance at29.33%, and commercial insurance at 22.6%. The mother's pension insurance type was also distributed fairly evenly among the four types, with urban employee pension insurance at 30%, institutional pension insurance at 27.5%, urban and rural resident pension insurance at 27.5%, and commercial insurance at 5%.

Table 2. Statistical analysis of variable interpretation and description (n=513)

Variable	Variable interpretation	Average value	Standard deviation
Dependent variable			
Dependent variable 1 Whether to pay attention to whether the target's parents have pension	*1=Very concerned about it* *2=Relatively concerned about it* *3=Not very concerned about it* *4=Completely indifferent to it*	*2.127*	*1.220*
Dependent variable 2: If you have children and he/she has reached marriageable age, do you care about whether the parents have pension insurance when your	*1=Very concerned about it* *2=Relatively concerned about it* *3=Not very concerned about it* *4=Completely indifferent to it*	*1.383*	*1.044*

children choose to marry			
Independent variable			
Gender	*1=Male 2=Femal 3=Undisclosed*	*1.440*	*0.499*
Age	*1=20-25 years old 2=25-30 years old 3=30-35 years old 4=35-40 years old 5=40-45 years old 6=45-50 years old 7=50-55 years old 8=55-60 years old*	*3.426*	*2.115*
Is there pension insurance?	*1=Yes 2=No*	*1.645*	*0.658*
Have child or not?	*1=Yes 2=No*	*1.977*	*0.512*
Type of household registration	*1=Non-agricultural household registration 2=Agricultural household registration 3=Undisclosed*	*1.359*	*0.904*
Single parent family or not?	*1=Yes 2=No*	*1.158*	*0.406*
Do parents have pension insurance	*1=Both parents have pension insurance 2=Father has a pension but mother doesn't 3=Mother has a pension but Father doesn't 4=Neither parent has a pension*	*1.359*	*0.997*

3.2 Analysis of empirical results

First of all, this study uses Pearson correlation analysis to study the relationship between quantitative data, whether there is a relationship, the closeness of the relationship, and so on. Then it analyzes the relationship between each Y and each x, and whether there is a significant relationship between Y and X; The correlation was positive or negative. The closeness of the relationship can also be explained by the size of the correlation coefficient. Finally, the analysis is summarized.

3.2.1 Impact analysis on whether to pay attention to whether the target's parents have old-age insurance

Table 3. Pearson result of factors on whether to pay attention to whether the target's parents have pension insurance

Whether to pay attention to whether the target's parents have pension insurance		
Gender	*Correlation Coefficient*	*-0.040*
	P value	*0.372*
Age	*Correlation Coefficient*	*-0.861*
	P value	*0.000*
Type of household registration	*Correlation Coefficient*	*0.304*
	P value	*0.000*
Single parent family or not?	*Correlation Coefficient*	*0.296*
	P value	*0.000*
Does the respondent have pension insurance	*Correlation Coefficient*	*-0.551*
	P value	*0.000*
Do both parents have pension insurance	*Correlation Coefficient*	*0.373*
	P value	*0.000*
**P<0.05 * * P<0.01*		

As shown in Table 3, the correlation coefficient between respondents' concern about their prospective partners' parents having pension insurance and their gender is -0.040, with a p-value of 0.372>0.05, indicating no correlation between the two. However, there is a significant negative correlation (-0.861) between this factor and respondents' age at the0.01 level, meaning that as age increases, respondents place more importance on whether their prospective partners' parents have pension insurance. The correlation coefficient between this factor and urban-rural household registration type is 0.304, indicating a significantly positive correlation at the 0.01 level, suggesting that urban household registration respondents are more concerned about this factor than rural household registration respondents. Similarly, the correlation coefficient between this factor and whether respondents come from single-parent families is 0.296, indicating a significantly positive correlation at the 0.01 level, meaning that non-single-parent family respondents place more importance on whether their prospective partners' parents have pension insurance than single-parent family respondents. Respondents with pension insurance have a significantly negative correlation (-0.551) with this factor at the0.01 level, indicating that they place more importance on whether their prospective partners' parents have pension insurance than those without pension insurance. Finally, the correlation coefficient between this factor and whether respondents' parents have pension insurance is 0.374, indicating a significantly positive correlation at the 0.01 level, meaning that respondents whose parents have pension insurance place more importance on whether their prospective partners' parents have pension insurance than those whose parents do not have pension insurance.

3.2.2 Analysis on "if you have children and he/she has reached the marriageable age, do you care about whether the parents of the target have pension insurance when your children choose to marry"

Table 4. Pearson estimates of "whether respondents pay attention to whether the parents of their children have pension insurance if they have children

If you have children and he/she has reached marriageable age, do you care about whether the parents have pension insurance when your children choose to marry		
Gender	*Correlation Coefficient*	*-0.104**
	P value	*0.018*
Age	*Correlation Coefficient*	*-0.396***
	P value	*0.000*
Type of household registration	*Correlation Coefficient*	*0.905***
	P value	*0.000*
Single parent family or not?	*Correlation Coefficient*	*0.887***
	P value	*0.000*
Does the respondent have pension insurance	*Correlation Coefficient*	*-0.360***
	P value	*0.000*
Do both parents have pension insurance	*Correlation Coefficient*	*0.753***
	P value	*0.000*
**P<0.05 * * P<0.01*		

As shown in Table 4, the correlation coefficient between respondents' concern about whether their prospective partners' parents have pension insurance when their children are of marriageable age and their gender is -0.104, indicating a significantly negative correlation at the 0.05 level, suggesting that women place more importance on this factor than men. The correlation coefficient between this factor and respondents' age is -0.396, indicating a significant negative correlation at the0.01 level, meaning that as age increases, respondents place less importance on whether their children's prospective partners' parents have pension insurance. The correlation coefficient between this factor and urban-rural household registration type is 0.923, indicating a significantly positive correlation at the0.01 level, meaning that urban household registration respondents place more importance on this factor than rural household registration respondents. There is a significant positive correlation between this factor and whether respondents come from single-parent families, indicating that single-parent families place more importance on whether their children's prospective partners' parents have pension insurance than non-single-parent families. Respondents without pension insurance have a significantly negative correlation (-0.360) with this factor at the 0.01 level, indicating that they place more importance on whether their children's prospective partners' parents have pension insurance than those with pension insurance. Finally, the correlation coefficient between this factor and whether respondents' parents have pension insurance is 0.753, indicating a significantly positive correlation at the 0.01 level, meaning that respondents whose parents do not have pension insurance place more importance on this factor than those whose parents have pension insurance.

4 Conclusions and discussion

The study found that children whose parents had pension insurance were more popular in the dating market. As far as age is concerned, with age, the importance of pension insurance for themselves and the parents of the next generation of marriage partner has gradually increased. As far as gender is concerned, although as far as the respondents themselves are concerned, whether they attach importance to the pension insurance of the parents of marriage partner does not show significant differences between different genders, women pay more attention to whether the parents of marriage partner have pension insurance when choosing marriage partner for the next generation. As far as family types are concerned, on the issue of "whether to pay more attention to whether the parents of marriage partner have old-age insurance", the study found that single parent family respondents pay more attention to whether the parents of marriage partner have old-age insurance than non-single parent family respondents, The study found that respondents whose parents had pension insurance paid more attention to whether their parents had pension insurance than those whose parents had no pension insurance.

Generally speaking, this research not only fills the gap in existing studies on the influence of parents' participation in pension insurance on their children's marriage market selection but also provides scientific evidence to promote the marriage rate and fertility rate in China.

References

1. National Bureau of Statistics(2022) Number of marriages in mainland China in the past decade. Retrievd from http://www.stats.gov.cn/
2. Douglas H. Wrenn, Junjian Yi, Bo Zhang, *House prices and marriage entry in China, Pages* 118-130, ISSN 0166-0462, https://doi.org/10.1016/j.regsciurbeco.2018.12.001. (Regional Science and Urban Economics, Volume 74, 2019)
3. Pei Jinsong, Jiao Meng. A study on the substitution effect of social insurance on children's old-age care for the left-behind elderly in rural areas: Based on micro-data of CHARLS [J]. Journal of Hebei University (Philosophy and Social Sciences Edition),2020,45(05):125-135.
4. Huang, Wei, and Chuanchuan Zhang. *The Power of Social Pensions: Evidence from China's New Rural Pension Scheme.* (American Economic Journal: Applied Economics, 13 (2): 179-205.DOI: 10.1257/app.20170789 2021)
5. Zou J, Fu X, Gong C, Ye LY. *Household Registration, Old-Age Insurance, and Consumption: Evidence from China.* (Systems. 10(5):179. https://doi.org/10.3390/systems10050179, 2022)
6. Luciano Fanti, Luca Gori (2012) Economic growth and stability with public pay-as-you-go pensions and private intra-family old-age insurance, Research in Economics, Volume 66, Issue 3, Pages 219-229, ISSN 1090-9443, https://doi.org/10.1016/j.rie.2012.04.003.
7. Wu Yufeng, Hu Jingbo, Nie Jianliang. *Research on the impact mechanism of basic endowment insurance for urban and rural residents on the health performance of rural elderly people* (Social security research, No.79 (06): 10-22, 2021)
8. Ou Yangxin, Zhang Yingliang. *Rural social pension insurance, income gap and welfare inequality* [J]. (Agricultural technology and economy, no.334 (02): 78-93. Doi: 10.13246/j.cnki. Jae. 2023. 02.003. 2023)
9. Cao Yuan, Fang Yu, Xiao Gandan et al. *Pension insurance system, family fertility decision-making and social welfare* [J]. (Southern Finance, no.542 (10): 72-86. 2021)

The Impact of Gender and Inequality Perception on Subjective Well-being——An Empirical Study Based on World Values Survey

Yu Hu *

Capital University of Economics and Business, Beijing, China

Abstract: Based on micro-data from the World Values Survey, this paper investigates the effects of gender and perceptions of gender inequality on subjective well-being using ordinal logistic regression, and tests the robustness of the findings using OLS regression and other methods. It is found that men have higher perceptions of gender inequality than women, and China has the highest perception of gender inequality compared to Japan, Germany and the USA. The regression results suggest that even though perceptions of gender inequality have a positive effect on subjective well-being, Chinese men's subjective well-being is still significantly lower than women, and this difference may be due to the social pressures and expectations faced by men. In a society with a low perception of gender equality, both men and women are victims.

1 Introduction

The history of feminist is the history of women's pursuit of "Legal equality, citizen equality and educational equality". According to the Fourth Survey on the Social Status of Chinese Women, the proportion of women in higher education has surpassed that of men, and they are becoming more and more involved in economic and social development. At the same time, however, a significant proportion of respondents still hold unequal gender perceptions. Nearly half of the men did not oppose or agree with the view that "men should be socially oriented and women should be family-oriented", and more than 30% of the respondents did not oppose the view that "wives should be submissive to their husbands at home[1]". Compared to the diverse development of feminism in Western countries, feminism in China is still in its formative stage, and there is still a long way to go.

It has been found that unequal gender perceptions have an impact on subjective well-being. Diener (2000) found that there is a gender difference between the degree of sexism and well-being, with lower levels of sexism associated with smaller gender differences in well-being at the national level[2]. Findings by Perry (2013) suggest that sexism increases the risk of low well-being and that gender perceptions inequality has a significant negative impact on African American women's well-being[3]. In existing domestic research, based on four surveys conducted by the World Values Study in China in 1990, 1995, 2001 and 2007, Li Lei (2017) et al. found that gender inequality perceptions negatively affected men's well-being and that social pressure from gender discrimination reduced men's subjective well-being[4]. Based on data from World Values 1981 to 2014, the findings of Huang Shaoan and Guo Junyan (2019) showed that there was a significant negative effect of gender inequality perceptions on well-being, The more unequal an individual's gender concept is, the lower his subjective well-being will be. Also the findings of this study support the study by Li Lei et al in 2017, which concluded that there are gender differences in the effect of gender inequality perceptions on well-being, with men who hold gender inequality perceptions having lower levels of well-being[5].

Building on existing research, this study uses the latest seventh wave (2017-2022) of the World Values Survey to select survey data from four different countries in China (2018), Germany (2018), Japan (2019) and the United States (2017) to describe the differences in perceptions of gender inequality across the four countries. Furthermore, this study analyzes the impact of gender inequality on subjective well-being in Chinese society, as well as the extent of this impact varies across subgroups such as education, age and gender, through ordinal logical regression.

2 Literature review

2.1 Measurement and influencing factors of subjective well-being

In recent years, as well-being has increasingly become the focus of social psychology research, the measurement of well-being has made great strides. According to the differences in measurement indicators and definition methods, well-being can be roughly divided into objective well-being and subjective well-being[6]. Objective well-being measures people's well-being based on objective conditions, starting from the external environment, health status and socio-economic factors

*Corresponding author: 15010484285@163.com

that affect the quality of life. Subjective well-being, on the other hand, places more emphasis on the importance of psychological factors and measures the level of well-being through respondents' descriptions and personal experiences[7]. Based on the study of the mechanisms underlying subjective well-being in a broad sense, two major research paradigms have emerged: subjective well-being and psychological well-being. Subjective well-being as a research paradigm focuses on people's satisfaction through material life, while psychological well-being focuses more on individual self-development and self-worth, emphasising the definition of well-being in terms of psychological satisfaction[8].

This study focuses on subjective well-being, which is a measure of well-being based on the subjective experience of the individual, and is the respondent's overall assessment of the quality of their life. While it is true that the subjective assessment of one's well-being cannot be separated from objective indicators such as economic level, education level and health status, this study focuses more on the respondents' own perceptions and evaluations of their well-being from a subjective perspective.

Regarding the factors influencing subjective well-being, according to previous studies, they can be divided into two parts: objective conditions and subjective psychology. Most of the objective factors affecting subjective well-being focus on socio-economic status, employment experience, human capital such as education level, and external environmental factors. For example, Peng Kaili and Yang Chen (2021) found that an increase in objective income level significantly contributed to farmers' subjective well-being[9] ; Lei Xiankai (2022) et al. showed that non-farm employment experience had a positive impact on farmers' subjective well-being[10]; Liu Xiaoxi (2022) et al. pointed out that community aesthetics, natural environment and fitness environment had a positive impact on residents' subjective well-being[11]. Compared to the influence of objective conditions on subjective well-being, there are fewer studies that use psychological perceptions as independent variables, and the studies that have been conducted include the influence of self-efficacy, perceptions of fairness, and individual traits on subjective well-being. For example, Peng Liping (2022) pointed out that self-efficacy and identity play a mediating role between volunteering participation and subjective well-being[12]; Qu Sha and Liu Xiaobin (2022) confirmed that perceptions of fairness have a positive effect on subjective well-being, and there are gender differences in this effect, with women's perceptions of fairness having a greater impact on subjective well-being[13]. In addition, some personal traits also have an impact on subjective well-being[14].

In summary, studies on the factors influencing subjective well-being have been characterised by an emphasis on objective conditions rather than psychological factors, and few scholars have explored the relationship between individuals' gender perceptions and subjective well-being, which may be related to the difficulty of measuring individuals' subjective factors. Based on the latest world values data from 2017 to 2022, this study analyses the impact of individuals' gender perceptions on their subjective well-being from a micro-psychological perspective. That is, do perceptions of gender inequality influence individuals' evaluations of their well-being? Under which factors does this influence differ?

2.2 The impact of gender inequality perceptions

Feminism has been brought to China by successive feminist movements in western countries. Scholars are increasingly concerned about the social constructiveness of gender inequality between men and women has stimulated research in the field of gender equality, in which the impact of gender inequality has received widespread attention.

In terms of educational attainment, Sadker (1995) have shown that teachers' classroom processes discriminate between boys and girls[15]; Li Kehan (2021) et al. have further shown that an increase in the socio-economic status of the father's generation reduces gender differences in the educational attainment of the offspring[16]. In the context of employment, there has been a significant amount of scholarly discussion on gender segregation in the labour market as a result of gender inequality perceptions. Tong Mei (2014) highlights that occupational gender segregation is an important indicator for measuring gender inequality[17]. Cheng Cheng (2015) et al. explain the gender gap in labour market earnings from the perspective of social capital, pointing out that female workers have lower returns to social capital than men[18]. In terms of political participation, Pan Zechuan and Xie Yan (2018) pointed out that there are gender differences in political participation awareness. Compared with men, women have a lower awareness of political participation[19].

As can be seen from the above, much of the research has focused on the impact of perceptions of gender inequality on educational attainment, employment and political participation, but less on its impact on individuals' experiences and perceptions of life, insufficient research in this field provides space for this study. In fact, as the two key variables in this study, perceptions of gender inequality and well-being, are both subjective psychological factors, making the issue of measuring and defining the variables particularly important. Gender equality refers to men and women having equal opportunities and being able to participate equally in social life, while gender inequality refers to the unequal opportunities and rights of men and women. Individuals who hold the concept of gender inequality tend to believe that men perform better in the political and economic spheres, as a manifestation of male superiority. Therefore, this study measures the extent to which individuals have a sense of male superiority in six areas: politics, education, leadership, right to work, income, value, as a reflection of their perception of gender inequality.

https://doi.org/10.1051/shsconf/202317801018

3 Theoretical model and hypothesis

Gender and subjective well-being

Are men happier, or are women happier? Research on gender differences in subjective well-being has yielded mixed findings.

Most overseas studies (Costa et al. 2001; Hansson et al. 2005; Smith and Reise 1998) point out that women have more negative emotions and lower well-being[20] [21] [22]. Similarly, Baltes et al. (1999) found that women tend to rate their subjective health lower than men, which is strongly associated with well-being[23]. Stevenson and Wolfers (2008) showed that objective measures of quality of life for American women have improved over the past 35 years, but their subjective well-being has tended to decline compared to men[24]. In contrast, studies have also shown that women have more positive emotions and higher life satisfaction[25]. Diener (1999) emphasises that for women, positive emotions are as easy to express as negative emotions and are not less important[26].

However, it is important to note that most studies on gender differences in subjective well-being have come from the US and Western European countries, with fewer relevant Chinese studies, and the findings do not match those of European and American studies. A community study by Liu Bin (2012) et al. showed that there were gender differences in the contagion of well-being, with well-being being more likely to be transmitted to women[27]. Li Lei (2017) et al. showed that women's well-being was significantly higher than men's in China, and that men's lower well-being may be due to higher social expectations[4]. A study by Huang Shaoan and Guo Junyan (2019) further illustrates that the perception of "male superiority" leads to lower well-being ratings for men[5].

In summary, regarding the effect of gender on subjective well-being, overseas studies are biased towards the conclusion that women's sense of well-being is low, while existing domestic studies are more likely to suggest that gender pressure on men leads to lower well-being compared to women. Given the inconsistency in the findings of the current academic research, this paper proposes two opposing hypotheses.

Hypothesis 1a. Men have significantly higher subjective well-being than women.

Hypothesis 1b. Men's subjective well-being is significantly lower than that of women.

3.1 Perceptions of gender inequality and subjective well-being

Most of the studies on gender inequality and subjective well-being have focused on factual studies rather than conceptual studies. In other words, most of the studies tend to analyze the impact of gender inequality on individual well-being through objective indicators, but less on the relationship between the perception of gender inequality and well-being. Regarding the former, the academic community has formed a more unified conclusion that the higher the level of gender inequality in a country, the lower the subjective well-being of individuals will be[28]. As for the latter, there is no consensus on the findings. Qian Ge (2017) concluded that, controlling for variables such as country, gender and income, the stronger an individual's perception of gender equality, the lower his or her subjective well-being will be[29]. In contrast, domestic scholars such as Huang Shaoan (2019) showed that the higher an individual's perception of gender inequality, the lower their probability of achieving high-level well-being[5].

Based on this, this paper proposes two mutually exclusive hypotheses to test the impact of individual perceptions of gender inequality and subjective well-being.

Hypothesis 2a. The higher the individual's perception of gender inequality, the lower his or her subjective well-being will be.

Hypothesis 2b. The higher the individual's perception of gender inequality, the higher their subjective well-being will be.

4 Data and variables

4.1 Data

This paper uses the World Values Survey (WVS) database, which has so far conducted seven waves of surveys covering changes in people's values in over one hundred countries from 1981 to 2022[30]. The main part of this paper uses data from the 2018 World Values Survey for China to analyse the relationship between gender differences in subjective well-being, perceptions of gender inequality and subjective well-being in Chinese society by constructing an ordered logistic regression model using stata15.1. In addition, data from four countries, China (2018), the United States (2017), Germany (2018) and Japan (2017), were selected for a multi-country comparison of gender inequality perceptions, containing a total of 7411 valid samples to reflect the degree of equality and differences in gender perceptions in China.

4.2 Variables

4.2.1 Dependent variable

The dependent variable in this paper is subjective well-being, which is measured by a subjective question, "Taking all things together, would you say you are:", which gives four options, "very unhappy" , "unhappy ", "happy" , "very happy" assigned a value of 1-4.

4.2.2 Independent variables

(1) Gender. Gender is a dichotomous variable, taking a value of 1 for females and 0 for males.

(2) Perception of gender inequality. This variable is measured through six dimensions: politics, education, leadership, right to work, income and values. Issues include political priority (Men make better political

leaders than women do), education priority (A university education is more important for a boy than for a girl), management priority (Men make better business executives than women do), value priority (Being a housewife is just as fulfilling as working for pay), work priority (Men should have more rights to a job than women) and income priority (If a woman earns more money than her husband, it's almost certain to cause problems) . Each question gave four options, "strongly disagree", "disagree", "agree", "strongly agree " assigned a value of 1-4.

4.2.3 Control variables

Based on the factors influencing well-being in previous studies, age, years of education, income level, health status, number of children, household savings status in the past year, and subjective social status were selected as control variables in this study.

5 Data analysis and results

5.1 Descriptive statistics

5.1.1 Descriptive statistics of variables

Table 1 Description of the variables used in the study.

Interval variables	Mean	SD	Max	Min	n
Gender	0.55	0.5	1	0	2909
Gender inequality perceptions_standard	0	1	4.07	-2.74	2909
Gender inequality perceptions	14.05	2.94	26	6	2909
Years of education	10.22	4.33	22	0	2909
Number of children	1.40	0.91	7	0	2881

Nominal variables	Frequency	Percentage	Cumulative percentage	n
Age				2909
16-24	307	10.55	10.55	
25-34	507	17.43	27.98	
35-44	598	20.56	48.54	
45-54	676	23.24	71.78	
55-64	513	17.63	89.41	
≥65	308	10.59	100	
Income level				2909
Low	1,078	37.06	37.06	
Medium	1,752	60.23	97.28	
High	79	2.72	100	
Health status				2909
Very good	18	0.62	0.62	
Good	196	6.74	7.36	
Fair	925	31.80	39.15	
Poor	1,007	34.62	73.77	
Very poor	763	26.23	100	

Table 1 (continued) Description of the variables used in the study.

Nominal variables	Frequency	Percentage	Cumulative percentage	n
Household savings (last year)				2909
Save money	977	33.59	33.59	
Just get by	1,303	44.79	78.38	
Spent some savings	404	13.89	92.27	
Spent savings and borrowed money	225	7.73	100	
Subjective social status				2909
Upper class	5	0.17	0.17	
Upper middle class	71	2.44	2.61	
Lower middle class	1,241	42.66	45.27	

Working class	1,043	35.85	81.13
Lower class	549	18.87	100

This paper mainly uses data from the WVS in China in 2018. After excluding the missing values of the main variables, there are 2881 valid samples for the variable "number of children" and 2909 valid samples for the other variables. In the independent variable, women accounted for 55% of the sample, slightly higher than men. The maximum value of gender inequality after standardisation was 4.07 and the minimum value was -2.74. The standard deviation of gender inequality before treatment was close to 3, indicating that people's perceptions of gender inequality varied significantly. In addition, the control variables selected for the study included years of education, number of children, age, income level, health status, household savings status in the past year and subjective social status, with the specific categories and statistics shown in Table 1.

5.1.2 Gender differences in perceptions of gender inequality

Table 2 Statistics and t-test of gender inequality between men and women (N = 2909).

Gender	n	Mean	SD	Min	Max	P-value
Male	1319	0.077	1.001	-2.738	3.728	0.0002
Female	1590	-0.063	0.995	-2.738	4.068	(3.7687)

This paper uses standardised gender inequality perceptions to analyse the differences between men and women. In Table 2, the mean value of perceptions of gender inequality for women was -0.063, which was lower than that of men at 0.077, and according to the t-test, there was a significant difference between men and women in terms of perceptions of gender inequality, meaning that women had lower levels of gender inequality and were more inclined to support equal gender perceptions. In terms of maximum values, the maximum value of gender inequality perceptions is higher among women than men, indicating that there are still those who hold the idea of "male superiority over women" among women.

5.1.3 Country differences in perceptions of gender inequality

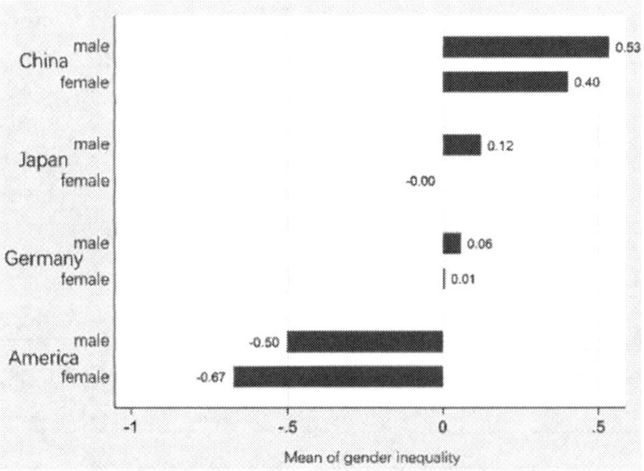

Fig.1 Bar chart of country differences in perceptions of gender inequality

Drawing a bar chart based on the standardised perceptions of gender inequality, it was found that among the four countries: China, Japan, Germany and the United States, China had the highest perceptions of gender inequality, followed by Japan and Germany, and the United States had the lowest perceptions of gender inequality. The finding may be related to the year in which the seventh wave of WVS was surveyed in different countries (2018 in China, 2019 in Japan). This differs from the findings of Huang Shaoan and Guo Junyan (2019) who found that gender inequality perceptions were highest in Japan and second highest in China. Although there are significant differences in perceptions of gender inequality across countries, in all four countries, men have higher perceptions of gender inequality than women, suggesting that men hold more gender inequality than women and are bound by the idea of "male superiority" shown in Fig.1.

 https://doi.org/10.1051/shsconf/202317801018

5.2 Impact of gender and inequality perceptions on subjective well-being

Table 3 Ordered logistic regression model of gender, inequality perceptions and subjective well-being.

Variables	Model 1	Model 2	Model 3	Model 4
Gender (F=1)	0.263***	0.200**	0.277***	0.280***
	(3.471)	(2.589)	(3.466)	(3.463)
Gender inequality perceptions	0.029*	0.114***	0.113***	0.112***
	(2.224)	(3.524)	(3.338)	(3.309)
Years of education		0.103*	0.113*	0.099*
		(2.424)	(2.559)	(2.218)
Number of children		0.09	0.06	0.059
		(1.951)	(1.130)	(1.119)
Years of education_Gender inequality perceptions		-0.010**	-0.010***	-0.010**
		(-3.228)	(-3.328)	(-3.230)
Age			-0.062	-0.057
			(-0.470)	(-0.436)
Age squared			0.042*	0.041*
			(2.332)	(2.275)
Health status			0.899***	0.855***
			(18.827)	(17.713)
Income level			0.284***	-0.04
			(3.712)	(-0.445)
Subjective social status				0.357***
				(5.855)
Household savings (last year)				0.145**
				(3.025)
N	2909	2881	2881	2881
R^2	0.003	0.009	0.092	0.100

Note: * $p < 0.05$, ** $p < 0.01$, *** $p < 0.001$, t-values in parentheses.

In Table 3, regarding the effect of gender and inequality perceptions on well-being, this paper uses a nested model from model 1 to model 4 by gradually adding control variables, considering the robustness of the regression results. Model 1 contains only the dependent variable well-being and two independent variables, gender and inequality perceptions. Models 2 to 4 include control variables step by step to test the robustness of the regression results, and model 4 is a full model with all control variables.

The effect of gender on subjective well-being was consistently positive and significant in models 1 to 4, indicating that women had higher subjective well-being than men. This result further supports the findings of Li Lei et al. (2017) and Huang Shaoan, Guo Junyan (2019), confirming hypothesis 1b that men have significantly lower subjective well-being than women. Regarding the effect of gender inequality perceptions on subjective well-being, gender inequality perceptions in Model 1 had a positive effect on subjective well-being at the 5% level, and this effect became more significant with the inclusion of control variables in the model. This result suggests that higher perceptions of gender inequality are associated with higher subjective well-being, that is the stronger the individual's perceptions of gender equality, the less likely they are to have higher levels of well-being, which means that hypothesis 2b is confirmed.

In addition, the results for the other control variables were as follows: years of education, health status, subjective social status and family savings in the past year have significant positive effects on subjective well-being. What's more, age showed a U-shaped relationship with well-being, which means that as age increases, well-being decreases first and then increases. These findings are intuitive and consistent with previous research, so this paper will not focus on them. In addition, the interaction of education years and gender inequality perceptions is consistently and significantly negative, although the coefficient is small, still suggesting that compared with the uneducated respondents, longer years of education inhibit the positive effect of gender inequality perceptions on subjective well-being.

Robustness test of regression results

In research on well-being, life satisfaction is often used as a reference indicator instead of well-being[31].

In the World Values Survey, life satisfaction is measured by asking respondents "All things considered, how satisfied are you with your life as a whole these days?", with answers ranging from 1 to 10 indicating the least satisfied to the most satisfied level. Referring to the robustness test of Li Lei (2017) et al, in this paper, life satisfaction was used as a proxy for well-being by first conducting OLS regressions on life satisfaction, while further dividing life satisfaction into four categories and conducting ordered logistic regressions to compare the differences between the OLS and ordered logistic regression results with life satisfaction as the dependent variable and the regression results with well-being as the dependent variable (the results are not in the text presented and can be obtained by performing the do file). The results show that the former is generally unchanged from the latter, both in terms of the positive and negative regression coefficients and the degree of significance, and this result helps to confirm the robustness of the paper's findings.

6 Conclusion and Discussion

This paper uses data from the Wave 7 (2017-2022) of World Values Survey to compare the national differences and gender differences of gender inequality through stata15.1 analysis, and uses the survey sample of WVS in China in 2018 to conduct an ordinal logical regression to analyze the impact of gender and inequality perceptions on subjective well-being. The findings of the study are as follows.

Firstly, there are differences in perceptions of gender inequality by gender and country. In terms of gender differences in perceptions of inequality, men's perceptions of gender inequality are higher than women's in different countries, and this result is highly significant in China. Studies have shown that higher levels of gender inequality not only undermine women's rights, but also force men to face higher social expectations and pressures, thus reducing their subjective well-being[5]. At the same time, some women still hold gender inequalities and suffer from the idea that men are inferior to women. In terms of country differences in perceptions of inequality, China has the highest perceptions of gender inequality, followed by Japan and Germany, and the United States has the lowest perceptions of gender inequality. The results show that Chinese society still has a long way to go in fostering gender equality. However, perceptions of gender inequality do not change spontaneously. In order to cultivate and improve the concept of gender equality, it is not only necessary to awaken women's awareness of equality, but also to improve the concept of gender equality in society as a whole, especially among men, through education and cultural environment. This is because an equal gender perspective will not only benefit women, but also men.

Second, in Chinese society, men's subjective well-being is significantly lower than that of women. This result is consistent with related studies analysing the six waves (1981-2014) of world values data, suggesting that this finding has a high degree of stability. Why are men, who are dominant in all aspects, instead more unhappy? The higher the individual's perceptions of gender inequality, the higher the subjective well-being will be, but why are more men who hold more gender inequality less likely to be happy? It is worth considering this paradox. Some studies have shown that men's lower well-being ratings are due to higher social pressure and workload, especially for men who hold gender inequalities and are bound by "masculinity", where discrimination against women and social expectations prevent men from experiencing well-being[4]. However, this study attempted to include the interaction of gender and inequality perceptions in the regression model, although the coefficient for men (β=-0.029) suggests that men with gender inequality are less likely to experience higher levels of well-being, the results were not significant (p=0.29). Therefore, the above results are not included in the text. Although gender differences in the effect of gender inequality on well-being could not be demonstrated, the findings of previous studies and the analysis in this paper have shown that both men and women are victims in a society with low gender equality perceptions.

Third, individuals' perceptions of gender inequality have a significant positive effect on subjective well-being. Conversely, in a society with a strong perception of gender inequality, the stronger an individual's perception of gender equality is, the lower his or her subjective well-being will be. This finding is not difficult to understand. In fact, previous research has shown that the higher the level of gender inequality in a country, the lower the subjective well-being of individuals will be. In terms of perceptions, the more individuals in a society hold gender equality perceptions that are ahead of their social environment, the less likely they are to be happy, while the more individuals hold gender inequality perceptions, the more likely they are to have higher levels of well-being.

It is worth noting that the study found that the positive effect of gender inequality perceptions on subjective well-being was weaker the longer the years of education. This result suggests that years of education play a moderating effect in the effect of gender inequality perceptions on subjective well-being. This paper suggests that the reason for this may be that the longer an individual has received education, the more likely his new ideas and thoughts are to conflict with the original concept of gender inequality, thus weakening their positive impact on subjective well-being. Due to the focus and data limitations of the study, this finding could not be explored further in this paper, and subsequent research could consider this aspect for more in-depth study.

Due to the characteristics of the data and the limitations of the author's ability, this study has many shortcomings. First, the measurement of the dependent variables in the study is more than simple, so the follow-up study can consider refining the measurement indicators of subjective well-being. Second, the study only uses cross-sectional data from the World Values Survey in China in 2018, which cannot be compared on a time series, and the follow-up study can be further expanded in terms of the breadth of time and the width of countries. Thirdly, due to the limitations of the data, the study uses more nominal variables than interval variables, which may affect the regression results to a certain extent, and the robustness of the findings needs to be further tested.

Reference

1. Main data of the fourth China Women's Social Status Survey[J]. Women's Studies Series, 2022(01):1+129.
2. E. Diener, "Subjective Well-Being: The Science of well-being and A Proposal for A National Index", American Psychologist, Vol. 55, No. 1, 2000, 34-43.
3. Brea L. Perry,Kathi L. H. Harp,Carrie B. Oser. Racial and Gender Discrimination in the Stress Process: Implications for African American Women's Health and Well-Being[J]. Sociological Perspectives,2013,56(1).
4. Li Lei,Liu Pengcheng,Sun Tran. Men and women,who is happier[J]. Statistical Research,2017,34(07):82-93.
5. Huang Shaoan,Guo Junyan. The impact of gender inequality perceptions on well-being--an empirical analysis based on data from the World Values Survey[J]. Social Science Front,2019(11):35-42.
6. Pedro Conceição, Romina Bandero, "A review of the literature on subjective well-being research", translated by Lu Yanhua, Foreign Theoretical Dynamics,No.7, 2013.
7. Geng Xiaowei,Wang Huiping,Zhang Feng. Research on the measurement of subjective well-being[J]. Psychology Exploration,2013,33(03):266-270.
8. Xu Xiaobo,Sun Chao,Wang Fengyan. Spiritual well-being:concepts, measurement, related variables, and interventions[J]. Advances in Psychological Science,2017,25(02):275-289.
9. Peng Kaili,Yang Chen. The mechanism of socio-economic status on farmers' subjective well-being--an empirical analysis based on mediating variables of cultural consumption behavior[J]. Journal of Hunan Agricultural University (Social Science Edition),2021,22(06):37-45.
10. Lei Xiankai,Luo Mingzhong,Ding Bachao. The influence of non-farm employment experience on farmers' subjective well-being and mechanism analysis[J]. Journal of Hunan Agricultural University (Social Science Edition),2022,23(02):36-44.
11. Liu Xiaoxi, Dai Junming, Shen Yifeng, Fu Xuelian, Li Xiaomei, Yu Yang, Zhao Qiowen, Gao Junling, Fu Hua. The influence of community environmental factors on residents' subjective well-being: an example from Pudong New Area, Shanghai[J]. Environmental and Occupational Medicine,2022,39(07):769-774.
12. Peng Liping. The effect of volunteering participation on college students' subjective well-being: the chain mediating role of self-efficacy and self-identity[J]. Chinese Journal of Clinical Psychology,2022,30(05):1126-1129.
13. Qu Sha,Liu Xiaobin. Fairness perception, social capital and residents' subjective well-being - an empirical test based on CGSS2017 data[J]. Investment Research,2022,41(05):25-36.
14. Yang Huifang,Guo Yongyu. The relationship between personal striving, personality traits and subjective well-being of college students[J]. Psychological Development and Education,2008(03):58-64.
15. Mara Sadker and David Sadker, Failing at Fairness: How Our Schools Cheat Girls, New York: Touchstone, 1994, 1-14.
16. Li Kehan,Lu Miaomiao,Sun Tao. A study on the influence of paternal socioeconomic status on offspring's educational attainment and gender differences[J]. Dongyue Series,2021,42(05):47-58.
17. Tong Mei. A study on occupational gender segregation[J]. Journal of Lanzhou University (Social Science Edition),2014,42(03):135-140.
18. Cheng Cheng,Wang Yixuan,Bian Yanjie. Gender earnings differentials in the Chinese labor market:A social capital explanation[J]. Population Studies,2015,39(02):3-16.
19. Pan Zequan,Xie Yan. A study on Chinese women's awareness of political participation from the perspective of gender consciousness - empirical findings based on the Hunan sample in CGSS2010 data[J]. Journal of Xiangtan University (Philosophy and Social Science Edition),2018,42(03):115-119.
20. Costa Paul T,Terracciano Antonio,McCrae Robert R. Gender differences in personality traits across cultures: robust and surprising findings.[J]. Journal of personality and social psychology,2001,81(2).
21. Anna Hansson,Pernilla Hillerås,Yvonne Forsell. Well-Being in an Adult Swedish Population[J]. Social Indicators Research,2005,74(2).
22. Smith L L,Reise S P. Gender differences on negative affectivity: an IRT study of differential item functioning on the Multidimensional Personality Questionnaire Stress Reaction Scale.[J]. Journal of personality and social psychology,1998,75(5).
23. Baltes, M. M., et al. Men and women in the Berlin aging study. In P. B. Baltes & K. U. Mayer (Eds.), The Berlin aging study: Aging from 70 to 100, 1999, 259 -281.
24. Betsey Stevenson,Justin Wolfers. well-being Inequality in the United States[J]. The Journal of Legal Studies,2008,37(S2).
25. Fujita F,Diener E,Sandvik E. Gender differences in negative affect and well-being: the case for emotional intensity.[J]. Journal of personality and social psychology,1991,61(3).
26. Diener, Ed,Suh, Eunkook M.,Lucas, Richard

https://doi.org/10.1051/shsconf/202317801018

E.,Smith, Heidi L.. Subjective well-being: Three decades of progress.[J]. Psychological Bulletin,1999,125(2).

27. Liu B,Li L,Mo Jiao. Is well-being contagious[J]. World Economy,2012,35(06):132-152+155-160+153-154.
28. Clemens Tesch-Römer,Andreas Motel-Klingebiel,Martin J. Tomasik. Gender Differences in Subjective Well-Being: Comparing Societies with Respect to Gender Equality[J]. Social Indicators Research,2008,85(2).
29. Qian Ge. The effect of gender equality on well-being: Statistical modeling and analysis.[J]. Health care for women international,2017,38(2).
30. World Values Survey website: https://www. worldvaluessurvey. org/WVSContents. jsp
31. Ronald Inglehart,Roberto Foa,Christopher Peterson,Christian Welzel. Development, Freedom, and Rising well-being: A Global Perspective (1981 -2007)[J]. Perspectives on Psychological Science,2008,3(4).

On the Copyright of Content Generated by Artificial Intelligence

Tang Han*

Law, College of Humanities and Development Studies, China Agricultural University, Beijing 100083, China

Abstract. While generative artificial intelligence participates in human creative activities on a large scale, it cannot be a natural human subject or an anthropomorphic subject because of its lack of rationality. However, copyright law encourages the creation of works, and under certain conditions, artificial intelligence generated content is conducive to the realization of the legislative purpose of copyright law and satisfies the objective criteria of originality of works, and therefore, such content has copyrightability. Under this premise, the attribution of artificial intelligence generated content copyright needs to be discussed. At present, there are designer's, user's and owner's views on its attribution. According to the current copyright law, in principle, the copyright of artificial intelligence generated content belongs to the user, but in special cases, it belongs to legal persons, employers, trustees and other subjects, unless otherwise agreed by the parties.

1 Introduction

With the rise of the fourth industrial revolution, big data, cloud computing, artificial intelligence and other technologies are becoming increasingly mature and deeply involved in human activities. [1] Among them, artificial intelligence is vigorously developed and gradually deeply involved in human creative activities. For example, as early as 2015, the timeliness and accuracy of news articles generated by artificial intelligence, such as "Xiaonan" of the Southern Metropolis Daily and "Zhang Xiaoming" of Today's Headlines, are completely comparable to those of professional news industry practitioners. In 2017, "Sunshine Lost the Glass Window," a collection of poems written by Microsoft's artificial intelligence "Xiaobing," was published, which once again aroused widespread concern and discussion in all walks of life. Now, a new generation of artificial intelligence products, represented by Chat Generative Pre-trained Transformer (hereinafter referred to as ChatGPT), has appeared, which automatically generates high-quality text by "learning" a large amount of linguistic data, so that the machine is able to understand the language like a human being. [2] The Copyright Law of the People's Republic of China (hereinafter referred to as the Copyright Law) is a law that encourages the creation and dissemination of works, and the wide application of artificial intelligence has generated a large amount of content. The copyrightability and other issues of such content should be responded by Copyright Law.

2 Clarification of the Connotation of Artificial Intelligence

As early as the 1950s, a definition of artificial intelligence had already been proposed. Marvin Minsky, the "father of artificial intelligence", defined artificial intelligence as "the science of making machines do things that humans need intelligence to do ". [3] With the development of artificial intelligence technology, experts in the theoretical and practical circles at home and abroad have discussed the definition of artificial intelligence. Wu Handong, a famous and renowned scholar of intellectual property law in China, believes that artificial intelligence is based on data storage, which, with the help of machine algorithms, aims to simulate certain thinking processes and intelligent behaviors of human beings through computers; [4]The European Commission's "European Report on Artificial Intelligence" issued by the European Commission defines artificial intelligence as "systems that possess intelligent behaviors, analyze the surrounding environment and automate it to a certain degree in order to achieve a specific purpose." In summary, this paper argues that artificial intelligence refers to artificial systems that mimic human intelligence and interact with their surroundings to automate specific goals that would normally require the human mind, with no specific purpose of their own. Generative artificial intelligence technology is a form of artificial intelligence technology that refers to models and related technologies that have the ability to generate content such as text, images, audio, and video.①

① See Article 22, paragraph 1, item 1 of the Interim Measures for the Administration of Generative Artificial Intelligence

* Corresponding author: thang7107@163.com

3 Artificial Intelligence Cannot Be Granted Legal Subject Qualification

On February 16, 2017, the European Union legislative proposal stated that "in the long term, complex autonomous robots could have the status of e-persons, thereby acquiring a special legal status and being compensated for the damage they cause. E-personhood may apply in cases where the robot makes independent autonomous decisions or interacts with third parties in an independent manner." Subject qualification, or personality, is the qualification granted by law to a person who possesses certain conditions to become a legal "person" and thus a bearer of rights, obligations and liabilities in a legal relationship. [5] The above legislative proposal of the European Union has triggered the discussion of the international community on whether artificial intelligence should be given the qualification of legal subject. Based on the philosophical perspective and the perspective of private law theory, this paper argues that artificial intelligence cannot be given the qualification of legal subject.

3.1 Philosophical perspective on the inability of artificial intelligence to become a natural person

At present, the definition of the legal subject of natural persons of various countries is based on the profound understanding of the human nature by the German classical philosophy, among which the Kantian philosophy is representative. [6] In Kantian philosophy, rationality, self-consciousness and freedom take an important place, and rationality is the core of it. Kantian philosophy believes that human beings have two indispensable attributes. On the one hand, as an animal with flesh and blood and desires, human beings belong to nature. They are subject to the laws of nature, and in a passive state of domination. On the other hand, as creatures with advanced thinking skills, humans are rational beings. They are capable of thinking to understand how the world works and applying knowledge and technology to transform the environment. This enables human beings to transcend the constraints of the laws of nature, to shape their own development and that of society with reason as their guide, to create norms such as law and morality, and to self-regulate and become free and self-disciplined ethical subjects. [6]

3.1.1 Artificial intelligence is not rational

On the whole, according to Kant, human action consists of two aspects. The first is the relationship between human beings and things. There are two kinds of beings in the world: irrational "things", which exist on the basis of natural will, have only relative value, and can only be regarded as "means". The other is the rational "man", whose essence is emphasized as "the end in itself" and has absolute value. [7] The second is the relationship between human beings, who, because of their absolute value, are ends in themselves, not means. Therefore, in their actions, people should respect others and consider both others and themselves as ends.

In terms of the relationship between humans and things, artificial intelligence is essentially a set of artificial systems composed of data, algorithms and programs, and is a scientific and technological achievement created by human beings for the ultimate purpose of survival, reproduction and development. [8] In other words, artificial intelligence is not a rational being, which is in the state of being dominated by human beings, and is an object, a tool and a means of accomplishing the creator's purpose, rather than the purpose itself. In terms of the relationship between human beings and human beings, the relationship between human beings and artificial intelligence only exists between human beings and objects but not between human beings and human beings.

3.1.2 Artificial intelligence does not have self-awareness

At the cognitive level, human rationality is reflected in the fact that human beings are able to recognize the world through consciousness and thus legislate for nature. The process by which human beings come to know the world is the process by which internal reason unifies external perceptual images. During this process, knowledge is formed dynamically through the use of unique thinking ability of human beings. ①This unique thinking ability belongs to human consciousness and originates from human reason. It is precisely because of this thinking ability that human beings have become the subjects of cognitive activity. However, all things in the world have become the objects of cognition because they do not have consciousness. Therefore, human beings have the ability to legislate for nature.

However, in terms of the stage of development of artificial intelligence, John R. Searle, an American philosopher, explored whether artificial intelligence should have cognitive abilities in a psychological and philosophical sense. He divided artificial intelligence into weak artificial intelligence and strong artificial intelligence, thinking that the former is only a tool to assist research, while the latter is an intelligent product that can be equivalent to human mind after appropriate programming, with "human-like" autonomous consciousness and rational thinking. [9] At present, we are still in the era of weak artificial intelligence. The current stage of artificial intelligence is still essentially an auxiliary tool used by human beings to complete the creation of works. The process of generating content by artificial intelligence is not equivalent to the thinking process of human beings. In addition, from the point of view of the source of knowledge, the knowledge base of artificial intelligence relies on human input. And its

Services.

① In Kant's view, all knowledge begins with experience, but does not come directly from experience. See [German] Kant, Critique of Pure Reason, translated by Deng Xiaomang, schooled by Yang Zutao, People's Publishing House, 2008 edition, p. 16.

generation of content relies on the analysis of existing data, which is the product of being trained by researchers. Artificial intelligence can only recognize and understand things through algorithms predetermined for them by humans or through training on large amounts of data. Their cognitive abilities are highly restricted, with limitations in their ability to understand abstract concepts and unstructured information. For example, while artificial intelligence can recognize words and sentences in text, it cannot truly understand the emotional expressions or deeper meanings of them. [6]Therefore, artificial intelligence does not possess the unique thinking ability of human beings, and thus does not produce self-awareness, does not possess rationality, and is essentially an object of human cognition.

3.1.3 Artificial intelligence does not have freedom

Because artificial intelligence does not have self-awareness, it can be assumed that it does not have the freedom to make choices based on free will. At the practical level, human rationality is reflected in the fact that human beings are free and able to transform the world in a dynamic way, thus legislating for themselves. Human free will manifests itself in illogical choices, where individual perceptions and desires lead to different decisions. Human beings are able to detach themselves from immediate sensations and desires, transcend the impulses of the senses and restrain themselves, and think and act with a broader perspective, pursuing what is more beneficial to the overall interest, rather than being confined to the immediate interests and desires of the present. [10] Thus, human beings are able to legislate for themselves and limit their behavior. Law requires the interaction of sentiment and jurisprudence, and this desire and choice of human beings influences the choice of the guiding ideology of jurisprudence, the design of the legal system, and the practical application of and compliance with legal norms. It can be said that human freedom is crucial to the development of law.

However, the judgment of artificial intelligence is the result of logical calculations. It strictly adheres to the built-in procedural settings, and will only use pure utilitarian logic to make judgments. It will not be able to weigh the advantages and disadvantages in complex situations. It cannot reason, cannot express meaning, and thus cannot perform legal acts. So it is meaningless to give it the status of a natural person as a legal subject.

In short, in the Kantian philosophical perspective, only human beings can become "natural persons". Artificial intelligence as a human creation can only be the object of cognition. The status of the subject and object can not be exchanged. It can be seen that artificial intelligence does not have the philosophical basis to become the subject of law.

3.2 Artificial intelligence cannot be fabricated as a legal subject from the perspective of private law theory

Based on the purpose of their own development, human beings can give the necessary things the subject status, so as to better show the properties of the means of things, reflecting the value of human beings. [11] Because of the human economic and social development needs, human beings endow legal persons with legal subject qualification. Therefore, some scholars propose that artificial intelligence can be given legal subject status according to this legal drafting techniques. This paper believes that the above view is a matter of debate. The reason why the legal person is fabricated as a legal subject is based on the ancient Roman people and personality separation legislation technology, the legal person itself and its beneficial. On the contrary, in terms of artificial intelligence, although there is now legislative technology, whether it has substantive and beneficial conditions is worth exploring.

3.2.1 Artificial intelligence has no substantive basis for being anthropomorphized as a legal subject

From the point of view of the legal subject itself, according to the current legal provisions, in order to be a legal subject, there are two requirements, one is that the subject of law should have "autonomy" and "self-discipline". [12] This means that the legal subject is able to decide on its own actions and comply with the law and ethical norms on its own; secondly, the legal subject is able to independently enjoy its rights, obligations and responsibilities in accordance with the law. [13] According to these requirements, the formulation of a legal person as a legal subject meets the requirements, but artificial intelligence does not meet the above criteria and cannot be formulated as a subject of law.

First of all, the legal person is a collection of people and property. The legal person is the organization, and is essentially a collection of natural persons. The will of the legal person is embodied in the will of the group of natural persons behind it. The implementation of the legal person's behavior, rights, obligations, and liability can not be separated from the natural persons. Therefore, the legal person can embody the rationality of natural persons, with the characteristics of autonomy and self-discipline. However, artificial intelligence is a collection of algorithms and programs, which is an intellectual achievement of human beings. It exists independently of natural persons after its design is completed. Not only does it not have its own rationality, but also fails to reflect the rationality of natural persons.

Secondly, as independent legal entities, legal persons have their own property and enjoy the right to corporate property. This means that legal persons can use their property to assume legal liability and pay debts. However, artificial intelligence is a human creation that does not have property, and it is unable to use its property to assume legal liability. Even if the relevant property

https://doi.org/10.1051/shsconf/202317801019

management system and liability assumption system for artificial intelligence are formulated, the ultimate property manager and liability fulfiller will be a natural person.

3.2.2 Beneficial conditions for artificial intelligence not to be proposed as a subject of law

From an economic and social point of view, firstly, the proposed establishment of legal persons as subjects can promote the development of legal persons themselves. Specifically, the provision of limited liability of legal person reduces the investment risk of investors, stimulates their investment willingness, and then promotes the emergence of more legal persons, which can form a virtuous circle. Secondly, the legal person emerges due to the development of commodity economy. This legal fiction is conducive to the development of commodity economy and brings economic benefits. Specifically, in order to incentivize investment and protect the interests of investors, a large number of legal persons emerge, which promotes the development of commodity economy. Making the legal person as the legal subject can avoid the complexity caused by the majority of its individual members as the subject of the legal relationship, and improve the efficiency of civil and commercial activities. The establishment of the subject qualification of legal persons has brought significant convenience to the development of civil and commercial activities.

However, granting artificial intelligence the status of a legal subject by legal fiction does not bring economic benefits. First of all, artificial intelligence itself is an intellectual achievement of mankind. Its own can get the protection of intellectual property law and its designers can be inspired by intellectual property law so as to create more and more high-quality artificial intelligence products. Giving artificial intelligence the proposed legal subject qualification can not motivate designers to create. Secondly, artificial intelligence is a product of the development of science and technology, and its emergence greatly improves the productivity of human beings. In other words, the purpose of artificial intelligence is to assist human labor, not to be participated in civil and commercial activities. Being proposed as a legal subject can not make it personally participate in civil and commercial activities, either can not bring vitality to the economic market.

In summary, artificial intelligence has no substantive basis and beneficial conditions to be proposed as a legal subject.

4 Artificial intelligence generated content is copyrightable

The issue of whether artificial intelligence generated content (hereinafter referred to as AIGC) is copyrightable is currently controversial in the academic community. There are mainly three different views: the first viewpoint takes a negative attitude towards the above issue, believing that AIGC has so far been generated by algorithms, programs and templates, which cannot show the unique personality of the creator. Therefore, it does not conform to the definition of creation under the copyright law, and should not be granted copyright protection. [14] The second viewpoint takes a positive attitude and thinks that the above issue can be judged by applying the originality standard. [15] The third viewpoint holds a neutral attitude towards this, arguing that the content generated by artificial intelligence neither originates from human beings nor requires human beings to pre-set the rules. Although it is difficult to interpret this kind of generated content as a work in the sense of the current copyright law, with the continuous development of the industry, when artificial intelligence can create independently from human beings, then it is necessary to revise the current system and take AIGC as the object of copyright regulation. [16] This paper argues that AIGC is copyrightable for the following reasons.

4.1 Copyrightability of AIGC is consistent with the legislative intention of copyright law

The purpose of the copyright law is to safeguard the legitimate rights and interests of copyright holders, to ensure that copyright holders are able to enjoy personal and property rights and interests in the works they create, and to encourage people to participate in creative activities through the protection and incentive mechanism for creativity. At the same time, the copyright law encourages the creation and dissemination of works that are beneficial to the construction of socialist spiritual and material civilization, conveys socialist core values and promotes social progress and civilization, as well as the prosperity of socialist culture and science. ①It can be seen that the Copyright Law encourages creation as a means to achieve the objective of the prosperity of socialist culture and science.

In terms of the collection of existing works, nowadays, with the support of powerful data and algorithms, artificial intelligence has an extremely large amount of information reserves, which can accurately copy and disseminate information resources at a low cost, assisting people in efficiently obtaining existing information resources for the creation of works, shortening the time for human beings to create the same or similar works, and improving the efficiency of human creativity. It helps human beings to save more time and energy for the innovations in ideas, styles, and other aspects, making it possible for people to create more works of different styles and types, satisfy people's diverse spiritual and cultural needs, and prosper the socialist cultural and scientific undertakings. In this sense, artificial intelligence works can liberate human beings from low-value repetitive labor, give more creative freedom to ordinary users, and help human beings develop more difficult and more valuable fields. [17]

At the level of work creation, AIGC exists objectively, and it is valuable to the subsequent creation of human beings, providing new ideas and revelations for human

① See section 1 of the Copyright Act.

beings, thus stimulating human beings' unlimited creativity and imagination, and broadening people's ways of thinking and horizons. If we do not recognize the copyrightability of the valuable intellectual achievements of artificial intelligence just because of its "inhumanity", it will lead to the fact that some high-quality intellectual achievements cannot be protected, which is not conducive to the researchers and developers to continue to optimize the algorithms and improve the performance of the artificial intelligence in order to generate more valuable intellectual achievements. It is not advantageous to stimulating innovation and the prosperity of cultural endeavors. Therefore, the determination that the content generated by is copyrightable is in line with the legislative intent of the Copyright Law.

4.2 AIGC can satisfy the elements of a work

According to article 3 of the Copyright Law, it can be seen that there are four elements of a work: first, it must be an intellectual achievement in the fields of literature, art or science; second, it must be an expression of an idea rather than an idea in itself; third, it must be capable of being expressed in a certain form; and fourth, it must be original. Of these, the first three are easier to judge, and originality is the key to determining whether an intellectual product constitutes a work.

Theoretically, the term "originality" has two meanings. One is uniqueness and the other is creativity. One the one hand, uniqueness refers to independent creation by the creator, with the intellectual achievements originating from the creator himself rather than others. One the other hand, creativity refers to the fact that the work must achieve a certain degree of creativity. The originality standard can be divided into subjective and objective standards, with the former requiring that the work be closely connected with a natural person, i.e., insisting on "independent creation" as a necessary element for the composition of the work; and the latter emphasizing the differences among works, i.e., insisting on "originality" in the object of the work. [18] Independent creation essentially embodies the relationship between the work and the author. It emphasizes that the author obtains the right to the work through the act of independent creation, the significance of which lies in determining the attribution of the right to the work and the source of the copyright holder's right to the work. It does not prove that the creation is a work. In other words, taking the relationship between the work and the author as an element of the work is not in line with the basic logic of the law, and such a move may lead to confusion between the connotation of the work and the attribution of the work. [19]Therefore, it is more appropriate to adopt the objective criterion of originality. In addition, the criterion of originality comes from social evaluation other than that of the author. The identity of the creator should not be a constitutive condition for the protection of a work. Whether AIGC can be a work, we should consider whether such generated content is "substantially similar" to the work created by others, and whether it can be recognized by the "general public" based on human readers. [20]

In practice, it has been difficult to distinguish between AIGC and human-created intellectual achievements with originality in form. It means that under the premise of adopting the objective standard of originality, if there are differences in expression between AIGC and works created by human, and if they have gained public acceptance, the generated content has the possibility of being recognized as a work. Currently, courts have recognized that AIGC is s a work. For example, in the case of Tencent v. PCCW, PCCW Technology published an article written by Tencent's artificial intelligence Dreamwriter on its website without Tencent's permission. The court held that the article was created by the creator according to his personal choice and arrangement, and that its external manifestation as well as the generation process satisfied the conditions for written works under the Copyright Law. Thus it can be concluded that the article in question was a written work protected under the Copyright Law.①

In summary, AIGC helps humans gather information, gives them inspiration, and improves the efficiency of human creativity. It can be copyrightable when it meets the standard of originality.

5 Attribution of Copyright in Content Generated by Artificial Intelligence

If AIGC meets the originality standard, it can be a work, so it is worth exploring the issue of its copyright attribution. Currently, the attribution of AIGC mainly involves the public domain, designer, owner, user, collaborative work, orphaned work, and other points of view.

5.1 Discussion of current attribution patterns

5.1.1 Public domain theory

The value of copyright law lies in balancing the interests of copyright owners with the public interest of society. On the one hand, it grants the copyright a certain period of protection, so that the copyright owner can gain income through monopolistic utilization of the copyright within this period, and then motivate people to create more works. On the other hand, it strictly restricts the period of protection, so that the works enter into the public domain for the free use of the whole society after the expiration of the period, in order to maintain the sufficiency of intellectual achievements in the public domain, and to ensure the continuous enhancement of the overall innovation ability of human beings. [8] There are only two cases in which a work enters the public domain: one, when the copyright exceeds the term of protection; and the other, when the copyright is not owned by the owner. According to the public domain theory, the

① See Civil Judgment No. (2019) Yue 0305 Minchu 14010 of the People's Court of Nanshan District, Shenzhen City, Guangdong Province.

inclusion of AIGC in the public domain not only helps to solve the problem of copyright ownership, but also balances the interests of copyright holders and the public interest, reduces the cost of innovation, and incentivizes innovation.

This paper argues that placing AIGC in the public domain may lead to an imbalance of interests. First of all, AIGC is not a thing without a copyright owner. Its designers, owners, and users may obtain the copyright of such generated content under certain conditions. They more or less paid labor for the generation of AIGC . Putting AIGC in the public domain will make their labor can not get equal benefits, which is not conducive to motivate their investment in artificial intelligence. Besides, AIGC objectively exists, which contains great innovative value. If the AIGC is recognized as a work, and then placed in the public domain, at this time, all the people of the AIGC have the right to use, without rights to prevent others from using. People tend to overuse competitively without restrictions on their use behavior, resulting in the depletion of the public domain of the innovative resources and waste, as well as the endless pursuit of personal interests. which is at the expense of the public interests of society , finally leading to a "tragedy of the commons".

5.1.2 The designer theory

Some scholars argues that copyright in AIGC rests with the designer of the artificial intelligence based on the following three main arguments, all of which have certain shortcomings.

First, AIGC benefits from the intellectual labor of artificial intelligence designers. The algorithm setting, creation training, and code modification of artificial intelligence rely on the creative labor of its designers. However, the artificial intelligence product is the intellectual achievement of the designer, who enjoys intellectual property rights over the achievement and has the right to control whether the artificial intelligence product is marketed as well as the revenue. Improtantly, the research results have already been property rightsized, and the existing intellectual property rights protection mechanism is sufficient to provide incentives for them. So the designer does not need to claim copyright over AIGC.

Moreover, the designer may choose not to publicize the artificial intelligence. For example, Pamela Samuelson argues that if copyright in the content generated by the artificial intelligence is not available, it is possible for the designer to protect the artificial intelligence he or she has developed as a technological secret, sign his or her name to the AIGC, and copyright the content in his or her own name, thus de facto enjoying copyright in the AIGC. [21] However, it is significant to regulate the signature of AIGC, and it is not reasonable for a designer to obtain copyright on AIGC in his or her own name. False or fraudulent attribution of AIGC will not only be detrimental to the social evaluation mechanism associated with attribution rights, but will also magnify the public safety risk of artificial intelligence-generated disinformation, leading to a serious social crisis of trust. [22] Compared with human beings, artificial intelligence generates information more efficiently, and the form of its generated content is indistinguishable from human-created content. The probability of artificial intelligence generating false information and confusing humans is extremely high. The proliferation of false information, which is difficult for people to distinguish, will make them skeptical of any information, resulting in a serious crisis of confidence. Once the information is signed by a human as the author, and in the form of the signature is the same as the correct information, such information is not easy to be recognized or detected by the public, but on the contrary, if it is signed by an artificial intelligence, people will have reservations about the authenticity of the content it generates. Currently, Amazon, Anthropic, Google, Inflection, Meta, Microsoft, and OpenAI have committed to the Biden Administration to make a series of voluntary commitments to protect users, including agreeing to security testing and adopting a new watermarking system to ensure that users are able to distinguish AIGC, thereby reducing the risk of fraud and misinformation. ①

Furthermore, AIGC can be considered as a derivative work. As an example: some researchers have argued that AIGC can be considered a product of artificial intelligence software and can be viewed as an extension or deduction of the original software work created by the designer. In the event that it is not possible to take the artificial intelligence as the right holder, the designer can be regarded as the initial creator of the content, owns the copyright of the content, and can extend the copyright of the derivative work generated by the artificial intelligence. [23] However, copyright law protects expression that is original, and AIGC is not inherently a derivative work of an artificial intelligence program. Derivative work is a work based on another work. Although the deducer adds original expressions in the process of creation, these expressions are not sufficient to make the derivative work completely independent of the original work. [24] In other words, the creation of a derivative work continues the basic elements of the original work. The designer has a copyright in the software required for the artificial intelligence to function, but the AIGC is not merely a translation of the designer's code. [25] The core expression of an artificial intelligence program is its code, while the original expression of its generated content is its combination of words, art, music, etc. The latter is not a modification of the former. Due to the inherent specialization and complexity of artificial intelligence technology, as well as the exclusionary business policies adopted by technology companies out of the need to pursue profits, it is difficult for people to distinguish the original algorithms and code from the generated content of an artificial intelligence, and it is impossible to recognize an artificial intelligence program from its generated content to recognize the original expression of the artificial intelligence program.

①http://www.chinanews.com.cn/gj/shipin/cns-d/2023/07-22/news965396.shtml

Therefore, AIGC does not conform to the logic of a derivative work.

In summary, the artificial intelligence designer is not entitled to copyright in the generated content when the user inputs instructions to the artificial intelligence and makes the necessary arrangements for the eventual generation of the AIGC.

5.1.3 Owners and users theory

The owner argument applies by analogy to the corporate works requirement, arguing that AIGC represents the will of the owner, therefore such generated content, once it meets the criteria of a work, is copyrightable. The copyright should be vested in the owner of the artificial intelligence. [15] The user argument argues that the user of artificial intelligence should enjoy the copyright of AIGC based on usage agreement, the tool attributes of artificial intelligence, the user's input of information, and the legislative purpose of the Copyright Law. [18] Some scholars also regard AIGC as commissioned works. The process of creation by users of generative artificial intelligence actually involves a contractual relationship between the user and the artificial intelligence developer. In this case, the artificial intelligence developer hires a technical team to run the artificial intelligence program to create the work according to the user's requirements, and the user pays the developer in exchange for the contractual agreement. The user is the "customizer" who provides specific requirements and instructions and expects the artificial intelligence to generate content to meet his or her creative needs, while the developer is the "contractor" who is responsible for running the artificial intelligence program and delivering the generated work to the user. Based on the contract, the user acquires the copyright of the AIGC by agreement with the artificial intelligence developer. [26] The above views have some desirability. All of them affirm the instrumental attribute of artificial intelligence, and the idea of analogical application also has some value. However, the identity change of the artificial intelligence owner and the user based on the employment contract and entrustment contract relationship has been ignored, and the above viewpoints have a certain one-sidedness.

In addition, there are scholars who apply the provision of service works by analogy, arguing that the user or owner of the artificial intelligence is entitled to copyright. On the one hand, the artificial intelligence achieves the result desired by the user and thus it enjoys the right of attribution. This allows AIGC to be distinguished from human-created content, and the artificial intelligence's creative contribution to it is noted. On the other hand, the fact that the user enjoys a copyright on the AIGC in addition to the right of attribution helps to stimulate creativity in human subjects, and enables the innovation stimulating effect of copyright law to be realized. [27]This paper argues that such a view is debatable. First of all, as mentioned above, the subject qualification is the qualification of a person with certain conditions to enjoy the rights and assume the responsibilities and obligations. [28] The right of authorship is the personal right to indicate the author's identity and sign his name on the work. Artificial intelligence does not have the qualification of legal subject, and thus cannot enjoy the rights stipulated in the law. Therefore, granting it the right of authorship does not have the legal basis. Besides, due to the lack of rationality, even if the artificial intelligence is given the right of authorship, it cannot fully exercise it. For example: the artificial intelligence can not independently decide whether to sign its name on the content it generates, can not decide the way of signing, can not prohibit others from signing their own works, and can not exercise the right of authorship to require others to sign their names when they use their works. Its right of authorship is ultimately exercised by the person who enjoys the right to domonate the proceeds of the AIGC.

5.1.4 Collaborative work theory

The collaborative work theory argues that there are two steps in the generation of content by artificial intelligence. First, the artificial intelligence designer creates the artificial intelligence product. And then, the artificial intelligence user generates the content using a pre-existing program. They work together to make the artificial intelligence creation, thus recognizing it as a collaborative work. [29]

However, this paper argues that a cooperative work refers to a work jointly created by two or more authors. Collaborative works need to satisfy the following elements at the same time. Firstly, there should be an agreement to co-create between the co-authors. Secondly, in the process of creation, the co-authors always carry out the intention of cooperation and consciously adjust their respective creative styles so that the works are articulated and harmonized with each other. Furthermore, the creators should intend to share the copyright as co-authors. Finally,the collaborative works should reach the standard of originality as stipulated in the copyright law. Specifically with regard to AIGC, first, there is no co-authorship agreement between the artificial intelligence designer and the user. What the designer designs is only the program of the artificial intelligence, which does not include the design of the generated content. And even if the user inputs the same command, the content generated by the artificial intelligence based on the built-in program may not be the same. Second, the designer has no control over the content generated by the user according to the program, and cannot consciously adjust the style of the content generated by the artificial intelligence to make the final cooperation results harmonious and unified as a whole. The AIGC ultimately depends on the user's choices and selections when inputting instructions. Again, the designer does not have the subjective intent to share the copyright of the AIGC with the co-authors. In other words, it is the designer's job to design the artificial intelligence product in accordance with the requirements, and the incentives he or she can be subjected to have already been enjoyed at the time of the completion of the product design. Finally, after the designer has built in the artificial intelligence program, he or she cannot control or decide whether the generated content meets the originality standard. And not all AIGC meets the originality standard. In summary, it is not reasonable to recognize AIGC as a collaborative work.

5.1.5 Orphan's work theory

Orphan works are works whose copyright holders cannot be contacted through best efforts to find them. [27] Copyright, other than the right of authorship, is exercised by the owner of the original work. After the identity of the author is determined, the copyright is exercised by the author or his heirs. ①Some scholars believe that treating AIGC as orphan works can effectively solve the problem of the lack of AIGC rights subjects. This is conducive to balancing the interests of artificial intelligence investors and users of generated content, improving the efficiency and value of the utilization of AIGC, and encouraging the owners of generated content to claim their rights. [30] The main goal of the orphan works system is to promote the dissemination and utilization of works, while taking into account the interests of copyright holders.

With regard to the issue of AIGC, firstly, in the process of dissemination and use of AIGC, it is difficult to define the "original owner", and ultimately it is still necessary to determine the attribution of rights to the designer, owner and user of the artificial intelligence. Secondly, even if the protection model of orphan works is utilized for its protection, it is not a long-term solution. AIGC is in fact not an orphan work, the generation of such content as an orphan work is only a temporary protection measure to avoid that when it can be a work, it is difficult to find the attribution of copyright rights and difficult to be protected by the copyright law, and then fall into a chaotic situation, which will jeopardize the development of the artificial intelligence industry, etc. In the end, it is still necessary to design the system of attribution of copyright for the generated content of artificial intelligence. [31] Therefore, the use of the orphan work model for its protection is still somewhat insufficient.

5.2 Proposals for attribution models

Thus, it appears that all the above views are one-sided. To summarize, this paper holds that artificial intelligence is a tool to assist human creation, and in principle, users of artificial intelligence enjoy the copyright of AIGC, unless otherwise provided for in the copyright law or agreed upon by the parties. This is also in line with the mode of attribution of works under China's copyright law: the principle of belonging to the author, supplemented by special provisions, and excepted by contractual agreement." [32]

5.2.1 The principle that copyright in AIGC belongs to the user

Specifically, our copyright law provides that the natural person who creates a work is the author. ②The so-called creation is an intellectual activity that directly produces a work. ③This also means that merely organizing others to create works, providing material and technical conditions, gathering information or providing ideas is not a creative act. The act of creation should be based on the criterion of whether or not substantial intellectual activity has been invested. If the user intervenes in the process of generating the content of the artificial intelligence by inputting commands, and the artificial intelligence carries out auxiliary work such as collecting information and providing ideas for human beings, but does not directly participate in the creation of the work, the user enjoys the copyright.

The reasons are as follows: first, artificial intelligence relies on users to generate content. Artificial intelligence cannot actively generate content, and it relies on the user's information input, content qualification, and generates intellectual results according to its algorithms, rules, templates, etc. It requires the user's input behavior to start its system. Users fixed AIGC, and it can be expressed in a certain form, to meet the elements of the work. Secondly, the user participates in AIGC training. The user uses the artificial intelligence for the purpose of completing the creation, and the artificial intelligence has the attribute of a tool to assist the creation. The input information reflects the user's choice and judgment. Finally, the user is responsible for the infringement of AIGC. The user has the ultimate power over the AIGC, can decide whether to use it and whether to benefit from it. He or she is the actual biggest beneficiary of the AIGC, who should bear the responsibility for copyright infringement. To summarize, at this point, it is possible to use legal drafting techniques to treat the user of the artificial intelligence as the author and therefore entitled to copyright.

5.2.2 Special provisions for copyright attribution of AIGC

The copiright of AIGC is attributed to special subjects if it belongs to special types of works. In China's current copyright legal system, special provisions mainly refer to corporate works, service works and commissioned works.

First of all, the Copyright Law stipulates that a legal person or an unincorporated organization is regarded as the author of a work that is created by a legal person or an unincorporated organization under the auspices of, on behalf of, and for which the legal person or unincorporated organization assumes responsibility. ④Therefore, for artificial intelligence like Dreamwriter which is used by a company, then the copyright of its generated content still belongs to the company.

① See Article 13 of the Regulations for the Implementation of the Copyright Law

② Section 11 of the Copyright Act.

③ Article 3 of the Regulations for the Implementation of the Copyright Law

④ Section 11 of the Copyright Act.

Secondly, if an employee creates a work using the legal person or unincorporated organization's artificial intelligence in order to complete the working tasks, the user's creation at this time is a service work. The copyright of AIGC should be determined in accordance with Article 18 of the Copyright Law. The copyright of the general work belongs to the user, and the legal person or unincorporated organization enjoys a two-year priority right of use within the scope of business; the legal entity or unincorporated organization owns the copyright of the special work, and the user only enjoys the right of authorship.

Finally, if there is a commissioning relationship between the user and other subjects, on the basis of which the commissioned parties participate in the training of AIGC and utilize it to create the work, the AIGC is a commissioned work at this time, and the attribution of its copyright shall be determined in accordance with the following rules. The ownership of copyright shall be agreed upon a contract between the commissioning and the commissioned parties. In the absence of such a contract or of an explicit agreement in the contract, the copyright in such a work shall belong to the commissioned party. ① If the copyright of the commissioned work belongs to the commissioned party, the commissioning party shall have the right to use the work within the scope of the agreement. If the two parties have not agreed on the scope of use, the commissioning party may use the work free of charge within the scope of the specific purpose of the commissioned creation.②

5.2.3 Exceptions to the Agreement on Copyright Ownership of AIGC

Private law pursues the freedom and equality of human beings. [33] Autonomy of Will is the embodiment of the concept of freedom and equality in private law and the basic principle of civil law. Copyright law is a special law of civil law, which should also follow the basic principle of civil law. The essence of Autonomy of Will emphasizes the autonomy of the individual and the importance of free will, which gives the subject of the right to freedom of action, and considers that the individual has the right to make choices, decisions and agreements independently. It means that the individual can participate in shaping the legal relationship with others based on his/her own will and interests, and embodies the respect and protection of human beings. [34] Therefore, a contract can be concluded between the designer, user and owner of artificial intelligence based on free will, and a clear agreement can be made on the specific matters of copyright attribution of AIGC. At this time, when there is an agreement between the parties, the agreement of the parties shall be applied in priority.

① Article 19 of the Copyright Act.
② Article 12 of the Interpretation of the Supreme People's Court on Several Issues Concerning the Application of Law to the Trial of Cases of Civil Disputes over Copyright.

6 Conclusion

The clarification of the legal subject status of artificial intelligence, the identification of works and attribution of rights of AIGC will contribute to the healthy development of the artificial intelligence industry as well as the growth of literary and scientific undertakings. Although artificial intelligence does not have the status of a legal subject at present, a large amount of its generated content is original and indistinguishable from human intellectual achievements in form. If it is not protected by copyright, it will be deterimental to incentivizing creativity and is not in line with the purpose of the legislation of the copyright law. Scholars have expressed their own views on the attribution of copyright. And it is necessary to further test the advantages and disadvantages of them in practice. The intellectual property law aims to balance nterests. In order to better cope with the impact of artificial intelligence on copyright law, when it comes to the attibution of AIGC rights, we should take into account the interests of multiple parties. and the logic of existing copyright law.

References

1. Jin C. Y., Xing H. Challenges and responses to the application of blockchain in digital music copyright management[J]. Science and Technology Management Research,2022,42(9):143-151.
2. Jing T. T., Zhang H.,Li F. N. Trends in human-computer coupling from ChatGPT[J]. Media,2023(10):29-31.
3. Minsky M. Semantic Information Processing[M]. Cambridge: MIT Press, 1968: 5.
4. Wu H. D., The question of patent law for artificial intelligence-generated inventions[J]. Contemporary Law,2019,33(4):24-38.
5. Ma J. J., Lecture Notes on the Theory of Personality and Personality Rights [M]. Beijing: Law Press,2009:36.
6. Liu H. H., On the Legal Status of Artificial Intelligence[J]. Politics and Law,2019(1):11-21.
7. Immanuel Kant. Foundations of the Metaphysics of Morals [M]. Sun S. W.,Translation. Beijing:Kyushu Press,2006:83-84.
8. Xu J. L., Copyright Attribution of Artificial Intelligence Generated Works[J]. Journal of Jinan (Philosophy and Social Science Edition),2023,45(4):37-49.
9. Searle J R . Minds, Brains and Programs[J]. The Behavioral and Brain Sciences, 1980(33): 417-423.
10. Cui S. L., On the differentiation and expansion of the subject qualification of private law [M]. Beijing:Law Press,2009:118.
11. Zhang B. H., The Dual Role Dilemma of Rights Capacity and the Reconstruction of Subject Qualification System[J]. Jurist,2014(2):12-29,176.
12. Li Y., Li X. Y., Exploration of the copyright problem

of artificial intelligence generated objects under the viewpoint of Kantian philosophy[J]. Journal of Law,2018,39(9):43-54.

13. Hu Y. H., The concept of legal subject and its characteristics[J]. Research on Law,2008(3):3-18.
14. Wang Q., On the Characterization of Content Generated by Artificial Intelligence in Copyright Law[J]. Legal Science (Journal of Northwest University of Politics and Law),2017,35(5):148-155.
15. Xiong Q., Copyright Recognition of Artificial Intelligence Generated Content[J]. Intellectual Property,2017(3):3-8.
16. Liu Y., A preliminary study on the copyright law protection of artificial intelligence generated objects[J]. Intellectual Property,2017(9):44-50.
17. Cong L. X., Li Y. L., Identification of Works and Copyright Attribution of Generative artificial intelligence--Taking ChatGPT's Work Application Scenario as an Example[J/OL]. Journal of Shandong University (Philosophy and Social Science Edition),2023(04):1-11[2023-08-11].
18. Cai Z. J., Che Y. L., Discussion on Legal Personality Recognition and Nature of Generated Content of ChatGPT[J/OL]. Journal of Xinjiang Normal University (Philosophy and Social Science Edition):1-8 [2023-08-11].
19. Dai Z., On the Concept of Works in Copyright Law[J]. Friends of editors,2016(5):89-94.
20. Wu, H. D., The Copyright Law of Artificial Intelligence Generated Works[J]. Chinese and foreign law,2020,32(3):653-673.
21. Samuelson P. Allocating Ownership Rights in Computer-generated Works[J]. University of Pittsburgh Law Review, 1986, 47: 1191.
22. Yao Z. W., Shen Y., The Risk and Regulation of Untrue Attribution of Artificial Intelligence Creations[J]. Journal of Xi'an Jiaotong University (Social Science Edition),2020,40(1):133-140.
23. Li Z. H., Justification of Copyright Protection and Copyright Attribution of Artificial Intelligence Creations[J]. Friends of Editors,2018(07):80-87.
24. Butler T L. Can a Computer be an Author? Copyright Aspects of Artificial Intelligence[J]. Hastings Communications and Entertainment Law Journal, 1982(4): 743.
25. Tao Q., On the Protection of Artificial Intelligence Generated Results by Copyright Law - Evidence of Data Processor's Right as Neighboring Right[J]. Law,2018(4):3-15.
26. Jin C. Y., Xing H. T., Study on Copyright Attribution and Infringement Liability Principles of Artificial Intelligence Publications[J]. Publishing and Distribution Research,2021(9):73-81.
27. Wang Q., "A Brief Introduction to the Design of the Orphan Works System[J]. China Copyright,2013(1):30-33.
28. Ma J. J., Lecture Notes on the Theory of Personality and Personality Rights [M]. Beijing: Law Press,2009:36.
29. Zhu M. Y., Design of copyright attribution system for artificial intelligence generated objects[J]. Journal of Shandong University (Philosophy and Social Science Edition),2019(01):118-126.
30. Liu Q., Liu Z. Y., Research on the protection mode of orphan works created by artificial intelligence[J]. Journal of Anyang Normal College,2018(4):37-44.
31. Li W. M., Positive Determination of Artificial Intelligence Intellectual Achievements in Copyright Law--Discussion with Prof. Wang Qian[J]. Oriental Law,2018(3):149-160.
32. Cao X. M., Legislative Improvement of China's Copyright Attribution Model[J]. Jurisprudence,2011(6):81-89.
33. Bender Lüthers et al. General Theory of German Civil Law [M]. Yu X. M., et al, Translation. Beijing:Law Press,2017:13.
34. Bettina Hüliman-Goppel et al. Swiss civil law: basic principles and human law [M]. Ji H. L., Translation. Beijing:China University of Political Science and Law Press,2015:20-21.

Research on Hangzhou's "Courtesy to Pedestrians" Policy from the Perspective of New Institutionalism

Kunyang Zhou*

Harbin Business University, Administrative Management ,Harbin , China

Abstract: Since 1970, new institutionalism has gradually become an important means and basis for analyzing various systems. This article takes the "Courtesy to Pedestrians" policy in Hangzhou as an example, and from the perspective of new institutionalism, studies the implementation process of this policy. It is found that it constructs legal elements through regulatory elements, constructs new moral standards through normative elements, and stabilizes the achievements of the above two elements through cultural cognitive elements, To achieve the goal of smooth policy implementation, And provide reference for the government to implement new policies more smoothly and achieve organizational goals more efficiently.

1. Research background

Since the reform and opening up, China has gradually moved towards marketization, industrialization, and urbanization, and people's production and living standards have greatly improved. At the same time, in pursuit of convenience, the number of private cars has been increasing year by year. This phenomenon has brought not only resource and environmental issues, but also pedestrian safety issues. The Traffic Safety Law of the People's Republic of China clearly stipulates that motor vehicles should slow down when passing through pedestrian crossings, and should stop and give way when encountering pedestrians passing through pedestrian crossings. However, there are still car accidents at pedestrian crossings, such as the "racing case" in Hangzhou, which has attracted high attention from the Chinese people. However, as of December 2022, Hangzhou has achieved "four consecutive championships" in the creation of a national civilized city. At the same time, as the first city to implement the "courtesy to pedestrians" policy, Hangzhou's successful experience can also be learned and implemented by many cities in China.

2. New institutionalism

In the 1970s and 1980s, the field of Western social sciences "rediscovered" the position and role of institutional analysis in explaining practical problems, thus forming a new paradigm of institutional analysis. In the perspective of new institutionalism, institution is defined as a multi-level stable social structure composed of social symbols, social activities, and material resources[1].

New institutionalism includes regulatory elements, normative elements, and cultural cognitive elements. Among them, regulatory elements are mandatory and can forcibly regulate people's behavior; Normative elements are considered from the perspective of the social environment of public life, and the social environment during a certain period of time also requires that everyone adhere to the code of conduct and moral requirements; The cultural cognitive element is the social consensus reached by the public based on their understanding, acceptance, and support for policies. These three elements are interdependent and interact with each other, forming a trinity of building a good social system.

Firstly, from the perspective of policy composition, it is the interaction between regulatory elements and cultural cognitive elements. By imposing mandatory rules on the behavior of the public and punishing those who violate the rules and regulations, a good social atmosphere is formed, which affects the original behavior habits of the public and guides them to accept the system or policy. Secondly, between regulatory elements and normative elements, regulatory elements are enforced and enforced by the public through mandatory means, but normative elements tend to be more flexible in terms of methods and methods. They guide the public to gradually recognize and accept through an informal means, forming a universal and correct value system to promote the effectiveness of regulatory elements, formulate relevant laws and regulations, reward and punishment mechanisms, and maintain Improve and develop this value system. Finally, between cultural cognitive elements and normative elements, the common understanding and behavioral activities of the public gradually form normative elements, which are understood and accepted by the public, gradually integrated into life, and form cultural cognitive elements in the mind and heart.

* Kunyang zhou
E-mail: 740745627@qq.com

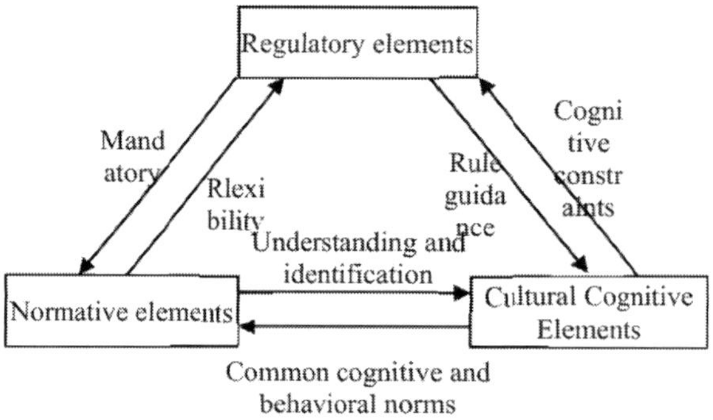

Figure 1 Elements and Interrelationships of New Institutionalism

3. Analysis of the Policy of "Courtesy to Pedestrians"

3.1. Policy Origin

In 2009, a shocking "70 yard racing case" occurred across the country. Hu killed Tan Zhuo while racing on Wen'er West Road in Hangzhou, causing him to die on the spot. Hu was sentenced to three years for traffic accidents. On August 27th of the same year, he was revoked his driver's license by the police for driving an illegally modified vehicle that was severely speeding on the main roads of the city. The zebra crossing is a guardian line for pedestrian safety. In order to prevent similar situations from happening, a traffic safety management action centered on "yielding pedestrians" has been comprehensively carried out in Hangzhou.

3.2. Implementation Process of the "Courtesy to Pedestrians" Policy

Hangzhou Public Transport Group was the first to propose the idea of "yielding to pedestrians". In 2005, the company's articles of association included "slowing down and giving way to pedestrians at pedestrian crossings", highlighting the rules and regulations that buses must slow down and allow pedestrians to cross pedestrian crossings. However, although the regulations are very clear, relying solely on the self-discipline of bus drivers to achieve the goal is difficult to truly ensure the safety of pedestrians in traffic. After the "Racing Case", Hangzhou Public Transport Group began to pay more attention to the implementation of the "Courtesy to Pedestrians" rules and regulations. The bus group dispatched inspectors to inspect the roads. Drivers who do not comply with regulations will be deducted 3 to 15 points depending on the violation, and the monthly safety reward of 300 yuan will be cancelled; If the violation exceeds 15 points, an additional 30000 kilometers of safe driving will be deducted, and the mileage will be reduced depending on the level of safe service; Drivers who violate the rules 5 times within a year will not only be deducted all their safety bonuses, but also be suspended from work to participate in company safety training for 200 yuan. Drivers who violate the rules multiple times may even be dismissed. After the implementation of measures such as institutional norms, training and education, assessment and punishment, and inspection by the municipal public transportation authorities, polite pedestrians in front of pedestrian crossings have gradually become a good habit and become a conscious action of tens of thousands of bus drivers, setting a standard in Hangzhou Pedestrian Street.

After the public transportation group took action, the Hangzhou Traffic Management Bureau and the Hangzhou Taxi Association called on taxi drivers to supervise each other, strengthen industry self-discipline, and promote the concept of "giving way to pedestrians". The Hangzhou Transportation Administration and the Hangzhou Taxi Association jointly launched the "Thousand Taxi Passed the Zebra Cross" event, calling on taxi drivers throughout the city to strengthen industry self-discipline, mutual supervision, and ensure the safety of citizens' travel. During the event, illegal "taxi drivers" will be exposed through the "industry exposure platform", and the company where the "taxi" belongs will be notified as an important basis for star rating of taxi drivers. Each taxi company has also formulated relevant rules and regulations, and has taken disciplinary measures such as criticism education, safety training, and withholding bonuses against the exposed "taxi drivers".

At the same time, 100000 households in Hangzhou have signed self-discipline agreements for civilized travel, allowing more and more private cars to gradually develop good habits of civilized travel in the community. And by organizing civilized teams such as traffic vanguards, road safety guides, youth, community residents, and driving schools, volunteer work is carried out in key periods and sections of road traffic. In cooperation with the media, a series of activities such as "zero accidents on the road, quality is loved by everyone in Hangzhou", "no crossing of zebra crossings and red lights" are carried out, allowing more people to participate in the "comity" ranks.

3.3.A Study on the Policy of "Courtesy to Pedestrians" from the Perspective of New Institutionalism

From the origin to the implementation of the policy of "giving way to pedestrians", there are three levels of regulatory elements, normative elements, and cultural cognitive elements in the new institutionalism, as follows

3.3.1. Regulatory elements in the policy of "giving way to pedestrians"

Without the protection of legal systems, rights are only a dispensable gift. Only by elevating "courtesy to pedestrians" to the level of laws and regulations can they serve as the basis for traffic management departments to ensure road safety and pedestrian rights[2]. Issue status can be enhanced as a result of several influences, such as the flow of policy proposals, the passing of legislation, explicit signals of commitments from political authorities, or changes in the political stream. Conversely, issue status may decline as a consequence of slow pace in the generation of policy proposals coming from the alternative-specification process, policy interference effects, or the absence of political authorities' commitment[7].

Therefore, in terms of regulatory elements, "courtesy to pedestrians" has evolved from the rules and regulations of Hangzhou bus and taxi companies to becoming a part of the "Three Year Plan for Building a Traffic Civilization Demonstration City in Hangzhou". On October 30, 2015, the 32nd meeting of the Standing Committee of the 12th Hangzhou Municipal People's Congress passed the "Hangzhou Civilized Behavior Promotion Regulations", And in November of the same year, the "courtesy of pedestrians" was included in the regulation. Hangzhou has gradually evolved the concept of "giving way to pedestrians" from the rules and regulations of buses to a reward and punishment system with complete implementation, supervision, and rewards and punishments. It has adopted a relatively tough approach to "cram" this policy into the lives of drivers, forcing them to comply with it. In terms of implementation areas and law enforcement equipment, it is constantly limited to iterative updates, and the reward and punishment system has been strengthened. In addition to fines, the driver's license must also adopt a deduction system. As the initiator of using the "courtesy to pedestrians" standard to punish motor vehicle drivers in Hangzhou, the Supreme People's Court has also recognized its legality in the form of guiding cases[3].

3.3.2Normative elements in the policy of "giving way to pedestrians"

The general public is the main force of social governance, and the indifference or "explosion" of public social participation in social governance can lead to disorder in social governance. The government needs to adopt reasonable organizational mobilization mechanisms to guide it[4]. Always adopting tough measures will stimulate the public's resistance, thus stimulating the public's sense of social responsibility and forming a new social moral standard is an important step in promoting the policy of "giving way to pedestrians". Through the activity of "evaluating taxis before the thousand taxi zebra crossing", the social responsibility of drivers is stimulated. By using mutual supervision mechanisms, an invisible constraint is formed between the bus and taxi groups, while illegal drivers influence the evaluation,Encouraged drivers' competitiveness and created a good social atmosphere for the new social moral standards. The government's strategy of prioritizing public and private use has set an example for various types of drivers, making the "courtesy to pedestrians" policy more widely accessible to the public

3.3.3Cultural Cognitive Elements in the Policy of "Courtesy to Pedestrians"

Culture can be understood as a complete conceptual system that dominates human behavior with ideology as its core. On this basis, the cultural system can be summarized as a series of rules composed of policies, regulations, codes of conduct, and moral norms that regulate human cultural activities[5].

At this level, the government has improved laws and regulations and established new moral standards in society, enabling drivers and the public to achieve cognitive consensus and behavioral activities. The frequency of people giving way to cars and "racing cases" in Hangzhou has become lower and lower, and the policy of "giving way to pedestrians" has been implemented to break the mentality of pedestrians avoiding vehicles due to fear, reconstruct the concept of pedestrians being the first among drivers, and achieve cognitive consensus, Reshaping the civilized landscape before the zebra crossing, pedestrians will experience the benefits of the policy and comply with the relevant requirements of the policy when converting to driver status, thus achieving consistency in behavior and cognition. In summary, see Table1.

Table 1 Research on the Policy of "Courtesy to Pedestrians" from the Perspective of New Institutionalism

type	content
Regulatory factors	1. Out of self-interest and with the protection of rules and regulations: Hangzhou Public Transport Group has incorporated the requirement of "giving way to pedestrians" into the company's rules and regulations; Clear rewards and punishments, improve the punishment mechanism for violations by bus drivers and the subsequent training system 2. There are laws to follow and law enforcement is based on: Hangzhou City has included "courtesy to pedestrians" in the "Hangzhou Civilized Behavior Promotion Regulations"; Clarify the amount paid in violation of regulations; Improve traffic police law enforcement tools

Normative factors	1.Establish and improve a public supervision mechanism: The Transport Administration and the Taxi Association jointly launched the "Thousand Pre evaluation Taxi Taxis at the Zebra Cross" event, aimed at mutual supervision; Establish a new ethical standard: for taxi drivers who violate regulations, they will be exposed through an "industry exposure platform" and use this as an important basis for evaluating star rated taxi drivers 2/Play a leading role: prioritize public affairs before private ones, drive private cars to sign relevant agreements, and jointly abide by the standards of courtesy for pedestrians
Cultural Cognitive Factors	1.Recognition of the benefits of giving way to pedestrians: Data fully indicates that the yield rate before the zebra crossing is as high as 93.9%, providing citizens with sufficient safety 2.The prospect of urban development is promising: the implementation of the "courtesy to pedestrians" policy has verified the civilized city of Hangzhou, and all types of vehicle drivers are able to consciously abide by it

4. Summary

Although the implementation of the "courtesy to pedestrians" policy faces a fixed group of drivers, there is a huge difference in the number and acceptance ability of the group. Therefore, Hangzhou prioritizes public transportation such as buses, gradually drives taxis and private cars from point to area, and improves relevant laws and regulations. It adopts mandatory methods to regulate drivers, supplemented by flexible reward mechanisms, and the two rely on each other, while using flexible means, Promote people to reach consensus and achieve the goal of establishing reward mechanisms. Institutions simplify political life by ensuring that some things are taken as given. Institutions provide codes of appropriate behavior, affective ties, and a belief in a legitimate order[6].

The regulatory elements, normative elements, and cultural cognitive elements in new institutionalism are effective methods for the government to implement policies, which can form a good interaction and restraint relationship between the public, society, and the government, ultimately achieving the desired effect of the policy. Therefore, before promoting policies with a wide audience, the government can refer to the three elements of new institutionalism, adopt flexible and tough methods, and update the original social consensus, In order to promote new policies more smoothly.

Reference

1. Li Xiangyu, Peng Qian. Institutional Analysis of the Implementation of the "Three Child" Fertility Policy and Supporting Measures from the Perspective of New Institutionalism [J]. Journal of Guangzhou University (Social Sciences Edition), 2022-21 (01): 125-135
2. Wang Xin. How to Effectively Play the Role of External Supervision in "Courtesy to Pedestrians" [J]. Legal and Social Science, 2017 (35): 234-235. DOI: 10.19387/j.cnki.1009-0592.2017.12.238
3. Zeng Fanyan The Implementation Path of the "Courtesy to Pedestrians" Norm - Analysis of the First "Zebra Crossing Penalty Case" in China [J]. Administrative Law Research, 2019 (03): 64-77
4. Yang Fengyin, Li Mengting, Yang Jinwen. A Study on the Linkage Mechanism between Government Governance and Social Autonomy: An Analysis Based on the Social Governance Practice of "Courtesy to Pedestrians" [J]. Journal of the Party School of the Hangzhou Municipal Committee of the Communist Party of China, 2018 (05): 74-80. DOI: 10.16072/j.cnki.1243d2018.05.012
5. Jia Shiqi, Qi Shuyu. The Inner Logic of the Construction of Socialist Cultural System with Chinese Characteristics: Based on the Perspective of New Institutionalism Theory [J]. Research on Governance Modernization, 2021,37 (05): 72-79
6. Tempel A , Walgenbach P .Global Standardization of Organizational Forms and Management Practices? What New Institutionalism and the Business-Systems Approach Can Learn from Each Other[J].Journal of Management Studies, 2010, 44(1):1-24.DOI:10.1111/j.1467-6486.2006.00644.x.
7. Barzelay M , Gallego R .From “New Institutionalism” to “Institutional Processualism”: Advancing Knowledge about Public Management Policy Change[J].Governance, 2006, 19(4):531-557.DOI:10.1111/j.1468-0491.2006.00339.x.

Improving people's well-being: The path to realize the modernization of social governance at the grassroots level

Fuyan Hong [1,*],Jiaqi Li [2]

[1]Harbin University of Commerce, School of Finance and Public Administration, 1979,Harbin, China
[2]Harbin University of Commerce, School of Finance and Public Administration,2001,Harbin, China

Abstract. The grass-roots social governance type national governance modernization work is an important part. The traditional governance concept and governance means have been unable to meet the needs of the current grass-roots development and construction. Therefore, it is of great significance to study the realization path of promoting the modernization of grassroots governance.

1 Introduction

Grassroots governance is the cornerstone of state governance, the long-term rule of the Embassy Party, the long-term stability of the country and the vital interests of the broad masses of the people. Promoting the modernization of the system and capacity of governance at the grassroots level is a basic project to realize the modernization of the national governance system and capacity, and it is also a concrete manifestation of the effectiveness of national governance.At the Fourth Plenary Session of the 19th CPC Central Committee, it is necessary to accelerate the modernization of the national governance system and governance capacity, and strive to form a more mature and stereotyped socialist system with Chinese characteristics. The 20th National Congress of the CPC proposed to comprehensively promote the great rejuvenation of the Chinese nation through Chinese-style modernization, and stressed that the modernization of the national governance system and governance capacity should be basically realized by 2035.

2 The content of promoting the modernization of grassroots governance

In the process of promoting the modernization of grassroots governance, common political governance, economic governance, and ecological governance are key construction contents, which must be paid attention to and emphasized in the development of grassroots governance.

2.1 Political governance

At present, the innovative work of grassroots social governance modernization is to clarify the power relationship between the government and the market, and between the government and society, and this issue contains many contents. Specifically, the work of political governance mainly includes the construction of grassroots rural public affairs, the construction of grassroots community governance systems, institutional publicity, etc., and these contents need to be improved in the development of grassroots social governance work to ensure the orderly development of governance work.[1]

2.2Economic governance

The economic development of local governments at the grassroots level is obviously backward, and the source of township income is single. One of the key tasks of grassroots governance is to improve economic development. Therefore, economic governance is also a part of the modernization of grassroots social governance. The content of economic governance includes the economic industry, the basic economic production mode, and all the relevant content about the economic source. In the governance process of the modernization of grassroots social governance, relevant departments need to guide and mobilize economic subjects to actively participate in grassroots governance work in order to complete the construction of economic governance.[2]

2.3Ecological governance

Ecological governance is also an important part of promoting the modernization of grassroots social

* Corresponding author: 1803118264@qq.com

governance. To do a good job in ecological governance, we can truly realize that green water and green mountains are golden mountains and silver mountains, and it is necessary to strengthen green ecological construction, reduce air pollution, crop pollution, soil pollution and other work.

3 Promoting the modernization of social governance at the grassroots level

In the process of promoting the modernization of grassroots social governance,it has been found that the environment for social governance at the grassroots level still needs to be improved, the participation of social entities is insufficient, the means of social governance are not rich enough, and the infrastructure is relatively weak.

3.1 The environmental conditions at the grassroots level need to be improved

In the process of promoting the modernization of community-level social governance, there is still much room for improvement in ecological environment and infrastructure development. Spring air pollution control work in rural areas is still not deep enough to promote, domestic garbage is piled up at will, and there is a lack of awareness of garbage reuse and centralized treatment.

3.2 Lack of participation of grassroots social entities

In the process of promoting the modernization of community-level social governance, there are still problems of poor implementation, resulting in insufficient participation of main bodies in the construction of this work in some areas. For example, in remote areas, some social subjects do not have a deep understanding of social governance, and the degree of cooperation with the social governance means carried out by relevant departments is not high, resulting in the development process of grassroots social governance work is slow.Moreover, there is also a low enthusiasm of local enterprises to participate in social governance, and the interaction between government and enterprises has not yet formed a good one, which affects the effectiveness of grassroots social governance.

3.3 Insufficient means of social governance

The progress of information and communication technology is an important basic condition for promoting social transformation and achieving the goal of modernizing national governance capacity. In the past, in the process of grassroots social governance, due to the constraints of the natural environment of innate conditions, many historical rural culture, politics, education and other social governance contents were difficult to integrate with the outside world, resulting in remote rural areas still based on traditional human management, and the phenomenon of single governance mode was more obvious. At the same time, with the development of information technology, digital governance can also empower grassroots social governance. Digital governance is an institutional arrangement and continuous process under the conditions of digital technology, led by the government, and the digital platform and enterprises, social organizations, online communities, individual citizens and other multiple subjects to participate in relevant affairs.[3]

3.4 Relatively weak public infrastructure

Public infrastructure is a necessary condition for social governance at the grassroots level. Grassroots organizations pay more attention to achieving economic growth, and invest most of the funds in the construction of various undertakings and industries, thus neglecting the construction of grassroots public infrastructure, resulting in problems. Such as relatively backward public medical equipment, insufficient educational resources and basic conditions, and relatively single cultural basic platforms in some areas.

3.5 The shortage of talents for grassroots governance

In recent years, the state has vigorously encouraged fresh graduates to participate in the construction and development of grassroots and use what they have learned to bring new ideas to grassroots work. However, the common problem with most students who go straight to grassroots jobs after graduation is that they lack the necessary work experience and need to be led by experienced staff when faced with difficult problems on the job.

4 Ways to promote the modernization of grassroots social governance

At present, through the efforts of the Party, the state and all the people, the construction of social governance at the grassroots level has achieved remarkable results, but there are still many deficiencies in the construction process, and relevant departments need to start from the actual deficiencies in the construction, from strengthening grassroots social governance, mobilizing the participation of multiple subjects, integrating media technology, improving infrastructure and other aspects to seek countermeasures to help the modernization of grassroots social governance, to achieve the maximum efficiency of grassroots social governance.

4.1 Focus on community-level social and environmental governance

In view of the problem that the social environment governance work in the process of grassroots social governance is not optimistic, the relevant departments should start from this aspect in the future and constantly improve the system and structure of grassroots social governance. Specific departments optimize and integrate social resources at the grassroots level to achieve social governance. For example, natural rural resources can be transformed into ecological parks or pilot areas for green crops, so as to improve the phenomenon of rural ecological governance and environmental governance. At the same time, grass-roots governments should also focus on the selection of talents, and introduce college graduates, entrepreneurs and veterans and other groups through various channels to serve as supervisors of social and environmental governance in grass-roots areas to ensure the implementation of daily social governance work.[4]Grassroots governments should focus on building high-quality professional social governance teams. With the development of new urbanization, the related rural social governance issues involve more aspects and complexity compared to urban community governance, which further tests the professional ability, comprehensive coordination, and the implementation and guarantee ability of policies such as uniting the masses organized by village and town governments in social governance.

4.2 Mobilize multiple entities to participate in governance

As pointed out in the report to the 20th CPC National Congress, strengthening and innovating social governance cannot be achieved without joint efforts, joint governance and shared benefits, and the coordinated participation of diverse social entities. Therefore, in the process of grassroots social governance, relevant departments need to fully mobilize the main forces of social enterprises, unit organizations and other main forces to jointly build a new pattern of social governance. Relevant departments can create three governance schemes according to actual needs: gap filling mode, collaborative mode and alternative mode. In the vacancy filling model, society and the market are the lead, and the government maintains and coordinates social relations; In the collaborative model, the government, the market and social entities each have the responsibility to divide labor, jointly design social governance norms, and negotiate public affairs; In the alternative model, the government is the leader, and the market and society play an alternative role to solve social problems and provide public services. These three models need to be cross-applied according to local conditions in order to optimize the effect of subject participation in governance.Give full play to the construction of rural social groups, change the single work center of social groups, expand the coverage of social groups' work, and allow more types of social groups to participate in social governance through volunteers, industry associations, united front groups and other relevant group organizations. Build an interactive platform, increase feedback channels, so that the public can feedback their real ideas to government departments, which can be targeted and purposeful when formulating policies, and at the same time, policies can be conveyed to the people in the first time to protect the legitimate rights and interests of the people.

4.3 Actively integrate into the media technology

In recent years, the role of cyberspace has been emphasized in national governance. Taking the entire network system and community as a unit, utilizing the entire internet for information sharing can significantly improve the transmission speed of social sentiment and public opinion. At the same time, information can be made public and transparent to the maximum extent, enabling people to timely understand the process and results of event resolution, allowing them to supervise the true implementation of various national policies and policies, and safeguarding the interests of the people.The development of information technology has brought conditions and opportunities for the promotion of grassroots social governance. In the future, we should actively integrate media technology into the work of social governance and innovate social governance means. For example, in grass-roots towns and villages, new media technologies can be used to establish and integrate media centers to carry out columns such as rural tourism and rural culture construction, and conduct investigations and records on the status quo of rural social governance on the ground. Or arrange financial media staff to carry out social governance theme shooting in the countryside, close to the actual life of grassroots people, and enrich the means of social governance communication. Digitalization is the leading force of Chinese modernization, and digital governance is the only way to modernize China's grass-roots social governance. First, improve the digital governance capacity at the grassroots level, establish digital thinking, grasp the balance between government leadership and the participation of multiple subjects, grasp the balance between administrative logic, market logic, technical logic and value logic, grasp the balance between vertical governance and horizontal governance, decentralized governance and intensive governance. The second is to enhance the digital feasibility of the grassroots, and increase the learning of digital literacy in basic education, higher education and recognition education. The third is to enhance the digital development capacity of the grassroots, promote the transformation of the industrial model from the supply channel east to the demand channel east through digital empowerment, increase the construction of digital human capital, and strengthen digital skills training. Fourth, enhance the capacity of grassroots digital services, encourage the empowerment of digital platforms and means, so that remote areas and low-income people have access to better social services such as livelihood security.[5]

4.4 Improve public infrastructure

At present, there are still problems in the construction of public infrastructure in grassroots social governance, and relevant departments should focus on improving public infrastructure and doing a good job in logistical support. Specifically, the construction of public service facilities such as health care, education work, and production operations in townships and other areas can be increased, and cash infrastructure, appliances and supplies can be purchased, and professionals can be arranged to carry out regular maintenance.

4.5 Professional training of grassroots governance talents

Grassroots social governance personnel are the closest personnel for the people to communicate and contact, and at the same time, they are also the bridge and link between the people and the government. For the government and social organizations at all levels, the existing grassroots staff have the ability to play their own roles in their original jobs, but regular operational training is still needed for grassroots staff. Each policy needs to be explained and trained by professionals at the grassroots level, and at the same time, it is necessary to expand the training at the knowledge level, and in the face of complex social problems, it is necessary to cultivate a variety of professionals, so that the work can be improved in all aspects.The development of economic and social development and people's livelihood is fundamentally at the grassroots level, the most solid force support for social governance is also at the grassroots level, we must grasp the grassroots and lay the foundation as a long-term plan and a strategy to consolidate the foundation, China's social governance system and governance capacity modernization innovation construction is also being continuously improved, according to the goal and purpose of serving the people, in order to continuously meet the people's continuous yearning for a better life, transform into the idea of serving the people, strengthen districts and counties, do enough townships and towns, cover villages, and effectively consolidate the root tomb of social governance, Based on modern social governance services, we will continue to promote rural revitalization.

5.Summary

With the development of new-type urbanization, social governance at the grassroots level has entered a new stage of development and is also facing new challenges. Therefore, while actively promoting the modernization of grassroots social governance, it is necessary for relevant departments to constantly innovate governance methods and explore multiple paths to realize modern governance, so as to provide reference and reference for the realization of modern governance system and governance capacity. [6]

Acknowledgments

This paper is one of the phased achievements of the National Social Science Foundation's general project "Research on Coupling Mechanism of Economic and Social Green Transformation and Local Government Green Governance in Northeast China" (21BZZ045).

References

1. Pan Kaihu. Exploration on the path of Promoting the modernization of grassroots Social governance in Western China [J]. New West,2023,No.561(02):93-95.]
2. Liu Y J. Realistic dilemma, concept transformation and path optimization of grassroots social governance in Western border areas [J]. Guangxi Ethnic Studies,2021(03):26-35.]
3. Li Tao. Promoting the modernization of grassroots Social Governance by Digital Governance [J]. Chinese Civil Affairs,2023,No.745(04):17.
4. P. Halla, A. Merino-Saum, C.R. Binder,How to link sustainability assessments with local governance? – Connecting indicators to institutions and controversies,Environmental Impact Assessment Review,Volume93,2022,106741,ISSN0195-9255,https://doi.org/10.1016/j.eiar.2022.106741.
5. Boon-Kwee Ng, Chan-Yuan Wong, Mary Grace P. Santos,Grassroots innovation: Scenario, policy and governance,Journal of Rural Studies,Volume 90,2022,Pages1-12,ISSN0743-0167,https://doi.org/10.1016/j.jrurstud.2022.01.004.
6. Wang Daguang. Practical Problems, Mechanism Analysis and Innovation Prospect of Public Participation in Grassroots Social Governance [J]. Teaching and Research,2022(04):45-55.

COVID19 Impact on Child Maltreatment: Evidence from Abuse and Neglect Investigations in Texas

Yitian Zhang*

University of Toronto, Department of Economics, 27 King's College Cir, Toronto, ON M5S 1A1, Canada

Abstract. COVID-19 has brought challenges to the society in various aspects, as one of the most vulnerable members of society, children's lives have also been significantly affected by it. This study aims to address the child maltreatment impacted by the COVID-19 pandemic. Leveraging county level data sets from Texas Department of Family and Protective Services and United States Census Bureau, XGBoost method and fixed effect model was used to investigate the most important economic, demographic, and social factors. It is found that population of 16-year-old or over, the rental vacancy rate, the population of 16-year-old or over that commutes to work by walking, and population of 16-year-old or over that works in agriculture, forestry, fishing and hunting, and mining industry is positively associated with total number of child maltreatment cases. On the other hand, the total population in labor force, the female population not in labor force, the average of public cash assistance (in dollars), the average time (in minutes) commuting to work, and the population whose household contains 2 units/rooms are negatively associated with total number of child maltreatment cases. Also, a single-difference model was implemented to identify that the COVID-19 pandemic was associated with 7.6% increase in total number of child maltreatment cases.

1 Introduction

The outbreak of the COVID-19 pandemic has brought unprecedented challenges to societies in North America, affecting every aspect of human life. Among the various concerning issues, child maltreatment emerged as a pressing concern during this turbulent time. The confinement measures, school closures, and economic hardships brought on by the pandemic have disrupted the protective networks for many children, exacerbating their vulnerability to maltreatment. As families faced heightened stress levels and uncertainties, risk factors for child maltreatment surged, prompting a closer examination of this critical issue. This study explores the effects of COVID-19 on child mal- treatment cases through an extensive analysis of existing literature, statistical modeling, and well as causality inferences. We will examine how changes in demographic, economic, and social factors contributed to child maltreatment cases. Additionally, the pandemic impact on number of child maltreatment cases was examined. The significance of this research lies in its potential to inform policymakers, healthcare professionals, and community organizations about the urgent need for attention in child protection and maltreatment prevention during stressed time. By understanding the complex factors at play, we can develop predictive strategies to identify children under risk of maltreatment and provide protective environments for children during times of crisis.

2 Literature Review

Child maltreatment is a complex issue that is influenced by a variety of social, economic, and demographic factors. Research has identified several key factors that can contribute to child mal- treatment, including poverty, a high number of children per adult resident, population turnover, and a high concentration of female-headed families [1]. During times of stress, such as the COVID-19 pandemic, these factors can interact in ways that lead to an even greater impact on child maltreatment. The COVID-19 pandemic has presented many challenges that had a significant impact on the mental health of children and adolescents [2]. Social isolation and loneliness have been shown to have negative effects on the mental health of young people during the pandemic [3]. Parenting has also been affected by the pandemic, with many parents reporting increased levels of stress [4]. This increased stress can increase the risk of family violence [5]. Healthcare workers have also been affected by the pandemic. Many were facing mental health challenges because of their work during this time [6]. Children and adolescents w also experiencing mental health challenges because of the pandemic [7], and school closures have had a negative impact on their mental health [8]. In addition to these studies, there are other reports and research articles that have documented the impact of the COVID-19 pandemic on child maltreatment cases. For example, a report by the Center for Innovation in Child Maltreatment Policy, Research and Training at Washington Univer- sity

* Corresponding author: yitian.zhang@mail.utoronto.ca

in St. Louis discussed how the pandemic has impacted child abuse and neglect. The report highlighted how the loss or reduction of job and financial stability, health, housing, and childcare can increase parental stress, which is strongly associated with increases in child abuse and neglect [9]. A systematic review found that compared to the period before the pandemic, some studies identified an increase in child maltreatment while others found a decrease or no difference [10]. In conclusion, child maltreatment is influenced by many factors and is a complex issue. The COVID-19 pandemic has presented new challenges and has had a significant impact on children and families. This study will investigate the social, economic, and demographic factors that affect child maltreatment and will analyse the impact of the pandemic on child maltreatment while controlling for these factors.

3 Methodology

3.1 Data

This study leveraged the county level economic and social factor data sets from United States Cen- sus Bureau and CPI 3.8 Abuse/Neglect Investigations - Alleged and Confirmed Victims data from Texas Department of Family and Protective Services. The United States Census Bureau provides county level demographic, economic, and social related factors for Texas. The Texas Department of Family and Protective Services provides data on victims in investigations that have been concluded by county from FY2013 to FY2022. We use the total number of cases (including confirmed and unconfirmed cases) as the dependent variable. A confirmed case in a completed investigation is a child who has been identified as a victim in at least one allegation with a disposition of reason to believe. An unconfirmed case in a completed investigation is a child who was alleged to be a victim in at least one allegation with a disposition of unable to complete, unable to determine, or ruled out. The two data sets were merged by year and county. Therefore, we utilized county-level panel data with 221 counties over the time series from 2013 to 2021 in Texas.

After extracting the data, linear interpolation or extrapolation is used to handle missing values. To avoid the interference of outliers, the variables are subjected to a 1% trimming process, which removes extreme values beyond the 1% percentile and 99% percentile. The natural logarithm transformation is applied to all variables (including dependent variable) for re-scaling and lowering standard deviation.

3.2 Model

This study first leveraged the XGBoost method to select the importance of variables, that is, to score the importance of variables. The XGBoost model is an algorithm based on gradient-boosted decision trees, which is known for significantly improving the scalability and generalization of vari- able selection. In each training iteration, the algorithm creates a new decision tree model in the direction of the gradient of the previous model's loss function and prunes it during the tree con- struction. On the other hand, the XGBoost algorithm adopts a distributed forward additive model, directly fitting the residuals of the upper-level trees after each iteration of generating weak learners, thus, enhancing variable selection accuracy through ranking and weighted quantiles.

After preliminary feature selection, a Hausman test is conducted to determine whether to use a fixed effects model or a random effects model. The results indicate a P-value lower than 0.01, suggesting that a fixed effects model should be used instead of a random effects model. The final model is expressed in the following format:

$$\log(Total_{c,t}) = \alpha + \sum_{i=1}^{n} \beta_i \log(X_{c,t}) + \theta_c + \epsilon_{c,t}$$

where c and t refers to county and year respectively, $Total_{c,t}$ refers to the total number of confirmed and unconfirmed child maltreatment cases in county c and year t, $x_{c,t}$ refers to the explanatory variable, θ_c represents the individual fixed effects (county-specific fixed effects), and $\epsilon_{c,t}$ representsthe error term.

Lastly, we would use single-difference model to study the COVID-19 pandemic impact on the total number of confirmed and unconfirmed child maltreatment cases. We used $Covid_{c,t}$ as a time dummy variable. As an indicator for pandemic, it equals 1 if the data point is after the year of 2020 (including 2020) and equals 0 otherwise. The model is expressed in the following mathematical format:

$$\log(Total_{c,t}) = \alpha + \alpha_1 Covid_{c,t} + \sum_{i=1}^{m} \beta_i control_{i,c,t} + \theta_c + \epsilon_{c,t}$$

where α_1 represents the main causal relationship that is being attempted to identify. A positive andstatistically significant α_1 indicates a significant positive effect of the COVID-19 pandemic on the dependent variable. $Control_{i,c,t}$ is a control variable, which is statistically significant explanatory variable from the fixed effect model constructed previously. θ_c represents the individual fixed effects (county-specific fixed effects), and $\epsilon_{c,t}$ represents the error term.

4 Results and Empirical Analysis

Based on the feature importance ranking generated from XGBoost model, we selected control variables and choose the most relevant ones for predicting the dependent variable to fit into the fixed effect model. Table 1 summarizes the top 20 variables with the highest importance score.

Table 1: Importance Score Ranking

Ranking	Variable Name	Weight	Average Gain	Average Coverage	Sum of Gains	Sum of Coverages
1	Population 16 Years and Over	282	60931606	10	1.72E+10	2723
2	Unemployed	72	1.2E+08	60	8.66E+09	4322
3	Armed Forces	49	39127	52	1917226	2563
4	Mean Cash Public Assistance Income (Dollars)	46	10198	2322	469093	10653
5	Mean Travel Time to Work (Minutes)	45	1291	238	58078	10696
6	Rental Vacancy Rate	41	5347	324	219210	13265
7	Mean Supplemental Security Income (Dollars)	39	2880	277	112337	10795
8	Walked	37	133336	140	4933447	5204
9	In Labor Force	35	121	7	4219	254
10	Mean Social Security Income (Dollars)	34	727	253	24718	8591
11	Not in Labor Force	31	8172863	46	2.53E+08	1426
12	Agriculture, Forestry, Fishing and Hunting, and Mining	30	3830	300	114907	9012
13	Average Household Size of Renter-occupied Unit	29	3332	446	96639	12933
14	Mean Retirement Income (Dollars)	28	5411	256	151512	7178
15	Median Earnings for Male Full-time, Year-round Workers (Dollars)	27	15890	278	429043	7510
16	Worked at Home	27	38211	349	1031697	9417
17	Other Means	27	454	71	12252	1928
18	Boat, RV, Van, etc.	26	7786	336	202438	8736
19	2 Units	26	27453	346	713766	8992
20	Public Transportation (Excluding Taxicab)	25	5695396	89.56	1.42E+08	2239

After fitting the dependent variable on the top 20 explanatory variables with the highest importance using fixed effect model, the regression outputs are summarized in table 2, and corresponding variable explanations are summarized in table 3.

Table 2: Fixed Effect Model Output

Variable Name	Estimated Coefficient
Constant	1.51 (1.42)
log(population16yearsandover)	2.18*** (3.45)
log(inlaborforce)	-1.15*** (-3.31)
log(notinlaborforce)	-0.75*** (-2.59)
log(meancashpublicassistance)	-0.01** (-2.04)
log(meantraveltimetowork)	-0.25*** (-4.56)
log(rentalvacancyrate)	0.04*** (4.13)
log(walked)	0.017* (1.94)
log(agricultureforestryfishingetc)	0.15*** (5.83)
log(units_2)	-0.03*** (-2.90)

Note: t-statistics in parentheses
*p<0.1; **p<0.05; ***p<0.01

Table 3: Variable Explanation

Variable Name	Explanation
log(population16yearsandover)	Logarithm of population that is 16-year-old or over
log(inlaborforce)	Logarithm of total population in labor force
log(notinlaborforce)	Logarithm of female population not in labor force
log(meancashpublicassistance)	Logarithm of average cash public assistance income (dollars)
log(meantraveltimetowork)	Logarithm of average number of minutes to travel to work
log(rentalvacancyrate)	Logarithm of rental vacancy rate
log(walked)	Logarithm of population that is 16-year-old or over and commute towork by walking
log(agricultureforestryfishingetc)	Logarithm of population that is 16-year-old or over and work in agri- culture, forestry, fishing and hunting, and mining industry
log(units_2)	Logarithm of total number of units/rooms in the household

From table 2, it can be observed that the coefficients of variables log(population16yearsandover), log(walked), log(rentalvacancyrate),and log(agricultureforestryfishingetc) are significantly positive, which indicates that these indicators have a significant positive impact on the dependent variable. On the other hand, the coefficients of log(inlaborforce), log(notinlaborforce), log(meancashpublicassistance), log(meantraveltimetowork), and log(units_2) are significantly negative, which indicates that the above-mentioned indicators have a significant negative impact on the dependent variable. Also, we calculated some descriptive statistics for all variables, which are summarized in table 4.

Table 4: Summary Statistics

Variable	Number of Observations	Mean	Standard Deviation	Minimum	Maximum
log(Total)	1,957	5.61	1.56	1.61	10.00
log(population16yearsandover)	1,957	9.86	1.52	6.22	14.25
log(inlaborforce)	1,957	9.29	1.57	5.63	13.87
log(notinlaborforce)	1,957	8.99	1.47	5.36	13.17
log(meancashpublicassistance)	1,957	7.81	1.13	-2.84	26.83
log(meantraveltimetowork)	1,957	3.11	0.25	2.46	3.60
log(rentalvacancyrate)	1,957	2.00	0.62	0.00	3.17
log(walked)	1,957	5.12	1.62	0.00	9.68
log(agricultureforestryfishingetc)	1,957	6.64	1.00	3.97	9.57
log(units_2)	1,957	5.17	1.81	0.00	9.51

The output for single-difference model is summarized in table 5. It can be seen that the estimated coefficient of $Covid_{c,t}$ is significantly positive.

Table 5: Single-difference Model Output

Variable Name	Estimated Coefficient
Covid	0.076***
	(3.542)
log(population16yearsandover)	2.238***
	(3.557)
log(inlaborforce)	-1.122***
	(-3.227)
log(notinlaborforce)	-0.746**
	(-2.577)
log(meancashpublicassistance)	-0.009*
	(-1.930)
log(meantraveltimetowork)	-0.237***
	(-4.264)
log(rentalvacancyrate)	0.038***
	(4.003)
log(walked)	0.017*
	(1.930)
log(agricultureforestryfishingetc)	0.137***
	(5.362)
log(units_2)	-0.026***
	(-2.887)
Constant	0.539
	(0.493)

Note: t-statistics in parentheses
*p<0.1; **p<0.05; ***p<0.01

5 Discussion and Conclusion

Based on the fixed effect model results, we can see that holding all other variables constant, when the population of 16-year-old or over, the rental vacancy rate, the population of 16-year-old or over that commutes to work by walking, and population of 16-year-old or over that works in agriculture, forestry, fishing and hunting, and mining industry increases by 1%, the total number of child maltreatment cases is estimated to increase by 2.18%, 0.04%, and 0.15% on average respectively. Also, holding all other variables constant, when the total population in labour force, the female population not in labour force, the average of public cash assistance (in dollars), the average time (in minutes) commuting to work, and the population whose household contains 2 units/rooms increases by 1%, the total number of child maltreatment cases is estimated to decrease by 1.15%, 0.75%, 0.01%, 0.25%, and 0.03% on average respectively. When controlling for those variables that have a significant impact on the dependent variable, based on the single-difference model output, we found that the COVID-19 pandemic was associated with 7.6% increase in total number of children maltreatment cases. There are literature suggesting that the COVID-19 pandemic has had a significant impact on child maltreatment cases. For example, a study found that child maltreatment often increases after natural disasters and economic downturns like the Great Recession [11]. According to a report by the International Labour Organization (ILO), the economic crisis caused by the COVID-19 pandemic is expected to contribute to global unemployment of more than 200 million people next year, with women and youth workers worst-hit [12]. At the same time, it is found that unemployment, food insecurity, and financial stress can increase parents' risk of child maltreatment [13]. Additionally, stress is known to increase the risk of child abuse and neglect [14]. These studies provide strong evidence that the COVID-19 pandemic has had a significant impact on child maltreatment cases, with increased levels of stress among parents being a major contributing factor. Moreover, it is found that the COVID-19 pandemic has been referred to as a "shadow pandemic" of domestic violence [15]. Another study found that parenting stress and the risk of child maltreatment increased during the COVID-19 pandemic [16]. Nevertheless, a paper discussing the potential impact of containment measures on child well-being, found that some pandemic mandates, such as movement restrictions and the closure of schools, services, and businesses increased the likelihood of violence against children and women, while cutting them off from social and professional support systems [17]. Finally, the patterns of parent stress across COVID-19 were found to increased [18]. These studies provide further evidence of the significant impact of the COVID-19 pandemic on child maltreatment cases and highlight the need for policymakers and practitioners to take these findings into account when developing interventions and support programs to address this issue.

Given the urgent need for attention of child maltreatment during stressed time such as COVID-19 pandemic, there are some suggestions to resolve this issue. For example, in Canada, one recommendation is to strengthen the surveillance and research framework for child maltreatment by adding questions about maltreatment to national health and victimization surveys, such as the upcoming Canadian Health Survey on Children and Youth [1]. This would provide valuable information on trends over time and among different subgroups, allowing for the development and evaluation of interventions to prevent child maltreatment. Similarly, based on our study, it is suggested that child protection agencies or social organizations should focus more on parents' mental health as well as the family financial situations during stressed time, which are closely related with child maltreat- ment. Lastly, it is suggested that the child protection agencies in North America, such as Child Aid Society (CAS) in Canada, Federal Agencies Associated with Preventing Child Abuse and Neglect in the United States, could cooperate more with medical institutions to aid in early identification and intervene when abuse and neglect are suspected. For example, the Canadian Paediatric Society has released a statement on "Medical neglect: Working with children, youth, and families" that outlines how health care providers can promote the best interests, well-being, and safety of children or youth at risk for medical neglect [19]. The statement highlights the importance of identifying and addressing any barriers that may prevent a parent from ensuring their child receives the care they need. Similarly, in the United States, the Child Welfare Information Gateway provides information on "Medical Responses to Child Abuse and Neglect," which discusses the crucial role that medical professionals play in assessing and documenting injuries that may be due to abuse or neglect [20].

To sum up, this study leveraged a data set from county level in Texas with rich number of social, economic, and demographic variables with a wide range of time including the stressed time as comparison. It is found that the COVID-19 pandemic has a significant impact on the child maltreatment cases, which led to increasing number of cases. It is suggested that policymakers, social workers, and organizations should pay sufficient attention to parents' mental health, financial stress, employment situations as well as stress resolutions to prevent and mitigate the harm of maltreatment brought to children. At the same time, it's important to note that child maltreatment is a complex issue with many contributing factors. It's difficult to determine a direct causal relationship between these variables and child maltreatment rates without conducting further research.

References

1. Gonzalez, A., Afifi, T. O., & Tonmyr, L. (2021). Completing the picture: A proposed framework for child maltreatment surveillance and research in Canada. Health Promotion and Chronic Disease Prevention in Canada, 41(11), 392-397. https://doi.org/10.24095/hpcdp.41.11.07
2. Fegert, J. M., Vitiello, B., Plener, P. L., & Clemens, V. (2020). Challenges and burden of the coronavirus 2019 (COVID-19) pandemic for Child and adolescent mental health: A narrative review to highlight clinical and research needs in the acute phase and the long return to normal- ity. Child and Adolescent Psychiatry and Mental Health, 14(1). https://doi.org/10.1186/ s13034-020-00329-3
3. Loades, M. E., Chatburn, E., Higson-Sweeney, N., Reynolds, S., Shafran, R., Brigden, A., Linney, C., McManus, M. N., Borwick, C., & Crawley, E. (2020). Rapid systematic review: The impact of social isolation and loneliness on the mental health of children and adolescents in the context of covid-19. Journal of the American Academy of Child & Adolescent Psychiatry, 59(11). https://doi.org/10.1016/j.jaac.2020.05.009
4. Brown, S. M., Doom, J., Watamura, S., Lechuga-Pena, S., & Koppels, T. (2020). Stress and Parenting during the Global Covid-19 Pandemic. https://doi.org/10.31234/osf.io/ucezm
5. Campbell, A. M. (2020). An increasing risk of family violence during the covid-19 pandemic: Strengthening Community Collaborations to save lives. Forensic Science International: Reports, 2, 100089. https://doi.org/10.1016/j.fsir.2020.100089
6. Greenberg, N., Docherty, M., Gnanapragasam, S., & Wessely, S. (2020). Managing mental health challenges faced by healthcare workers during covid-19 pandemic. BMJ, m1211. https: //doi.org/10.1136/bmj.m1211
7. Golberstein, E., Wen, H., & Miller, B. F. (2020). Coronavirus disease 2019 (covid-19) and Mental Health for Children and Adolescents. JAMA Pediatrics, 174(9), 819. https://doi.org/ 10.1001/jamapediatrics.2020.1456
8. Lee, J. (2020). Mental health effects of school closures during COVID-19. The Lancet Child & Adolescent Health, 4(6), 421. https://doi.org/10.1016/s2352-4642(20)30109-7
9. Carter, D, Odama, A, Obi, N, & Lang, H. (2021). COVID-19 and its Impacts on Child Abuse and Neglect.. Report submitted for TPS Child Maltreatment class, Brown School. Retrieved from

https://cicm.wustl.edu/items/covid-19-and-child-abuse-and-neglect-brief/

10. Rapp, A., Fall, G., Radomsky, A. C., & Santarossa, S. (2021). Child maltreatment during the COVID-19 pandemic. Pediatric Clinics of North America, 68(5), 991-1009. https://doi.org/ 10.1016/j.pcl.2021.05.006
11. Seddighi, H., Salmani, I., Javadi, M. H., & Seddighi, S. (2019). Child abuse in natural disasters and conflicts: A systematic review. Trauma, Violence, & Abuse, 22(1), 176-185. https://doi. org/10.1177/1524838019835973
12. International Labour Organization. (2021). Ilo Monitor: Covid-19 and the world of work. 7th edition. COVID-19 Pandemic in the World of Work: ILO Monitor: COVID-19 and the world of work. 7th edition. (2021, January 25). https://www.ilo.org/global/topics/coronavirus/impacts-and-responses/WCMS_767028/lang--en/index.htm
13. Brooks-Gunn, J., Schneider, W., & Waldfogel, J. (2013). The great recession and the risk for child maltreatment. Child Abuse & Neglect, 37(10), 721-729. https://doi.org/10.1016/j. chiabu.2013.08.004
14. Yang, M.-Y. (2015). The effect of material hardship on Child Protective Service involvement.

    Child Abuse & Neglect, 41, 113-125. https://doi.org/10.1016/j.chiabu.2014.05.009
15. Jain, M. (2021). The "Shadow pandemic" of domestic violence. BMJ. https://doi.org/10. 1136/bmj.n2166
16. Wu, Q., & Xu, Y. (2020). Parenting stress and risk of child maltreatment during the covid- 19 pandemic: A family stress theory-informed perspective. Developmental Child Welfare, 2(3), 180-196. https://doi.org/10.1177/2516103220967937
17. Bhatia, A., Fabbri, C., Cerna-Turoff, I., Tanton, C., Knight, L., Turner, E., Lokot, M., Lees, S., Cislaghi, B., Peterman, A., Guedes, A., & Devries, K. (2020). Covid-19 response measures and Violence Against Children. Bulletin of the World Health Organization, 98(9). https://doi.org/10.2471/blt.20.263467
18. Adams, E., Smith, D., Caccavale, L. J., & Bean, M. K. (2020). Parents Are Stressed! Patterns of Parent Stress across COVID-19.

    https://doi.org/10.21203/rs.3.rs-66730/v2
19. Canadian Paediatric Society. (2018) Medical neglect: Working with children, youth, and families. Paediatrics & Child Health, 23(7), 468-472. Retrieved from https://cps.ca/en/ documents/position/medical-neglect
20. Child Welfare Information Gateway. (2021). Medical Responses to Child Abuse and Neglect.

    U.S. Department of Health and Human Services, Children's Bureau. Retrieved from https://www.childwelfare.gov/topics/responding/collabresponse/medical/

https://doi.org/10.1051/shsconf/202317801023

Re-training vocational education to nurture craftsmen for the new era

Chuanli Wei*, Depeng Lv

Harbin University of Commerce, Harbin, 150028, Heilongjiang Province, China

Abstract.A great nation rises, craftsmanship builds dreams. The working class occupies the position of the leading class in China, as the main force of reform and opening up as well as the construction of socialist modernisation. The construction of the new era of socialist economy with Chinese characteristics cannot be separated from the new era of craftsmanship, however, to carry forward the spirit of craftsmanship, excellent professional knowledge and skills is an indispensable quality of the workers group. Therefore, we should focus on the development of vocational education, comprehensively improve the level of vocational education, provide more skilled personnel support for regional modernisation, breed new-age craftsmen in line with the development requirements of the times, motivate the majority of young people to take the road of skills to become a success and serve the country with their skills, accelerate the construction of an army of knowledge-based, skilled and innovative workers, and provide strong talent support for the comprehensive construction of a modernised socialist country.

1 Content of vocational education to cultivate craftsmanship in the new era

The so-called craftsmanship refers to the attitude of pursuing excellence in creation or work, and it is also the embodiment of professional ethics and character, as well as a kind of professional value orientation and behaviour of practitioners. For individuals, craftsmanship is a kind of doing a thing, loving a thing, specialising in a thing, and doing a good job of a thing, down-to-earth and serious, persistent, meticulous, fine polishing spirit. For enterprises, craftsmanship is a kind of corporate culture of sticking to the speciality, making high quality products, creating technology, building standards, persevering, being more competent and pioneering. To the society, it is a social culture of cooperation, contract observance, creditworthiness, promotion of coordination, division of labour, win-win cooperation and perfect improvement. Advanced vocational colleges focus on cultivating students' vocational skills, and end with training students to eventually find employment in the industry. At this time, it is necessary to integrate craftsmanship into teaching, so that students can have a correct and affluent mindset, learn theoretical knowledge on the premise of having a good quality of professionalism, and take the initiative to learn and develop their skills and specialities, in order to cultivate excellent technical talents in all aspects.

1.1. Love and respect for work

The cultivation of artisan spirit in higher vocational colleges is an urgent demand for high-quality economic and social development and people's better life, as well as the internal pursuit of building a high-quality higher vocational education system in the new era. [1]To love one's job is to respect and love one's profession, as well as to be committed to and pursue one's work. The talents trained by modern vocational education are not only working machines, but also people with dreams, ideals and beliefs. They may not be able to become the strong elite of the society, but they will surely become the pillar of social progress. They are dedicated to their work, highly love the work they are engaged in, have a sense of responsibility for their work, not only to take responsibility for their work seriously, but also to strive for excellence, not a hair out of place. The root of craftsmanship is love and dedication to work, and the ultimate goal to be achieved by vocational education to cultivate craftsmanship is also love and dedication to work. Therefore, in vocational education, we should strengthen the concepts of "people-oriented", "student-centred", "job competence-based" and "quality improvement as the core", and take love and respect for work as the goal of training talents in modern vocational education. Vocational education personnel training objectives. [2]

* e-mail:1649104896@qq.com

https://doi.org/10.1051/shsconf/202317801023

1.2. Striving for excellence

Vocational education students should have the quality of excellence. Craftsmanship requires students to be very responsible for their professional work, to pay attention to all details, to pursue perfection and to be rigorous and meticulous. Craftsmen create things with their craftsmanship expertise, and in the continuous improvement and breakthrough of their profession, they perform the exquisite skills of "being able to do what others can't" by virtue of the pursuit of excellence.[3] Respect and promote the spirit of craftsmanship is a fine tradition in China since ancient times. Since the founding of New China, the Communist Party of China has always insisted on promoting the spirit of craftsmanship in the process of leading the people to carry out the construction of socialist modernisation, creating a large number of craftsmen in pursuit of the ultimate, the pursuit of excellence. As students of the new era, we should abandon impatience, quiet mind, not on the surface of things, and do not just ask for a pass, but to do better, especially in vocational education and training, we always have to instil this ideology of striving for excellence.

1.3. Innovative breakthroughs

Under the pressure of free market competition and market transformation and upgrading, "Craftsmanship" is given a new connotation oriented to innovation. To cultivate a large number of big country high skill new era craftsmen talents, modern vocational education needs and students of higher vocational colleges and universities need to have innovative consciousness and innovative ability, the formation process of craftsmanship is the process of forming innovative consciousness and innovative ability of students in higher vocational colleges and universities, so higher vocational colleges and universities need to focus on cultivating the innovative ability of students, but cultivating the innovative consciousness and ability of students is not a one-off, it needs the colleges and universities However, the cultivation of students' innovative consciousness and ability cannot be achieved overnight, which requires institutions to have a long-term vision and planning, and focus on cultivating students' innovative ability. Higher vocational colleges and universities should also strengthen practical teaching and the construction of practical training bases, so that students can continuously improve their problem-solving ability in practice, thus enabling students to gradually develop innovative and creative abilities. Finally, vocational schools should provide students with more opportunities and platforms for social contact, so that students can get in touch with the society, understand the society, and cultivate outstanding talents with innovative consciousness and skills.

1.4. Working together

In today's society, the wave of globalisation is constantly flashing, and countries are paying more attention to win-win cooperation with each other, no matter in economic, political, cultural and social aspects. Therefore, the team spirit of collaboration is indispensable in the new era of vocational education and training. The youth of the new era should not only have a strong ability, but also understand the meaning of cooperation, and recognise the importance of win-win situation, and work together in the process of mutual cooperation to complement each other's strengths and weaknesses, in order to achieve the effect of one plus one is greater than two. As one of the aspects of craftsmanship, the team spirit of collaboration and co-progress requires us to focus on competition and innovation in the cultivation process, while at the same time inculcating the team spirit all the time, so that the majority of young people will have the team spirit and become a trustworthy reserve army.

2 Strategies for vocational education to accelerate the breeding of craftsmen for the new era

With the development of society, the status of education is getting higher and higher, but it is not difficult to find out that the development of vocational education is slow through Fig.1, but the development of vocational education is the basis for cultivating craftsmen in the new era, so the following strategies are proposed.

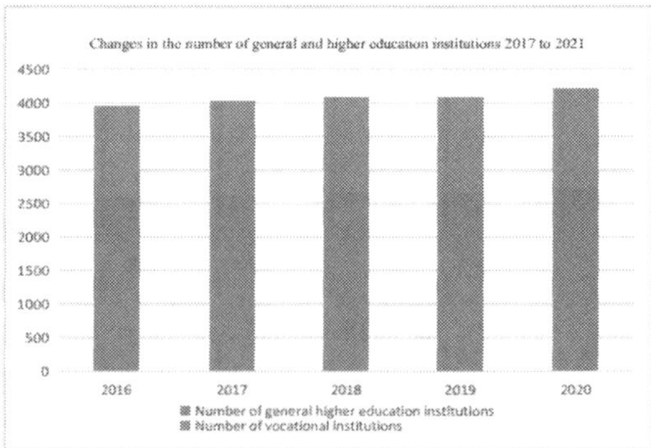

F ig. 1.Changes in the number of general and higher education institutions 2017-2021

2.1. Increasing innovation in the vocational education system

craftsmanship spirit and professional skills have a high degree of internal correlation, so it can not achieve the corresponding effect if they are cultivated separately. [4] The development of modern vocational education requires theoretical innovation and the integration of craftsmanship and talent training. From the perspective of school philosophy, modern vocational education must take moral education as a fundamental task, focus on cultivating students' innovative spirit and practical ability, and integrate the cultivation of craftsmanship into the whole talent cultivation process with the fundamental goal of students' comprehensive development. From the concept of theoretical teaching, modern vocational education must change the traditional concept of theoretical education, to cultivate "craftsmanship" as an important guarantee to achieve the goal of talent cultivation in schools; from the concept of practical teaching, modern vocational education, the traditional teaching concept of the curriculum "knowledge-based", "ability-based"and so on. From the perspective of practical teaching concept, modern vocational education traditional curriculum teaching concept of "knowledge-based" "ability-based" concept needs to be changed, to the student as the main body, practice-oriented, guide students to participate in the whole process of curriculum teaching. From the viewpoint of talent cultivation objectives, modern vocational education should integrate the cultivation of craftsmanship into the whole process of talent cultivation, and take the cultivation of students, social demand, recruitment demand and school professional construction objectives as the talent cultivation objectives.

2.2. Increase the supply of vocational education policies

The development of vocational education has become an important strategic choice for all countries in the world to meet the challenges of economy, society, population, environment and employment and to achieve sustainable development. [5] In the new era, we should attach great importance to vocational education and place vocational education in a more prominent position in economic and social development and educational reform and innovation. There are many policies related to vocational education in recent years, so increase the policy support for vocational education, turn vocational education and ordinary education into the same type of education, award vocational education degree certificates with the same value as ordinary university degree certificates, improve the Chinese model for the development of modern vocational education, continue to increase the supply of policies, accelerate the improvement of the modern vocational education system, and construct a diversified pattern of schooling as well as modern governance system, injecting a strong force into the path of Chinese-style modernised vocational education.

2.3. Improvement of the social status of skilled personnel

Improve the political treatment of skilled workers in many forms. In the view of many people, even if the income of skilled workers is high, they still do not enjoy the full respect of society. Only by improving the political treatment and social status of skilled workers can we encourage more people to join the skilled labour force. To advocate the "craftsman spirit", we must actually improve the status and income of skilled workers. The development of enterprises cannot be separated from scientific and technological innovation talents. Only by helping scientific and technological talents to solve their concerns, their hearts can focus on the development of the enterprise. Higher wages will enable employees with excellent skills to enter their positions and give them a strong sense of acquisition, happiness and achievement. The wages of highly skilled workers should be higher than the national average. The wages of skilled workers in some special precision manufacturing industries are those of ordinary civil servants and even higher than those of university professors. In formulating talent policies, the status of skilled workers should be fully respected, and skilled workers should be encouraged institutionally, so that there will be an endless stream of such practical workers, and the outstanding craftsmanship of skilled workers will be used to create the world-renowned "Made in China".

2.4. Enhancing the capacity of the vocational education teaching force

In school practice, there is a tension between the ideal educational policy goals of student self-management and autonomy and the lack of teacher time and attention to students.[6]Vocational education is a complex educational activity, containing the unity of opposites between teachers and students, schools and enterprises, education and industry and other elements. [7] Gradually raise the level of academic knowledge for teacher training. In the examination of degree qualifications and the management of student enrolment programmes, universities in the field of vocational and technical education, especially those engaged in vocational and technical teacher education for a long period of time, will be supported. Support in-service teachers to upgrade their academic qualifications and make appropriate policy support. Explore diversified paths for educating people. Improve the vocational education teacher training system and create high-quality teacher training bases, while strengthening the construction of existing teacher training centres. Promote the construction of digital learning platforms for vocational teachers. Start a teacher learning and training version of the National Vocational Education Intelligent Education Platform, and share it with all teachers. Promote the construction of national vocational university teacher innovation groups. Map survey and research, strengthen process management, and at the end of the year, test the first team and systematically organise the construction results. The most important thing is to create a good atmosphere of respect for vocational teachers in the whole society. On the basis of strengthening the cultivation of "dual-teacher" teachers in higher vocational colleges and universities, we will further improve the operation mechanism of school-enterprise co-operation in building "dual-teacher" teacher training bases, so that "dual-teacher" teacher training bases will become the base for the formation of students' technical skills and the formation of students' technical skills and the formation of students' technical skills. On the basis of "dual-teacher" teacher training efforts, further improve the operation mechanism of

https://doi.org/10.1051/shsconf/202317801023

school-enterprise cooperation to build "dual-teacher" teacher training bases, so that "dual-teacher" teacher training bases become an important base for the formation of students' technical skills and professionalism. [8]

2.5. Carry out in-depth school-enterprise co-operation

Change and innovate the idea and mode of school-enterprise cooperation, carry out joint school running, take into account and balance the interests of higher vocational colleges and universities, enterprises and students, mobilise the enthusiasm of enterprises to participate in the education and training of enterprises into the education and training of enterprises as a role in the process of vocational education, play the cultivation advantages of the enterprise and according to the different attributes of the enterprise, the positioning of the terminal students, the formation of a healthy production line, so that different attributes of the students to learn and apply to the enterprise to get deeper progress after entering the enterprise. The students with different attributes can learn to use what they have learnt and make deeper progress after entering the enterprises. We should supervise and regulate the cooperation between the two sides in the system, learn from the vocational education models of developed countries, increase the government's financial and policy support for school-enterprise cooperation, and endeavour to build a set of reasonable and healthy education models and management systems for the joint progress of vocational colleges and enterprises, which will not only make contributions to the progress of the society, but also contribute to the strengthening of the bilateral construction.

Colleges can build practical training bases according to actual needs and carry out practical training teaching for students by introducing talents from enterprises. Practical training teaching integrates and intersperses theory and practice, and adjusts the teaching method and class time according to actual needs. In understanding the interests of students, according to the characteristics of each student to guide students to independent learning, stimulate students to explore the professional field of interest. You can continue the "old with new" mode in the enterprise, using the apprenticeship system for teaching, selecting experienced technicians in the enterprise to lead the class for practical operation, so that students can deeply feel the spirit of craftsmanship in practice, and this spirit is passed on, and gradually bring the spirit of craftsmanship into the industrial sector, the line of influence, and eventually Promote the formulation of industry standards and norms, and set up spiritual benchmarks.

3 Conclusion

Craftsmen are an important force for industrial development, and craftsmanship is an important spiritual source for innovation and entrepreneurship. We should endeavour to cultivate more high-quality technical and technological talents, heavenly craftsmen and craftsmen of the times, and provide strong personnel guarantee for the comprehensive construction of a modern socialist country. System innovation, policy supply, to increase investment, improve the spirit of craftsmanship, improve the social status of technical and functional talents, for the comprehensive construction of socialist modernisation of the country, for the realization of the Chinese dream of the great rejuvenation of the Chinese nation to provide a strong pillar of human resources and functions. The cultivation of craftsmanship in modern vocational education is analysed, and specific cultivation paths are proposed, aiming to provide theoretical reference for the reform and development of vocational education.

References

1. Li D ,Tahir S . The Effect, Dilemma and Countermeasures of Cultivating Students' Craftsman Spirit in Higher Vocational Colleges in the New Era[J]. Curriculum and Teaching Methodology,2023,6(17).
2. X. Chen, Vocational education and human-centredness. Mod. Vocat. Educ. **1**, 1 (2018)
3. J.L. Tu, Research on the cultivation path of craftsmanship in modern vocational education. Contin. Edu. Res. 93-97 (2023)
4. Qing J ,Chunshang W . Research on the Cultivation Path of the Integration of Craftsman Spirit and Professional Skills[J]. SHS Web of Conferences,2023,157.
5. Ministry of Education, China Vocational Education Development White Paper. China Vocat. Tech. Educ. 154 (2022)
6. Maija A N ,Markku J . Educating self-governing learners and employees: studying, learning and pedagogical practices in the context of vocational education and its reform[J]. Journal of Youth Studies,2020,23(9).
7. D.Y. Xu, Z.J. Chen, Teacher team construction of vocational education in China:experience, problems and prospects. South. J. Vocat. Educ. 11, 74-80 (2021)
8. J.L. Tu, Research on the cultivation path of craftsmanship in modern vocational education. Contin. Edu. Res. 93-97 (2023)

Research of Public News Headlines' Bias in the Context of New Media: Using "Weibo" Hot Search Headlines as a Case Study

Xinyi Shen*

School of Journalism and Communication, Hangzhou City University, Hangzhou, Zhejiang, 310015, China

Abstract. In recent years, the criteria for creating popular search headlines on Weibo have made it a socially controversial source of information. Using content analysis as the primary research method, this study seeks to systematically dissect the presentation of news topics within Weibo's popular search headlines. Through a careful examination of news content, trends, and patterns, it aims to reveal the existence and dynamics of biases in news selection and presentation. Furthermore, this paper investigates the dissemination of public news and its complexity while exploring both internal and external factors contributing to deviations in the publication of public news headlines. It also strives to enrich our understanding of the significance of this platform and its potential impact on Chinese society. Insights gained from this research contribute to our understanding of the evolving patterns of news dissemination within the Chinese internet ecosystem. It underscores the importance of critically evaluating the role of digital platforms in shaping public discourse and information consumption, with a focus on bias issues and their broader societal implications.

1 Introduction

In the age of Web 4.0, where information dissemination is increasingly mobile-centric, the internet's ever-expanding landscape of information sources has reshaped how individuals' access and engage with news and content. Notably, Weibo, a prominent social media platform, has emerged as a central player in shaping the narrative and direction of news and information within the Chinese media landscape. Through trending topics and viral headlines, Weibo has demonstrated its capacity to shift the focus and direction of news events, often reconfiguring the cause-and-effect relationships surrounding these events. This phenomenon inadvertently sparks controversies and propels debates, thereby influencing public opinion and discourse.

In China, the 50th Statistical Report on the Development Status of the Internet in China, which was published by CNNIC, indicates that as of June 2022, China had 1.051 billion Internet users, 1.047 billion mobile phone users, 788 million users of online news, and 29.5 hours of per capita weekly Internet access. Because there are so many people using the Internet, the Internet news user group makes up approximately 75% of all Internet users. There are now 1.051 billion Internet users in China, released is the 50th Statistical Report on the Growth of the Internet in China. It's also important to discuss and conduct tests on the problem of content bias in digital and data-driven news distribution media behind algorithms and manual audits as the primary way of tweeting popular topics.

This study aims to explore and analyses the trends in Weibo headlines from the previous year, utilizing the lens of communication science. By conducting a comprehensive analysis of these trends, the study seeks to shed light on the ways in which Weibo shapes public news communication in China and reaffirms moral values within this new media landscape.

2 Current literature review

Ardèvol-Abreu A remarks that social media can be a source for mainstream and citizen-created news, but also for hybrid information containing the attributes of both and also finds that the perception of media bias negatively impacts news consumption across all media platforms, leading individuals to seek less biased alternatives, but it acknowledges the complexity of trust development in the context of social media [1]. Entman R M stated that that media bias can affect political power dynamics by strengthening favoured officials and weakening others, and integrating concepts from media studies has the potential to yield empirical evidence and improve normative recommendations for enhancing media's contribution to democracy [2]. Toni G. L. A. Van der Meer, Michael Hameleers & Anne C. Kroon emphasize the alignment of negativity in news reporting with citizens' selection bias, driven by confirmation bias and congruence with their political beliefs, suggesting that ease of processing source cues is not the sole determinant of news exposure [3].

One discusses how social media serves as a source of diverse news types and its role in influencing individuals

* Corresponding author: 1064354566@qq.com

to seek less biased alternatives, emphasizing the challenge of trust development in the social media context and focuses on the political consequences of media bias and advocates for integrating media studies concepts to enhance the media's role in democracy. Someone highlights the alignment of negativity in news reporting with citizens' selection bias, driven by confirmation bias and political beliefs, challenging the idea that source cues alone determine news exposure. These viewpoints collectively highlight the complexity of media bias and its impact on news consumption and politics.

According to Zhu Hongjun and Zhou Kui, the term "neutrality" is frequently used by new media platforms as an excuse to avoid their social obligations to the media. These platforms' dubious content, lack of verification, commercial-type algorithmic tendencies, and ethical responsibility for the use of technology are all signs of their "pseudo-neutrality"[4]. "It contradicts the value of the new media platform (Weibo) itself to disseminate the amount of value by attributing all of the public opinion issues brought on by the names of Weibo hot search titles to the calculation and pushing of big data algorithms while ignoring the positioning of people as the "gatekeepers," externalizing social responsibility, and attributing all of the issues to technology [4]. Zhang Yunfeng identifies two primary traits in the creation of news headlines: the urge to get readers to click and share, or to make information more readable by using colloquial language and the frequent appearance of "headline parties" which frequently grab readers' attention by using exaggerated language. Readership and conversation are typically sparked by contentious headlines, and editors' subjective consciousness is preoccupied with the headlines' ability to influence public opinion [5]. In response to the overwhelming volume of information on Weibo, Yang Ning believes that most users lack the time and energy to actively choose and can only passively choose to learn about recently occurred events by perusing the popular subjects in the Weibo hot search list. Weibo headlines are the primary source of the public's grip the way of propagating ideas as a result, and they are extremely susceptible to the effect of relevant algorithms and artificial gatekeepers in the fragmented time when users collect news information [6]. The stereotypes are aggravated by violent symbolic attacks on the "stigmatization" of women, which are primarily distinguished by the target of the attacks being mainly women and the content of the attacks being mainly sexualized slander and degradation, according to Chen Qi [7]. Hou Yufei writes in his article "A glimpse of the "one-sided" phenomenon of gender issues on microblogging platforms" that media platforms (microblogs) have recognized that gender issues are quite topical and have increased the proportion of gender issues in the selection of news events such as hot searches, and that the publication of such social events is strongly correlated with the setting of the platform's intervention agenda [8]. More frequently than "a man," "a woman" is the subject of headlines that deflect attention and present contrasting arguments to elicit debate. According to Song Mengyuan, prejudiced discourse structures are the root of regional stigma because of cognitive differences and under-expression, which causes all verbal views of a place to be transformed into rumors, prejudices, etc. Big data crawling or individual access to information frequently defaults to a quantity-based strategy, often resulting in the broadcast of pertinent false information that is both verifiable and challenging to test [9].

The discussion delves into the critical issue of content bias within digital news distribution, raising questions about the concept of "pseudo-neutrality" employed by new media platforms. These articles also further examine the pivotal role of headlines in news consumption, emphasizing their use of colloquial and exaggerated language to capture readers' attention. Gender-related stereotypes in news coverage and regional stigma stemming from prejudiced discourse structures should be solved as well. Likewise, it notes the challenges posed by big data strategies, often leading to the dissemination of false information, highlighting the complexities of today's digital media landscape. Overall, this paragraph offers a comprehensive overview of the intricate interplay between technology, media, and societal dynamics in the digital age.

3 News headlines' biased textual analysis

Fragmented semantic concepts that emerge more frequently and are more consistently in nature are distilled into categories with implications to broaden the scope and dimensionality of the study perspective. A total of 204 contentious hot search titles that appeared on the "Weibo Hot Search List" from October 2021 to October 2022 were chosen as the sample for analysis to pursue the accuracy and comprehensiveness of the textual content. These hot search titles were manually categorized and refined to create a coding system that reflected the "Weibo Hot Search List" in October.

The percentage of 35 headlines that dealt with "relationship between men and women" among them reached 17.2%. For instance, "Man driving drunk during internship insists he is a woman," "Woman told she has a boyfriend, not recruited," "A boy in Shanghai builds a family amusement park for his girlfriend," and more. Only three of these headlines, usually concerning criminal instances that jeopardize social security, clearly demonstrate a bias in favour of positive reporting. The analysis found that 28.6% of the headlines were based on the victim's side of the event, with only one male and the rest female, and that they frequently used terms like "was" and "were" to emphasize how vulnerable women were in the incident and spark conversation, for example, "The headline "Woman burned, scarred and amputated by boyfriend" skilfully employs a sequential relationship to distract the public's attention and generate a conversation on "what did the woman do to deserve the petrol? ", emphasizing what the woman experienced rather than what the guy did. This is an ingenious use of the sequential relationship to draw attention away from the debate about "what the woman did to deserve the petrol" and focus it on "why the man threw petrol." Many news headlines, or 25 out of the 71.4%, did not use passive subjects, although they were nonetheless narrative in nature. Examples include "Drunk man beats up female neighbor in lift" and "30-year-old female master's degree student cheated out of $580,000 for online dating in a pig killing dish. "The headlines accurately and succinctly describe what happened, and they identify a particular group of people with the qualifiers "drunk" and "female master," attaining the goal of objective assertions without sacrificing the discussion of the subject. Weibo's topical angle about male-female relationships is still primarily negative news-based now, with the intention of challenging male-female relationships while exposing social difficulties. There is no shortage of headlines that are meant to draw attention or purposefully obfuscate active and passive relationships, even though mainstream newspapers often convey facts objectively in a "subject-verb-object" style.

The comparison of geographical regions and the corresponding preconceptions is the major issue with the headlines of "geographical" hotspots. For instance, the headlines "No chicken can escape Guangdong" and "No sheep can leave Gansu alive" automatically give the idea that all residents in Gansu enjoy eating chicken and sheep, respectively. With such regional headlines, which are obviously subjective, accounting for 11.1% of Weibo's coverage of local events, such extreme emotions tend to reinforce preconceptions of a particular region and are not favourable to cultural understanding and integration between regions. With more than 200,000 talks, the percentage of such headlines was 78.6%. The headlines were frequently brief, easy to muddle with behavioural relationships, and not obviously biased. Although the headlines are usually brief and opaque behavioural relationships without any clear reporting bias, they tend to trigger public debate and big-topic debates between various geographic areas at the first level. Positive values frequently shift when headlines like "eye-catching" and "curiosity hunting" appear. Hot searches for "Boss promises programmer 15 days of annual leave after getting off" and "Bride throws bouquet, 7 bridesmaids scream and dodge" totalled 23, or 11.8% of the total, and reported information that threatens social security in the form of hilarious jokes, for instance. He's in love, "Bride throws bouquet, 7 bridesmaids scream and dodge," and "The bride throws bouquet, 7 bridesmaids scream and dodge" are just a few examples of the subjective search headlines. Several male pals were drugged and knocked out by a drug dealer to commit molestation.

Big data algorithms and people play two "gatekeeper" roles in the creation of Weibo hot searches in the current setting of the new media era of big data. According to the findings of the textual analysis of Weibo headlines, hot search headlines that objectively state news events are still the mainstream drafting direction, while factors that primarily affect public psychology and the scope of public opinion are not the main targets. This is because the central government and the General Administration of Radio, Film, and Television have a strict control over online public opinion. Unavoidably, some headlines do need to be altered in terms of formulation. To start, they should stop using the victim as the main subject and switch to using the vulnerable as the angle of traffic attraction. This would draw more public attention to the headlines. According to the definition of symbolic violence, it was French sociologist Pierre Bourdieu who first put forth the concept. Bourdieu claimed that symbolic violence is distinct from other types of violence because it is a "gentle, undetectable form of violence... that is not so much forced as it is coerced." It is more freely chosen than it is accepted. To give the audience another option, new media platforms like Weibo first screen publicly available news material once, compile it, and create catchy headlines. This study examines whether media groups generate headlines based on "objective statements" or on the bias of information diffusion between primary and secondary screening, both automatically and manually. In summary, the trend of information distribution and communication is being driven by the flow of new media, and high levels of debate and retweets have emerged as one of the standards for determining if information is reliable. From the standpoint of symbolic violence, audiences actively choose contentious and unusual news stories as a useful entry point into the public discourse, which to some extent also results in the platforms themselves using their position as "gatekeepers" to meet social needs and other requirements by "embellishing" news content that would otherwise seem commonplace and ordinary. This has, to some extent, resulted in the active use of its function as a "gatekeeper" within the platforms to "embellish" news material that appears routine and every day, as opposed to the active use of "symbolic violence."

4 The Role of Symbolic Violence in Media Headlines

According to the content analysis, the concept of "symbolic violence," as initially articulated by French sociologist Pierre Bourdieu, carries profound implications when applied to the analysis of media headlines [10]. Bourdieu posited that symbolic violence differs from other forms of violence as it operates subtly, almost imperceptibly, coercing rather than forcing individuals. It is a form of violence more freely chosen than accepted, making it particularly insidious [10]. In the context of contemporary new media platforms like Weibo, the process of headline creation serves as an illuminating example of symbolic violence at play. These platforms function as gatekeepers of information, holding significant sway over the narratives that permeate the public sphere. They possess the authority to curate, synthesize, and frame headlines, thereby shaping what captures the public's attention, instigates debates, and fuels discourse.

As revealed by the 204 textual analysis of Weibo headlines, the prevailing trend still leans toward objective reporting of news events. However, the dynamics at play extend beyond mere factual dissemination. The headlines crafted often play with the positioning of subjects, subtly shifting the focus from the perpetrator to the victim or the vulnerable. This strategic choice influences the public's perception of events and serves as a form of symbolic violence. Crucially, this manipulation of headlines does not occur in a vacuum. Audiences actively participate in this complex ecosystem by selecting contentious or unusual news stories as entry points into public discourse. They are drawn to narratives that spark controversy or intrigue, thereby reinforcing the gatekeeping role of these platforms. It is a symbiotic relationship where social demands and media influence converge, resulting in the embellishment and amplification of seemingly mundane news content.

In summary, the study of symbolic violence in media headlines is an ongoing journey, one that continues to reveal the intricacies of power and influence in our information-saturated age. The evolving landscape of information distribution and communication, driven by new media platforms, underscores the significance of headlines as gatekeepers of public discourse. The active use of symbolic violence, where narratives are subtly shaped and perspectives are influenced, has become an inherent part of media headline creation. This analysis contributes to a deeper understanding of the mechanisms at play in the digital age, where news headlines are not just conveyors of facts but powerful tools for shaping public opinion and discourse.

5 Prospects and conclusion

According to the meticulous analysis of the texts under scrutiny, news headlines continue to adhere steadfastly to the indispensable standards mandated for information dissemination in China. Firstly, the confirmation that news headlines in China align with the established standards of impartiality. In an era where misinformation and biased reporting have garnered significant attention globally, this reaffirmation of journalistic integrity in China serves as a model worth emulating. Moreover, the emphasis on objectivity and effectiveness within the communication industry raises questions about the potential global applicability of these practices. Could the principles observed in China's news headline creation serve as a blueprint for journalism practices worldwide? This sparks a dialogue about the cross-cultural exchange of ideas and best practices in the field of journalism, an avenue ripe for further exploration and collaboration.

Understanding whether editorial choices can indeed shape the agenda of public conversations will not only deepen our comprehension of media's role but also offer insights into the mechanisms of information dissemination in the digital age. However, as we delve into this complex web of influence, it raises pertinent ethical and societal questions. What are the responsibilities and obligations of these gatekeepers in an era where information is a potent tool? How do these manipulations affect our collective values and beliefs? It is incumbent upon us to scrutinize the impact of symbolic violence in media headlines and to reflect on the evolving dynamics of information dissemination in our increasingly interconnected and digitized world.

In conclusion, according to the analysis of the texts used in this study, news headlines are still created in accordance with the necessary standards for information dissemination in China, which include being impartial, fair, true, and credible and not actively biased in favor of one side or the other. The communication industry has a positive side because it is objective and effective at spreading comprehensive and highly accurate information. Future analysis and research will examine the connection between the editors' choice of hot headlines "which does not reflect their subjective consciousness" and the volume of discussion and buzz they produce.

References

1. Ardèvol-Abreu A, Gil de Zúñiga H. Effects of editorial media bias perception and media trust on the use of traditional, citizen, and social media news[J]. Journalism & mass communication quarterly, 2017, 94(3): 703-724.
2. Entman R M. Framing bias : Media in the distribution of power[J]. Journal of communication, 2007, 57(1) : 163-173.
3. Van der Meer T G L A, Hameleers M, Kroon A C. Crafting our own biased media diets : The effects of confirmation, source, and negativity bias on selective attendance to online news[J]. Mass Communication and Society, 2020, 23(6) : 937-967.
4. Zhu Hongjun, Zhou Kui. Pseudo-neutrality: an examination of the gate-keeping mechanism and social responsibility of information aggregation platforms [J]. Journal of Nanchang University (Humanities and Social Sciences Edition),2017,48(05):102-107.
5. Zhang Yunfeng. An analysis of the innovation strategy of television livelihood news in the era of media integration [J]. Today's Media,2022,30(11) :51-53.
6. Yang Ning. Research on the problems of microblogging hot list under the perspective of communication science[J]. News Communication,2017(12) :121-122.
7. Chen Q. "Victim's original sin theory" : symbolic violence and public opinion failure in the online commentary of sexual assault cases[J]. Contemporary Communication,2018(06) :77-81.
8. Hou Yufei. A glimpse into the phenomenon of "one-sided" gender issues on microblogging platforms[J]. DOI : 10.15997/j.cnki.qnjz.2022.22.031.
9. Song Mengyuan. Exploring the formation factors of regional stereotypes based on the perspective of stigma[J]. Modern Communication,2020(08) :250-251.
10. Bourdieu P. Pascalian meditations[M]. Stanford University Press, 2000.

Influence of different types of advertising on the advertising conversion rate in college students

Yuxin Li*

Earth, Soc, Env Sustainability, Liberal Arts & Sciences, University of Illinois Urbana-Champaign, U.S., Urbana city

Abstract. The conversion rate is important measurable data as a metric of the effectiveness of advertising. The higher the conversion rate is, the more people who saw the advertising would take action. Many factors could affect the advertising conversion rate. In this paper, we aimed to find whether the format (video and poster) of advertising presentations impacts the advertising conversion rates among college students. The researcher designed an online survey that was distributed to college students, and respondents were asked to evaluate eight products' ads with both video and poster types. In addition we investigated five potential factors, including the level of impact social media has on advertising, gender, preferences of activities, time spent on different advertising channels, and time from viewing the ads before buying products, that would influence people's conversion rates. The results suggested that the conversion rates between video and poster are different ($p<0.05$). However, all five elements had no correlations with the conversion rate.

1 Introduction

Due to the widespread use of electronic products nowadays, online shopping plays a significant role in purchasing goods for customers. Mover Shakers (2021) conducted a survey which found that the ease of making purchases online motivates about 47% of respondents, while 42% appreciate the variety of goods available. Customers tend to learn information about the products from online advertising [1].

Online advertising is becoming one of the most important methods to attract potential customers. According to Maiya (2020), online advertising significantly impacts how customers perceive products and is regarded as the most important factor in predicting consumer purchasing behavior [2]. Liu et al. (2012) also claimed the rapid growth of online advertising that with an anticipated $80 billion in spending in 2011 and an expected 17% annual growth rate, online advertising is quickly emerging as one of the primary advertising channels [3]. To estimate the success of advertising, people use it to help businesses evaluate the effectiveness of their advertising campaigns. A lot of previous research discussed how to increase the conversion rate. Yan et al. (2009) suggested the behavioral targeting tactic could make the campaigns more effective. They discovered that by correctly segmenting consumers for behaviorally targeted advertising, ads CTR (Click-Through rate) could be boosted on average by as much as 670% [4]. On top of this, Beales (2010) also indicated that behaviorally targeted advertising is more effective than traditional run-of-network advertising since it provides consumers with more useful adverts [5]. Pandey et al.'s (2011) research showed that behavioral targeting can be significantly enhanced by focusing on hand-tuned segments [6]. Other studies have focused on improving click-through rates at the big data level in the personalization text. Researchers are focused on how the general market management strategy has affected advertising conversion rate and how behavior targeting is a significant determinant.

The click-through rate will considerably rise in the big data era if advertising is positioned with purpose and tailored to the user's preference settings. According to the research of Schwartz, Bradlow, & Fader (2017), focusing on click-through rates in the marketing funnel instead of direct conversion would result in a 10% loss in client acquisition, suggesting that personalizing ads could drive higher actual conversions [7].

However, these methodologies focus more on a macro marketing angle, which require a substantial investment of data, time, as well as financial resources in order to obtain reliable evidence to suggest the effective strategies. As (Lu et al., 2017) suggested that due to a number of issues, such as extremely scarce conversions, delayed feedback, communication gaps between the platform and the third party, etc., CVR prediction is quite difficult [8]. Alternatively, we could approach the same goal of improving the conversion rate from a narrow perspective by examining which advertising features and design characteristics can boost click-through rates. Examining how various campaign creatives could significantly influence consumer behavior, Braun & Moe (2013) found the targeted person's history of ad impressions is the most effective way, with the date of an increase of 13.8% in expected conversions and a 12.7% rise in expected visitors [9]. Some other studies have researched how appealing advertisements are to users regarding whether they utilize direct or metaphorical expressions. According to Ding et

* Corresponding author: yuxin17@illinois.edu

al. (2022), metaphorical advertisements stimulated higher tourism intention by the theoretical framework of ELM. For high-popularity posts, metaphorical advertisements draw more visual attention than straightforward advertisements [10].

Research focusing on Analyzing what types of formats work well to increase click-through rates offers a direct way to assess how they affect conversion rates, and the results are more explainable. However, there is little research investigating the impact of ad presentation formats (video, poster, text, etc.) on conversion rates. Therefore, this study's main goal is to identify the type of advertising presentation that will increase conversion rates.

The precise goals are to: (1) assess the difference between different advertising formats (video and poster) in their impact on conversion rates and (2) investigate possible variables as confounders that might affect conversion rates. Researching the impact of different types of advertising on conversion rates provides insights for advertisers to attract more customers by creating proper formats of advertising products. And if they can deliver the predicted more effective ad formats with the same cost investment to get more conversions and thus increase the business's profit.

Factors that influence conversion rates

A. Factor 1: [the impact level of social media on purchasing decisions]

B. Factor 2: [gender disparity]

C. Factor 3: [the usual channels people have access to advertising]

D. Factor 4: [different preference of activities]

E. Factor 5: [how long of making a choice of purchasing the products after seeing the advertising]

2 Method

The research aimed to compare whether different advertising presentation formats will have different impacts on conversion rates and investigate possible impact factors on advertising conversion rates among college students. Participants included 36 individuals (35 students, one other) aged 18-34 years (21 males, 15 females) in different educational occupations ranging from undergraduate students (80.6%), graduate students (11.1%), Ph.D. students (5.6%), and other (2.8%). The participants were given the link to a Google form to submit their answering questionnaire online.

In section one, the questionnaire included eight advertising products (electric devices brands (2), food and drinks (3), car (1), and skin care (1)) with two advertising contents (video, poster). The order of two different advertising rates was randomly allocated.

After viewing the content, participants needed to rate each on a 10-point Likert scale (1 = completely unlikely in purchasing, 10 = very likely in purchasing) of each item. Before rating the scale, we set a question of "Have you completed viewing the video/photo?" to ensure our participants finished watching the video and poster. In section two, participants must answer four multiple-choice questions about potentially related variables that may influence the advertising con-version rate. The four variable questions were:

1. How much do you think social media has an impact on your purchases?
2. Which of the following advertising channels did you usually see or hear the advertisement?
3. Which one do you prefer to engage in?
4. If you purchased as a result of the advertisement, how soon after seeing/ hearing the advertisement did you make the purchase?

3 Results

A two-sample t-test was used to test whether the conversion rates between video and poster are different. Conversion rates between video and poster are different (paired t test showed a significant difference, $p<0.05$). The conversion rate for video is 47.14%, and the conversion rate for poster is 35.97%. The conversion rate for video is significantly higher than the conversion rate for poster, demonstrating the audience in the sample is more attractive under the form of video advertising presentation.

Spearman correlation test and chi-squared test were used to find what other factors would influence the conversion rate. In the spearman correlation test, the extent to how much social media influences consumers ($ps > 0.05$ for both poster and video) and gender ($ps > 0.05$ for both poster and video) disparity demonstrated there were no significant evidence to suggest that there is association with two types of conversion rates with large p-value. In the chi-square test, we found factors of variety preferences of activities ($p = 0.73 > 0.05$), the frequently using advertising channels($p>0.05$), and how long from viewing the ads before buying products ($p>0.05$) indicate that there is also no enough evidence to show the association relationship with conversion rates. (See Table 1)

Table 1. Results Of P-value From Both Spearman Test And Chi-square Test.

Spearman test	p-value result	
	Conversion Rate of Video	Conversion Rate of Poster
the extent to how much social media influences consumers	p-value = 0.2171	p-value = 0.3095
Gender	p-value = 0.9874	p-value = 0.5706
Chi-square test	p-value result	
	Conversion Rate of Video	Conversion Rate of Poster
variety preferences of activities	p-value = 0.2171	p-value = 0.3095

the frequently using advertising channels	p-value = 0.9874	p-value = 0.5706
how long from viewing the ads before buying products	p-value = 0.6893	p-value = 0.2410

4 Discussion

From the results, we could know that different presenting forms between video and poster have different impacts on the conversion rate, and the video format of advertising is more attractive for the audience compared to picture dynamic form.

The difference between the impact of video and poster may stem from several factors. Firstly, video content may attract more attention due to motion and constant novel stimulation. More attention to the content of the ad is likely to boost the liking of the content, thus increasing the conversion rate.

Secondly, the video content in this study took longer time to view than the poster, since we did not limit the view time of the poster. Thus the difference between the mere exposure time between the two could introduce the difference in conversion rate. It would be interesting to test a condition where the participants were forced to view the poster for the full duration of the video, and compare the conversion rate. However, this might be challenging since it would be an unnatural task for the participants to do. Future researchers might come up with smart ways to test this hypothesis.

Thirdly, video involves multiple modalities of sensory inputs such as visual and auditory inputs, whereas the poster only involves visual input. Future research could focus on whether adding auditory input alone can boost the conversion rate.

Additionally, the analysis of four p-values among assumed four correlation factors which might potentially impact the conversion rate are all larger than the significance level (α=0.05), showing there is no significant evidence to say there is correlation between the factors and conversion rate. Based on the different type of data, we used two various analysis method. For example, we use the spearman test to conduct the numeric variables of gender (male =1, female = 5) and 5-point Likert scale (1 = None, 5 = Greatly) on how respondents' choice of purchasing would affect by social media, while we use the Chi-squared test to analyze the possible correlation between the conversion rate types and categorical data among different activities preference, preference of advertising channels, and time for purchasing after watching the advertising introduction. The results, which were all against our alternative hypotheses, might be due to the following reasons: 1. The sample size (n = 36) was small to represent the population parameter cases. 2. After evaluating eight cases of different advertising products, the respondents would be too tired to carefully finish the question part in section two, increasing the response bias.

The results may be different if we change the subjects from college students to currently employed people with more purchasing power. Compared to college students, people who are employed may have better skills in managing their wealth and can have more independent selections on commodities based on their preferences. Due to fewer restrictions, they certainly can decide whether they want to buy the products or not through watching advertisements to increase or decrease higher chances of conversion rates. Our results of conversion rates between video and poster may be more apparent, and the four possible variables are likely to have an impact on the final results.

In addition, to reduce response bias and enhance the accuracy of results, we could explore more effective and survey techniques by boosting the subjects' engagement. Though we added the video format of advertisements to bring dynamic visual experiences within our questionnaire, respondents were still needed to rate each advertisement in the traditional questionnaire format for a total eight different products. Also, we cannot strictly guarantee they have finished every video or poster. In this scenario, we are considering whether or not it would be feasible to replace the conventional online text format of the questionnaire to an interactive video form questionnaire. The final results could be more accurate by the video type of questionnaire for the following reasons: 1) The video form of the questionnaire could capture respondents' immediate feelings and allow respondents to rate directly. In our traditional form questionnaire, respondents need to revisit the questionnaire interface after watching the video form advertisements, increasing the potential distractions of their judgments. 2) The video questionnaire could give the respondents access to the following product question only after viewing the previous video or poster entirely. In this case, we could enhance the competence of our final results and ensure the validity of each subject's results. The video questionnaire will indeed require us more energy to create due to its complexity. Still, this new type of questionnaire could ensure a higher assurance of the final results, especially when we investigate the outcomes, including video-based questions. To minimize distractions during the response period and enhance each answer's accuracy, it is worth it for future research to design suitable types of questionnaires.

About the final results, we are surprised by finding no correlation between the impact of social media on people's purchasing attitudes and the conversion rate of two forms of ads and assume the formats of video conversion rate should affect their purchasing conversion rate more. Since there is more video-based content on social media, people are supposed to be exposed to more information by watching brief videos instead of pictures. Therefore, in the trend of social media, people may access more advertising via short videos by the main social media like TikTok or YouTube. However, the p-value was higher than the significance level (α=0.05), which was out of our expectations. For the improvement in the future research, the researchers could perform the research as an experiment, decreasing the distractions and have more reliable results.

5 Conclusion

Based on the analysis results from collecting data, it can be concluded that the advertising presentation formats are important to the advertising conversion rates. The results indicate that the two different types (video and poster) of advertising have influenced the conversion rate. Moreover, by comparing the conversion rate between video and poster, we found that the conversion rate of video format is higher than that of poster format, suggesting that customers may be more attracted by video-format advertisements and more likely to purchase.

This research clearly illustrates in the process of creating advertising, advertisers need to consider which forms of advertising can maximize product presentation so as to attract more potential users through this form. Furthermore, while this research designed a questionnaire combined with video and poster using Google form, future research could design a new type of video questionnaire to increase the validity and accuracy of the results. However, this research also raises the question of whether it is possible to show the product in a more three-dimensional way by combining pictures and videos to further increase the conversion rates. For future research, we can consider the effect of other forms of advertising (short videos, emojis) on conversion rate.

References

1. Marketing Charts. (2021, June 30). What Motivates Consumers to Buy Online? | Movers & Shakers.
2. Maiya, U. (2020, October). (PDF) IMPACT OF ONLINE ADVERTISING ON CONSUMERS BUYING BEHAVIOUR: A STUDY WITH REFERENCE TO KARKALA TALUK. ResearchGate.
3. Liu, Y., Pandey, S., Agarwal, D., & Josifovski, V. (2012). Finding the right consumer. Proceedings of the Fifth ACM International Conference on Web Search and Data Mining.
4. Yan, J., Liu, N., Wang, G., Zhang, W., Jiang, Y., & Chen, Z. (2009). How much can behavioral targeting help online advertising? Proceedings of the 18th International Conference on World Wide Web
5. Beales, H. (2010). The Value of Behavioral Targeting. Www.semanticscholar.org.
6. Pandey, S., Aly, M., Bagherjeiran, A., Hatch, A., Ciccolo, P., Ratnaparkhi, A., & Zinkevich, M. (2011). Learning to target. Proceedings of the 20th ACM International Conference on Information and Knowledge Management.
7. Schwartz, E. M., Bradlow, E. T., & Fader, P. S. (2017). Customer Acquisition via Display Advertising Using Multi-Armed Bandit Experiments. Marketing Science, 36(4), 500–522.
8. Lu, Q., Pan, S., Wang, L., Pan, J., Wan, F., & Yang, H. (2017). A Practical Framework of Conversion Rate Prediction for Online Display Advertising. Knowledge Discovery and Data Mining.
9. Braun, M., & Moe, W. W. (2013). Online Display Advertising: Modeling the Effects of Multiple Creatives and Individual Impression Histories. Marketing Science, 32(5), 753–767.
10. Ding, X., Feng, P., Wang, J., & Lin, M. (2022). Metaphorical or Straightforward? Com-paring the Effectiveness of Different Types of Social Media Advertising. Frontiers in Neuroscience, 16.

Research on the reform of integrated talent cultivation of middle and high school undergraduate in the context of 1+X certificate

Desheng Zeng[1], Xiaodan Chen[1], Shuanglong Pang[1,*], Xue Jiang[2], Chunyang Chen[2] and Jun Nie[3]

[1]Guangdong Innovative Technical College, School of Information Engineering, 523960 Dongguan, China
[2]Guangdong Innovative Technical College, Middle Vocational Department, 523960 Dongguan, China
[3] Guangdong University of Science & Technology, School of Computer Science, 523083 Dongguan, China

Abstract. With the deepening of industrial transformation and upgrading and reform of vocational education, the demand of society for high-level vocational technicians has increased. Combining with the "20 articles of vocational education" and "1+X" certificate system pilot project, based on the existing system of secondary and higher vocational education in our college, we jointly carry out the reform of "integration of courses and certificates" with Guangdong University Of Science & Technology. According to the characteristics of the school's majors, the vocational skill level certificate of "intelligent computing platform application development" was selected as a pilot reform, and the X certificate was fully utilized as the basic measure and standard for practitioners to measure their competence, so as to build an integrated talent training system of middle and high school, and to realize the organic connection between the curriculum systems of middle school, high school and undergraduate levels. This will help cultivate more excellent technical skills talents for the Guangdong-Hong Kong-Macao Greater Bay Area and is expected to provide a reference for building a perfect vocational education system.

1 Introduction

1.1. Background

With the industrial transformation and upgrading, the development of digital economy, the increased use of automation and intelligence in industry and other kinds of industries, the demand for low-level skilled workers is decreasing year by year, and the demand for high-level technical skilled personnel is rising year by year. The demand for technically skilled personnel in various industries is becoming more and more urgent, and the important status and role of vocational education is becoming more and more prominent. However, compared with developed countries, China's vocational education still has a system construction is not perfect, the system is not sound enough, the participation of enterprises in the school is small, and the supporting policies for the growth of technical and skilled personnel are not perfect, and the quality of schooling and talent training varies across the country, and the actual situation in Guangdong Province, the level of talent training in vocational institutions in the Pearl River Delta region is much higher than that in eastern Guangdong and northwestern Guangdong, etc. The actual situation in Guangdong Province is that the level of talent training in vocational institutions in the PRD region is much higher than that in the eastern and northwestern Guangdong regions[1,2].

At the same time, as Guangdong Province is pushing forward the reform of classified examination, the policy of "separation of general high school and secondary vocational school" at the secondary school level is gradually implemented, and more and more students will be trained by secondary vocational education. Higher vocational and undergraduate education has become a clear pathway for these students to advance in their studies. In the existing vocational and technical education system, an integrated pattern of progression of qualifications is built to meet the urgent needs of society and parents for the career development of secondary vocational school students[3].

1.2. Significance

(1) Take the theme of building a digital economy in the Greater Bay Area as a research opportunity. In recent years, the national, Guangdong, Hong Kong and Macao governments at all levels have continued to formulate and improve the policy framework for the construction of the Greater Bay Area, while the rise of the digital economy has brought new growth points for the development of the Greater Bay Area in Guangdong, Hong Kong and Macao. Technology and industrial innovation in many fields, such as big data, artificial intelligence, intelligent manufacturing, and financial technology, will become an important driving force and strategic support for the construction of the Greater Bay Area. Based on the demand for talents in the economic

* Corresponding author: pangshuanglong@vclimbers.org

and social construction of the Guangdong-Hong Kong-Macao Greater Bay Area, this study fits the demand for ICT talents in the digital transformation of the Greater Bay Area, with the main goal of cultivating high-quality talents in the application development of intelligent computing platforms, which helps to provide important support and talent guarantee for the development of the digital economy in the Greater Bay Area[4].

(2) Build a segmented integrated talent training system of "integration of courses and certificates". Fully reflecting the concept of "X certificate", we adopt segmentation to build the curriculum system of middle and senior vocational level, and integrate the content of primary, intermediate and senior certification syllabus, core courses and practical training courses into the professional curriculum system according to the requirements of "X certificate" in the talent training target. In this paper, we have taken a look at the curriculum system of Guangdong, Hong Kong, Macao and Taiwan, which has been integrated into the professional curriculum system to obtain skill level certificates in phases.

This paper conducts a study on the demand for technically skilled personnel in the social development and economic construction of the Guangdong-Hong Kong-Macao Bay Area, based on the development needs of the Dongguan regional economy and society, and in line with the current national policy of vigorously developing vocational education. Combined with the construction plan of our school, we make full use of the specialized curriculum system of "1+X" certificate to gradually build the characteristics of the vocational education system of intermediate, higher vocational and local vocational through training, and form a competitive brand of vocational education.

2 Problems of the current vocational education system of middle and high school

2.1. Inadequate articulation between major levels

At present, in the integration cultivation process of secondary school, higher vocational school and undergraduate school, the articulation of majors is a crucial part, and it is necessary to do a good job of stage articulation of majors in order to realize the organic articulation between each level, otherwise students cannot study the same majors coherently, which is easy to cause the fault of knowledge and skills. Secondly, there are certain differences in the training objectives and orientation of students due to the differences in the corresponding school orientation and school foundation of different segments.

2.2 Inadequate curriculum articulation

Due to the lack of unified combing of curriculum system among current secondary, higher vocational and undergraduate institutions, institutions work separately and do not pay attention to the organic docking of curriculum. In the process of student training, there are often problems such as repeated learning of certain knowledge, or a certain course repeatedly appearing, and even some knowledge or skills have a discontinuous phenomenon. The effect of technical skills training is far from the expected effect [5].

2.3 Lack of mechanism system

At present, among the three types of institutions, there is still an obvious problem that our country lacks guarantee mechanism for the articulation of middle and high school, and the governments in different places also lack the support policy for the articulation of middle and high school, and lack the motivation for the articulation, and the perfection of the work in all aspects is different, and there is great resistance from all aspects, and there is no way to form a synergy to promote the articulation of middle school, high school and undergraduate[5-6].

3 Integrated talent training reform ideas

3.1 Sorting out the knowledge points in the X-certificate standard

Led by the school, the X-certificate standard analysis team is formed by the front-line technical backbone of industry and enterprises, professional leaders and key teachers to analyze the "intelligent computing platform application development vocational skill level standard", decompose all the knowledge points covered by the skill requirements in the vocational skill level standard, and sort all the knowledge points by the principle of similarity. All the knowledge points are classified and sorted by the principle of similarity, and the same category is sorted by the progressive relationship of the difficulty of knowledge points, and the knowledge points are improved by the pre-paving knowledge and the subsequent extended knowledge settings. In addition, the transitional knowledge should be added when the knowledge before and after is connected, so as to form a knowledge system that can be studied by students at all stages[1,2].

3.2 Restructuring of talent training plans at each level

With the document "Guiding Opinions on the Formulation and Implementation of Professional Talent Training Programs in Vocational Colleges and Universities" (No. 13 of Education Vocational Training [2019]) issued by the Ministry of Education, the talent training program is a normative document for institutions to organize teaching activities and arrange teaching tasks, as well as the basic basis and implementation carrier of talent training, which also makes the important position of the talent training program in the process of talent training clearer. In each academic year, the professional leader should organize the front-line technical backbone of industries and enterprises and professional backbone teachers to adjust the professional talent training program based on the established professional training objectives and the overall requirements of the state for vocational education talent training, combined with the newly constructed curriculum system, so as to achieve the optimization of the dual participation of schools and enterprises in the talent training process.

3.3 Restructuring the curriculum

Curriculum system is the most important carrier to achieve the goal of talent training, and it is the key to guarantee and improve the quality of education. The curriculum system is mainly composed of specific course objectives, course contents, course structure and course activities. As a type of employment-oriented education, vocational education has a comprehensive implementation of the concept of technical skills training, each major should be based on professional training objectives, service orientation and employment orientation, to build a curriculum system suitable for serving the needs of students to grow and become successful[1,3].

3.4 Clarify the main occupational job groups and job responsibilities

Intelligent computing platform application development occupational skill level is divided into three levels: primary, intermediate and advanced, and the three levels are progressive. Through research and analysis, as well as reference to the relevant specifications of the intelligent computing platform application development vocational skill level standard, we clarify the main job groups and job responsibilities as follows: infrastructure management, IDE integrated development environment management, application software and artificial intelligence application product development and testing departments for IT Internet enterprises, traditional enterprises and institutions of digital transformation, and government, etc., engaged in intelligent computing platform Build, platform management, IDE integrated development environment deployment and system testing, data management and analysis processing, basic function application development and artificial intelligence application product development testing, etc., according to business needs to complete offline online integrated development environment deployment, management, system testing, as well as data analysis and processing, artificial intelligence algorithm optimization and implementation, artificial intelligence application product development testing the job tasks include[7].

3.5 Clarify occupational skills requirements

Through the above analysis of the main professional job groups and job responsibilities, and with reference to the grade standard, we divide the vocational skills related to "intelligent computing platform application development" into four modules: platform construction, platform management, data management and application development.

The following is a typical "platform management" as an example, the main vocational skills for the three levels, mainly as follows:

(1) In the primary certificate, for the "platform management" module, the main task is "system management". Firstly, they can independently complete the daily operation and maintenance management operations of the storage system of the intelligent computing platform using the equipment operation and maintenance management tools provided by the vendor, such as: system status monitoring, log collection, daily inspection, etc. Secondly, the vendor's equipment operation and maintenance management tools can be used to independently complete the daily maintenance and management operations of the artificial intelligence dedicated servers (such as GPU accelerated servers, Kunpeng general-purpose computing servers, and Sunrise heterogeneous computing servers) of the intelligent computing platform, such as: equipment inspection, log collection, and equipment status indicator identification, etc. Finally, you will be able to apply the specifications and skills of writing system operation and maintenance management documents to assist senior technical support staff to sort out and improve the network topology diagram, system operation and maintenance management and other related documents of intelligent computing platform system.

(2) In the intermediate certificate, for the "platform management" module, the two main tasks are "system management" and "system testing".

In the task of "system management", firstly, according to the design requirements of the business, you can use the system management tools provided by the product vendor to independently complete the operation of storage resource expansion and upgrade of the intelligent computing platform. Secondly, according to the design requirements of the business, you can use the system management tools provided by the product manufacturer to independently complete the system operation status monitoring and inspection, performance analysis and optimization, security reinforcement, and fault analysis of the artificial intelligence dedicated servers (such as GPU accelerated servers, Kunpeng series servers, and Rising series servers, etc.). Finally, you will be able to use project documentation tools and templates to independently organize and write O&M reports and technical support documents for intelligent computing platform systems.

In the task of "system testing", first, be able to use the common methods and tools for system troubleshooting, independently analyze the causes of common failures, and put forward improvement suggestions and methods and measures. Secondly, according to the system troubleshooting manual provided by the vendor, use troubleshooting tools or the system's own troubleshooting commands or functions to deal with system emergencies and ensure the stable operation of the system.

(3) In the advanced certificate, the "platform management" module also contains two tasks: "system management" and "system tuning".

In the task of "system management", firstly, according to the requirements of business design, you can use the system management tools provided by the product vendors to realize the overall status monitoring, resource management and system tuning of the intelligent computing platform. Second, according to the project change guidance, you can participate in and complete some of the intelligent computing system changes related to the implementation of operations. Finally, you can use document development tools or templates to independently complete the preparation, optimization and archiving of documents related to the operation and maintenance management of the intelligent computing platform system.

In the task of "system testing", firstly, you can use problem management tools to track and solve the problems of business-related systems and realize the closed-loop management of problems. Secondly, to provide technical support for major incidents, emergencies and major changes according to the emergency handling process and specifications for major incidents. Lastly, the ability to use documentation development tools or templates to independently complete the preparation, optimization and archiving of routine documentation for system tuning and testing.

4 Building an integrated talent training system

Due to the differences in the orientation and infrastructure of secondary, higher vocational and undergraduate institutions, there are problems of knowledge and skills discontinuity and repetition of teaching contents in the cultivation of secondary and higher education talents. In this paper, based on the rules of talent cultivation and teaching, we adopt a segmented approach to integrate the requirements of the corresponding "X-certificate" examination syllabus for primary, intermediate and advanced levels into the curriculum system of each stage, and design the following way to build an integrated training system. Firstly, in the intermediate stage, the study and certification of the primary "X certificate" will be completed; secondly, in the senior stage, the study and certification of the intermediate "X certificate" course will be completed; finally, in the undergraduate stage, the study and certification of the senior "X certificate" course will be completed. Finally, at the bachelor's degree stage, we will complete the study and certification of the advanced "X Certificate" course. Take "Intelligent Computing Platform Application Development Vocational Skill Level Standard" as an example, its main applicable professional situation.

Therefore, in the pilot process of integrated talent cultivation reform in our school, according to the existing professional construction of the school, the middle-level stage major is selected as: software information and service; the senior-level stage is: computer application technology; the undergraduate stage is: software engineering. The process of building integrated talent training system is mainly divided into the following steps.

4.1 Clear training objectives

The main focus is on the infrastructure management, IDE integrated development environment management, application software and artificial intelligence application product development and testing departments of IT and Internet enterprises, traditional enterprises and institutions transforming to digitalization, and government, etc., to cultivate technical skills talents with good professional ethics and humanistic qualities, with enterprise employment as the guide and job skills and comprehensive quality as the core. The program aims to cultivate technical and skilled talents with good professional ethics and humanistic qualities.

4.2 Clarify the relationship between X-certificate in major, curriculum and credit

The development of X-certificate originates from the typical work tasks corresponding to vocational positions (groups), and the demand for knowledge and skill points of typical work tasks is closely related to the cultivation of corresponding professional core skills in schools, but there is not a unique correspondence between professional core skills and majors. In the design of our integrated talent training system for secondary, senior and undergraduate schools, X-certificate is used as the foundation for consolidating students' sustainable development, developing vocational skills level credentials for expanding employment and entrepreneurial skills and alleviating structural employment conflicts. For example, the knowledge and skills involved in the vocational skills level certificate of intelligent computing platform application development are embodied in the form of core skills for professional talent training in the relevant majors of middle-level, senior-level and bachelor's degree.

4.3 The study stage of secondary school

In the secondary stage, that is, in the learning stage of software information and service majors in our school, secondary school students are required to become junior technical skill talents of "intelligent computing platform application development" after completing the study and certification, master the basic knowledge of intelligent computing platform application development, and be able to complete intelligent computing hardware and software platform and development environment deployment, as well as daily management of the development platform and basic application function development test according to the business configuration requirements. They can complete the deployment of intelligent computing software and hardware platform and development environment, as well as the daily management of the development platform and the development and testing of basic application functions according to the business configuration requirements.

4.4 The study stage of senior

In the senior-level stage, that is, the study stage of computer application technology in our school, senior-level students are required to become intermediate technical skill personnel of "intelligent computing platform application development" after completing the study and certification, and be able to complete the deployment and management of offline integrated development environment, as well as the basic processing of data, product development and testing of artificial intelligence primary applications according to the requirements of business management, They will be able to complete the deployment and management of offline integrated development environment, basic data processing, and product development and testing of artificial intelligence primary applications according to business management requirements.

4.5 The stage of undergraduate

In the undergraduate stage, that is, the students of computer application technology in our school, after completing their studies, they are promoted to Guangdong Institute of Science and Technology, and as students in the undergraduate stage, they are required to become advanced technical skills talents of "intelligent computing platform application development" after completing their studies and certification, and they can complete the tasks of cloud integration development environment Students are required to complete the deployment, management and system testing of the cloud integration development environment, as well as the advanced processing of data, optimization of artificial intelligence algorithms and development and testing of advanced application products according to business needs.

5 Conclusion

Firstly, in the process of current VET talents cultivation, the principles of wholeness, standardization, practicality and operability of X-certificate related standards are used to carry out integrated cultivation of middle-grade, high-grade and undergraduate talents, avoiding the problems of non-connection of majors and non-connection of courses in the process of talents cultivation in each stage, avoiding repeatedly offering the same courses for many times, effectively using the learning time of each stage and cultivating qualified VET talents. The X-certificate system strengthens the training and evaluation of students' vocational ability and enhances students' vocational ability.

Secondly, in this paper, we try to explore the integrated talent training model based on X-certificate to build secondary, higher vocational and undergraduate education, which helps to establish a perfect vocational and technical education system, in meeting the increasing demand of enterprises for high-level talents, and also helps to provide the urgent demand of society and parents for vocational and technical talents to upgrade their qualifications.

Acknowledgments

This article was funded by the Provincial High-level Professional Construction Project of Guangdong Education Department: Computer Application Technology Major Group of Guangdong Innovative Technical College, No. GSPZYQ2021046 and Guangdong Provincial Education Science Planning Project: Research on Talent Training Mode of "Post-Course Competition Certificate" in Higher Vocational Internet of Things Application Technology, NO. 2022GXJK691.

References

1. Ding ZG & Guo YNA. (2020). The implementation path and guarantee of 1+X certificate system in higher vocational institutions. China Vocational and Technical Education (10), 53-56.
2. Li Q, Lu W & Yin XJ. (2019).1+X certificate system: progress, problems and countermeasures. Journal of the National Academy of Educational Administration (12), 18-25.
3. Liu, WJ. (2019). Research on curriculum reform of vocational education under the 1+X certificate system. Vocational Education Forum (07), 47-53.
4. Zeng, DS., Jiang, X., Shao, S., Fu, J. & Wang, SY. (2023). Research on the reform of higher vocational information professional talent training. Industrial and Information Technology Education(03),11-15+25.
5. Yi XS & Jiang X. (2021). An analysis of the current situation of talent cultivation in China's middle and high school articulation. Journal of Qiqihar Normal Higher College (04), 89-91. doi:10.16322/j.cnki.23-1534/z.2021.04.031.
6. Huawei ICT Academy. (2022). Huawei 1+X Vocational Skill Level Certification. Huawei 1+X Vocational Skills Level Certification. https://e.huawei.com/cn/talent/cert/#/oneXcert?navType=oneXCert
7. Occupational Skill Level Certificate Information Management Service Center. (2020). 1+X portal. Occupational skill level certificate information management platform. https://vslc.ncb.edu.cn/

Digital Exploration of Audio and Video in Archive Management Based on Big Data Era

Fang Liu*

School of Finance, Shandong Technology and Business University, Yantai, China

Abstract. With the development of the era of artificial intelligence and big data, archive databases have rapidly developed relying on the "information highway", and digital media such as audio and video have become an indispensable part of people's daily lives. Especially in fields such as culture, education, and scientific research, the importance of digital archive management has gradually been recognized and valued by more and more people. However, due to various reasons, the development of digital audio and image processing has always been lagging behind. Only digital information can provide users with comprehensive services of various information. This article uses the QuestionStar app to understand the actual status of audio and image through questionnaire surveys, Provide practical basis for future research and practice of archive digitization.

1 Introduction

Digital archive management originally originated in European and American countries. It is a technical means of digitizing traditional physical archive data to achieve long-term preservation, convenient retrieval, and shared utilization. With the continuous progress of computer and network technology, digital archive management is increasingly valued and has become an essential part of major institutions, enterprises, and government departments. Through digital means, multimedia resources such as audio and video can be transformed into forms that can be recognized and processed by computers. This not only facilitates storage and transmission, but also greatly improves the efficiency of archive management. The digital archive management system can intelligently classify and annotate massive audio and image resources, enabling users to quickly locate the required resources, greatly improving the efficiency of processing and utilization. Digital archives can be remotely accessed and transmitted through the network, enabling cross spatial and cross temporal information sharing. The digital archive management system can provide a unified interface and specification, making it convenient for users to search and call. Users can search and obtain the required audio and video resources in the digital archive according to their own needs and permissions, achieving the reuse and recreation of information. The promotion and popularization of digital archive management can not only accelerate the creation and dissemination of digital audio and video, but also bring more opportunities and development space for academic research, education and training, cultural heritage, and other fields.

We distributed 340 questionnaires to some enterprises and universities in Shandong Province through the "Survey method for the practical scope of digital information" and all of them were returned. The questionnaire recovery rate is 100%, but the effective response rate is 81.2%. Through analysis, audio and video have been widely applied and developed in the field of digital archive management, but the construction of archive databases is a very challenging project and lacks overall planning. The digitization of audio and video archives varies greatly among different types of carriers, and the processing of audio and video archives has always been lagging behind due to various reasons.

2 Limited digital conversion technology for audio and video

Digital audio image conversion is a great technological innovation, and audio image digitization is the core technology in the field of archive management. The digitization technology of audio and video archives [1] can effectively collect the content of traditional audio, image, video and other files, and store the original information obtained through digital conversion technology in the archive database, making it convenient and efficient for query, retrieval, reference and other operations. Audio and image digitization can be processed, stored, and transmitted by computers. The development of digital audio image conversion is still subject to some limitations, which have had a certain impact on the quality and application range of digital audio. Through the organization and summary of data, Table 1 clearly reflects the limiting factors of audio and video in information archive management, Table 1 clearly reflects the limiting factors of audio and video in information archive management, Lack of technical support is currently a common problem and

*201513369@sdtbu.edu.cn

the first issue to be addressed in the digitization of audio and video in archive management.

Table 1. The influencing factors of electronic audio image conversion

Serial number	Information Types of electronic document	Proportion
1	Lack of technical support	55.2%
2	Carrier function is not powerful	52.4%
3	Lack of professionals	50.1%
4	Leaders do not pay attention	26.3%

1.The accuracy of audio image digital conversion technology has always been a focus of academic attention. The process of digitization requires the use of a series of technical means and equipment, such as digital scanners, sound acquisition cards, etc. Sample continuous sound signals. Two basic parameters are used for sampling: first, how many audio samples are collected per second; The second is the sample size. As shown in the formula: (fs)=2f (f is the highest frequency of the sampled signal). When sound and video are received by the acquisition device, they are converted into electrical signals and transmitted to the computer through an audio and video interface. After receiving the electrical signal, we need to convert it into a digital signal. Due to the limitations of sampling rate and quantization depth required for digitization, digital audio cannot perfectly restore the original analog audio signal. Therefore, in the process of digital audio conversion, there may be quality issues such as distortion and noise, which not only affect the listening experience and objective performance of audio, but also have adverse effects on audio in different application scenarios.

2.The format issue of audio image digital conversion is also a difficult problem to solve. Traditional analog audio and video technology is stored on magnetic tape or film, which is not easy to save. Digital audio and video can be saved in optical and magnetic storage media, and digital audio and video can be reprocessed for errors. However, the diversity and incompatibility of digital formats have become a major challenge in the digital conversion of audio and video, and there are incompatibilities between different audio and video formats. This not only leads to low conversion efficiency, but also brings great difficulties to the application and sharing of audio and image digital conversion.

The energy consumption issue of audio and image digital conversion technology is also worth our attention. Currently, most digital audio image conversion chips and algorithms require a significant amount of computational resources and energy. Usually, an AD converter is required, which converts electrical signals into digital signals and stores them in a computer. The final step in digitizing audio and video is encoding. In this process, the digital signal is converted into a digital encoding format. Currently, the most popular audio encoding formats are MP3, AAC, WMA, etc., while the most popular image encoding formats are MPEG-2, H.264, AVC, etc. These encoding formats can compress the original digital signal, thereby reducing storage space consumption. These digital audio image conversions take a long time and also consume a large amount of computer storage and energy resources. In scenarios that require a large amount of digital audio and image processing, these resource consumption may also become an issue that cannot be ignored.

The development of digital audio and image conversion technology, while facing a series of limitations, also requires continuous technological innovation and breakthroughs. Only by addressing these limitations can we promote the further development of digital audio conversion technology. Therefore, in future research on digital audio conversion technology, we need to pay more attention to technological innovation and breakthroughs, continuously improve the quality and application range of digital audio, and achieve wider application of digital audio in various fields.

3 Lack of platform support for digital audio and image archive management

Audio image digitization platform refers to a platform that digitizes audio image archives and stores, manages, and transmits them through a network [2]. It is implemented based on digital archive management. This platform can digitize traditional materials such as audio and video tapes, convert them into digital format files, and store them on the network. This method can greatly improve the storage capacity and sharing efficiency of archives. The construction of an audio and image digitization platform needs to consider specific application requirements. Show in table 2, it can be seen that, for different types of files, different digitization methods need to be adopted to ensure the accuracy and completeness of digital archives. At the same time, it is also necessary to consider factors such as network bandwidth and storage capacity to ensure the efficiency of platform usage.

Table 2. Support for Electronic Document Information in Universities

Serial number	Electronic Document Support Platform	Proportion
1	Information Supporting Situation of electronic document	42%
2	Digital information network construction	32.2%
3	Cooperative Information Technology Company	33. 1%
4	Lack of funds	23.4%

Audio and video archives are easily saved, disseminated, and managed in digital form. In terms of digital archive management for audio and video archives, we need to prioritize two aspects: the technical details of the digital process, and the systems and services for managing digital archives. Firstly, the technical details of the digitization process are crucial for achieving digital archive management of audio and video. The acquisition process requires the use of high-quality acquisition equipment, which converts audio and video into digital data through high-fidelity digital technology. Through the editing process, we can edit, organize, and organize data to make it easier to manage and use. In some cases, we also need to perform post processing on audio and video to ensure the quality and integrity of the data. The transmission, storage, and retrieval processes also require the use of high-end technology to ensure the efficiency and security of data transmission[3], storage, and retrieval. Secondly, the system and services for digital archive management are equally crucial. The management platform for digital archives needs to have strong functional performance, including data management, data retrieval, data integration, data retention, data output, and other aspects[4]. The security and guarantee of digital content require the use of high-level encryption technology and key management technology to ensure that archive information is not stolen or lost. Advanced data backup and recovery technologies are required for the backup and recovery of digital archives to ensure that materials are not accidentally lost。

The digital platform can quickly convert archive files into digital format and store them, reducing the time and workload of manual processing. Digital archive management can also facilitate the access and utilization of archive data, which has increased efficiency several times compared to traditional paper archive management methods. It can effectively protect archives from tampering or leakage, and ensure the integrity and security of archive data. Digital archive management can promote archive sharing and cooperation, especially for archive sharing across multiple departments and regions. The digital platform can achieve the sharing of archival data, reduce the situation of repeated collection by multiple departments, improve the efficiency of archival sharing, and better contribute to the process of social informatization.

4 The challenges faced by digitizing audio and video

Digital archive management of audio and video also faces many challenge.The main challenge lies in the information security and reliability of digital archives. Once digital archive materials are damaged, they may have irreparable impacts on multiple fields such as academia, culture, history, and so on. Therefore, digital archive management must strengthen the security and technical guarantee of archive materials to ensure the integrity, credibility, and availability of digital archives.

In addition, digital archive management also needs to address technical issues such as data storage, data transmission, and data format conversion. Especially for multimedia data such as audio and video, there are more issues that need to be addressed. Multimedia data not only has a large amount of data and complex formats, but also requires effective compression and encoding to meet the needs of data transmission and storage. In addition, it is necessary to overcome the compatibility and sustainability issues of digital archives to ensure the long-term preservation and use of archival data shown in Table 3.

Table 3. Factors affecting digital audio and imagings

Serial numbe	Factors affecting digital audio and imaging	Proportion
1	Capacity requirements for digital archives	55.2%
2	Classification and indexing	45. 6%
3	Copyright security	33.4%
4	Sharing and utilization	32.5%

With the convenience of digitization and online dissemination, the problems of piracy and infringement have become increasingly serious. Copyright protection of digital audio and video is also an important challenge. Huan (2019) [5]Another issue discovered in digital archive management is the issue of copyright protection. Sheng (2003) [6] Copyright is an important aspect in digital archive management. Yang (2017) [7] Discovering that digitization of archives requires copyright protection and avoidance of infringement. We can use digital watermarking technology and copyright authentication technology to protect copyright and ensure the legality and reliability of digital archives. Digital archive management needs to find effective ways to protect copyright, prevent the spread of piracy, and maintain the interests of legitimate content and the healthy development of the industry.

The construction and maintenance of databases are also crucial for the digital management of audio and video archives. The digitization of audio and video archives requires support for multiple data formats and access to terminal devices. Therefore, how to build and maintain an efficient database is the core issue of digital archive management [8].

In digital archive management, the application of big data is becoming increasingly important [9]. With the continuous growth of data volume, the information in the database becomes increasingly complex, so more advanced technologies need to be adopted for processing [10]. Big data technology can provide more accurate and efficient processing methods for digital archive management, providing us with more accurate, comprehensive, and reliable historical and cultural information.

5 Conclusion

In short, the implementation of digital archives is conducive to the modern transformation of traditional archive protection and management, providing new paths and directions for social development and cultural inheritance. Digital archives, as an important component of information construction, play an irreplaceable role. The promotion of digital archives construction also requires cooperation and joint promotion from all walks of life to achieve sustainable development and innovation in digital archive management. By strengthening copyright protection, data security, data efficiency, and data sharing, digital archive management can be made more comprehensive and reliable.

Acknowledgments

We are grateful for the 2018 Shandong Province Higher Education Humanities and Social Sciences Research Program Project "Research on the Impact and Promotion Strategies of the Use of Modern Instant Communication Tools on College Students' Psychological Harmony - Taking WeChat as an Example"(J18RA157)

References

1. He,F.(2015)Exploring the Ways to Carry out the Digitalization of Enterprise Archives in Construction.J.Corporate culture,2:123-123.
2. Guan, Y.(2019)Research on Innovative Services for Township Archives Management under the New Situation.J.File management,35:47-47.
3. Norris M, Oppenheim C. (2007)Comparing alternatives to the Web of Science for coverage of the social sciences'litera-ture.J.Journal of Informetrics, 1(2):161-169.
4. Tahamtan I, Afshar A S, Ahamdzadeh K. (2016)Factors affecting number of citations: a comprehensive review of the litera- ture.J.Scientometrics, 107(3):1195-1225.
5. Huan, Y. (2019) On the copyright protection of digital archives.J.Science and wealth, 35:30-30.
6. Sheng, L.(2003)Problems to be solved in the digitization of archives information.J.Anhui archives,6:15-17.
7. Yang, R.(2017)The protection of archives digital copyright in the context of informationization and the countermeasures.J. Corporate culture,6:243-243.
8. Jie, X(2016)Research and Discussion on the Digital Construction of Archives.J.File,11:103-103
9. Li(2020), Analysis of the Application of Big Data in Social Security Archives Work.J.Urban construction archives, 2:36-37
10. Ning(2016), Research on the Reform of Computer Science Education in Universities under the Background of Big Data.J.Digital Communication World,5:302-302

 https://doi.org/10.1051/shsconf/202317802005

How Mass Media Influences U.S. Political Polarization— A Comparison Study of CNN and Fox News

Jinhe Yang*

School of English and International Studies, Beijing Foreign Studies University, Beijing, China, 100089

Abstract. This paper presents a study on how mass media, specifically CNN and Fox News, have influenced the polarization of American politics concerning environmental policies. The study used a qualitative approach to analyze the content of CNN and Fox News broadcasts related to environmental policies. The results show that CNN and Fox News have significantly contributed to the polarization of American politics, especially on issues related to public policy. While CNN emphasizes the need for environmental protection and regulation, Fox News focuses on the economic consequences of such regulations. These differing views have influenced the political attitudes of viewers and created a divide among the American public on environmental policies.

1. INTRODUCTION

As a terminology in the field of political science, "political polarization" is currently one of the main concerns also in the field of social and behavioral sciences, as "different societies worldwide seem to have undertaken a path of political conflict recrudescence, intergroup ideological distance, and growing intergroup animus", which has been linked to a range of negative outcomes, such as reduced trust in government, increased political instability, and decreased social cohesion. Over the years, the U.S.A. has concentrated on the factual existence and implicit influence of political polarization that has pervaded. As a sophisticated phenomenon with a wide range of causes and corresponding detrimental consequences, in recent years, the mass media with specific political biases and stances in the United States has been a topic of concern and debate, even if they coverage, critics argue that many of these organizations have a clear political bias that shapes their reporting.

As media has been "a ubiquitous part of modern life" [1] and has inoculated profound influences on many aspects of society, including politics, "many of the current options are extraordinarily social, dramatically increasing people's capacity to form bonds with individuals and groups that would otherwise have been entirely inaccessible" [2]. However, concerns have been raised about the role of mass media in contributing to political polarization, which can have detrimental effects on democracy and social cohesion.

This paper aims to use qualitative research methods including textual analysis, discourse analysis, and case study on two conventional mainstream media based on their previous and published newsletter on American environmental policies—namely CNN and Fox News, to focus on the mechanism of media in shaping public attitudes towards environmental policies and contribution to political polarization in the US, and to reveal the conceptualization of polarization in politics in the U.S.A.

2. LITERATURE REVIEW

2.1 Conceptualization of Political Polarization

According to Kubin & von Sikorski [3], the concept of political polarization has been discussed in political science since the 1950s, when scholars began to examine the effects of party systems and electoral behavior on political conflict. Political scientists typically distinguish between two levels of political polarization: elite and mass. Elite polarization focuses on the polarization of the political elites, like party organizers and elected officials. Mass polarization focuses on the polarization of the masses and the public, most often the electorate or the public.

The studies on political polarization have evolved to reflect the changing political and social contexts in the United States. In the 1950s and 1960s, Morgan Kelly [4] claimed that American politics was characterized by a lack of polarization and ideological clarity among both elites and the masses. They suggested that a moderate consensus prevailed in the post-war era and that parties were weak and overlapping in their policy positions. In the 1970s and 1980s, however, other scholars challenged

*yjhchristopher@bfsu.edu.cn

this view and pointed out the increasing ideological divergence and partisan alignment among elites and masses. They attributed this trend to various factors, such as the civil rights movement, the Vietnam War, the rise of new social movements, and the realignment of party coalitions.

In the 1990s and 2000s, scholars further examined the sources and implications of political polarization in the US and other countries. They explored how "institutional factors (e.g. electoral rules, media systems, and campaign finance), psychological factors (e.g. motivated reasoning, social identity, and confirmation bias), and contextual factors (e.g. globalization, immigration, and terrorism)" influenced political polarization among elites and masses [4][5]. They also analyzed how political polarization affected democratic performance and outcomes, such as policy-making, representation, accountability, participation, and trust.

The literature on political polarization has also developed different theoretical frameworks and methodological approaches to understand and measure such phenomenon. Some scholars have adopted a normative perspective and argued that political polarization was either beneficial or detrimental to democracy. Others have adopted a descriptive perspective and focused on explaining how political polarization varied across time, space, and levels of analysis [5]. Some scholars have used quantitative methods and relied on survey data, roll-call votes, or text analysis to measure political polarization. Others have used qualitative methods and employed case studies, interviews, or discourse analysis to examine political polarization.

2.2 Cause and Effect of Political Polarization

Firstly, social identity has been recognized as a cause of political polarization. Kreiss [6] argued that social identity is a significant factor contributing to political polarization, which is constructed by social groups, particularly journalists in this study, within a specific socioeconomic context. This construction of social identity has important implications for the legitimacy of institutions in the eyes of the public. Journalists are a distinct group of people in society, and their social identity is related to other groups in society. The social identity of journalists is historically specific, contextually relevant, and politically significant, which indicated the significance of social identity to the phenomenon of political polarization (p.27). According to Mason's [7] research, individuals who are highly sorted into partisan groups are more likely to engage in political action, regardless of their policy attitudes. This is especially true when they feel threatened or angry (p. 99).

Media consumption is another important factor contributing to party polarization. People today have access to a wide range of news sources, many of which have partisan leanings. This can lead to the formation of so-called "echo chambers", where individuals are exposed only to viewpoints that align with their pre-existing beliefs [8]. Mason [9] suggests that when people consume information from mass media, they can gain insight into the extent to which their political behavior is influenced by irrational or unfair factors (p. 17). Prior [8] notes that despite the availability of a wide range of news sources, there is still divergence in people's viewpoints. This is because it is difficult to measure the average opinion when it comes to news consumption, which further complicates arguments on this topic (p. 285).

Finally, political elites can also contribute to polarization by adopting extreme positions and engaging in divisive rhetoric. According to Noelle-Neumann [9], political elites can contribute to political polarization by adopting extreme positions and engaging in divisive rhetoric. This can create a "spiral of silence," where individuals who disagree with the dominant discourse feel marginalized and are less likely to speak out. Political elites, like the presidents and leaders, may influence the polarization of the electorate (the mass public) by taking more extreme ideological positions and actions, and by deliberately fueling animosity and distrust toward the opposing side [8] [10]. They may also use social media to organize offline political activities, disseminate false or disinformation, and increase political polarization among the citizens [8]. Political polarization among the electorate may manifest as affective polarization (disliking and avoiding the other side), ideological polarization (adopting more extreme views), or behavioral polarization (engaging in conflict or violence) [29].

Political polarization may have negative consequences for democracy, such as reducing compromise, cooperation, and representation [3].

Regarding the consequences of this political phenomenon, another major effect is that it can lead to gridlock and an inability to make progress on important policy issues. This can lead to a sense of disillusionment and disengagement among citizens, who may feel that their voices are not being heard.

Another consequence is that political polarization can undermine the democratic process itself. When individuals become more ideologically entrenched, they may become less willing to compromise or engage in constructive dialogue with those who hold different viewpoints. This can lead to increased hostility and intolerance, making it more difficult to build consensus and make decisions in the public interest.

2.3 Media in Politics

Media can influence political polarization by exposing people to different or similar information and opinions about political issues and actors. Political polarization can also affect media use and preferences by motivating people to seek out or avoid certain sources of information and opinions. The literature on media and political polarization has examined various aspects of this relationship, such as the causes, consequences, and measurement of media exposure and political polarization. According to Humayun [11], media has become an omnipresent aspect of contemporary life and has had a significant impact on various aspects of society, including politics. Sunstein [12] suggests that many current media options are highly social and have the potential to significantly increase people's ability to connect with individuals and groups that were previously inaccessible. This can lead to the formation of strong bonds between people who may not have otherwise connected (p. 65). One of the ways that media has been shown to relate to political polarization is through the media's role in shaping public opinion. This effect was particularly strong among individuals who identified as politically engaged.

Selective exposure is also a kind of measure to affect political polarization, whereby individuals seek out news content that reinforces their pre-existing beliefs.

Media fragmentation has also been linked to political polarization. A study by Gentzkow and Shapiro [13] found that the rise of cable news channels and the internet has led to greater fragmentation of the media landscape, with individuals increasingly seeking out news sources that align with their political beliefs. This fragmentation has been linked to greater polarization, as individuals are exposed to fewer diverse viewpoints and are more likely to hold extreme political views.

Another question in the literature is whether media exposure contributes to or reduces political polarization. Some studies have argued that media exposure can increase political polarization by creating "echo chambers" or "filter bubbles" that insulate people from opposing views and reinforce their existing beliefs [14] [15]. Other studies have suggested that media exposure can decrease political polarization by exposing people to diverse and balanced perspectives that challenge their prior beliefs and foster mutual understanding [16] [17]. The effects of media exposure on political polarization may depend on various factors, such as the type, content, and quality of media sources; the frequency, duration, and intensity of media use; and the individual characteristics, motivations, and preferences of media users. Subsequently, how political polarization influences media use and preferences is also a concentration. Some studies have shown that political polarization can affect media use and preferences by leading people to select or avoid media sources based on their ideological congruence or incongruence [18]. Other studies have indicated that political polarization can affect media use and preferences by shaping people's perceptions of media credibility, bias, and trustworthiness. The effects of political polarization on media use and preferences may vary depending on the availability, accessibility, and diversity of media sources; the salience, complexity, and controversy of political issues; and the social norms, expectations, and pressures of media users

3. RESEARCH METHOD

3.1 Overall Design

Qualitative analysis methods were used to analyze the data collected through coding, which was used to categorize and organize the data [19]. Narrative analysis was used to examine how individuals construct stories and how these stories shape their experiences [20]. Media texts were analyzed using thematic analysis to spot trends in the way messages are put together and conveyed. The study employs conventional qualitative analysis techniques like thematic analysis, narrative analysis, and coding.

The research was searched from the archive website of CNN and Fox News in a specific period from Dec. 31, 2021 (after Joseph Biden came to power) to the up-to-date deadline of Apr. 1, 2023. After selecting the homogeneous category relating to "environment" from CNN and Fox News and searching under the keywords of the outcome was there emerged 27,201 newsletters on CNN and 25,500 from Fox News. A random selection was conducted: firstly, the newsletters were numbered from 1 to the end according to the time sequence; secondly, use the function of "=RAND ()" in Excel to decide which 50 passages will be selected for each media coverage, thus 100 passages in total, as samples of this research (see Annex 1). This step can ensure equity and randomicity the same number of newsletters would be chosen with validity and credibility.

Table 1. The Random Number

0.62514012	0.57714128	0.94487861	0.2246206	0.27312736
0.37975304	0.65199578	0.35163979	0.62873144	0.16828869
0.03082766	0.67873878	0.83775133	0.14490305	0.004924
0.2905721	0.30391524	0.10141453	0.97049255	0.9451935
0.18724918	0.13224971	0.0808627	0.12037682	0.79470079
0.09302146	0.34617002	0.37131865	0.01526678	0.21497434
0.79250378	0.86347053	0.65392736	0.5991139	0.3530141
0.29980965	0.3350924	0.80068398	0.02754698	0.63556279
0.89870829	0.16273102	0.23370692	0.20693771	0.24116763
0.85246079	0.07071243	0.62984441	0.72495058	0.18628898

This sample selection will take three decimal places of each value, for example, 0.62514012, and the 625th numbered article from each media outlet of CNN and FOX News will be selected as the sample, as shown in Table 1 and Table 2.

Table 2: The Number of Selected Passages

625	577	944	224	273
379	651	351	628	168
030	678	837	144	004
290	303	101	970	945
187	132	080	120	794
093	346	371	015	214
792	863	653	599	353
299	335	800	027	635
898	162	233	206	241
852	070	629	724	186

3.2 Sampling and Data Collection

Fox News and CNN were chosen as the targeted research object because they are two of the most widely recognized news networks in the United States, and they are known to have different political biases.

CNN is a news-focused cable television station that broadcasts in the United States. It covers a variety of news subjects, including politics, business, health, technology, entertainment, and foreign news. People in the United States and other countries have long regarded CNN as one of their primary news sources.

CNN is rated as having a center-left bias in the Ad Fontes Media media bias chart, which assesses news organizations based on their bias and dependability. This indicates that although the news outlet delivers factual information, its reporting can occasionally be swayed by a liberal point of view. CNN is still regarded as having very balanced reporting when compared to other news sources, nevertheless (see Figure 1).

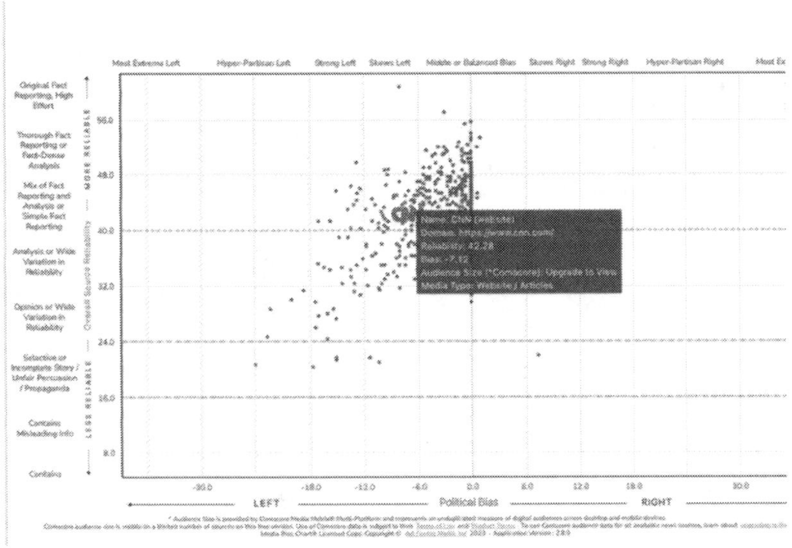

Figure 1: Interactive Media Bias Chart[21]

Fox News is a conservative news network that covers a variety of news subjects, including politics, business, health, technology, entertainment, and international news, according to Ad Fontes Media. According to Figure 2, Fox News is widely seen as having a significant right-wing slant. This indicates that the news agency offers information from a conservative standpoint and that conservative political ideology may occasionally impact its coverage.

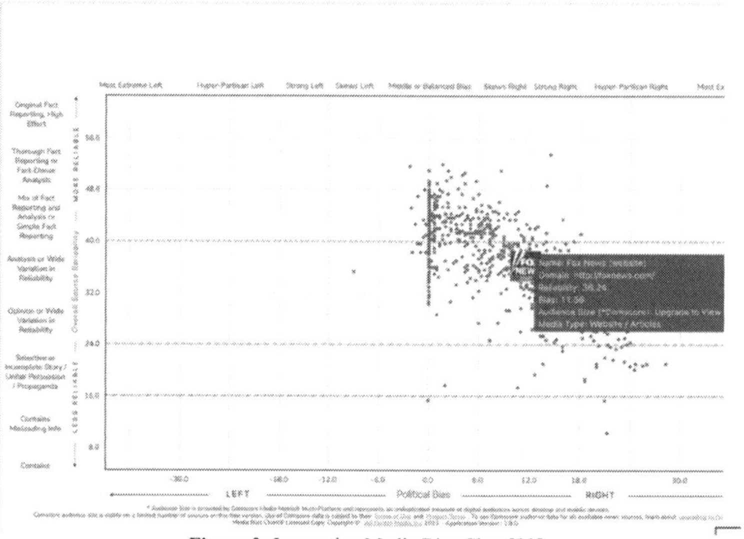

Figure 2: Interactive Media Bias Chart[22]

By examining media bias in Fox News and CNN, we can learn more about how media bias shapes public perception, how various news networks present political problems, and how media coverage might exacerbate political polarization. In conclusion, choosing these two networks as samples is a legitimate and practical method for researching media bias.

3.3 Coding and Thematization

Firstly, the open coding process was conducted by thoroughly reading the 100 passages, with the outcome gathering into the codebook (see Annex 1). Then, the axial coding procedure divided the open codes into three themes: political, environmental, and socio-economic perspectives. By examining and categorizing the codes into the three themes, codes like "Biden", "government", "presidency", "administration", "EPA", "policy", "proposal", "efficiency", "harm", etc. were thematized into "Political" perspective. Such thematization methods were conducted in the codebook.

4. FINDINGS

4.1 Media Coverage of Environmental Policies: Implications for Political Polarization and the Role of Ideological Leanings

The study found that both CNN and Fox News covered a wide range of environmental policies aimed at tackling climate change and promoting sustainability. These policies included efforts to transition to green energy, implement stricter air pollution controls, regulate waste management, expand national parks, and address issues such as rising sea levels, endangered species protection, and sustainable agriculture. By examining the relationship between media, environmental policies, and political polarization, this study provides insights into the complex dynamics at play. It underscores the importance of media responsibility and the need for a well-informed and engaged citizenry in fostering a more constructive and inclusive political discourse on environmental issues.

The study also found that these ideological tendencies are reflected in the framework, theme selection, and object selection, which may affect readers' interpretation and reinforce existing beliefs, ultimately leading to political polarization

4.2 Media Bias in Environmental Coverage: Framing, Topic Selection, and Object Option as Factors of Political Polarization

Firstly, the framing of news stories can be influenced by the ideological leanings of the news outlet. For example, CNN may emphasize the need for immediate and extensive action to address climate change, framing it as a pressing crisis requiring urgent solutions. On the other hand, Fox News may adopt a more skeptical stance towards climate change, focusing on the potential economic impact of environmental regulations. Additionally, the selection of topics covered by each network can reflect their ideological leanings. CNN might prioritize stories that highlight the negative consequences of environmental degradation and emphasize the importance of government intervention and international cooperation.

Meanwhile, Fox News might emphasize individual freedoms, market-based solutions, and the potential economic costs associated with environmental regulations. Furthermore, the choice of objects and topics interviewed on these networks can contribute to biases. CNN's video programs may invite guests who support progressive environmental policies, such as climate scientists and environmental activists. In contrast, Fox News might feature guests who express skepticism towards climate change or advocate for limited government intervention, such as conservative think tank representatives or industry experts. These biases in framing, topic selection, and guest choice can shape viewers' interpretations of environmental policies and reinforce their existing beliefs. Viewers who align with the liberal or progressive spectrum may feel their perspectives are validated by CNN's coverage, while conservative viewers may find their viewpoints affirmed by Fox News. This reinforcement of preexisting beliefs can contribute to political polarization, as viewers become more entrenched in their ideological positions and less receptive to alternative viewpoints. It is important to recognize that media bias is a nuanced and complex issue. Biases can arise from various factors, including the perspectives of journalists, editorial decisions, and the target audience of a particular news outlet.

Additionally, biases can vary within networks, as individual journalists may have their own biases that influence their reporting. To mitigate the potential impact of biases in news coverage, media consumers should actively seek out diverse sources of information and engage in critical analysis. By exposing themselves to a range of perspectives, individuals can develop a more comprehensive understanding of environmental policies and avoid the pitfalls of echo chambers and confirmation bias.

4.3 Case Analysis

In the first quarter of 2023, the Environmental Protection Agency of the US government released the latest policy on Electronic Vehicles (EVs), and both CNN and Fox News have covered this policy while simultaneously manifesting their political biases in their passages, which have contributed to macro-political polarization from a micro perspective.

The article “The EPA's Plan to Accelerate the EV Revolution” written by CNN Politics discusses the importance of fuel economy regulations for electric vehicles in reducing greenhouse gas emissions and reaching climate goals. The article implicitly states that it supports the EAP’s new policy on EVs, and in terms of emotional color, the words used in the article are focused on straightforwardly presenting information and analysis, without overtly positive or negative connotations. However, some words and phrases such as "critical for achieving climate targets" and "potential to significantly reduce emissions" suggest a positive view of the policy.

Moreover, the article “EPA proposes new tailpipe rules that could push EVs to make up” also written by CNN supports the policy of implementing tougher car pollution standards and promoting electric vehicles to combat climate change, as it emphasizes the benefits of reducing emissions and mentions the support of President Biden and environmental groups.

To judge the emotional color of the words in the article, one can look for language that conveys a positive or negative tone towards the policy or the people involved. For example, positive language used in CNN’s passages includes words like "benefit," "progress," and "success," while negative language might include words like "harm," "failure," and "controversy."

The article from Fox News titled "EPA administrator claims 'absolutely no punishment' for those who don't buy electric vehicles" suggests a critical stance towards the policy of promoting electric vehicles. The use of words such as "claims" and "punishment" may imply skepticism or doubt about the policy.

Regarding the article “Biden unveils toughest-ever car emissions rules in a bid to force”, based on a narrative analysis of the article, it appears that Fox News is critical of the policy announced by President Biden to set the toughest-ever car emissions rules to force electric vehicle purchases.

First, the headline uses the word "force," which has a negative connotation, rather than a more neutral term like "encourage." The article also frames the policy as a "radical" and "expensive" plan that will impact consumers and the auto industry. Additionally, the article includes quotes from Republican lawmakers and industry representatives who criticize the policy. Furthermore, the article includes phrases such as "green agenda" and "climate crusade," which suggest that Fox News views the policy as part of a broader liberal environmental agenda rather than a pragmatic solution to reduce emissions

5. DISCUSSION

5.1 Media Possess Ideological Bias to Political Polarization

CNN has often been criticized for having a liberal or progressive bias in its coverage. Critics argue that the network tends to favor left-leaning perspectives and often presents news stories and analyses from a liberal standpoint [23]. This perception of bias has manifested by viewing CNN as promoting a particular political agenda and catering to a predominantly liberal audience [24], especially distinctive in the verbal usage, emotion, and comments on the environmental policies in this study. This may be seen in the language they employ, the subjects they decide to cover, and the way they frame those subjects in their writings. But it is also crucial to keep in mind that news organizations could exhibit variable degrees of prejudice toward various themes and situations.

On the other hand, Fox News has faced accusations of having a conservative or right-wing bias. Critics argue that the network promotes conservative viewpoints, employs conservative pundits, and frames news stories in a way that supports a conservative narrative [23]. Some viewers perceive Fox News as catering to a conservative audience and promoting a conservative political agenda [24], which could also be observed from the environmental policy commentaries selected. Based on the language and tone of these articles, it appears that Fox News is presenting a negative view of the policy and may be reflecting a conservative or right-leaning political bias.

These perceived biases can shape the content and framing of news stories, as well as the selection of topics and objects. The ideological leanings of CNN and Fox News can influence the way information is presented, leading to different interpretations of events and issues.

This can contribute to the formation of partisan attitudes and the reinforcement of existing beliefs among viewers [16].

It is important to note that perceptions of bias can vary among different audiences. Individuals with different political affiliations or ideological orientations may interpret the same news coverage differently, perceiving bias in different directions [24]. However, the perceived bi-polarized biases on the same environmental policy of CNN and Fox News have contributed to the perception that the media landscape is polarized, with each network serving as a platform for partisan viewpoints[25].

5.2 Information Filtering and Selective Exposure Deepen Manifestations of Political Polarization

Individuals tend to seek out and consume news that aligns with their existing beliefs and values, resulting in selective exposure to ideologically compatible sources [18] [26]. This filtering process leads to an echo chamber effect, where individuals are predominantly exposed to information that reinforces their preexisting views [12].

CNN and Fox News, as prominent news outlets, employ the environmental policies conducted in this study of Biden's Administration, which cater to their target audiences, thereby contributing to information filtering and selective exposure. Previous research has demonstrated that individuals who identify as liberals are more likely to consume news from CNN, while conservatives are more inclined to turn to Fox News [17]. These media outlets strategically frame and emphasize certain political issues, appealing to their respective partisan audiences [16] [26].

Furthermore, technological advancements, such as personalized algorithms and social media platforms, exacerbate the problem of information filtering and selective exposure. These algorithms track users' online behaviors and preferences, curating content that aligns with their past consumption patterns [27]. As a result, individuals are often presented with a narrow range of perspectives, reinforcing their existing beliefs and contributing to the echo chamber effect.

The consequences of information filtering and selective exposure are significant for political polarization. Individuals who engage in selective exposure tend to become more entrenched in their ideological positions, as exposure to opposing viewpoints is limited [26] [28]. This limited exposure to diverse perspectives can lead to an increase in hostility and a decrease in understanding between partisan groups [17] [25].

To address the challenges posed by information filtering and selective exposure, it is crucial to promote media literacy and encourage individuals to seek out diverse sources of information [7] [26]. Additionally, media organizations should strive for balanced and comprehensive reporting, presenting multiple perspectives on political issues [12] [29]. By promoting media diversity and providing individuals with the tools to critically evaluate and access a wide range of information, it is possible to mitigate the negative effects of information filtering and selective exposure on political polarization.

5.3 Political Polarized Media on Public Attitudes and Behaviors

The media, including CNN and Fox News, play a significant role in shaping public attitudes and behavior regarding political issues. Previous research has indicated that media exposure can have a profound impact on individuals' political beliefs and actions [30] [31]. In the context of political polarization, the influence of the media becomes particularly salient, as it can contribute to the widening divide between partisan groups.

Both CNN and Fox News have been found to shape the attitudes of their respective audiences through their editorial choices, framing of issues, and partisan commentary. Studies have shown that exposure to ideologically aligned media outlets can reinforce existing beliefs and attitudes, leading to increased polarization.

Moreover, the media's influence extends beyond shaping attitudes and beliefs to influencing public behavior. Research has demonstrated that media consumption can impact individuals' political participation and voting behavior.

However, it is important to acknowledge that the relationship between media exposure and public attitudes and behavior is complex and multidirectional. While the media can reinforce existing beliefs, it is also influenced by the preferences and demands of its audience [32]. Media outlets may tailor their content to cater to their viewership, resulting in a feedback loop where media and audience preferences reinforce each other.

Exposure to ideologically aligned media outlets can reinforce existing beliefs, deepen partisan divisions, and influence political participation. However, it is crucial to recognize the reciprocal relationship between media and the audience, as media outlets also respond to the demands and preferences of their viewership.

5.4 Limitations and further expectations

Although this study provides a preliminary comparative study of the impact of CNN and Fox News on political polarization in the United States, there are some limitations.

First, this study is limited to two media outlets, CNN and Fox News, and does not consider the impact of other news organizations on political polarization. Future research could consider more media outlets to get a comprehensive understanding of the media's contribution to political polarization.

Second, this study relied intensively on narrative and thematic analysis to analyze CNN and Fox News coverage. However, media consumption is a dynamic process that is influenced by individual experiences, social context, and other external factors. Further research could employ more diverse research methods, such as longitudinal studies, field observations, and quantitative methods, to gain insight into the interactions and influences between media and viewers.

In addition, this study could extend the exploration of the relationship between media and political polarization. In addition to the media's influence on viewers, further research could explore the feedback effects of political polarization on media coverage. For example, political polarization influences the media's stance and reporting style, thus creating an interactive relationship.

Future research could also delve into the impact of the diversity of media consumers and the trend toward customization of media on political polarization. People are increasingly inclined to choose media that match their positions and preferences when accessing news and information. Thus, the research could explore how factors such as individual differences, social factors, and technological developments interact with selective media access to further exacerbate or slow down the trend of political polarization.

6. Conclusion

CNN and Fox News are known to have different political biases, with CNN being perceived as having a liberal or left-leaning bias and Fox News as having a conservative or right-leaning bias. It is not uncommon for news articles to reflect the bias of the source, so the articles from CNN and Fox News may present different perspectives or interpretations of the same news story.

In terms of the specific articles provided, the CNN articles focus on President Biden's plans to reduce carbon emissions and promote electric vehicles. They discuss the impact of proposed regulations on the auto industry, the potential benefits of electric vehicles, and the challenges in transitioning to cleaner transportation. While the articles are critical of the Trump administration's rollback of environmental regulations, they also acknowledge the difficulty of achieving significant changes in a short period.

The Fox News articles also cover Biden's policies regarding electric vehicles, but they take a more critical stance, portraying them as a "power grab" and suggesting that the administration is punishing people who choose not to purchase electric vehicles. The articles are more focused on the perceived negative effects of the proposed regulations, such as increased costs and reduced freedom of choice.

Overall, based on these articles, it appears that the CNN articles present a more optimistic perspective on the potential benefits of electric vehicles and the importance of reducing carbon emissions. The Fox News articles, on the other hand, are more critical of Biden's policies and present a more skeptical view of the benefits of electric vehicles.

References

1. Humayun N. (2013, January 4). "The Impact of Social Media on Modern Society".
2. Sunstein, C. R. (2017). #Republic: Divided democracy in the age of social media. Princeton University Press
3. Banda, K. K., & Cluverius, J. (2018). Elite polarization, party extremity, and affective polarization. Electoral Studies, 56, 90–101.
4. Morgan Kelly. (2021, September 12). Political polarization and its echo chambers: Surprising new, cross-disciplinary perspectives from Princeton. Princeton University.
5. Kubin, E., & von Sikorski, C. (2021). The role of (social) media in political polarization: A systematic review. Annals of the International Communication Association, 45(3), 188–206.
6. Kreiss, D. (2019). The social identity of journalists. Journalism, 20(1), 27–31.
7. Gentzkow, M., & Shapiro, J. M. (2011). Ideological segregation online and offline. The Quarterly Journal of Economics, 126(4), 1799-1839.
8. Prior, M. (2007). Post-broadcast democracy: How media choice increases inequality in political involvement and polarizes elections. Cambridge University Press.
9. Mason, L. (2018). Uncivil Agreement: How Politics Became Our Identity. University of Chicago Press.
10. Wilson, A. E., Parker, V. A., & Feinberg, M. (2020). Polarization in the contemporary political and media landscape. Current Opinion in Behavioral Sciences, 34, 223–228.
11. Humayun N. (2013, January 4). "The Impact of Social Media on Modern Society".
12. Sunstein, C. R. (2017). #Republic: Divided democracy in the age of social media. Princeton University Press.
13. Gentzkow, M., & Shapiro, J. M. (2006). What Drives Media Slant? Evidence from U.S. Daily Newspapers.
14. Bail, C. A., Argyle, L. P., Brown, T. W., Bumpus, J. P., Chen, H., Hunzaker, M. B. F., Lee, J., Mann, M., Merhout, F., & Volfovsky, A. (2018). Exposure to opposing views on social media can increase political polarization. Proceedings of the National Academy of Sciences, 115(37), 9216–9221. https://doi.org/10.1073/pnas.1804840115
15. Kubin, E., & von Sikorski, C. (2021). The role of (social) media in political polarization: A systematic review. Annals of the International Communication Association, 45(3), 188–206.
16. Levendusky, M. S. (2018). Political polarization and the American public. Annual Review of Political Science, 21, 563-578.
17. Wojcieszak, M., & Mutz, D. C. (2022). How group discussions with like-minded others can increase political polarization. Proceedings of the National Academy of Sciences, 119(3), 1-6.
18. Garrett, R. K., Carnahan, D., & Lynch, E. K. (2013). A Turn Toward Avoidance? Selective Exposure to Online Political Information, 2004–2008. Political Behavior, 35(1), 113–134.
19. Gibbs, G. (2007). Analyzing Qualitative Data. SAGE Publications, Ltd.

20. Taylor, S. J., Bogdan, R., & DeVault, M. L. (2016). Introduction to qualitative research methods: A guidebook and resource (Fourth edition). John Wiley & Sons, Inc.
21. interactive chart. Ad Fontes Media. (2023, July 8). https://adfontesmedia.com/interactive-media-bias-chart/#elementor-action%3Aaction%3Dpopup%3Aopen%26settings%3DeyJpZCI6IjQ3MDgyIiwidG9nZ2xlIjpmYWxzZX0%3D.
22. Interactive chart. Ad Fontes Media. (2023a, July 8). https://adfontesmedia.com/interactive-media-bias-chart/#elementor-action%3Aaction%3Dpopup%3Aopen%26settings%3DeyJpZCI6IjQ3MDgyIiwidG9nZ2xlIjpmYWxzZX0%3D
23. DellaVigna, S., & Kaplan, E. (2019). The Fox News effect: Media bias and voting. Quarterly Journal of Economics, 134(1), 1503-1544.
24. Iyengar, S., & Hahn, K. S. (2021). Red media, blue media: Evidence of ideological selectivity in media use. Journal of Communication, 71(1), 22-40.
25. DellaVigna, S., & Kaplan, E. (2019). The Fox News effect: Media bias and voting. Quarterly Journal of Economics, 134(1), 1503-1544.
26. Stroud, N. J. (2020). Polarization in the news environment: Examining audience selective exposure and media trust. Journal of Communication, 70(5), 680-702.
27. Tsfati, Y., & Cappella, J. N. (2018). The third-person effect: A meta-analysis of the perceptual hypothesis. Mass Communication and Society, 21(2), 197-224.
28. Bakshy, E., Messing, S., & Adamic, L. A. (2015). Exposure to ideologically diverse news and opinion on Facebook. Science, 348(6239), 1130-1132.
29. Prior, M. (2013). Media and political polarization. Annual Review of Political Science, 16, 101-127.
30. Jones, A. B. (2018). The effects of media consumption on political attitudes: A meta-analysis. Mass Communication and Society, 21(2), 173-194.
31. Smith, R., & McDermott, R. (2020). Media exposure and political polarization: Evidence from the introduction of Fox News. American Economic Journal: Applied Economics, 12(3), 224-251.
32. Stroud, N. J. (2020). Polarization in the news environment: Examining audience selective exposure and media trust. Journal of Communication, 70(5), 680-702.

The Influence of Advertising Language on Adolescents' Consumption Perspectives from the Perspective of Behavioral Economics

Nianhe Wu *

Nianhe Wu, Shanghai Starriver Bilingual School, 2588 Jindu Road, Minhang District, Shanghai, China.

Abstract: Advertising leverages linguistic concepts and metaphorical techniques to enhance the persuasiveness and expressiveness of products. In the current era of widespread electronic devices and online shopping platforms, advertising exerts an unprecedented influence on consumer behavior, particularly among adolescents at the peak of their cognitive development. Adolescents are more susceptible to the impact of advertising language due to their unique psychological characteristics. This paper combines theories such as behavioral economics and mental accounts to analyze the underlying causes of irrational consumption behavior among adolescents within the context of China's distinct mental account structure. The study aims to examine the influence of different advertising approaches on adolescents' consumption attitudes from a social-psychological perspective. Additionally, it proposes reform policy recommendations, including strengthening government regulation, enhancing educational guidance, promoting social governance, and conducting empirical research, to contribute to the creation of a civilized and harmonious developmental environment for adolescents.

1. Introduction

Adolescents are a distinct group between children and adults, possessing unique needs[1]. Scholars like Erik Erikson have emphasized that adolescents experience identity and role confusion during this developmental stage, requiring the development of their identities and self-awareness, including aspects such as gender identity, political inclinations, and religious beliefs. Concurrently, adolescents undergo rapid cognitive growth and the development of complex thinking abilities. They begin to question the beliefs inherited from previous generations and challenge social relationships[2]. Consequently, adolescents have been regarded as "miniature adults"[3].

At its core, advertising is a language-based medium. Cao Zhiyun defined advertising language as "the language used in advertisements, encompassing all linguistic and textual information found within advertisements, such as speech, words, sentences, texts, punctuation, and textual graphics" [4]. In the rapidly evolving era of the Internet and continuous innovations in television media, advertising has grown significantly and diversified across various platforms, becoming an integral part of people's lives. Presently, advertising silently influences adolescents, often yielding negative effects. Many advertisements employ metaphorical techniques to induce adolescents into engaging in irrational consumption. For instance, F. Ziesmer noted that adolescents are emerging as a valuable consumer segment due to economic recession and cultural changes[5]. Research in 2013 reported that adolescents, as the largest consumer group in the United States, contribute approximately $150 billion annually. The primary goal of this survey is to investigate how advertisement linguistics affects adolescent consumption behavior from a behavioral economics perspective. In this research, we designed a survey that embodies the questions about how often juveniles (age 17-22) consume luxury brands, how often the advertisements attract them, and to what extent they agree with the connection between advertisements and the consumption of affection[6].

In contemporary society, advertising strategies have gradually diversified. Businesses, driven by profit, tend to adopt extreme and class-oriented commercial advertising language and image designs. They highlight differences in class, race, and gender through advertising language, continuously influencing adolescents' thoughts, and lifestyles and even inducing them to adopt erroneous life values. Existing research primarily focuses on the consumer psychology of high school and college students, with limited attention given to studying NFT (Non-Fungible Token) products within the metaverse context[7]. Previous research has predominantly explored linguistic, social psychological, linguistic, and advertising psychological aspects, paying comparatively less attention to the behavioral economic theories encompassed behind advertisements[8]. Hence, this paper designed a survey that seeks to analyze the influence on adolescents' consumption attitudes from a social-psychological perspective and discuss corresponding countermeasures.

*Email: wu_nianhe@126.com

2. The Inevitability of Adolescents' Irrational Consumption

(1) The Influence of Businesses Utilizing Digital Products and the Internet to Enhance Adolescents' Irrational Consumption

Advertising language serves as a key means of communication in advertising, as it can evoke emotional resonance among the audience and guide their consumption behavior. Advertisers aim to induce adolescent consumers by carefully designing advertising language that associates consumer products with their emotions, familial ties, romantic relationships, and other values. Adolescents, in a special period of life development, are characterized by their inclination to think critically, quickly adapt to new things but possess weaker discernment abilities. When exposed to advertising language, they are easily influenced by various factors, such as emotional needs, social identification, and peer pressure. Under the influence of sensory impulses and emotional factors, adolescents are susceptible to losing rational analysis and making consumption decisions devoid of logical judgment, ultimately leading to irrational consumption behavior.

Over the past two years, businesses have focused on marketing non-fungible tokens (NFTs) in the non-metaverse realm. NFTs represent unique cryptographic systems that captivate adolescents, primarily due to the metadata associated with NFTs, which bestows product ownership and transaction transparency. For instance, when significant enthusiasm and buzz surrounds Alibaba's NFT product series, most consumers highly prioritize it. This indicates that companies should regularly interact with their target markets to create a community composed of potentially interested customers, a task facilitated by social media platforms. Businesses attract adolescents to join these communities through social media by sharing information, pictures, videos, lotteries, and other content related to NFT products. Subsequently, businesses interact with potential teenage customers by providing professional consultations, answering questions, collecting feedback, and engaging in other activities to continually reinforce adolescents' trust and loyalty. These efforts lead to heightened sales and promotional outcomes. For example, Gucci and the gaming platform Reblox collaborated to launch a co-branded virtual equipment collection via an electronic game social platform, offering lower prices than Gucci's physical products. Adolescents obtain virtual luxury goods through NFTs, satisfying their desires for comparison and vanity, fueling irrational consumption.

(2) The Influence of Mental Accounts on Adolescents' Irrational Consumption

Nobel laureate Richard Thaler first introduced the mental account theory. Mental accounts are the cognitive process in which individuals mentally categorize, record, and evaluate their wealth. This process is influenced by individuals' estimations and expectations for different accounts (Thaler, 1985) [9]

Building upon this theory, researchers such as Li Aimei and Sun Daqiang have established a unique mental account structure specific to Chinese individuals[10-11]. They have noted that Chinese mental accounts exhibit distinct cultural attributes, and advertising language, in conjunction with these cultural attributes, creates a unique mental account for Chinese consumers. The study confirms that Chinese mental accounts have three fundamental components: income, expenses, and savings. Income is further divided into regular work income, irregular additional income, and income from business investments. Expenses are categorized into essential living expenses, family development and personal growth expenses, emotional connection expenses, and leisure and entertainment expenses. Savings accounts are divided into security-oriented protection accounts and risk-oriented investment accounts.

According to Professor Li Aimei's model, we conclude that consumers' expenditures on luxury goods are based on their leisure and entertainment expenses. If advertisers bind advertising language with emotions such as family affection and filial piety, consumers' mental accounts for expenditures shift from leisure and entertainment expenses to emotional connection expenses[10]. For example, the advertisement for DR diamond rings, with the language "A man can only customize one ring in his lifetime," combines love and brand value. According to DR's prospectus, the highest-selling FOREVER series of simple luxury items in the first half of 2021 cost approximately 4,000 yuan to produce, yet they are priced at 16,000 yuan, achieving a gross profit margin of 72.58%. In contrast, other domestic jewelry giants such as Zhou Dasheng and IDO have an average gross profit margin of around 40%. Behind the high cost and low profit is the reduced price sensitivity of consumers when it comes to symbols of love. Couples perceive that they are paying for love. These studies reveal the correlation between consumers' mental accounts and irrational consumption.

Swatch, a world-renowned fashion watch brand, captures the emotional spending of the adolescent demographic by infusing fashion with the concept of meeting personalized requirements. Swatch conveys the artistic charm of individuality, youthfulness, and excitement to teenagers. Due to characteristics such as a younger age structure, lower cultural literacy, weaker defense mechanisms, and a strong desire for comparison, adolescents are easily influenced by the "non-substitutability" effect. This effect states that wealth belonging to different accounts cannot be completely substituted for one another. Adolescents pursue the maximization of emotional satisfaction, leading to a cognitive illusion that emotional connection expenses are not luxury expenses. Consequently, numerous irrational consumption decisions are made. (see Table 1)

 https://doi.org/10.1051/shsconf/202317802006

Table 1: Psychological Account Structure for Chinese Individuals Propose

ACCOUNT CLASSIFICATION	PSYCHOLOGICAL ACCOUNT SUBCATEGORIES	SPECIFIC EXAMPLES OF PSYCHOLOGICAL ACCOUNTS
INCOME ACCOUNT	Regular work income	Salaries, bonus
	Irregular additional income	Windfall, inheritance, gifts
	Income from Business Investments	Loans, fixed asset investments
EXPENSE ACCOUNT	Essential Living Expenses	Food,transportation, communication, healthcare, clothing, etc.
	Family Development and Personal Growth Expenses	Professional development, vehicle expenses, expenses for children
	Emotional Connection Expenses	Respecting parents, emergency expenses, donations, social expenses
	Leisure and Entertainment Expenses	Luxury goods, entertainment, beauty and wellness, travel and vacations
SAVINGS ACCOUNT	Security-oriented Protection Account	Current deposits, fixed deposits, cash
	Risk-oriented Investment	Account Stocks, options, bonds, precious metals

(3) The Influence of Celebrity Endorsements on Adolescents' Irrational Consumption

Adolescents are known for their self-expression, rebellious nature, and pursuit of fashion and trends. Therefore, celebrity culture is one of the most admired cultures among adolescents. Commercial companies utilize celebrity endorsements not only to enhance brand awareness but also to project the positioning of products in the market, target audience, and brand values. Celebrity endorsers have a dual nature, and the replacement of untrustworthy endorsers can have an impact on the brand image, leading to changes in adolescents' perception and trust in the brand, thereby influencing their consumption decisions. Therefore, when selecting endorsers, businesses often invest time in assessing whether the endorsers' image and reputation align with the brand's image and values to avoid negative effects resulting from endorser replacements.

At this point, celebrities possess unique charm and leadership that give rise to the so-called "halo effect"[12] When they serve as endorsers, the intangible halo effect is transferred to the products they endorse, imbuing the products with a halo as well. Businesses leverage the celebrity effect to establish a connection between adolescents and the endorsed products based on the celebrities' image and values. Subtly and unconsciously, this stimulates their desire to purchase the products, making them consumers of the endorsed brands. For example, luxury brand Dior chose celebrity Liu Yifei as its ambassador because she embodies the brand's core values of "continuously pursuing excellence, exuding feminine charm, and being refined and elegant." This choice targets adolescents who identify with Liu Yifei's values but may still have immature thinking, thereby leading to irrational consumption.

It is important for adolescents and their parents to develop media literacy and critical thinking skills to discern the intentions behind celebrity endorsements and make informed consumption choices. Additionally, businesses should take ethical considerations into account when selecting celebrity endorsers, ensuring that their values align with the brand's values and that they serve as positive role models for adolescents. By promoting responsible and transparent marketing practices, the influence of celebrity endorsements on adolescents' irrational consumption can be minimized.

3. Questionnaire Design: Investigating the Influence of Luxury Product Advertising on the Irrational Consumer Behavior of Chinese Youth

This questionnaire aims to investigate the luxury consumption behavior among Chinese youth aged 14 to 22 years old. A total of 86 participants from junior high schools, high schools, and universities in Shanghai participated in the survey. The data revealed that 88.37% of respondents have either shown interest in or purchased luxury products. In the context of luxury advertisements pervading social media and the internet (100% of respondents agreed to encountering luxury advertisements on platforms such as WeChat and Douyin), 92.11% of participants were found to be influenced by attractive advertisements when making luxury purchases. The study further found that 88.57% of respondents in this group were influenced by models and celebrity endorsements when making luxury purchases. This result aligns with the primary emphasis of luxury advertisements, where 73.68% of survey participants pointed out that such advertisements often highlight the brand as the ultimate choice for specific groups (e.g., actors). Moreover, all respondents (100%) agreed that luxury advertisements associate the brand with social class or specific emotions (e.g., love, filial piety), and 55.26% of survey participants admitted that they consume luxury products to highlight their own social status, while 76.32% of respondents revealed that they buy luxury products for their relatives due to the influence of advertisements. In conclusion, luxury product advertisements utilizing celebrity endorsements and associating the brand with social class or specific emotions have led to irrational consumption behaviors among a significant portion of Chinese youth.

4. Policy Recommendations from the Perspective of Behavioral Economics

In order to address the issue of excessive consumption among adolescents influenced by advertising, this paper proposes the following policy recommendations from the perspective of behavioral economics:

(1)Establish regulations and strengthen supervision

In response to emerging areas such as the digital economy, it is important to improve relevant laws and regulations on advertising supervision, promote the issuance of departmental regulations, and encourage local legislation. Establish a regular inspection mechanism for advertising supervision and increase efforts to investigate and penalize illegal activities such as the dissemination of false information. Further urge online platforms to fulfill their responsibilities and ensure the authenticity and reliability of product information through careful review and verification.

(2)Strengthen education and provide positive guidance

Promote the inclusion of financial literacy education in the school curriculum, improve the interactive education mechanism between schools and families, and guide adolescents in establishing sound psychological accounts and understanding basic financial knowledge and money management skills. Simultaneously, enhance critical thinking skills and logical reasoning abilities through general education, helping adolescents better cope with the advertising and temptations of businesses.

(3) Social governance and create a conducive environment

Utilize various channels such as the internet, television, radio, and live events to intensify the promotion of rational and moderate consumption. Guide adolescents to develop proper consumption values, establish good consumption habits, and assist them in distinguishing between genuine needs and excessive consumption. Additionally, help them understand that consumption is not the sole means of satisfying their needs and encourage the pursuit of meaningful values.

(4) Conduct empirical research and strengthen the foundation

Conduct in-depth analysis of the psychological account structure of Chinese consumers and its impact on consumption behavior. This research serves as a theoretical basis and research foundation for government policy formulation, as well as aids businesses in accurately identifying target consumer groups and developing more effective marketing strategies.

In conclusion, by strengthening government supervision, creating a favorable environment, raising consumer awareness, and conducting further research, we can ultimately achieve the goal of regulating market order, protecting the healthy development of adolescents, and safeguarding the legitimate rights and interests of consumers.

References

1. Erikson, E. H. (1950). *Childhood and Society*. W. W. Norton & Company.
2. Widick, C., Parker, C. A., & Knefelkamp, L. (1978). Erik Erikson and Psychosocial Development. *New Directions for Student Services*, 1978(4) 1978, 1-17. https://doi.org/10.1002/ss.37119780403
3. Grinder, R. E. (1969). The Concept of Adolescence in the Genetic Psychology of G. Stanley Hall. *Child Development,* 40(2), 355. https://doi.org/10.2307/1127408
4. Cao, Z. (1992). Advertising Language Art. *Hunan Normal University Press.*ISBN:9787810311335
5. Ziesemer, F., Hüttel, A. & Balderjahn, I. (2021). Young People as Drivers or Inhibitors of the Sustainability Movement: The Case of Anti-Consumption. *J Consum Policy* 44, 427-453. https://doi.org/10.1007/510603-021-09489-x
6. Schudson, M. (1984). *Advertising*, The Uneasy Persuasion: Its Dubious Impact on American Society. Routledge.
7. Xiang T(2022). Research on the design application of Chinese traditional cultural elements in stop-motion animation advertising. *Zhejiang University of Science and Technology for Master's Degree.* DOI: 10.27840/d.cnki.gzjkj. 2022.000078
8. Zeng X. (2021). Emotional comfort, social interaction flexibiblity, and value realization: young people's consumptuion of mystery boxes and the re-imagination of youty subculture. *Journal of Fujian Normal University(Philosophy and Social Sciences Edition)*,226(01):133-141.
9. Thaler, R. (1985). Mental Accounting and Consumer Choice. Marketing Science, 4(3), 199-214. https://doi.org/10.1287/mksc.4.3.199
10. Liu, H., Li, A., & Wang, H., Wei, H. (2012). The Effect of Promotion Types on Consumers' Purchase Decisions: From the Perspective of Construal Level Theory. *Acta Psychologica Sinica*, 44(8), 1100-1113. https://journal.psych.ac.cn /xlxb/CN/Y2012/V44/I8/1100
11. Sun, Daqiang. (2008)"An Exploratory Study on Mental Accounting Based on the Context of Chinese Culture." Ph.D. thesis, *Beijing Normal University*. https://d.wanfangdata.com.cn/thesis/Y9027749
12. Nisbett, R. E., & Wilson, T. D. (1977). The halo effect: Evidence for unconscious alteration of judgments. *Journal of Personality and Social Psychology*, 35(4), 250–256. https://doi.org/10.1037/0022-3514.35.4.250

A Study of the Application of New Media and Technology to Innovative Design Education in the Science of Learning

Jiajun Chen[1], Shiyan Luo[1], Jiaen Lai[2], Leyi Zhu[1], Peijuan Cai[1], Jiaying Zeng[1] and Ya Zhang[1, *]

[1] Zhongkai University of Agriculture and Engineering,He Xiangning College of Art and Design, Guangzhou, China
[2] Guangzhou Institute of Science and Technology, Guangzhou, China

Abstract. This paper proposes an innovative design education model that effectively combines learning science theory, new technology, and new media to address the problems of serious homogenisation and disconnection from social requirements in design education. The model is based on the Chinese context, in accordance with the three principles of innovative design education summarized in Tim Brown's design thinking, and utilizing the learning science theory, new technology, and new media. Analyzing the relationship between learning science theory, new technology, and new media, providing a reference for the full application of new technology and new media in Chinese modern design education, and promoting the application of learning science theory in Chinese modern design education constitute the construction of the model.

1 Introduction

As a result of the fourth industrial revolution, new technologies and media are reconfiguring the education model further. In the 1990s, embodied cognition became the core of the second generation of cognitive science, the teaching paradigm shifted from "teacher-centered" to "student-centered," and learning science emerged to design and promote learning innovations for scientific comprehension [1, 2]. However, the status quo of art and design education in Chinese colleges and universities in the contemporary new era is still dominated by design theory education, resulting in a significant disconnect between graduates and social design needs. Due to technical and interdisciplinary issues, it is difficult for design students to produce innovative designs through the curriculum, and learning science theories is rarely employed in design education.

This thesis focuses primarily on the learning science theory, analyzes the relationship between the theory development process and new technology and new media, summarizes the three major principles of innovative design education in China's design education context, constructs a classroom model of innovative design education, applies the findings of learning science research to the practice of design education, and offers a new way of thinking.

2 Overview of the science of learning

Learning science is rooted in cognitive science, and its primary research focuses on cognitive theory, methodology, sociology, and other facets. Traditional cognitive science does not include the body in its research scope, thereby ignoring the interaction between humans as an organic whole and their physical and mental environments [3].Contextuality became one of the foundational concepts of "second-generation cognitive science" in the 1980s, which asserts that cognition is situational because embodied intelligence is embedded within the constraints of natural and social environments[4]. With the internal drive for educational change and the promotion of technology [5], learning science was born from the second generation of cognitive science and gradually developed into a new discipline, which reflects the teaching and learning process, the cognitive subject (students) to the classroom environment, context of the deepening degree of dependence and interaction, learning is no longer merely skill learning, but should be the students continue to select their own learning strategies. "Applied Learning Science" refers to the use of knowledge of how people learn to enhance the effectiveness of instructional design and assist learners in learning authentic tasks; it is the application of the findings of learning science's investigation into how people learn to teaching and assessment.

3 Connotation of Innovative design education

The combined use of new technology and new media has broadened the scope and limits of art and design education and teaching methods. Teachers and students, students and pupils, use various media and disciplines to investigate the diverse expressions of design, which helps to cultivate "design thinking" and conforms the

* Corresponding author: michelleguyu@126.com

design to the current economic development while presenting a fresh perspective. Tim Brown believes that the prerequisites for creativity are the social and spatial environments, and that design thinking is a holistic, human-purpose and human-fundamental means of solving innovation's problems.

Based on this understanding, in the context of China's practice of "observing the right and innovating, seeking the truth and making use of it", innovative design in our country should be a design that implements the new development concept, integrates into the new development pattern with the domestic macrocycle as the main body and the domestic and international double cycle promoting each other, conforms to the trend of digital science and technology, and solves the challenge of innovation and is rich in the deep meaning of design. Therefore, in order to enable design students to shift from the "technical" to the "creative", innovative design education should follow the following three principles.

3.1 Student-centered and accurate educational objectives

From the perspective of students, the abundance of high-quality design education resources has increased the flexibility and selectivity of students' independent learning. From the perspective of educators, educators have ideal standards for their own teaching classrooms, and they will subjectively determine what type of teaching situation is most appropriate for design majors. Nevertheless, the actual classroom is still "teacher-centered" one-way knowledge transfer, which is frequently quite distinct from the ideal standard. "Student-centred" refers to the teaching concept of focusing on student growth, student learning, and learning outcomes [6], i.e., according to the teaching goals that educators wish to achieve, they will continue to design the teaching and summarise and reflect on the teaching, prescribe the right medication, and move further towards the goal so as to allow students to construct in different contexts, understand knowledge, and achieve knowledge coherence. A comprehensive educational objective should include: 1. what has been learned 2. applying what has been learned 3 how to evaluate academic performance [7], so that students can comprehend the transition from memorization to knowledge creation after learning. Therefore, innovative design education classroom should be based on solving the problem of what kind of innovative design talents need to be cultivated in the new era for the establishment of objectives, focusing on the four aspects of design professional theory, design professional skills, interdisciplinary knowledge and moral quality of the new era, so that students internalise their knowledge during the teaching process, and teachers are also improved in the process of teaching.

3.2 Utilizing innovative technologies and media for design innovation

Regarding art and design, art is an abstract reflection of human thought activity and experience, whereas design is a deliberate creative act. Modern design education should actively address the innovative challenges of the AI era and assume responsibility for comprehending and communicating the meaning of design as well as cultivating creative abilities.

In 2017, "Deer Class" delivered 400 million images to the group's e-commerce companies during the Double 11 Shopping Festival, overturning the traditional design process and demonstrating the need for innovative design education to keep up with the cutting edge of technology. Platforms such as Ali Data, Tencent Cloud, and FanSoft provide multi-screen data presentation solutions for individuals and businesses to analyze problems using multiple technologies such as big data and artificial intelligence, which is a concrete example of the combination of design and new technologies. Shenzhen University has utilized WeChat, a new media, to communicate in a creative manner. APP is not limited to social activities, but also art and design education services, through the establishment of the University of Shenzhen WeChat brand culture shop sales of cultural and creative products at the same time, the design education and practice, so that the output of the course has a clear destination, to stimulate the creative passion of the students, so that the cultural elements can be further refined and re-designed to meet the needs of innovation. The program is also an excellent illustration of how new technologies and media can assist designers in creating innovative designs. The new technology and new media serve to complete the design, and the design responds to the new technology and new media in such a way that the new technology and new media can be presented and used in various ways.

Innovative design education should utilize all types of educational resources provided by big data, build an innovation and entrepreneurship platform, introduce teaching courses or interdisciplinary courses of new technologies in a timely manner, provide students with the means to independently learn new technologies and interdisciplinary knowledge, realize the freedom of learning and the direct presentation of design education's results, and reduce the number of teaching activities. Ineffective teaching practices resulting from instructors' inability to keep up with new technologies or lack of knowledge in other fields.

3.3 Integration of industry and education to acclimate to the "circular" economy's new pattern of development.

The Fifth Plenary Session of the 19th Central Committee of the Communist Party of China proposed the major strategic plan of "accelerating the construction of a new development pattern with the domestic macrocycle as the main body and the domestic and international double cycle mutually promoting each other." To realize the integration of industry and education, traditional design

https://doi.org/10.1051/shsconf/202317802007

education should seize this opportunity to reform and convert into an innovative design education classroom. Through the construction of an innovation and entrepreneurship platform with abundant educational resources, increasing the degree of school-enterprise cooperation, so that students can understand and fundamentally grasp the development of enterprise demand orientation, improving the transformation of design and innovation results at the same time, the enterprise also quickly receives the innovative design of its services, establishing a "small cycle" on the campus, and then outputting in line with the results of this cycle. This will establish a "small cycle" on campus, which will then produce "circular economy"-aligned design works, aid the new development pattern of the domestic large cycle, and enhance the competitiveness of art and design talents.

3.4 New technologies and media for creating "environment" in science theory learning

In 2007, the "4E+S" theoretical model of cognition was proposed for the first time in the learning sciences, where Embodied, Embedded, Enactive, and Extended cognition occur in a particular context emphasized by Situated Cognition. And the process of every action is intrinsically dependent on its physical and social environment[8]. In actuality, the combination of new technologies and new media has always been conducted in three core areas of the field of learning sciences: (1) research on the design of learning environments, (2) research on the underlying mechanisms of learning, and (3) research on learning analytics techniques [9]. By applying to real classroom situations (VR, WeChat, etc.), new media directly influence the traditional educational environment and context, which is no longer limited to the directly perceived physical world, and enrich the biological channels through which we receive information (visual media, auditory media, olfactory media, etc.)[10]. Through resource integration and sharing as a driving force for learning, new technologies subvert the individualisation of cognition and provide means of collecting and processing pedagogical data, enabling educators to realize effective and comprehensive pedagogical assessment and communication, and to create a variety of learning contexts according to the learners, which is in line with the current "student-centred" concept of education. It aligns with the current "student-centered" philosophy of education.

Based on the above summary of innovative design education, its relationship with learning science theory and new technology and new media is sorted out to provide a theoretical premise for the construction of innovative design education model, whose relationship is shown in Figure 1.

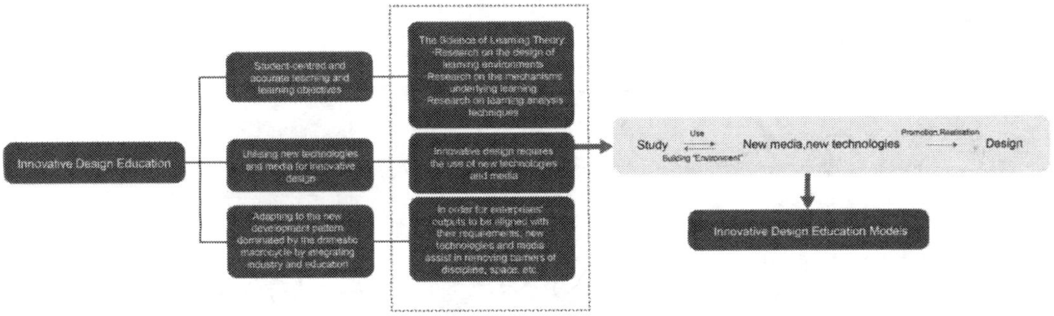

Figure 1. A compilation of innovative design education in relation to the study of science theories and the introduction of new technologies and media.

4 Constructing an innovative classroom paradigm for design education

Based on the feasibility analysis of learning science theory combined with new technology and new media applied to innovative design classroom, following the cyclic relationship of "teaching-learning-assessment", combining the three cognitive processes of "selecting, organising, and integrating" of meaningful learning in the theory of learning science with the three assessment modes, constructing the innovation and entrepreneurship platform, and integrating it into the education system,as shown in Figure 2 and Table 1.

Table 1. Innovation and Entrepreneurship Platform Specific Requirements.

Missions	Objectives	Problem solving
Publication of pre-coursework	Determine the technical and disciplinary knowledge necessary for the design based on the project context, design requirements, design objectives, and the student's design vision.	1.Who should this course be designed for? 2.What is the intent of the design of this course output? 3. What techniques and related disciplines must students master in order to implement the design concept (including depth, specific chapters, etc.)?
Provide technical and multidisciplinary resources, as well as communicate with businesses.	Provide pertinent and helpful technology (IoT, data analytics, 3D printing, VR, AR, etc.) and interdisciplinary resources for the student's design's completion and future development.	1.Can students utilise technology and interdisciplinary resources in design? 2.How should students' access to technology and interdisciplinary resources be incorporated into design? 3.Do the technology and interdisciplinary resources provided to students positively influence their future paths?
Provide a venue for the display of works and facilitate school-business collaboration.	The labour is creative and meets societal needs.	1.Is the design of this course's final project innovative? 2.What societal needs are fulfilled by the design of this course's final product?

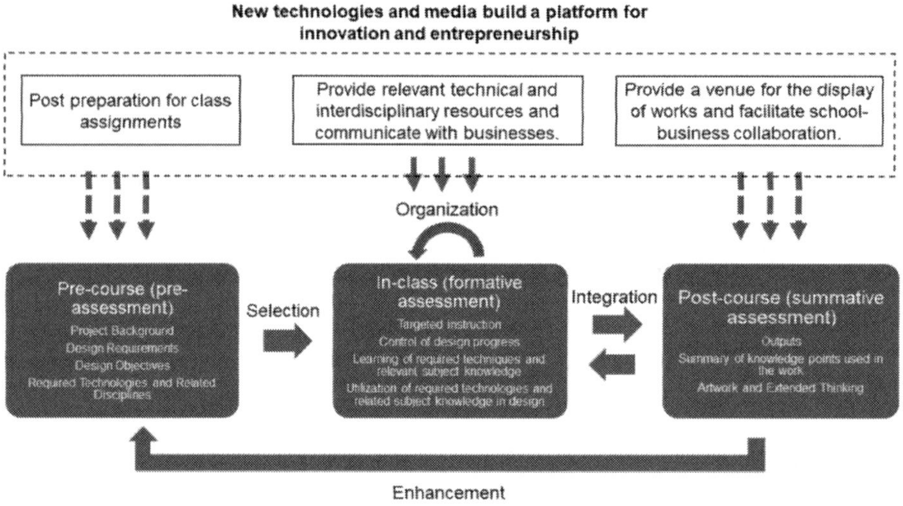

Figure 2. Innovative Design Education Model.

4.1 Using assessment theory and new technologies and media to guide cognitive processing in the science of learning

Assessment, teaching, and learning are a cyclical system in the innovative design education classroom, and assessment should not be a distinct task tacked on after the classroom. The learning objectives of a course not only lead to student success, but also enhance group effectiveness and cohesion, and effective assessment can reveal whether students internalise knowledge to construct meaning and accurately move towards the set learning objectives. Cognitive processing is the process through which learners select pertinent material during learning, organise the selected material and form a coherent representation, and integrate the selected material with the original knowledge activated in long term memory [7]. In design education, teacher-student communication is of the utmost importance; teachers can control the progress of students' works through evaluation; they can influence students' cognition from the social world; they can prevent students from producing "ineffective design"; and they can adjust design professional classroom teaching methods in a timely manner. Presently, the emergence of WeChat applets and big data education platforms enables the implementation of comprehensive evaluation.

The innovative design classroom carries out pre-assessment, formative assessment and summative assessment in the three phases before, during and after the class, combining the three assessment methods with the process of choosing, organising and integrating the three meaningful learning processes. In pre-assessment, pre-course tasks are released by the platform to make students pay attention to the information related to the specialised courses of this design and make the first step of effective selection based on the given information; in formative assessment, students make consistent mental representations of the selected information and the corresponding technologies and interdisciplinary resources provided by the platform to form the students' knowledge structure corresponding to the design; and in summative and formative assessment, each time the new technology and new media are actually intervened to help students to connect the represented knowledge with the original knowledge extracted from long term memory.

By completing the corresponding teaching assessment according to the requirements of different time points in the curriculum, teachers can adjust their teaching strategies in time, so as to make students go through the process of transforming from sensory memory to long-term memory through a complete cycle, so that students can complete the process of pre-study, learning and reviewing as well as informal learning outside the classroom in school, and to help students to reduce irrelevant cognitive processing, make good use of the basic cognitive processing, and promote the generation of cognitive processing.

4.2 The Function of New Technologies and Media in the Classroom for Innovative Design Education

New technologies and new media should not be used separately in innovative design education classrooms, but rather in tandem. The innovation and entrepreneurship platform is the result of combining new media and new technology. From the teacher's point of view, the combination of new technology and new media is a tool for teaching, through the construction of the platform, the introduction of new technology, interdisciplinary and other curriculum resources, and to build a bridge for school-enterprise co-operation, to show the final results of the teaching. From the student's point of view, the combination of new technology and new media is a tool for learning and completing the design, which demonstrates the final results of the teaching. And by applying new technology and new media in reverse to the design process.

5 Conclusion

This paper analyses the current situation of design education and new technology and new media, summarises the concepts of innovative design and innovative design education in the Chinese context, and constructs a classroom model of innovative design education by combining new technology and new media under the guidance of the theory of learning sciences, which offers a new approach to the reform of design education and a real shift to a "student-centered" teaching classroom. The model represents a new approach to designing education reform and a genuine transition to a "student-centered" classroom. The model is currently constructed from the general direction of design education; however, it can also be used in design education of a specific profession or specific design skills, etc., which is not described in detail in the text; however, future scholars can be based on the design education from all directions and perspectives of the design education to explore and further improve the model, the theoretical model into the design education, to promote the application of the model. In the future, scholars can investigate and further improve the model from a variety of perspectives, implement the theoretical model into design education, promote the application of learning science theory in design education, and find a suitable path for the development of design education in China so as to better meet the social demand for innovative talents and innovative design.

References

1. LI, B., *An Effective Design for the Reform of Learning Science to Teaching Practice: A New Mission of UbD.* Theory and Practice of Education, 2021. **41**(31): p. 59-64.
2. Kolodner, J.L., *The Learning Sciences: Past, Present, Future.* Educational Technology, 2004. **44**(3): p. 34-40.
3. Jingjing, Z., et al., *The Connotation and Application of "4E+S" Cognition in the Learning Sciences* Modern education technology, 2021. **31**(8).
4. Sawyer, R.K., *The Cambridge Handbook of the Learning Sciences.* Vol. XI. 2014, New York: Cambridge University Press.
5. Manli, L., et al., *From Cognitive Sciences to Learning Sciences:the Past,the Present and the Future.* TSINGHUA JOU R NAL OF EDUCATION, 2018. **39**(4): p. 29-39.
6. Juning, Z., *Leading the Reform: Organizing and Managing the SC Reform.* Research in Higher Education of Engineering, 2021(4): p. 8-22.
7. Mayer, R.E., *Applying the Science of Learning.* 2016, Beijing: China Light Industry Press Ltd. 30,56.
8. Yonghong, X. and L. Jianhui, *Cognition Beyond the Boundaries of the Brain:Situational Cognition and Its Answers to Questions about the Nature of Cognition.* Philosophical Trends, 2015(12): p. 89-98.

9. Jun-jie, S. and P. Lei-si, *Reflection on the Several Important Issues in the Development of Learning Sciences* Modern Educational Technology, 2018. **28**(1): p. 12-18.
10. Jian, Z., et al., *A Review of Research in Learning Science* Open Education Research, 2007. **13**(2): p. 15-20.

The Construction of Zeno's "Ideal Community"

Yaoye Wu *

Beijing International Studies University Political Party Diplomacy College, Chaoyang District, Beijing, China

Abstract: Within the annals of Western political thought, the essence and centrality of the Stoic School have played a vital role in the development of Hellenism and subsequent philosophical ideologies. Zeno of Kition, one of the founding figures of Stoicism, made substantial contributions to the formulation of early Stoic principles. Alongside his notable advancements in fields such as ethics, physics, and logic, Zeno's concepts of "the whole world" and "the citizen of the world" represented a pioneering breakthrough within the societal context of that era. These concepts transcended the prevailing class system and held epoch-making significance.

1.Preface

In 333 BC, Zeno was born into a merchant family in Kition, Cyprus. Despite being a Phoenician, he was exposed to Greek philosophy from an early age. His father would often bring back books from his business trips to Greece, which not only developed Zeno's reading and thinking habits, but also introduced him to the works of Socrates, which were renowned in their neighborhood for their wisdom.[1] As he grew older, Zeno started accompanying his father on trips to transport dyes from Phoenicia to Athens. However, a tragic incident occurred when their merchant ship sank, drastically altering Zeno's life trajectory.

In 312 BC, Zeno's ship embarked from Phoenicia with the purpose of conducting business with Bayrius. However, it tragically met its demise in the waters near Athens. Once he was rescued and reached land, Zeno found himself aimlessly wandering the streets of Athens, soaked from head to toe. In his hands, he held cherished Greek philosophical texts that he had cherished since his teenage years. Zeno's addiction to books was so profound that he tirelessly inquired about locations where he could engage with thinkers akin to Socrates. It was during this period that fate intervened, as the owner of a local bookstore noticed Clates. Coincidentally passing by, the owner motioned towards Clates and directed Zeno to "Follow him." From that moment forward, Zeno chose Clates as his mentor and delved deeper into the realm of Greek philosophy under his guidance.

He studied in Athens for twenty years, during which time Zeno not only absorbed the thoughts of cynicism but also actively engaged with other schools. Through these interactions, his ideological system matured gradually. After leaving Clates and other teachers, Zeno frequently engaged in philosophical discussions with like-minded individuals such as Ariston, beneath the colonnade of the palace located in downtown Athens. It was during this period that the Stoics emerged onto the stage of Western philosophy.

The Athenians held Zeno in great reverence, entrusting him with the key to their city wall for safekeeping. Furthermore, they bestowed upon him a golden crown and a bronze statue as tokens of their admiration. Throughout his lifetime, Zeno dedicated himself to exploration and approached academic pursuits and reasoning with utmost seriousness. His intellectual endeavors resulted in the authoring of 20 significant works, including "On Life Following Nature," "On Responsibility," "On Law," and "Ethics." Regrettably, the core essence of these profound insights has not withstood the test of time, as they remain lost to contemporary understanding.

Zeno possesses the commendable qualities of thriftiness and resilience. According to reports, he consumes food that does not require the use of fire for preparation and wears a remarkably thin cloak. Additionally, Zeno exhibits remarkable adaptability, and is not particularly fond of engaging in exhilarating activities. He holds the conviction that young individuals should embrace humility and strive to avoid arrogance. Moreover, he highlights the significance of focusing more on the advantages inherent within language rather than mere expression and vocabulary, with the aim of becoming more intellectually astute. Zeno emphasizes the importance of maintaining a dignified and proper demeanor at all times.

Virtue and dignity hold immense value for Zeno, as they are not only seen by him as the means to achieving happiness but also as happiness itself. This notion also remains consistent with Zeno's life, as he lived to the remarkable age of 89 without ever succumbing to illness. Zeno is credited with being the pioneer to introduce the

*18985224060@163.com

concept of "duty" into discourse.

2.The connotation of Zeno's "national thought"

Before discussing Zeno's concept of the state, it is essential to understand the teachings of Heraclitus, an ancient Greek philosopher. Heraclitus, who existed in the 5th century BC, proposed the theory of the origin of fire. According to him, the world consists of active matter and passive matter, with fire representing the active matter and an unspecified material representing the passive matter. Every element in the world is a combination of these two fundamental constituents. In Heraclitus' philosophy, fire itself embodies the concept of "logos" (which can be equated to "idea" and "God" in idealism). Therefore, the notion of "logos" is inherent in everything that exists. This "logos" is what gives every entity in the world its ability to function. By expanding our perspective, we can grasp that the entire universe is also governed by this "logos," meaning that the entire universe operates in accordance with an overarching "idea."[2]

The Stoics were heavily influenced by the ideas of Heraclitus as well. They believed that "logos" encompassed everything in the world and governed the entire universe. All things operated in accordance with "logos", and all individuals adhered to its laws. Building upon this foundation, Zeno presented the concepts of "the law of nature" and "the whole world" in his book "Ideal World" (originally known as "Politeia", to distinguish it from the works of Plato and Aristotle while incorporating the essence of Zeno's thoughts).

Let us begin by discussing the concept of "the whole world". The interpretation of the term "whole world" in this context differs from the cosmopolitanism currently being studied. However, I am confident that the former significantly impacts the latter. Furthermore, Zeno's philosophy has been translated by some as "world country" or "world city-state", but I contend that neither term adequately captures the essence of Zeno's concept. This is due to the fact that both country and city-state are politically biased, whereas Zeno's notion of "world whole" resembles more of a spiritual community.

As previously stated, all individuals are governed by the principle of "logos," thus belonging to a collective known as "the whole world." Within this global community, every person is deemed equal and referred to as a "citizen of the world." [3]

Simply stated, virtue is the sole good and its opposite is evil. Happiness, which fosters the flourishing of individuals, represents the ultimate objective, and it can be attained through the pursuit of virtue. [4]Zeno's notion of "the whole world" aims to establish a profound sense of human harmony. To this end, Zeno has established a citizenship criterion within this "whole world," wherein only individuals possessing "good moral character" are eligible to become citizens. Zeno's conceptualization of "The World as a Whole" overlooks distinctions based on race, class, and physiology, instead, employing morality as the sole criterion for differentiation.

It is evident that he embodies not only the notion of being a "citizen of the world," but also encompasses a comprehensive design of the entire world.

Firstly, it is prohibited to construct temples, courtrooms, and sports facilities within this area. Secondly, there is no circulation of currency within this zone. Furthermore, both men and women are expected to wear identical clothing, disregarding physiological distinctions. Nonetheless, Zeno did not elaborate further on the specifics of the lifestyle. [5]These three points delineate Zeno's departure from the city-state system, his pursuit of harmony, and his rejection of hierarchy. A more detailed explanation of these concepts will be provided subsequently.

The pursuit of a unified world may seem idealistic, but in reality, the systems, laws, and other elements among political entities such as city-states and countries vary greatly. Thus, the question arises: how can we strive for a cohesive global entity? As previously discussed, the world operates based on the law of logos, representing a natural law capable of reconciling divergent realities. According to Zeno, this natural law remains intangible, as citizens unconsciously adhere to and embrace it. It serves as the foundation for all concrete laws, existing within various ideological systems.[6] The essential requirements of natural law for citizens can be summarized as "responsibility" and "virtue." "Responsibility" entails aligning one's actions and words with the natural laws, while "virtue" represents the aim of observing these laws and fostering the harmonious development of the global community.

From Zeno's exposition on "the whole world" and "the citizens of the world," it becomes evident that this entity is fundamentally distinct from the city-states of that time. The public spaces abolished by Zeno were core elements of the city-states during that era. Additionally, the size of the world as a whole is not constrained but rather contingent upon the number of virtuous citizens within it. Virtue is the sole criterion for citizenship, thereby abolishing past divisions between Greeks and Manchus, as well as between free individuals and slaves in actual city-states. Zeno also places great faith in enlightenment, asserting that even the wicked can transform into virtuous individuals and attain the status of world citizens. [7]Moreover, the absence of currency circulation in Zeno's constructed world can be best understood as an indication of the state of self-sufficiency and harmony that has been achieved.

3.The origin of Zeno's "national thought"

3.1Alexander's Expansion

To fully grasp the essence of Zeno's ideas, it is crucial to delve into the historical backdrop of his time. As previously stated, Zeno was born in Phoenicia, Cyprus, but due to his father's professional engagements, he immersed himself in Greek philosophy, particularly the works of Socrates. Eventually, he journeyed to Athens, where he became a devoted student of the cynic Clates. Consequently, Zeno's thoughts bear a significant imprint

https://doi.org/10.1051/shsconf/202317802008

of Greek philosophy, exemplified by the "cosmopolitanism" inherent in cynicism. Nevertheless, Zeno's thinking also exhibits a notable departure from the conventional city-state-oriented mindset depicted in Plato's Republic.

In addition to the influence of Greek philosophy, the political circumstances of Zeno's era greatly impacted his conception of the entire world. Following a decade of intense warfare, Alexander consistently emerged victorious, conquering and expanding into other regions. The Middle East, and even India, became marked by the presence of Greek city-states. Realizing that sheer force was insufficient to ensure stability within this vast empire, Alexander implemented numerous policies aimed at fostering ethnic integration and establishing positive relationships between Greeks and non-Greeks. In an effort to dismantle Greece's inherent sense of superiority, Alexander also embraced the inclusion of two foreign princesses, setting a precedent for imitation. [8]Despite Alexander's extensive endeavors, he was unable to quell the chaos brought on by expansion. The resulting disorder had a profound impact on the social fabric, leading to moral decay and a deterioration of the world at large.

Under the circumstances of expansion and integration, a relevant concept emerged that greatly influenced traditional Greek philosophy and compelled the ancient Greeks to reconsider the potential for global unification.

3.2 Construction and Development of Zeno's Ethical Theory

According to the Stoics, philosophy can be categorized into logic, physics, and ethics. They liken logic to a trunk, ethics to flesh and blood, physics to the soul, and philosophy as a whole to an "animal" that integrates the first three.

Some people classify the logical part of this categorization as rhetoric and dialectics. The soul is initially devoid of knowledge and experiences, but perception, akin to a seal imprinted on wax, generates an impression which then solidifies into a memory, ultimately combining to form an experiential framework. Through the synthesis of sensation and imagery, ideas are formed, and if these ideas are rooted in experiences, they will yield common understandings. Scientific concepts emerge consciously and systematically, representing the "culmination of meticulous contemplation." Moreover, they argue that the yardstick for knowledge is the "self-evident presence of impression and concept." In essence, if an individual perceives the existence of a tangible object corresponding to a concept upon reading about it, that concept is considered scientifically valid and accurate. The Stoics give significant importance to formal logic, particularly syllogism.[9]

Ethics, often associated with logic, holds significant importance for the Stoics, akin to the significance of flesh and blood. The founder of Stoicism, Zeno, expounded extensively on ethical research, even during his time as a teacher in Clades, where he was exposed to the radical ethical ideals of cynicism. This exposure greatly influenced Stoicism, leading to the incorporation of ethical doctrines right from its inception. Stoicism holds human society in high regard while disregarding natural philosophy. Furthermore, according to Stoics, the relationship between man and the universe can be likened to that of a small spark and a raging fire, with the former being an integral part of the latter, exhibiting synchronization and coordination. Additionally, Zeno emphasized the eventual burning up and return to the original state of fire for all things, followed by a fresh start. According to Stoicism, the world unfolds according to this supreme law, with everything occurring in a predetermined manner for a specific purpose. This belief can thus be characterized as Stoic fatalism.

Therefore, it is essential for individuals to gauge their personal objectives based on the realization of the universe's purpose. [10]This necessitates adhering to God's will, abiding by societal norms, and fulfilling one's responsibilities. Such a moral existence, in their belief, leads to the pursuit of happiness. The early Stoic school of thought maintained that genuine joy and fulfillment in life could only be achieved through leading a virtuous and rational existence. Consequently, they advocated for the restraint of desires and the abandonment of irrationality. Whether one is wealthy or destitute, honored or disgraced, high-born or low-born, healthy or plagued by illness, the Stoics advocated for adopting an indifferent stance, aiming to achieve tranquility and free themselves from cravings. This early Stoic philosophy, which aligns with the natural course of destiny and demonstrates apathy, combines elements of both fatalism and asceticism. However, in contrast to the Epicureans, the Stoics place excessive emphasis on leading a simpler life and exhibiting indifference, ultimately succumbing entirely to idealistic fatalistic beliefs.

The Stoic concept of "cosmic tranquility" or "freedom from passion or indifference" is fundamentally guided by the principle of reason, known as the cosmic logos (which will be elaborated later). In the context of human society, the implementation of this principle manifests as "virtue is the only good." This idea emphasizes that such "good" can be attained through practice and education. Socrates viewed error as merely a specific instance of reason, whereas Stoics, who regard themselves as upholding the Orthodox Stoic tradition, expand on it to include various passions such as joy, fear, desire, and sadness.

Misinterpreting current kindness brings happiness, while misinterpreting future goodness incites desire. Misinterpreting current malevolence leads to pain, and misjudging future evils evokes fear. All forms of emotions are abnormal states of the soul, which ought to be not only controlled, but also eliminated.[11]

4.Breakthrough and enlightenment of Zeno's "national thought"

4.1Breakthroughs and Limitations of "The World as a Whole"

As previously mentioned, Plato's Republic and Cynics' Cosmopolitanism had a significant impact on Zeno's ideology.[12] Both Plato and Zeno extensively discussed the concept of the "state", yet they presented two distinct worlds. Plato's envisioning of the "state" remained confined within the framework of a city-state, lacking any transcendence of the existing Greek city-state system. His Republic was comprised of three classes: rulers, defenders, and laborers, each fulfilling their respective responsibilities. In contrast, Zeno's "world as a whole" represented a departure from the limitations of the city-state system, embodying the essence of a global community. This whole had no predefined boundaries, and all citizens were considered equal. Influenced by his mentor Clates, Zeno's ideas were further shaped by the principles of cynicism. One hallmark of cynicism is the proposition that the city-state deviates from its inherent nature, a notion also reflected in Zeno's conception of the World as a Whole.

However, it is worth noting that cynicism only pays attention to personal happiness, while Zeno's "The World as a Whole" pays attention to the community under "logos", and this community is a spiritual community, not a political unity, with morality and ethics as the link.

From the above discussion on Zeno's thought, it is evident that his concept of the "world as a whole" defies the prevailing principle and hierarchy of the Greek traditional thought, thus showcasing the value of Zeno's ideas. While the city-state system was dominant during that era, it also fostered division and isolation. Each city-state, as an independent political entity, established its own self-contained system encompassing citizens, institutions, and laws, consequently leading to the fragmentation and separation among city-states. The Peloponnesian War serves as a vivid example of the hostility and conflicts that occurred between these city-states. Within a multitude of city-states, the inhabitants boasted a sense of superiority over foreigners, thus restricting the boundaries of ancient Greek thought within the confines of the city-state system. However, with Alexander's extensive conquests, people began to realize the need for a new perspective, and it was Zeno who first proposed the idea of viewing the world as a whole. Additionally, Zeno believed that virtue was the sole criterion for distinguishing individuals, thus rejecting prevailing ancient Greek ideas such as slavery and the division of society into barbarians and non-barbarians. This revolutionary stance transcended the established class system of that time and held immense significance for its era.

However, it must be acknowledged that Zeno's World as a Whole is characterized by rationalism. Firstly, within this world, virtue has replaced race, nationality, rank, and physiology as the sole criterion for distinguishing individuals. However, this concept is disconnected from reality, and this moral community only exists in the realm of imagination. Secondly, it is unrealistic to attribute dominance to the natural law determined by "logos", as it fails to fulfill the role of a genuine law in maintaining social operations. Additionally, there are numerous contradictions in Zeno's ideology. For instance, while assuming that the world is governed by natural laws, the purpose of blindly striving for ultimate goodness and virtue becomes questionable. Similarly, Zeno's view on destiny suggests that people's fate is predetermined by "logos", yet he also acknowledges the significance of individuals' efforts in shaping their own destiny. When this perspective on destiny is applied to society, citizens find themselves trapped in a dilemma of either adapting to their fate or attempting to change it.

4.2"The World as a Whole" and "a community with a shared future of mankind"

Throughout the course of human development, we can see that the concept of "the whole world" is constantly enriched and improved in the social evolution. Both the attempt of supranational organization and the assumption of world government fully demonstrate the yearning for peace of mankind. However, the differences in ideology, nationality and geography still limit human exploration. We can't ignore these objective facts, but this does not mean a total negation of the idea of "the whole world".

Zeno's "world whole" is different from the present "world country". The core of the former is actually a moral community, which is not subordinate to modern political or economic fields. In order to achieve the harmony of the world and even the universe, human beings should respect each other, conform to nature and abide by morality, and take this as the way to get happiness. Up to now, the reasonable part of the core of the thought of "the whole world" is still applicable: global problems such as climate deterioration, epidemic situation and rising terrorism stand in front of all mankind, which cannot be dealt with by the meager strength of one country and one region. Only by letting go of prejudice and diluting differences can we join hands to better promote the development of human society. This is the significance of "a community with a shared future of mankind".

"The World as a Whole" holds that the link connecting the people of the world is "logos"; For the "a community with a shared future of mankind", this link refers to the fundamental interests of human survival. The background of the former is the expansion of imperial territory and national integration under Alexander's rule, and the resulting social disorder and chaos; The latter was born in the unprecedented changes in the world, and the change of international pattern and order added instability to the international community. The former has idealistic color and puts forward ideas for solving social problems at that time; The latter is a feasible scheme based on reality and contributing to the current human development.

5.Conclusion

From the above discussion, it can be seen that both "the whole world" and "a community with a shared future of mankind" have crossed the existence of objective obstacles, thus achieving a high degree of human harmony. Since ancient times, mankind has never stopped exploring world peace and stability, but how to face the current global challenges, I think we can get the answer from Zhi Nuo's "world as a whole" and "a community with a shared future of mankind".

Reference

1. Diogenes Laertius，Lives of Eminent Philosophers（Harvard University Press，1925），7，p32.
2. Wang Xiaochao's "A Brief History of Greek Philosophy: From Homer to Augustine", Shanghai Sanlian Bookstore, 2007 edition, page 204.
3. Diogenes Laertius，Lives of Eminent Philosophers（Harvard University Press，1925），6，p63～72。
4. Mark Aurelius, translated by Zhu Ruqing, "Reflections", China Social Sciences Press, 1988 edition.
5. Diogenes Laertius，Lives of Eminent Philosophers（Harvard University Press，1925），7，p80～130
6. Zhao Dunhua: "A Brief History of Western Philosophy", Peking University Press, 2004 edition, pages 90-94.
7. Cui Lina, "Zeno's" World State "Thought of the Stoic School," Research on Ethnic History, 2016, Issue 00, pp. 188-197.
8. Bertrand Russell, translated by Liu Changzhou: "History of Western Philosophy", Jiangxi People's Publishing House, 2017 edition, page 85.
9. Zhao Yongfeng's "Exploration of the Linguistic Philosophical Thoughts of the Stoic School", 2009 edition, p. 143.
10. Wendelban, translated by Zhan Wenjie, "History of Ancient Philosophy", Shanghai Sanlian Bookstore, 2014 edition, page 293.
11. Leon Loban, translated by Chen Xiuzhai, "The Origin of Greek Thought and Scientific Spirit", Guangxi Normal University Press, 2003 edition, page 367.
12. Xu Jian's "Paradox" in Zeno's "Political System", edited by Liu Xiaofeng in "Classical Studies" (Volume 1), East China Normal University Press, 2018 edition, pp. 31-48.P

A Study of the Korean Criminal Hesitation System in the Perspective of Comparative Law

Wang Jin[1]; Jin Aihua[1]; Xia Yikun[2*]

[1] Tianjin Normal University, Tianjin,China
[2] Criminal Investigation Police University of China,Shenyang,Liaoning,China

Abstract. Early intervention of criminal law in social governance has become an irreversible trend in China, which plays a significant role in preventing and controlling social risks, but the serious collateral consequences of penalties also lead to many new social contradictions. Korea's criminal hesitation system1 pays attention to the ethics of rules and regulations and has a strong color of subjectivism, which can better achieve the double effect of crime prevention and human rights protection. Our country should learn from the advantages of Korea, introduce the system of criminal deferral in due course, and expand the scope of application of the criminal deferral system in order to dissolve the contradictions caused by the consequences of the penalty. At the same time, it should also adhere to the concept of "criminal integration" and promote the systematic process of misdemeanour management in China.

1. Presentation of the problem

In recent years, the positive criminal law view was introduced into our country, emphasizing that the criminal law should be more actively involved in social problems, in order to realize the preventive function of the criminal law to intervene in advance and effectively prevent and control risks [1]. For example, the development of the financial market itself has a tendency to profit, involved in the crowd, the social public of financial investment is not prudent will lead to the unreasonable flow of capital or even touch the crime of illegal fund-raising, coupled with the popularization and application of the Internet, artificial intelligence technology, the number of victims of financial crimes and the amount of money involved in parallel and rapid growth, which is a huge impact on the formation of the security and stability of the community [2]. For this reason, the Amendment (XI) to the Criminal Law raises the statutory maximum penalty for the crime of illegally absorbing public deposits from ten years' imprisonment to fifteen years' imprisonment. In addition to this, the Criminal Law Amendment (XI) also added 18 new offenses, expanded the scope of criminal punishment, and at the same time lowered the age of criminal responsibility. We must admit that the position of adhering to the positive view of criminal law in the context of risk society has a very positive effect on the prevention and control of social risks, strengthening the governance of crimes against public legal interests and the protection of social order [3]. However, we also need to pay attention to is that in our country, the severity of criminal punishment is not only reflected in the penalty structure of mainly free punishment, but also lies in the seriousness of the consequences attached to the punishment, the stigma of criminals may be accompanied by a lifelong. For example, although the penalty for the crime of dangerous driving is not heavy, once the penalty is imposed, the offender will almost be put on a "blacklist", as stipulated in article 39 of our Labour Law, which provides that if a worker is held criminally liable according to the law, the employer may terminate the employment contract. Not only that, but the family of the offender will also be implicated by the criminal penalty. The Dongguan Municipal Compulsory Education Program for Children of New Guans under the Integration System stipulates that "100 points will be deducted if the child has received a criminal punishment in the past five years". This means that children of criminals will be seriously affected when enrolling in compulsory education. Some scholars have carried out statistics and found that the provisions on the collateral consequences of penalties stipulated in various types of normative documents in China have risen from 643 to 880 during the period from 2014 to 2018[4]. The Prison Administration of Sichuan Province has conducted a study on the recidivism of ex-prisoners, and conducted questionnaire statistics on 5505 recidivist offenders in 10 prisons around Chengdu, and the results showed that the recidivism rate of ex-prisoners in Sichuan was 69.38%. Among them, 32.41% reoffend because of "difficult to be accepted by neighbors". 84.9% reoffend because of "difficult to be accepted by neighbors". And 84.9% of the re-offenders "did not pay social insurance". Some insured criminals, due to serving their sentences, have had their social insurance, medical insurance, and occupational and practice licenses interrupted, making re-employment difficult; 34.33% of them were "discriminated against when looking for a job" after their return to the country,

* Corresponding author:xyk199703@126.com

and most businesses require a clean criminal record or a letter of no criminal record.[5] This kind of discrimination is not only a problem, but also a problem for criminals who have been convicted of a crime. The result of this kind of restriction is not conducive to the effective rehabilitation of offenders, not to mention the better reintegration into the society. [6]

For the above problems, a considerable number of scholars have called for the activation of China's crime system, as a response to the growth of the criminal population. However, unfortunately, the current criminal path in China is characterized by the mismatch between the causes of crime stipulated in the substantive law and the ways of crime stipulated in the procedural law, the insufficient typology of the causes of crime stipulated in the substantive law, and the single way of crime stipulated in the procedural law, etc. [7], which results in the failure of China's criminal system to achieve the desired effect. In addition to the establishment of Korean criminal law, such as deferred prosecution, the elimination of previous convictions and other traditional ways to get out of the crime, but also around the penal system itself to build a more complete criminal hesitation system, compared with China's criminal hesitation system, which is more flexible penal power to eliminate the consequences of the crime attached to it. The use of penal power can not be absolute rigidity, but should abide by the human nature of criminal law and moderate tolerance [8]. In this regard, we need to examine, compare and learn from the Korean criminal hesitation system, and try to improve our criminal hesitation system on the basis of respecting the basic conditions of our country as a new path to eliminate the social contradictions brought about by the concept of positive criminal law.

2. Examination of the Korean Criminal Hesitation System

According to the current Korean Penal Code, the Korean criminal hesitation system consists of four parts, namely, the criminal law suspension system, the sentence suspension system, the felony parole system, and the sentence reduction system.

2.1 Suspension of sentence

Suspension of sentence means that in cases where the defendant is sentenced to less than one year of imprisonment or a term of imprisonment, suspension of qualification® or a fine, the sentence may be suspended if there are mitigating circumstances and the defendant has shown significant repentance in applying the sentencing circumstances of Article 51 of the Korean Penal Code. There are three conditions for the application of the system of suspension of sentence: first, the defendant has committed a crime and has been sentenced to a lesser sentence of imprisonment, disqualification or fine, i.e., the defendant has been sentenced to less than one year of imprisonment, disqualification, or a fine. Second, according to Article 51 to consider the offender's age, character, intelligence and environment; relationship with the victim; the motive, means and results of the crime and post-crime circumstances, the defendant has extenuating circumstances[9]and will be repentant performance is significant. Thirdly, the prohibitive condition is that the defendant does not have a prior criminal conviction in which he or she has been sentenced to suspension of qualification or a heavier penalty. It is worth mentioning that according to Article 59, paragraph 2 of the Korean Penal Code, even in the case of multiple offenses, all or part of the sentence can be suspended.

This suspension of sentence is also not an absolute effect. Article 60 of the Korean Penal Code stipulates that a stay of sentence is effective for two years, and a person is deemed to have been exempted from prosecution after two years have elapsed since the date of the stay of sentence. Of course, if, during the period of suspension, the person whose sentence has been suspended is sentenced to another judgment that imposes a sentence of disqualification or more, or if it is found that he or she has a prior criminal record that imposes a sentence of disqualification or more, the sentence that has been suspended shall be pronounced. A suspended sentence may be pronounced on a person who has been placed under protection observation pursuant to Article 59-2 of the Korean Penal Code if the person violates the matters to be observed during the period of protection observation and the degree of violation is serious.

2.2 Suspension of Execution of Sentence

Suspension of execution of a sentence means that the execution of a sentence of imprisonment for a term of less than 3 years, a term of imprisonment, or a fine of less than 5 million yen ® may be suspended for a period of 1 year to less than 5 years if there are extenuating circumstances in consideration of the offender's age, conduct, intelligence, and environment, relationship with the victim, the motivation, means, and results of the crime, and the circumstances of the crime as stated in Article 51 of this Law.

Although the words "suspend" are used in both the provisional pronouncement of a sentence and the suspension of the execution of a sentence, they are fundamentally different. First of all, the two can not be applied at the same time. Suspension of the sentence is not yet pronounced on the defendant, the defendant does not know the criminal behavior of the penalty, while the defendant is suspended execution is to be pronounced on the sentence, the penalty they face belongs to the state of knowledge. Second, the two apply different levels of penalties. A suspended declaration of sentence is used only for imprisonment of less than one year, a term of imprisonment, suspension of eligibility, and a fine, whereas a suspended execution of sentence limits the applicable degree of penalty to imprisonment of less than five years, a term of imprisonment, and a fine of less than five million yuan. Finally, the conditions for expiration differ between the two. As mentioned earlier, the suspended sentence shall be pronounced if the person whose sentence has been suspended has been sentenced to another judgment that imposes a penalty above the disqualification level or if the person is found to have a prior conviction for a crime that imposes a penalty above the disqualification level. A suspended sentence may be pronounced if a person who has been placed under protective observation pursuant to Article 59-2 of the Korean Penal Code violates the matters to be observed during the period of protective observation and the degree of the violation is serious; and if a person who has been placed under suspension of sentence has been sentenced to imprisonment for a term of more than a year for committing an intentional crime during the period of suspension, and the sentence is finalized, the suspension of the sentence becomes effective.

In addition to this, the system of suspension of sentence has additional elements of revocation. This is mainly reflected in Korea's new law No. 5057 of December 29, 1995, which stipulates that the suspension shall be revoked if the sentence to which it applies was imposed on a defendant who committed a crime within three years after the execution of the final sentence of imprisonment for a term of imprisonment or a heavier sentence was completed or waived; or if the defendant violated an order to observe, engage in community service, participate in lectures, etc., that he/she was ordered to perform during the period of suspension of the sentence. The suspension may be revoked if the violation is serious.

2.3 Parole for felonies

Articles 72-76 of the Korean Penal Code provide for parole for felonies. The elements include: 1. the offender has been sentenced to a term of imprisonment or a term of confinement and the sentence is to be carried out; 2. the offender is of good behavior and has shown significant remorse; 3. the offender has served 20 years of a life sentence or more than one-third of a term of imprisonment; and 4. the offender has paid the full amount of the fine or the amount of the punishment4.

Parole also has a protection period. According to the current Korean Penal Code, the period of protection for a suspended sentence is 1 year, the period of protection for a suspended sentence is the period of suspension or the judge's discretion during the period of suspension, and the period of protection for parole is set at 10 years; however, the setting of protection observation is not absolute, and it can be dispensed with if the administrative agency that granted the parole deems that it is not necessary. The time limit of the protection period is the same as that of the parole period. If the offender is sentenced to imprisonment or more during the parole period and the sentence is determined, the parole lapses; if the driven person violates the provisions of the parole supervision or the matters to be observed under the protection observation, and if the degree of violation is serious, the parole may be revoked. It should be noted that the Korean Penal Code provides a negative element for the lapse of parole, i.e., if the offender is sentenced for a crime of negligence during the parole period, the parole is still valid. The revocation of parole, however, is not specified in terms of the consequences arising from an intentional or negligent crime. However, according to the method of interpretation of citing the weight to make clear the light, the statute of a certain matter will be pushed and other similar matters for interpretation and application [10], it should also be considered that the intentional crime will lead to the revocation of parole, because of the negligent crime, the negligent violation of the provisions of the supervision of parole or negligent violation of the protection of the observation of observance of the matters to be complied with, the parole shall not be revoked.

2.4 Reduction of sentence

Korea's sentence commutation system is stipulated in the Korean Pardon Act, a separate criminal law, and together with pardons and reinstatement of rights, it constitutes Korea's criminal pardon system. Korea's sentence reduction system is divided into general and special reduction, with general reduction meaning that the sentence is changed if there is no special provision, and special reduction meaning that the sentence is reduced, while the sentence can be changed if there are mitigating circumstances for the offender.

The application of mitigation is harsher than suspension of sentence, suspension of execution or parole for felonies. The main manifestations are:

In terms of the initiation of sentence reduction procedures, there are two types of applications for sentence reduction and recommendations for sentence reduction. The former requires the Prosecutor General, on his or her own initiative or on the basis of a report from a prosecutor of the Prosecutor's Office supervising the execution of a sentence or a report from the head of the correctional institution in which the offender is detained, to request the Minister of Justice to reduce the sentence of a specific person; the latter requires a prosecutor of the Prosecutor's Office supervising the execution of a sentence or the head of the correctional institution in which the offender is detained who intends to reduce the sentence of a specific person to submit a report with a specific explanation of the reasons for the recommendation to the Prosecutor General. A report specifying the reasons for the recommendation shall be submitted to the Prosecutor General, accompanied by the relevant statutory instruments.

With regard to the substantive discretion of commutation of sentence, it is necessary for the Minister of Justice to request the President on the received request for application for commutation of sentence or recommendation for commutation of sentence, and the commutation of sentence for a specific person shall be made by the President in the form of a presidential decree. In addition to the request to the President, the Minister of Justice is also required to request a review by the Pardon Board at the same time.

3.The Korean criminal hesitation system evaluation

The emergence and development of Korean law, had a long time by the influence of foreign law [11], in particular, a long time by the influence of Chinese Confucianism, emphasizing the moral basis of criminal law, to the modern era, in the continental law thought, especially the dominance of German thought in the case of Confucianism ethical morality has been denied, but the current Korean criminal law still retains the traditional rule of law thought and the color of the ethics of the rules and regulations. For example, in Article 250 of the Korean Penal Code, there are two paragraphs for the crime of homicide, the first of which reads, "Anyone who kills a person shall be punished by death, life imprisonment, or imprisonment for a term of more than five years", and the second of which reads, "Anyone who kills his or her own or his or her spouse's immediate family member shall be punished by death, life imprisonment, or imprisonment for a term of more than seven years". " As can be seen, the pursuit of the current Korean criminal law to play the role of the last barrier in maintaining social order and family order is obvious and sincere. The same is true for the criminal hesitation system. For example, Korea's current criminal law in the sentence suspension system, is considered to be "typical of the subjectivist criminal law thought as the theoretical basis" [12], subjectivism to the subjective danger of the perpetrator as the standard of conviction and punishment, externalized into the perpetrator repeatedly commit criminal acts of the danger of the performance of illegal acts [13]. Sentence suspended declaration system will be the offender's age, conduct, intelligence and environment; relationship with the victim; the motive, means and results of the crime, as well as the circumstances of the crime as the content of the investigation to determine the objective danger of the offender, the offender no longer pronounce a sentence for eligible offenders, to achieve the essence of the "elimination of the previous conviction", to a certain extent, can dissolve the penalty attached to the sentence. To a certain extent, this can eliminate the collateral consequences of punishment and play a special preventive role in criminal law. For example, in order to reduce the recidivism rate of juvenile delinquents and to help them spontaneously recognize their mistakes and return to society as soon as possible, the Korean Criminal Law introduced the protection observation system. According to the Korean Protection Observation Act, protection observation is a form of punishment in which the juvenile is not detained and is allowed to live normally in his/her community, but at the same time, he/she is subject to the relevant regulations and supervision by a protection observation officer. Protection observation is educational and restorative, and also has a mild punitive component, such as making up for the juvenile's misdeeds through a certain form of labor payment. [14]

However, the criminal hesitation system in the current Korean criminal law also has certain drawbacks. First, in terms of the procedural setting of the criminal hesitation system in Korea, Article 335 of the Korean Criminal Procedure Act stipulates that the suspension of punishment needs to be requested by the prosecutor to the court that has jurisdiction over the defendant's place of residence or last place of abode, and that the court needs to decide on it after asking the defendant or his or her representative for his or her opinion. In the current system of administration of justice in Korea, prosecutors have absolute superiority as both investigative and prosecutorial departments and can counter the decision of the court with their legal supervisory power. Suspension of sentence can only be initiated by the prosecutor's office, which means that even if the judge believes that the defendant has the conditions for suspension of sentence, he or she can only act as a "bystander" due to the lack of discretion, and whether or not the prosecutor applies the conditions for suspension of sentence when the defendant has the conditions for application depends entirely on discretion, and it is not a mandatory requirement, and the prosecutor has become "the judge behind the judge" in essence. In essence, the prosecutor has become the "judge behind the judge". The initiation of sentence reduction procedures is even more complex. In Korea, the procedure for commutation of criminal sentences is set out in the Korean Pardon Act, a separate criminal law, and is a form of pardon. In ancient times, pardons were used as a tool for emperors to maintain their rule, to show benevolence, and to win the hearts of the people, with a strong color of humanism. Nowadays, the pardon system has undergone a profound metamorphosis of the nature of jurisprudence, the maintenance of the overall interests of society and the realization of special preventive purposes as the fundamental follow. [15]Although the issuance of pardons by the head of state is one of the common practices in today's countries, it is clearly inappropriate for the president to issue orders for the commutation of criminal sentences in Korea, with the involvement of a number of state-level institutions in the foreground. There are three reasons for this: first, because the degree of clemency of commutation is lower than that of sentence suspension and sentence suspension, and the complexity of the procedure of the system with a lower degree of clemency than that of a higher degree of clemency is not conducive to the maintenance of the unity of the law and order; second, in the procedure of commutation of sentence, it is necessary to involve the Minister of Justice and the Prosecutor General, and ultimately the President to issue the Presidential Decree, but in fact, the top officials of the relevant national administration and the Head of State need to deal with a lot of affairs, and high-ranking officials are in charge of the President's decree. However, the fact that the top officials of the relevant administrative departments and the Head of State have to deal with a wide range of matters, the control of high-level officials over procedural decisions will inevitably lead to inefficiency. Third, Korea is a country with a separation of powers between the legislature, the judiciary and the executive, and although the Korean prosecutors undertake legal functions such as investigation and public prosecution, they are of the same nature as the Ministry of Justice and belong to the sequence of administrative organs as they are subordinate to the Ministry of Justice. The fact that the system of sentence reduction, which has judicial attributes, is entrusted to the executive branch gives rise to the suspicion of executive interference in the judiciary.

https://doi.org/10.1051/shsconf/202317802009

4.The significance of Korea's criminal hesitation system for our country

From the level of criminal substantive law, the criminal hesitation system in China includes probation, sentence reduction, parole, and pardon. Some scholars believe that, in the process of China's rule of law, especially in the context of the criminal policy of leniency and severity, the criminal hesitation system adhering to the concept of decriminalization, non-incarceration is undoubtedly the modern society of criminal reform a "good medicine. [16] However, in recent years, China's criminal legislation has gradually introduced the positive criminal law concept, which makes the criminal legislation and criminal justice in the participation of social governance further ahead. Although through the establishment of misdemeanor, thus making the criminal law of the punishment range greatly expanded, the punishment boundary gradually front, so as to deal with the current increasing social risk does have a certain legitimacy and necessity. The emergence of a large number of misdemeanor criminal cases exacerbated the strain on judicial resources, the expansion of the criminal population also brings many factors of social instability [17]. Therefore, it is necessary for us to further reflect on our criminal hesitation system.

4.1 Timely introduction of the criminal hesitation system

Combing through the existing criminal hesitation system in China's substantive law, whether it is the probation system or the sentence reduction system, or the parole or pardon system, we can't help but find that there exists a commonality, that is, it needs to be applied only after the defendant has been declared guilty. In practice, however, not all criminal cases, especially misdemeanor cases, all have to enter the criminal justice process and make a guilty verdict [18]. Perhaps, in China's existing criminal prosecution system, let the criminal case into the criminal justice process is more conducive to the investigation of the facts of the case, but also more in line with the work of China's judicial organs, but whether it must be made a guilty verdict is the judiciary can be discretionary, even if a guilty verdict, if you can be suspended declaration of leniency, can be effective in mitigating the rise in the criminal population of this social problem.

4.2 The appropriate expansion of the scope of application of our criminal deferral system

Some scholars have found that China's 2003-2005 period of reduction of the application rate of 25%, the application rate of parole for about 1% to 3%, the difference between the two is nearly ten times more than the current situation of judicial practice rely on the reduction of the sentence [19]. In the author's opinion, the reason for this, in addition to the technical limitations of the legal language, the judicial practice preference, there is another important reason is that the parole system applies the prohibitive conditions are too harsh. Article 81 of China's Criminal Code provides that recidivism and criminals sentenced to more than ten years of imprisonment or life imprisonment for intentional homicide, rape, robbery, kidnapping, fire prevention, explosions, the release of hazardous substances, or organized violent crimes shall not be released on parole. And the sentence reduction system does not have this provision, means that for this part of the offender directly exclude the application of the parole system, only to apply the sentence reduction. It is worthwhile for us to think about is that, our country parole system applies the substantive criteria for "recognized to abide by the prison rules, accept the education and reform, really repentant performance, there is no danger of reoffending", although the recidivism and the implementation of violent crimes perpetrators of the physical danger in the implementation of their criminal behavior and even accept the trial compared to other criminals higher, but it does not mean that they will be punished during the period of accepting the sentence, but this does not mean that they will be punished during the period of accepting the sentence. It does not mean that their rehabilitation effect is not good during the period of receiving the sentence. It is not feasible to fully adopt the Korean parole model in our country at present, but we can try to make moderate adjustments to the prohibitive conditions, such as retaining the basis of the substantive conditions for the application of parole, and setting the requirement of a longer period of time to be served for recidivism and offenders sentenced to more than ten years' imprisonment for violent crimes than for ordinary offenders, in order to incentivize offenders to accept rehabilitation better.

5. Conclusion

Misdemeanor legislation in the context of the risk society has its rationality, but the serious consequences of penalty attachment lead to the deviation between the actual and the deserved of criminal governance, and we must reflect on our criminal hesitation system with objectivism as the standard. In terms of comparative law, Korea's criminal hesitation system, which is based on the standard of subjectivism, deserves to be recognized for its full respect for social ethics, but its complicated institutional procedures have also become an obstacle to its effectiveness. China has an incomparably deep legal culture, as early as the early Northern Song Dynasty, there is a more complete record of the idea of prudent punishment of the "persuasion of prudent punishment of the text" and "prudent punishment of the aphorism". To this day, fewer arrests, more cautious prosecution, and more cautious detention have been formalized as a criminal policy, and the policy of leniency and severity is also deeply practiced in the process of China's rule of law reform. In the face of the new situation of criminal governance in China, we should draw on the strengths of different countries to improve our criminal hesitation system. Finally, it should be noted that, in addition to the participation of substantive law in criminal governance, procedural law is also essential, criminal hesitation system needs to be criminal justice procedures on the ground. Entering the era of misdemeanor, it should be based on the integration of criminal, rationalize the coordination and articulation between the criminal law and other criminal laws such as the criminal procedure law [20], and realize the systematic governance to cope with the dilemma of misdemeanor legislation.

Acknowledgments

This work was supported by the Seed Program for Korean Studies of the Ministry of Education of the Republic of Korea and the Korean Studies Promotion Service at the Academy of Korean Studies(AKS-2022-INC-2230003).

Appendix

(Note: ① Korea's local literature expressed "한국형사집행유예제도", directly translated as "Korea's criminal probation system". However, in order to avoid conflict of understanding with China's criminal probation system, taking into account the word practice in the academic community, so this paper is translated as "Korea's criminal hesitation system".

② According to Article 43 of the Korean Penal Code, the cessation of qualification sentence means: "A person who has been sentenced to death, life imprisonment or life imprisonment shall be deprived of the following qualifications: (1) the qualification to become a public servant; (2) the right to vote and the right to stand for election in public law; (3) the qualification to engage in the business of the public law which shall have the necessary conditions as prescribed by law; (4) the qualification to become the director of the juridical person, supervisor, general manager or inspector in relation to the affairs of the juridical person, (4) Qualification to be a director, supervisor, general manager, or inspector or property administrator of a juridical person in connection with the affairs of the juridical person."③The type of currency in this document refers to the basic unit of the Korean currency, the won, unless otherwise specified.④ Section material is a type of property penalty in the current Korean Penal Code. Article 47 of the Korean Penal Code stipulates that the amount of koji shall be from 2,000 won to 50,000 won or less.)

References

1. Fu Liqing. On the Positivist View of Criminal Law[J]. Politics and Law Forum,2019,37(01):99-111.
2. Xia Yikun. Interpretation of legislative modification of illegal fund-raising crimes:Taking the Criminal Law Amendment (XI) as a perspective[J]. Progress of Jurisprudence,2023,5(3):152-160.
3. Chen Zipei, Feng Weiya. The Path Choice of Attachment Consequence of Crime under Positive Criminal Law View[J]. Journal of Railway Police College,2022,32(03):91-97.
4. Wang Ruijun. The Function and Boundary of "Penal Attachment Sanctions"[J]. Jurisprudence,2021(04):44-60.
5. Sichuan Prison Administration Group. An analysis of recidivism of ex-prisoners in Sichuan Province[J]. Crime and Rehabilitation Research,2020(05):2-7.
6. Sun Guoxiang. The basic position of criminal law development in the new era[J]. Jurist,2019(06):1-14+191.
7. Shi Limei. Exploration of the path of criminalization under the vision of criminal integration[J]. Journal of Law,2023,44(04):92-104.
8. WANG Zhen. Rational Review of Criminal Hesitation System--Concept, Theory and System[J]. Frontier,2011(05):97-104.
9. Chen Zhijun. The Korean Criminal Code and Separate Criminal Laws [M]. Beijing:China University of Political Science and Law Press,2021:13.
10. [Japanese] Kawamura Yasu, Zhou Dongping, Xu Jing. Lifting the weight to make it lighter, lifting the weight to make it lighter and comparing it to the annex (below)[J].Legal History Translation Review,2021,9(00):202-228+2.
11. [KOR] Park Byung-ho. Korean law and ethics[J]. Jurist,1993(03):88-91.
12. Cho Byung-chi. An Important Contribution to Sino-Korean Legal and Cultural Exchanges--Preface to the Criminal Code of Korea and its Separate Criminal Laws[J]. Jurist,1996(06):88-89.
13. Xu Jing. Expansion and Response of Subjectivism in Criminal Law Legislation[C]/Shanghai Law Society. Shanghai Law Research Collection 2021, Vol. 23 - Southeast University Volume, 2021:9.
14. Liu W. Research on Restorative Juvenile Justice Reform in Korea [D]. Shanghai Normal University,2021.
15. Wang Yuhan. On the Pardon System [D]. Shanghai Normal University,2021.
16. Dong Bangjun. Research on Criminal Hesitation System[J]. Journal of Shandong University of Science and Technology (Social Science Edition),2010,12(06):29-35.
17. Xia Yikun, Li Ying. The legitimacy and path of procedural offenses in the context of misdemeanor legislation[J]. Journal of Party School of Nanning Municipal Committee of the Communist Party of China,2023,25(02):38-43.
18. Fan Chongyi. Theory and practice of misdemeanor governance under Chinese-style criminal justice modernization[J]. China law review,2023(04):191-201.
19. SUN Pengqing,ZHANG Yin. Structural deviation in the application of sentence reduction and parole and its systematic correction[J]. Research on Crime and Rehabilitation,2023(08):47-56.

20. PENG Wenhua,FU Liang. The goals and paths of criminal governance of crime in the context of structural change of crime[J]. Journal of the Chinese People's Public Security University (Social Science Edition),2023,39(02):12-25.

Research on the Operation Mechanism and Optimization Path of Wechat Gig Group

Yunlin Ye [1,*], Yonghui Deng [2,**]

[1] Harbin University of Commerce, Songbei District, Harbin, Heilongjiang, China
[2] Harbin University of Commerce, Songbei District, Harbin, Heilongjiang, China

Abstract: The rise of platform economy accelerates the diversification of labor market forms and makes the combination of labor and capital more diversified. Among them, wechat gig group meets the needs of low-end labor market and provides employment opportunities for homogenous low-end labor groups. In actual operation, wechat gig group, as a small urban labor market in the digital economy, has the advantages of transaction costs, the convergence of labor prices, and the gradual transformation of the working relationship between group owners and group members into community embedments. It is found that there are problems in the operation process such as information overload, man-post mismatch, gray level management that reduces the standardization of labor dispatch and information technology application may inhibit the employment of low-end labor, and puts forward optimization paths from community awareness, man-post dynamic optimization, interest perception and other ways to help the harmony of labor-capital relations in the platform economy period.

1. Introduction

The wide application of the Internet platform has accelerated the diversification and non typical characteristics of labor forms, made the relationship between labor and capital obscure and complicated, and enabled the rapid development of gig economy in the form of recruitment, temporary, part-time, piecework, emergency and hourly workers. According to a report released by the Institute of the School of Social Sciences of Tsinghua University, China's gig economy accounted for 2.64% of GDP in 2019, and it is expected that by 2035, the proportion of gig economy will reach 6.82%. It shows that the gig economy has been widely practiced in various industries.

Based on the market intermediary operation between digital platforms and workers, Yan Jinghua and Shi Xianmei (2021) believe that gig economy is a process in which workers use idle resources to provide certain specific services[1] in a certain period of time with the help of digital platforms. Zheng Qi and Yang Weiguo (2019) divided the gig economy into the new gig economy and the old gig economy, the latter referring to the temporary and project-based employment model, such as the employment model of "standing at a big post" to obtain employment opportunities; The former is based on the Internet platform, and relies on the application of digital knowledge and communication technology of high-tech companies to improve the matching efficiency of supply and demand[2] . The difference between the two lies in whether the medium relies on Internet technology. In the new gig economy, according to the scope of information acquisition and choice, it can also be divided into national gig economy platform relying on Internet gig app and regional gig economy information network relying on instant messaging software. At present, the concept commonly discussed in the theoretical community of the gig economy mainly refers to a new model of temporary on-demand matching between labor supply and demand sides achieved by internet platform enterprises relying on internet technology[3].

The gig economy has recently become a key theme in modern employment practice[4].However, the service areas of regional gig economy platforms are mostly based on intra-city services. Among them, wechat gig group is a labor market form and employment information network with smaller scale and more subdivided functions. It attracts homogenous low-end labor who are relatively older, have low education, low skills, can't keep up with information technology application, have limited employment channels, and take survival as their main employment appeal. This is also why Xie Fusheng and Wu Yue (2019) analyzed on-demand labor. The low-en[1]d blue collar workers described by Xie and[5] Wu. It is difficult for them to quickly match jobs depending on their own abilities. However, wechat gig group, built by strong network of relatives and friends, enables small gig resource integrators to play the role of market intermediary between employers and labor providers by releasing temporary job information, providing "one-stop" services

*Corresponding author: 18646304575@163.com
** 73dengyonghui@163.com

such as employment recruitment, management and delivery to jobs, so that they can quickly match jobs. To obtain the income on which they depend.

Therefore, as a self-created for-profit organization that takes the homogenous low-end labor market employment matching as its profit source, what problems exist in helping the employment of groups with employment difficulties? From the perspective of labor dispatch and information technology, it is of practical significance to propose the optimization of the operation path of this labor market intermediary

2.The analysis ofthe operating characteristics of wechat gig groups

In the operation of gig employment and employment, wechat gig group has adapted to the employment needs of gig workers who have limited employment opportunities and cannot obtain stable work groups. However, compared with the practitioners of the Internet app gig platform, the "flow" between types of work and positions is larger and more frequent.

(1)Wechat gig groups make the price of labor service converge

Wechat gig groups are characterized by a large number of homogeneous low-end labor services, no barriers to entry and exit markets, and complete information of superior human resources companies, indicating that wechat gig groups, as labor dispatch intermediaries, are in a perfectly competitive market. In reality, the owner of the wechat gig group has the right to decide the service price of the daily contract worker; However, when there is a service price difference between different group members who provide services for the same employer, it will be a phenomenon that the low-priced ones flow into the high-priced wechat group in the exchange of information. This phenomenon of low barriers to entering and leaving wechat groups will lead to the convergence of labor prices of the same employer and the same type of post in different wechat groups.

(2) The attributes of social platforms have transaction cost advantages

As a high-frequency and ubiquitous social platform, wechat gig group is the default work and task information release group for group members. It has become the core for group members to participate in the gig economy and can quickly connect and match according to labor supply and demand. Specifically, the group owner publishes the labor demand information, the group members share the employment information, and the group owner can easily and quickly communicate with the group members who have the intention to work. They can communicate their needs and discuss the progress of the project through private chat, voice call, video call and other ways, and can reach an employment agreement in a timely and effective manner to improve work efficiency. At the same time, it ADAPTS to the labor form of on-demand employment and cost-saving, and the "daily settlement" of salary reduces the worry of unpaid salary. Instant satisfaction is an important reason for wechat odd job incentive.

(3) The trust mechanism based on word of mouth is the bond of group activity

Due to their large scale, face-to-face interaction between people is difficult to achieve, and nationwide gig platforms mainly rely on users' resumes and evaluation systems to build trust, but this approach may have problems with information asymmetry and fake evaluations. In contrast, through the advantages of city-specific social networks, it is easier for the owner of a wechat gig group to obtain real feedback from the employer, establish a real evaluation of the group members' work performance, and achieve screening and connection with the group members. At the same time, the group members can establish a good reputation by actively participating in the work and obtaining good work evaluation. This word-of-mouth effect is not only a trust mechanism and external cognitive set, but also can bring economic returns when labor supply exceeds demand.

(4) Work relationships are linked to community embeddings

Since the industrial economy, along with the production of socialized employment into a stable social organization, there has been a stable business relationship, so that individuals have a stable social affiliation outside the family and other consanguineous relations. However, for those who rely on wechat gig groups for employment, they have strong flexibility and lack stable employer industry relationship. As Marx said, the "kind essence" of human being is a social existence, and it is first manifested in the labor nature of human beings, which is manifested in the collective production labor and social relations. And the wechat gig group builds an atmosphere and scene of collective labor. From the perspective of the relationship between group owners and group members, wechat gig groups can be summarized into three stages of development: instrumental rational interaction, virtual community interaction and acquaintance social construction.

In the first stage, that is, at the beginning of group construction and group chat, the interaction between the group owner and the group members is aimed at completing work tasks. The formation purpose of wechat gig group is instrumental rationality, which is a manifestation of "birds of a feather flock together, and people are divided into groups". The members of the group have high homogeneity in terms of social status, work ability and emotional identity. In the second stage, with the initial agreement between the group owner and the group members, when the group owner sends labor service providers to the employer and gets a cut, and the group members also get job opportunities and economic rewards, the weak relationship virtual community in the name of the wechat gig group begins to form. The third stage is that with the release of information, resource sharing and continuous completion of work tasks within the wechat group, the interaction frequency between the group owner and group members and among group members increases, as well as the increase of emotional identification and common memory, and the homogenous community attributes of the wechat gig group begin to appear[6] . Wechat gig group has become a social capital network including all members, and all members have formed a "virtual acquaintance society" [7] in which they exchange work information, share work experience, rely on each other, help each other and benefit each other. The boundary of group members' communication begins to expand from the work field to the non-work field, and individuals change from the original atomization state. To the collectivized state of mutual cooperation and partnership relationship. At this time, the "weak relationship" between the group owner and the group members, and the group members with high interaction frequency, will change from the "weak relationship" of instrumental network to the "strong relationship" emphasized by Granovetter, to the interpersonal communication with emotional identification and trust interaction, and expand the social network and sense of belonging of each member. The convenient wechat payment and leave mark functions not only improve the transaction efficiency, but also encourage the group members to continue to participate in odd jobs.

3. Problems existing in the operation of wechat gig groups

The gig employment of wechat gig group is mostly low-skill "U disk employment" [8] , which is in essence small-scale labor dispatch. The master of wechat gig group plays the role of lead worker or on-site customer service of human resources company, providing employment information for vulnerable groups who cannot get regular employment opportunities. However, there are the following problems in the labor market of wechat gig group, an Internet platform.

(1) Information overload increases screening costs

As more gig workers join the wechat gig group, the amount of information will also increase exponentially. Users may be faced with an influx of information, including job postings, job recommendations, and workers' self-introductions. Due to the diversity and quantity of information, users need to invest a lot of time and energy in sifting, processing and responding to the information. At the same time, information is often repeated or redundant, which brings cumbersome processing tasks to users, and it is easy to miss some important information. Users need to spend more effort to manage and process this information in order to find the right employer or worker that meets their needs.

(2) The mismatch between employees and posts leads to security risks

Although the information of wechat gig group is relatively sufficient and gig workers can find suitable positions according to their own positioning, wechat gig group cannot fully guarantee the real credibility and service quality of its members. For example, the information publisher does not provide accurate information; Gig workers have certain differences in their needs and abilities, and some group members may exaggerate their abilities or maliciously post false job information; Even the real abilities and job information may be affected by the bias of individual subjective evaluation, making it difficult to find a partner who fully meets the needs. As a small labor dispatch unit, wechat gig Group is a social platform. Compared with the formal employment mechanism, the employment relationship between the group owner and the group members in the wechat gig group circumvents legal protection. Once there is an accident, there will be a dispute of buck-passing.

In terms of the positions of the workers, diversification, complexity, and the result orientation of employment will lead to security risks. For example, in the same express sorting night shift recruitment, including a number of actual jobs, including loading, unloading, small parts sorting, unloading, flexible loading and unloading, etc. At the same time, the working time is generally more than 10 hours, the labor process has supervision of supervisors, night shift positions, etc., so that the age and physical strength of gig workers can not adapt to the needs of the post. However, there are accidents and safety accidents due to continuous long-term physical labor. The accident of an over age Courier in Ningbo, Zhejiang Province, which caused public concern in early 2023, exposed the shortcomings of the gig economy, such as the mismatch between people and jobs and the insufficient protection of workers' rights and interests.

(3) Grayscale management has lowered the standardization of labor dispatch

As we all know, the phenomenon of labor subcontracting exists in the construction industry, and it also exists in the gig economy. That is to say, in the relationship network of the wechat gig group, the wechat group owner and group members have the fact that they are subordinate to the human resources company, but they do not have the name of the subordinate gig workers, and the group members do not have clear evidence to prove the reality of this labor dispatch. When the employer issues the work tasks to the gig workers in a temporary and one-time way, the wechat group owner, as the leader, not only undertakes the task of recruiting and conveying workers to the post, but also standardizes the workers' behavior through pre-job training, attendance checking, standardized management, etc., which will reduce the employment cost and the user is happy to see its success.

However, the wechat group owner relies on the management of irregular labor dispatch to participate in the labor market recruitment and the enterprise management of the employer. When undertaking labor tasks, wechat gig workers have no fixed employer, unclear employment relationship, can not sign labor contracts with the group owner and superior human resources company, and can not enjoy social insurance.

Wechat gig workers are mostly middle-aged and elderly people, housewives over 40 years old, farmers who do not go out to work, and young slasher who earn pocket money by doing part-time jobs. The atomization and lack of unity of these people make wechat gig workers vulnerable groups in the gig economy. As a result, once occupational accident injuries occur, wechat gig workers will not be able to obtain reasonable protection. Wechat group owners will also be because of the dispute, denial and can not be compensated, in more cases can only be their own bear, the years of savings consumed, do not know whether they can continue to work. Therefore, wechat gig work is one of the most exploited flexible employment methods.

(4) The application of information technology may inhibit the employment of low end labor force

With the development of wechat gig groups, some wechat group owners have begun to use other ways, such as wechat mini programs and other applications, to register for work information and pay daily wages, rather than the traditional sense of private chat, reply and transfer of red envelopes. In this case, some workers who only know the basic wechat operation, or older workers will lose their job opportunities and problems such as salary collection. The continuous development and evolution of this situation will lead to the reduction of the audience of wechat gig workers and affect the effect of ensuring employment and stabilizing employment.

4. The optimization path of wechat gig group operation

(1) To develop and implement group announcements to form a sense of community

A sense of community is a shared spiritual bond that allows group members to feel a close sense of dependency and community[9] . Generally speaking, in a wechat gig group, the group owner will clearly inform the group members at the beginning of the group's establishment of the purpose of the group and what information should be shared to ensure a good communication order, and restrict or remove the members who send irrelevant content, advertising or junk information to avoid irrelevant information other than employment in the group. In the group setting, the group master can limit the right to speak, and ensure that only the filtered content is posted; Use the "group announcement" function to release important information, and put some commonly used information at the top, regularly clean up redundant information in the group, maintain the cleanliness and effectiveness of the information in the group, ensure that the group members quickly find valuable information, reduce the problem of information overload. When the group operation enters a stable stage, individuals will have a sense of identity and belonging to the group, and will have a sense of community, and will take the initiative to collect and disseminate relevant information of the group. Group members will also choose suitable positions according to their own skills, promote higher level of cooperation in wechat gig group, and then obtain more job resources from the employer and the superior human resources company. To enhance the scale and activity of wechat gig group.

(2) Dynamic optimization of person-post matching

Dynamic optimization of man-post matching, not only the online interactive information to be accurate and clear, but also need specific group operation measures to ensure. First, in gig groups, the group mainly ensures that the posted job information is clear and clear, and details the required skills and experience; Second, the group owner evaluates the work effect and labor feedback in the past, and "marks" the group members with poor work performance or safety risks to restrict their choice of work. To review and screen applicants to ensure that everyone has the ability to do the job. Third, in the group announcement, list the safety measures and precautions in the process of work, including personal protective equipment, the correct way to use tools and equipment, and the response measures in case of emergency, and make sure that all group members understand and abide by these regulations. Fourth, the group owner has an emergency contact group or a shared address book, so that in the event of an emergency or emergency, the relevant personnel can be quickly contacted. This can improve the efficiency of emergency response and ensure that problems are addressed in a timely manner. Fifth, if certain positions require specific skills, it is necessary to carefully review the working abilities and qualifications of the group members, so that problems can be solved before employment and the risk of accidents can be reduced.

(3) Regulate illegal labor dispatch with the perception of interests

The odd job employment of wechat gig group members is subsistence employment, which determines that it is difficult for them to safeguard their labor rights and interests according to their individual strength. Different pay for the same work, disguised charges, non-payment of social insurance and other phenomena will always occur[10] . What is more serious is occupational accident injury in the labor process. These atomized and non-solidarity people, while individually vulnerable, also determine that they cannot comply with Article 64 of the Labor Contract Law (2013). "Participate in or organize a trade union in the labor dispatching unit or the labor employing unit according to law", and Article 10 of the "Interim Provisions on Labor Dispatching" (2014) stipulates that "if the dispatched worker suffers an accident injury due to work in the labor employing unit, the labor dispatching unit shall apply for the identification of work-related injury according to law... The dispatching unit shall bear the liability of industrial injury insurance, but may agree on compensation measures with the employing unit ". Under the existing employment constraints and legal conditions, it is suggested that the dispatching party and the employing party should learn from the practice that the MT platform takeout rider has 3 yuan of accident insurance a day, and give the wechat group gig workers as much protection as possible to ensure their own profits and smooth production within the scope of operability.

(4) Strengthen the promotion of new technologies

They point out that information technology is a necessary tool skill for completing odd jobs in the modernization process of a country with a large population. For these older wechat gig group members with limited education level or lack of technical ability, if they can use mobile devices to complete registration, certification, order receiving and payment, they may get more job opportunities, and even apply Internet gig app to independently order and increase their income.

In the wechat gig group, the group master will help those older gig workers familiarize themselves with and master application skills such as using wechat mini programs. In addition, by sharing methods, guidance materials and other relevant information in the group, the group master can strengthen the publicity and popularization of the new way of job declaration and daily salary settlement in the wechat gig group, and master the operational ability of task participation and salary range. At the same time, the records of salary payment by wechat mini programs and other applications should be published regularly to ensure the fairness and transparency of work declaration and salary payment, and enhance their participation and satisfaction. It will not only help improve workers' employment opportunities and income levels, but also promote inclusive development of society.

5. Summary

As a social media platform connecting employers and workers, wechat gig Group provides additional employment channels for people who have difficulties in obtaining employment information, increasing employment opportunities and flexibility for gig workers. At the same time, wechat gig group also provides enterprises with more extensive and diversified human resources, enabling them to flexibly meet the employment demand, quickly find the right labor force, carry out flexible manpower allocation, improve competitiveness and adapt to changes in market demand.

In order to give full play to the social benefits of wechat gig group, its operation mechanism can be further explored and optimization strategies can be adopted to standardize its operation process and protect the rights and interests of gig workers. The development of wechat gig groups depends on laws and regulations, labor market development and gig practitioners' acceptance of the platform. While there are opportunities for development, it also requires stakeholders to work together to address potential problems and challenges and ensure the sustainable development of wechat gig work groups.

Acknowledgments

Supported by: Philosophy and Social Science Research Planning Project of Heilongjiang Province "Research on the Realization Path of Labor Benchmark Guarantee in New Employment Form"（Item number: 21JYE403）

References

1. Yan Jinghua, Shi Xianmei. (2021) Political Economy Analysis of the relationship between capital and labor in gig economy. Theoretical Monthly,08:35-43.
2. Zheng Qi, Yang Weiguo. (2019) Research perspective of Gig economy: A Review of Western classic Literature China Human Resources Development,01:129-137.
3. Horney.The Gig Economy:A Disruptor Requiring HR Agility[J].People and Strategy,2016(3) :20-28.
4. Taylor,M.,Marsh,G.,Nicol,D.,&Broadbent,P.(2017). Good work:The Taylor review of modern working practices.Retrieved May 14,2021,from:https://www.thersa.org/globalassets/pdfs/reports/good-work-taylor-review-into-modern-workingpractices.pdf
5. Xie Fusheng, Wu Yue. (2019) Is Gig economy a new employment relationship with win-win situation between labor and capital . Economist, No.246(06):5-14.
6. Song Qingyu, Liu Neng.(2023) Social Relations and operation mechanism of college students running teams: Based on the analysis of wechat group chat data. Youth Research, No.450(03):24-36+94-95.
7. Wu Zhenqi, Guo Chengcheng.(2023) From loudspeaker to low voice wechat group: Rural public reproduction and Social Governance Transformation: Based on a field investigation in a North China village . China Rural Observation, No.170(02):34-52.
8. Zhu Dajian.(2020) U-disk employment: Three major problems of gig economy in Chinese context [J]. Exploration and Contention, 07:9-12.
9. MCMILLAN D W, CHAVIS D M.(1986) Sense of community: A definition and theory. Journal of Community Psychology, 14(1):6-23.
10. Yao Xianguo, Feng Lobing.(2021) Protection and realization of labor rights and interests in Labor Dispatch . Zhejiang Journal, No.246(01):98-105.

Research on Operational Strategies of Video Platform Content Creators
—— A Case Study of Lifestyle Content Creators on Bilibili

Zihe Cao*

Ji Nan University JNU Journalism/Electronic Commerce Shenzhen, Shenzhen, China, 518053

Abstract: The rapid development of the internet has spurred the rise of new media. Within the ecosystem of new media, a substantial number of content creators, referred to as "UP zhu" or video uploaders on the Bilibili platform, have emerged, engaging in various media-related activities across different platforms and domains. These activities are inevitably accompanied by operational strategies, which constitute a pivotal aspect of business endeavors for companies and relevant practitioners, a concept that applies equally to Bilibili. This study takes lifestyle-oriented content creators on Bilibili as a case study, examining Vlogs' development, characteristics, and operational strategies on Bilibili. It analyzes the operational strategies of lifestyle-oriented Vloggers on Bilibili and concludes by summarizing existing issues among video platform content creators and proposing optimization strategies.

1. Introduction

The emergence of content creators is not a happenstance but an inherent phenomenon within the context of media convergence. Amidst the backdrop of media convergence, a myriad of media practitioners has surfaced, including media operators from governmental institutions, professionals from news organizations, and newcomers in the realm of self-media. Particularly, those within the realm of self-media leverage the expansive landscape of new media to partake in tasks such as information production, dissemination, and processing. Simultaneously amassing substantial followings and remarkable viewership, they also embark on "online operations," each exhibiting distinctive operational models, showcasing their prowess within the online milieu and capitalizing on their traffic.

Bilibili is a video platform that is currently beloved by the youth demographic and offers multiple domain modules for users to select, with Vlogs being one of them. In March 2017, Bilibili publicly introduced the Vlog module, which rapidly burgeoned in popularity and captured the adoration and attention of users. Lifestyle-oriented Vlogs are characterized by authenticity, interactivity, continuity, and professionalism. Currently, they hold a considerable proportion within Bilibili, encompassing both video quantity and the number of content creators. Furthermore, the audience and fan base for lifestyle-oriented Vlogs on Bilibili are vast and expansive.

2. Literature Review

There is a copious volume of research in this sphere. Numerous scholars have delved into media convergence, practitioners, and operational management. Additionally, several studies have specifically scrutinized Bilibili as a case study.

2.1 Operational Strategies

The subjects covered in studies concerning operational strategies vary between China and other countries. In China, the focus primarily gravitates toward economics, while overseas research is more inclined toward management studies and the operational environment.

In general, operational strategies denote measures taken by enterprises and pertinent practitioners in the competitive market environment. These measures are aimed at optimizing internal structures and carving out developmental space by comprehensively considering existing strengths and weaknesses. Operational strategies encompass policies, plans, and competitive approaches adopted by enterprises to achieve desired business objectives within a specific market environment[1]. This concept is equally applicable to the operational activities of content creators. There are diverse perspectives to examine operational strategies, with numerous associated theories, such as Pan Qiangsheng's integration of enterprise operational strategies, PEST analysis, Porter's Five Forces model, competitive strategy theory, and STP marketing theory[2].

Scholars have also explored operational strategies for

*15039779786@163.com

various new media, traditional media, and content creators within the context of media convergence. Zheng Xiaohui examined advertising operational strategies within the era of converged media for traditional broadcasting[3], while Hong Li, Zhang Shengxi, Che Li, and others explored newspaper advertising operational strategies within the backdrop of media convergence[4] [5] [6]. Zheng et al. scrutinized branding strategies for cultural variety shows using the television show "The Readers" as a case study. [7]

2.2 Media Convergence

In the present era, media convergence has evolved beyond being merely a prominent industry concept; it has ascended to become a noteworthy subject of academic discourse. Over recent years, research into media convergence has experienced a progressively steep ascent.

Du et al. contend that the foundation of media convergence lies in the amalgamation of regulatory frameworks, where the engine propelling this amalgamation is technological integration, and its trajectory is industrial amalgamation[8]. The engine driving media convergence is technological integration, with technological fusion serving as both the cornerstone and the central impetus. Sun posits that the ultimate amalgamation in media convergence resides in the fusion of technology and human involvement[9]. Moreover, Liu advocates that media integration transcends surface-level amalgamation, placing substantial emphasis on cultural assimilation. To achieve authentic success in media convergence, comprehensive cultural amalgamation is imperative[10].

In contemporary times, research conducted from the perspective of media convergence is proliferating and exhibits a notably diversified character. For instance, Yue Jinhong delves into the art of proficiently conducting in-depth reporting within the era of integrated media[11]. Li et al. scrutinize the predicaments surrounding music copyright protection in the realm of short videos within the context of media convergence and subsequently propose pertinent countermeasures. Through these investigations, the emerging trends in the production and dissemination of Chinese television dramas within the framework of media convergence are expounded upon. [12]

2.3 Media Practitioners and Media Operators

Hu Shutong suggests that research concerning the journalism profession largely converges around five domains: studies on the positioning of the journalism profession, research into role ethics and norms, investigations into role functionality, inquiries into role dislocation and deviation, and examinations of specific position journalists. [13]

In recent years, as new media continues to evolve, the landscape of media ecology has undergone significant transformations. Consequently, media practitioners have acquired new connotations and identity, and the research on this group has demonstrated diversification. Scholars have increasingly explored aspects like the professional development of media workers, their job satisfaction, work-related stress, and the roles of media leaders. In this study, we consider video uploaders on Bilibili as media operators.

2.4 Media Business Management

Currently, scholars predominantly examine media operational activities from the perspective of media business management, which can be broadly classified into three areas: operational strategies, transformation directions, and management reform[14]. In China, numerous scholars have authored works investigating media business management, such as Zhan Chengda's *Media Business Management*, Gao Fuan and Sun Jianghua's *Introduction to Media Management*, and Nai Peng's *Online Media Business and Management*.

Several scholars delve into media operational strategies. Notably, You Xiangyu analyzes CCTV anchors as subjects, employing a media business management approach to conclude that while CCTV anchors' product promotion display is professional and approachable, there is room for improvement in adhering to communication norms[15]. Zhang Yao examines iQiyi as a case study, dissecting its media operational management strategies and proposing that iQiyi should allocate more focus on users and technology, while considering cross-industry collaborations when appropriate[16].

2.5 Research on Bilibili

Based on the literature review, Bilibili exhibits unique characteristics in terms of its media and cultural attributes. Consequently, researchers take varied approaches, angles, and focuses while conducting studies on Bilibili.

Bu et al. posit that "Bilibili, with its robust content ecosystem, hosts a multitude of high-quality content creators. The content generated by these creators not only drives user growth but also significantly enhances user engagement and activity." [17] Huang Pingping, considering the backdrop of "slow-motion" development, analyzes Bilibili's slow-motion videos in terms of narrative features, interactive methods, and representative elements[18]. Zhou et al. empirically study the phenomenon of individual subtitles on Bilibili, contending that from a cultural rights perspective, individual subtitles promote fair dissemination of cross-cultural resources[19]. Zhai Chenhua explores the theme of UGC content marketing on Bilibili, investigating the distinctive branding features of UGC content marketing within the platform[20]. Yu Ke examines the case of Bilibili, dissecting the profit model of bullet screen video platforms. [21]

3. Analysis of Business Strategies of Lifestyle Vlog uploaders on Bilibili

Under the backdrop of new media, the business models and audience consumption patterns of lifestyle Vlog uploaders on Bilibili have evolved with the changing times. They are no longer confined to fixed templates and unchanging routines; instead, they exhibit characteristics of diversification and spontaneity.

3.1 Analysis of Interviews with Lifestyle Vlog Uploaders on Bilibili

To understand the operational strategies of video uploaders in creating Vlogs, how these strategies are implemented in their actual content creation, and the results and impacts these strategies yield, practical interviews were conducted. The interview subject selected for this study is the Bilibili video uploaders "Yi Zhi Dan Dan" , who currently has approximately 121,000 followers on Bilibili and around 400,000 followers across various platforms. This uploader is a typical representative of the lifestyle Vlog category, with iconic video collections like "Clever Life of Study Abroad" , "Living Aloneing" , "Random Sharing" , and "Clever Dorm Life" .

The interview content and results are outlined below:

a. Creative Strategies – Their Unique Vlog Creation Features

Firstly, creators of lifestyle Vlogs often emphasize authenticity and spontaneity, which contribute to the attractiveness of this category. As "Yi Zhi Dan Dan" shared during the interview: *"For me, presenting an authentic life is more important. It's interesting and is a form of existence. My life itself should be interesting – it's just my life."*

Furthermore, they consider user engagement essential in their creative process. Lifestyle Vlog creators pay extra attention to interacting with their fans, enhancing fan loyalty.

Moreover, many content creators are active across multiple platforms. For instance, the interviewee creates content not only on Bilibili but also on platforms like Douyin (TikTok) and Xiaohongshu (known as the Chinese version of Instagram). Due to the distinct nature of each platform, their operational strategies may differ accordingly.

b. Marketing Strategies – Monetization Models for Video Uploaders

Regarding marketing approaches and monetization, each video uploader adopts distinctive strategies. While summarizing marketing models for Vlogs, it was observed that video uploaders employ diverse marketing strategies, ranging from video content and fan communities to both online and offline avenues.

However, the interviewee also mentioned that many video uploaders are shifting towards short-form videos. As a result, the amount of content created on Bilibili has decreased. This trend reflects the migration of audiences to other platforms, impacting the number of advertisements video uploaders can secure on Bilibili and affecting their monetization potential.

c. Operational Strategies – Video Uploaders' Self-Management

The methods employed by video uploaders to manage their Vlogs, maintain favorable relationships with their fans, garner long-term attention, receive positive reviews, and achieve monetization require careful operational strategies. A synthesis of successful video uploaders' practices reveals alignment with the STP theory. They segment the market, then focus on cultivating their target market segment intensively. For instance, some video uploaders target student demographics, while others cater to working professionals, optimizing content based on their target audience's characteristics and preferences.

d. Problem Awareness Strategies – Identifying Existing Issues

Effective video uploaders continuously possess problem awareness, enabling them to promptly identify, rectify, and improve upon their existing issues. This proactive approach is essential for achieving higher traffic and growth. A review of various video uploaders' content indicates that alongside their engaging creations, they also encounter numerous challenges.

e. Platform Optimization Strategies – Improving Bilibili

The creative, marketing, and operational efforts of video uploaders rely on the opportunities and conveniences provided by the platform. As a collective entity, Bilibili should also maintain an optimization focus. Timely optimizations, in line with video uploaders' needs and actual circumstances, are vital for Bilibili's development.

3.2 Business Model and Monetization Methods of Lifestyle Vlog Uploaders on Bilibili

Analyzing interviews with lifestyle Vlog uploaders on Bilibili reveals that their business models and monetization methods are diverse and encompass the following aspects:

a. Advertising Revenue

Earning income through in-video advertisements is the most direct monetization method and a common approach among many video uploaders. Advertising placement includes direct methods like product introductions, highlighting product benefits, and urging viewers to make purchases. Indirect advertising placement involves incorporating advertised products as props, background elements, or decorations within the video or subtly mentioning them.

b. Knowledge Monetization

This advanced business model involves video uploaders selling their knowledge or skills, such as cooking or crafting, to fans. They offer tutorials and earn income in return. This approach requires substantial knowledge and professional skills to provide content teaching and sell courses.

c. Brand Collaborations

Video uploaders collaborating with brands typically possess a large fan base, industry recognition, and a positive reputation. This enables them to promote products or services in partnership with brands and receive collaboration fees. Currently, the number of video uploaders who can directly collaborate with brands is relatively small, the majority of whom are primarily the top creators within the industry

d. Fan Donations

For uploaders with a substantial fan base and high fan engagement, fan donations become a significant income source. These uploaders, appreciated for their quality content, receive support from fans while they watch Vlog videos. Fans utilize the platform's donation feature to reward creators.

e. Social E-commerce

Using live broadcasts or videos to showcase and sell products constitutes a social e-commerce business model, generating revenue from product sales. This typically involves advertisers paying for product placement and video uploaders earning "commissions" from fans' purchases through provided links.

f. Cross-Industry Collaborations

Some video uploaders engage in collaborative efforts with other uploaders, appearing together in videos, co-creating content, or participating in online interactions. This approach introduces novelty to fans and facilitates shared traffic. Collaborations with uploaders from different domains lead to jointly crafted content and collaboration-based revenue.

Certainly, different uploaders may adopt distinct business models or combine multiple approaches for revenue generation. Additionally, some uploaders increase fan loyalty and revenue by offering membership subscriptions and exclusive benefits.

g. Content Operations

Each video uploader's operational strategy varies based on their target audience, core focus, and video style. Lifestyle Vlog uploaders often start by creating foundational content that aligns with their personal style. After attracting an initial batch of followers, they engage in interactive forms like bullet screen comments, comments, and community engagement to discuss video ideas, foster positive interactions, and produce new content.

4. Audience Feedback Analysis Based on Bilibili

To gain a deeper understanding of audience feedback for lifestyle Vlogs on Bilibili, we conducted research encompassing analysis of audience feedback on the platform and interviews with users.

4.1 Analysis of Audience Feedback on Bilibili

Currently, Bilibili has attracted a significant number of popular video uploaders who produce videos in their respective domains. To illustrate this, we will analyze the audience feedback for two uploaders, namely "Li Ziqi" and "Ouyang Nana," to better understand their audience dynamics.

First, it is vital to delve into the audience feedback analysis for "Li Ziqi." Renowned for her culinary videos, Li Ziqi achieved the Guinness World Record for "Most Subscribed Chinese YouTube Channel" in February 2021, with a staggering 14.1 million subscribers. Beneath her videos, a plethora of feedback has been compiled, primarily positive and supportive. Examples include comments like *"Soothing and appetizing, makes me crave!"* and *"Watching her videos is so comforting. We need more creators like her."* However, there are occasional negative opinions such as *"Vey unreal. Real rural life is nothing like this"* and *"Who does chores so exquisitely? Overly packaged!"* Overall, Li Ziqi's successful business model is evident; she has cultivated a self-brand, achieved monetization, and garnered a positive reception.

Next, it bears significance to analyze the audience feedback for "Ouyang Nana." In comparison to Li Ziqi, Ouyang Nana maintains a slightly higher update frequency and offers a more diverse range of content, encompassing a broad spectrum of Vlog topics. Presently, Ouyang Nana's audience feedback exhibits polarization. Some audiences commend her for her authenticity and her ability to display a glimpse of her daily life as a celebrity. This satisfies curiosity about celebrity lives, and fans are inclined to embrace her recommendations, including products from her own clothing brand. Examples of positive comments include *"Love you, Nana! I'm going to buy it now!"* and *"Nana is one of my comforts during the time my preparation for the postgraduate examination."* Nonetheless, there are also negative voices, such as *"She clearly doesn't lack money, yet she seems so eager to make more, unattractive behavior."* This analysis of Ouyang Nana's audience feedback highlights that celebrity popularity can indeed attract traffic and followers as content creator. Yet, they also bear the burden of greater responsibilities and heightened scrutiny.

4.2 Audience Analysis

4.2.1 User Portrait

Based on Bilibili's user portrait report for the year 2022, it is evident that the user base of Bilibili exhibits a distinct trend towards youthfulness, with nearly 80% of users being under the age of 24. In terms of gender, males account for 57% of the user base, while females comprise 43%, indicating a higher proportion of male users. From a geographical perspective, Bilibili's primary user concentration is in prosperous coastal regions (such as Guangdong, Shandong, and Zhejiang) as well as in first-tier cities (Beijing, Shanghai, Guangzhou, Shenzhen). These regions benefit from favorable geographical locations and relatively developed economies, enabling users to be exposed to diverse cultures from Hong Kong, Macau, Taiwan, Japan, South Korea, Europe, and the United States. As a result, they possess a higher capacity to embrace subcultures and readily accept novel concepts. [22]

In light of this, a concise summary of the user portrait for lifestyle Vlogs on Bilibili can be outlined as follows:

a predominantly youthful audience, mainly comprised of students, with a slightly higher male-to-female ratio but generally balanced; an overall higher level of education; and a notable concentration in economically prosperous coastal regions.in addition , These engagements are self-presentation bricolage,self-identified branding,and community participation.[23]

4.2.2 Analysis of Audience Purchase Intent and Consumer Behavior

First and foremost, the youthfulness of the audience signifies a strong propensity for embracing novel experiences. The audience for lifestyle Vlogs on Bilibili is largely concentrated within the 18-24 age group, characterized by their heightened openness to new ideas. Consequently, they are not only receptive to the Vlog content itself but also inclined to engage with product endorsements and introductions within the content.

Furthermore, when examining geographic factors, residents of first and second-tier cities, as well as economically developed regions along the eastern coast, exhibit higher economic activity. This economic affluence translates to more disposable income available for product consumption. Moreover, these regions tend to display a more active mindset, making them more willing to accept innovative concepts and explore new products and ideas.

The audience for lifestyle Vlogs on Bilibili displays a notable inclination towards accepting advertisements. This phenomenon can be attributed to two primary factors. Firstly, Bilibili's advertising formats have departed from conventional methods, avoiding intrusive and forced approaches. Secondly, each content creator on the platform specializes in their own distinct field. This contextual relevance in advertising aligns with the interests of their fan base, making such promotions more appealing to fans. In the end,Bilibili must pay attention to the platform's interactive experience and user satisfaction in order to gain users' continuous subscription.[24]

4.3 Inspiration for Media Practitioners to Enhance Operational Strategies

The aforementioned audience interviews, feedback analysis, and user portrait analysis provide valuable inspiration for media practitioners to refine their media management and enhance their operational strategies.

a. Segment the Overall Market and Identify Target Audience

As content creators on the Bilibili, the first step should involve segmenting the Bilibili community according to one's unique style, creative attributes, and other defining characteristics. By doing so, content creators can pinpoint their specific target audience. This strategic alignment allows for more effective content creation, better traction with audiences, and the achievement of sustainable growth.the stronger happy emotion video has, and the video is popular. [25]

b. Analyze Market Characteristics and Formulate Operational Strategies

Once the target audience is determined, media practitioners should promptly analyze the distinctive features of their intended market. This includes understanding age groups, professional backgrounds, preferences, and other pertinent traits. Utilizing this comprehensive understanding, media practitioners can then develop appropriate operational strategies tailored to their specific audience.

c. Pay Attention to Audience Feedback and Optimize Operational Strategies

Timely attention to audience feedback plays a pivotal role. It offers a deeper insight into the audience's mindset and their opinions and suggestions regarding content and Vlogs. This ongoing assessment facilitates agile adjustments, enabling the optimization of operational strategies and ultimately fostering enhanced growth.

5. Existing Issues and Optimization Strategies for Video Platform Media Operators

At present, media operators encounter an array of challenges in their pursuit of effective media operation. In response, we have meticulously outlined these challenges and concurrently presented optimization strategies.

5.1 Analysis of Existing Issues for Media Operators

Through our summary and analysis of lifestyle Vlogs on Bilibili, it is evident that while these videos have numerous advantages, they also face several pressing issues that require attention. These issues can be categorized as follows:

a. Weak Innovation Awareness and Excessive Imitation

To maintain high-quality, engaging, and lasting content that appeals to fans, strong innovation and originality are essential. Many successful content creators have amassed large fan bases and high view counts due to their consistently innovative approach. When compared to other creators, they stand out by consistently presenting fresh and distinctive content that is both imaginative and unique. From the perspective of a creator's personal growth, the ability to continuously innovate and refresh their content while building upon their existing foundation is vital. This approach caters to users' curiosity, ensuring that their content remains engaging and relevant over time.

In contrast, some content creators lack innovation and resort to imitating outdated content and production styles. This leads to repetitive video themes and styles that struggle to capture users' attention. In more severe cases, creators directly copy other content or steal ideas from fellow creators, demonstrating not only a lack of innovation but also infringing upon others' legal rights.

b. Excessive Commercialization and Over-Promotion

As discussed above, one of the profit strategies for content creators is advertising revenue. They achieve this by incorporating advertisements within their videos or promoting related products, thereby earning advertising fees or commissions. When done in a way that doesn't compromise the audience's viewing experience, judiciously inserting advertisements or product promotions is acceptable. Properly timing these promotions and maintaining a balanced approach can generate income for content creators without compromising the quality of their videos.

However, some content creators excessively prioritize profitability, leading to severe commercialization of their content. They incorporate a high volume of unrelated advertisements or product promotions with little or no relevance to the video's content. These promotions occupy a significant portion of the video's duration, resulting in viewer dissatisfaction. Such behavior is akin to "killing the goose that lays the golden egg." While it may yield short-term gains, it is extremely detrimental to the content creator's development in the long run.

c. Inconsistent Content Quality and Proliferation of Low-Quality Videos

Due to the fact that anyone can upload videos, including their own Vlogs on Bilibili, there exists a significant disparity in content quality, with some content creators producing videos of lower quality. On a smaller scale, such Vlogs struggle to attract viewers, and on a larger scale, these low-quality videos undermine viewers' trust in Bilibili, ultimately impacting the platform's growth.

This low quality is evident in multiple aspects. Firstly, a lack of craftsmanship, encompassing issues like poor camera work, low resolution, and subpar editing techniques, leads to an unsatisfactory viewing experience. Secondly, the content might be unoriginal or even vulgar, including direct copying, offensive pranks and embarrassing acts. Such practices not only tarnish the image of the content creators themselves but also pollute the overall environment of the Bilibili.

d. Diminished Legal Awareness and Infringement of Others' Legal Rights

Lifestyle Vlogs often involve capturing daily routines and minor occurrences, which may inadvertently include other individuals or objects in the footage. This also extends to editing others' videos. However, this practice raises a critical concern—violation of others' rights.

Some content creators, in the process of filming videos, lack a sense of discretion and extensively capture footage of others. Worse yet, certain individuals might engage in surreptitious filming or recording of others without their knowledge, subsequently directly publishing the videos on online platforms. This constitutes a classic violation of others' rights, including privacy and likeness rights. In cases of serious offenses, content creators are liable to bear legal responsibilities.

e. Involvement in Illegal Activities and Disruption of the Online Environment

Certain content creators, while producing lifestyle Vlogs, might engage in illegal activities or violations of regulations to attract attention and gain viewership. This could include activities like illegal camping, trespassing, or pranks. From an individual perspective, such behavior poses personal safety risks for the content creators, as their actions might expose them to danger. There have already been cases of fatalities resulting from extreme challenges conducted in violation of regulations. From a legal standpoint, some of these actions have contravened laws and regulations, subjecting those involved to legal liabilities. In terms of the internet environment, these actions set a highly misguided example for viewers, leading to negative influences. This impact is particularly concerning for young and inexperienced individuals, who might be more prone to mimicry, thereby exposing themselves to potential hazards.

5.2 Improvement and Strategies for Media Practitioners' Operational Models on Video Platforms

To achieve success and long-term growth in the context of media convergence, media practitioners need to continually optimize their content creation, output quality, and improve their operational models.

a. Enhancing Content Production and Quality

With the rapid development of media convergence, content production has become a core competitive advantage. High-quality content serves as the most direct and fundamental force for attracting users. Therefore, media practitioners need to significantly improve their content creation capabilities and elevate their creative skills and quality.

Simultaneously, achieving a balance between depth and diversity is crucial. "Depth" refers to delving deep into a specific field, creating content with substance and insight that differentiates the creator from others. This uniqueness enables audiences to recognize the creator's distinctive contributions within that domain. On the other hand, "diversity" emphasizes avoiding rigidity and adherence to established norms. It involves offering diverse and comprehensive content on the media platform, continuously innovating, and meeting the evolving needs of users.

b. Increasing Technological Investment to Drive Transformation and Upgrading

As media practitioners, it is imperative to possess a proactive mindset and take action towards progress. This entails both investing in technology and undergoing transformative upgrading. In the rapidly changing landscape of new media, technology is continuously evolving. As media professionals, it is essential to promptly adapt to new technologies, learn about them, and incorporate them into operations by intensifying technological investments. Furthermore, attention should be given to transformation and upgrading based on actual network requirements, environmental needs, user demands, and more.

For those operating in the realm of new media, it is crucial to embrace new technologies. For instance, leveraging artificial intelligence technology for mining and analyzing web data can enrich the content of one's creations. Additionally, utilization of cloud computing technology can enhance information storage and transmission efficiency.

c. Strengthening User Communication and Implementing Timely Adjustments

Media practitioners should not operate in isolation or pursue a unilateral approach; instead, they should actively engage with user opinions and suggestions. It is essential to establish open channels of communication and interact with users proactively. Through these exchanges, media professionals can promptly grasp user needs and feedback, facilitating adjustments and optimizations according to user responses.

Enhancing user communication can be achieved through various means, including but not limited to engaging with comments sections, utilizing message boards, interacting with live bullet comments, and fostering interactions within fan communities. By participating in these online communication activities and engaging in discussions on relevant topics, media practitioners can gain valuable insights into the sentiments of their audience. This dynamic, two-way interaction not only enhances the user experience but also increases user engagement and loyalty.

d. Advancing Platform Collaboration for Accelerating Model Innovation

Media practitioners can foster collaborations with diverse platforms to drive innovation, amalgamate resources, and comprehensively enhance market competitiveness. For instance, collaborations with content production companies and advertising agencies can be pursued to develop products and services that align with user needs jointly. Simultaneously, proactive exploration of new business models and revenue streams, such as subscription-based services and e-commerce, is recommended.

Successful collaborations facilitate information exchange and method optimization, creating a win-win or even a multi-win scenario. Therefore, media professionals should take the initiative to engage in collaborations with peers and across various platforms. By doing so, they contribute to the elevation of content creation, expediting the innovation of operational models.

e. Prioritizing Information Security and Enhancing Data Protection

In the era of digital information, which has comprehensively arrived, issues such as information leaks and data breaches have become potential threats. Therefore, media practitioners must pay attention to information security and bolster data protection measures, while increasing investment in relevant technologies and management protocols.

Additionally, individuals should practice self-regulation and adhere to pertinent laws, regulations, and industry guidelines. Safeguarding users' personal privacy and information security is essential, making it imperative for media operators to operate in compliance with regulation.

References

1. Wu Xiaoyan. "Study on the Business Strategy of Shanghai Sunbright Industrial Co., Ltd." Master's Thesis, China University of Mining and Technology, 2022.
2. Pan Qiangsheng. "Research on Optimization of Business Strategy in H Company's Club Stores." Master's Thesis, East China Normal University, 2022.
3. Zheng Xiaohui. "Business Strategy for Traditional Radio and Television Advertising in the Era of Integrated Media." Journalist Cradle, Vol. 12, 2022, pp. 147-149.
4. Hong Li. "Analysis of Business Strategies for Newspaper Advertising in the New Media Era." Marketing Circles, Vol. 19, 2019, pp. 28+30.
5. Zhang Shengxi. "Business Strategy for Newspaper Advertising in the Context of Media Convergence." Press Outpost, Vol. 3, 2019, p. 2.
6. Che Li. "Analysis of Business Strategies for Newspaper Advertising in the New Media Environment." Journalism Communication, Vol. 21, 2018, p. 2.
7. Zheng Suhui, Xi Zhanhong, Lu Hongcheng. "Branding Operation Strategy of Cultural Variety Programs: A Case Study of 'The Reader'." TV Research, May 2022, pp. 80-82.
8. Du Zhongfeng, Luo Jing. "Discourse Changes and Inherent Logic of Media Convergence in China: A Perspective from Discourse Analysis." Editorial Friend, Vol. 1, 2020, pp. 12-18+29.
9. Sun Wei. "Cyber Humans: Media Convergence in the Post-Human Era." Shanghai Journalism Review, Vol. 6, 2018, pp. 4-11.
10. Liu Bing. "Cultural Integration: An Examination of Cultural Factors in the Process of Media Convergence." Editorial Friend, Vol. 1, 2018, p. 65.
11. Yue Jinhong. "How to Conduct In-Depth Reporting in the Era of Media Convergence: Review of 'Party Newspaper Path Exploration in the Age of Media Convergence'." Media, Vol. 8, 2023, pp. 97-98.
12. Liu Jiajia. "Discussion on the New Trend of TV Production and Communication in the Perspective of Media Convergence." Contemporary TV, Vol. 1, 2023, pp. 66-72.
13. Hu Shutong. "Research on the Occupational Roles of New Media Practitioners from the Perspective of News Professionalism." Master's Thesis, Shandong Normal University, 2017.
14. Liu Jing. "Valuation Research of Kuaishou, an Internet Media Enterprise, from the Perspective of Media Operation and Management." Master's Thesis, Henan University, 2022.
15. You Xiangyu. "Analysis of CCTV Anchors' Attempt at Live Commerce from the Perspective of Media Operation and Management." West China Broadcasting TV, Vol. 6, 2021, pp. 91-93.

16. Zhang Yao. "Analysis of iQiyi's Media Operation and Management Strategy." China National Exhibition, Vol. 2, 2020, pp. 234-236.
17. Bu Yanfang, Zhu Yuning. "Analyzing Bilibili's Creative Ecosystem from the Perspective of 'Luoxiang Talks about Criminal Law' and Its Communication Network." Journalism Lover, Vol. 2, 2023, pp. 48-52.
18. Huang Pingping. "Narration, Interaction, and Elements: An Analysis Based on Bilibili Slow Videos." Media, Vol. 8, 2022, pp. 47-49.
19. Zhou Meng, Chen Wanqian. "Promoting Cultural Fairness through Niche Communication: An Empirical Study on Bilibili's Individual Subtitle Phenomenon." Youth Journalist, Vol. 14, 2021, pp. 113-114.
20. Zhai Chenhua. "Brand Characteristics of UGC Content Marketing on Bilibili." Media, Vol. 12, 2021, pp. 74-75+77.
21. Yu Ke. "Profit Models of Danmaku Video Websites: A Case Study of Bilibili." Youth Journalist, Vol. 14, 2016, pp. 97-98.
22. Blue Lion Technology. "https://www.bilibili.com/read/cv18496411/," March 15, 2023.
23. Wang, XX,Picone, I.Mobilized, negotiated and balanced: Chinese school vloggers' platform engagements and layered identity construction on Bilibili.Journal of Youth Studies,Asian Anthropology 22:3, pages 213-216.
24. Moon, Sung, Bae, Seung-Joo, and Lee, Sang-Ho. (2022). A study of influencing factors on OTT Bilibili users' intention to continue media subscription. Journal of the Multimedia Society, 25(12), 1756-1767.
25. Z. Wu and E. Ito, "Correlation Analysis between User's Emotional Comments and Popularity Measures," 2014 IIAI 3rd International Conference on Advanced Applied Informatics, Kokura, Japan, 2014, pp. 280-283

Research on the Influencing Factors of Management Policy for Depressed College Students Based on Family-School Interaction

Yu Guo *, Xiaofang Zhang

Zhangjiagang Campus, Jiangsu University of Science and Technology, Suzhou, China

Abstract—Based on parents' behavioral intention, the satisfaction of parents is divided into the cognitive dimension of policy interpretation, cue mutual verification, natural language and emotional dimension of authenticity, altruism and interaction. Combined with the previous research on the theory of caring for depressed students, the satisfaction model of depressed students' parents is constructed. The data were collected through questionnaires and conducted using SPSS and AMOS. The results show that policy interpretability, cue mutuality and natural language have positive effects on parents' perceived usefulness. Authenticity, altruism and interactivity of affective dimension have positive influence on the perceived trust of parents. Parents' perceived usefulness has a positive effect on parents'satisfaction. Parents' perceived trust has a positive effect on parents'satisfaction.

1 Introduction

The capacity and level of society'sself-organization, or the maturity of civil society, is a prerequisite for a state governed by law. Whether authority can play its expected role depends on the strong supervision and constructive criticism of authority by the growing civil society. In an increasingly pluralistic society with highly differentiated social subsystems, governance begins with the right of citizens to autonomy, that is, to decide their own affairs.

In college student management affairs, the special attention of students with bad mood is an important means to prevent college emergencies. According to the article 25 of 《the Regulations on the Administration of Students in Ordinary Institutions of Higher Learning》, the circumstances of suspension include that the student applies for suspension or that the school considers that it should be suspended. In practice, some students who are depressed or in bad mood are asked to suspend school or ask for leave to go home. Some parents think that the students' negative mood is not enough to leave school, causing disputes between students and schools.

From the domestic court's cancellation of a university's "decision to revoke Cheng's compulsory suspension"[1]to Hunter College's "decision which forcibly not return to the dormitory",[2]the society expects schools to give more care and trust to students with bad emotions. Jeffrey Benson argues that teachers who work with students who have mental illnesses face day-to-day challenges and high burnout rates. Instead, the structures must change--by way of clinical consultations, mentorships, and in-school support systems--to best serve both teachers and their students. [3]Kevin Cokleysuggests college students' trust of faculty maybe particularly important for Black males and is likely a contributing factor to academic misidentifications.[4]

At present, some unreasonable management behaviors in colleges and universities cause the anxiety of students with bad mood to increase further. How to guide colleges and universities to have warm care for students, reduce students' anxiety and promote good interaction between home and school is the content of this paper, which has certain reference value for the actual student management work in colleges and universities.

2 Construction of a hypothetical model of satisfaction index

2.1. Theoretical review

At present, European and American scholars have developed and applied a large number of purchase satisfaction index models, among which the American purchase satisfaction and purchase intention index models are more widely used and more mature. [5]

At present, it mainly focuses on the satisfaction survey of students' parents in primary and secondary school education management. For example, Zheng Qing believes that teachers' caring behavior will not only directly affect junior middle school students' depression, but also indirectly affect junior middle school students' depression through redundant thinking and forgiveness. [6]

Most of the studies on the job satisfaction of college teachers are on the management of colleges and universities. In 2014, Liu Kai from Tibet University for Nationalities published the Construction of the

*Email: 34132671@qq.com.

Satisfaction Index Model of college students' Parents -- Taking Tibetan College students' Parents as an example in Journal of Changchun University of Technology (Higher Education Research Edition), which identified the satisfaction index model of college students' parents. Therefore, this paper will learn from the American customer purchase intention and satisfaction index model, and combine the adjustment of college student management. Purchase intention refers to whether the consumer is willing to buy the product at the equilibrium market price according to the consumer's monetary income. The influence of purchase intention is mainly studied from the perspectives of consumer attitude, maximum perceived value and minimum perceived risk. [7]

Through in-depth interviews and investigations with parents of college students, the team adjusted the relationship between variables of the American customer purchase satisfaction Index model. According to the concerns and feedback of parents of depressed students, and the satisfaction of home-school education and services, an initial hypothesis model of parental behavior index for depressed students is proposed. Based on the preliminary investigation, this study prepared an open questionnaire, and conducted a survey on 200 college students in Suzhou and Zhenjiang.By means of on-site distribution and two-dimensional code scanning, the survey was developed. They were invited to return home during the summer vacation to ask their parents to answer the questionnaire. At the same time, the parents of 32 students in colleges and universities were interviewed by telephone with the contents of an open questionnaire. From the above two ways, the survey could sort out the specific items of parental satisfaction for the management policy of depressed students in colleges and universities.

The influence of cognition on perceived usefulness refers to the accuracy of information transmission between families and schools. In home-school communication, two-way transmission of information is crucial.Parents often first understand the policy with the instructor. If they contact other teachers in the student work department, they can meet the needs of parents to obtain more policy information, and all kinds of information are the consensus view of multiple teachers to confirm each other, so as to give parents more reliable answers to their doubts. Furthermore, according to scholars, such as Wei Haiying ,Yang Guoliang, real-time interaction between sellers and consumers on the sales platform is an effective way to build trust. Similar to the above sales relationship, parents can not only understand the policy information from the text, pictures and documents provided by the school, but also perceive the authenticity and altruism of the school's suggestions through the verbal and non-verbal information of teachers, such as expressions and gestures. The school can reply and answer in time, and the above behavior will undoubtedly increase parents' trust in the school. [8]

The interaction between parents and schools is mainly based on cognition and emotion, which have a direct impact on parents' behavior intention in school suspension policy. "Policy interpretability", "cue mutuality" and "natural language" are selected as important factors of parental cognition. "Authenticity", "altruism" and "interaction" are selected as the three important factors of parental emotion. "Perceived usefulness" and "perceived trust" are selected as the internal state of parents' behavior. Under the joint stimulation of various factors, parents produce internal state changes of perceived usefulness and perceived trust, so as to respond to "parent satisfaction", and construct a reflection diagram of school suspension intention under home-school cooperation, as shown in figure 1.

2.2 .Structural model

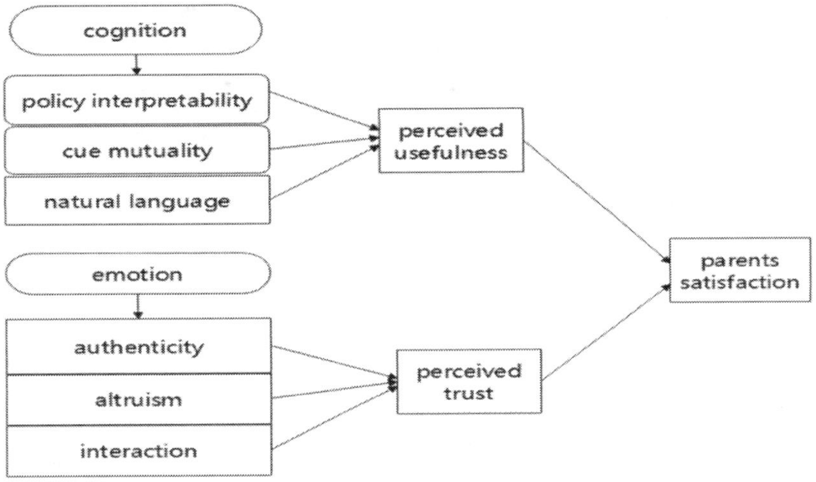

Figure 1. Diagram of school suspension intention under home-school cooperation

2.3.Research hypothesis

According to the constructed structural equation model, the following hypotheses are proposed:

H1: Student administrators' interpretation of university management policies can meet the needs of parents for answers to their doubts, and has a significant impact on parents' perceived usefulness;

H2: Counselors, class teachers and Students Office staff have the same interpretation of the policy, and there are words and cases to confirm it, which has a significant impact on the perceived usefulness of parents;

H3: Parents believe that following the school's advice is the best solution, which has a significant impact on parents' perceived usefulness;

H4: The information about students' symptoms and school management is true without any modification or concealment, which has a significant impact on parents' perceived trust;

H5: Schools really consider for parents and students, which has a significant impact on parents' perceived trust;

H6: Instant interaction between school and parents, barrier-free communication, has a significant impact on parents' perceived trust;

H7: Parents' perceived usefulness has a significant impact on parents' willingness;

H8: Parents' perceived trust has a significant impact on parents' willingness.

Table1. Latent variables and observed variables

variable	code	Measured item
Policy interpretability	py1	Parents believe that schools can provide professional solutions to students' problems
	py2	The school can understand the parents' meaning very well
	py3	Parents can understand the school's explanation very well
	py4	The policy and explanation conveyed by the school was exactly what I expected
cue mutuality	cl1	In the home-school connection, parents can get some idea of how to solve the problem through other similar cases
	cl2	In home-school contact, parents can be informed about how to solve problems through the policies provided by the school
	cl3	In the homeschool contact, various organizations and staff of the school have the same interpretation of the policy
natural language	nr1	The school staff spoke naturally, without a sense of overstatement or urgency
	nr2	In homeschool communication, parents listen to their teachers as if they were listening to their friends
Authenticity	tr1	Information about students is faithfully stated by the school
	tr2	Regarding the relevant policies of the school, the school truthfully states
	tr3	A complete and truthful statement about the school's similar case
altruism	al1	The program offered by the school is truly considered for parents
	al2	The programs offered by the school are truly considered for the students
	al3	According to this program, students' psychological condition can be improved
	al4	Students can improve their academic performance by following this program
interaction	ra1	For parents' questions, the school can answer in a timely manner
	ra2	Parents can inform the school of the student's condition in a timely manner
Perceived usefulness	pu1	Through home-school communication, problems can be solved
	pu2	The explanation of school policy is helpful for parents' decision
perceived trust	pt1	Parents trust the information provided by the school
	pt2	Parents believe that the solution provided by the school is the optimal solution
parent satisfaction	ps1	Parents are willing to solve problems as suggested by the school
	ps2	Parents are willing to interact with school staff and answer questions in a timely manner
	ps3	The interpretability of school policies has a significant impact on students' satisfaction in problem solving

3 Study design and sample analysis

3.1 Scale design and data collection

There are nine variables in this study, which are: policy interpretability, cue mutuality, natural language, authenticity, altruism, interactivity, perceived usefulness, perceived trust and parental satisfaction. Based on the characteristics of family-school communication, the collected scales were classified and improved, and after the inquiry of relevant experts and scholars, the measurement scales of various variables were developed, as shown in Table 1 -- latent variables and observed variables.

3.2 Data collection

Students at three schools who had or were currently depressed were asked to have one of their parents respond during the summer vacation. It is also explained that the results of this questionnaire are an academic survey and will not be related to the formulation of relevant policies of the school. Please fill in the questionnaire carefully. After the questionnaire is made on the website of "Questionnaire Star", it will be distributed in the form of links or two-dimensional codes. This questionnaire adopts Likert five-level scale with score values ranging from 1 to 5, and the degrees are "1" representing "strongly disagree", "2" representing "disagree", "3" representing "uncertain", "4" representing "agree" and "5" representing "strongly agree".

The selection of research objects in this study is based on two considerations: on the one hand, some parents of depressed students are reluctant to participate in this survey due to their economic and ideological situation, so they invite other elders of students in their family to be the survey objects, such as grandparents and siblings of students. On the other hand, due to the unbalanced urban economic development in different regions of China, the participation initiative of depressed students is limited. Therefore, through the efforts of the members of this research group and the opportunity of the psychological counseling center where the members are located, 237 questionnaires were randomly sent to three colleges and universities in Jiangsu Province, and 221 were recovered, with a recovery rate of 93.24%. Then, the questionnaires were sorted out. (There are three exclusion criteria:

Whether the subjects answered more than 90% of the questions in the questionnaire; whether there will be regular responses in the questionnaire.Finally 200 effective questionnaires were selected, and the effective recovery rate of the questionnaires was 84.38%, which was in line with the standard of statistical analysis.Sample characteristics of parents' satisfaction with depressed students are shown in Table 2.

Table2. Sample characteristics of parental satisfaction of depressed students

Statistical variable	category	Number of people	percent	Statistical variable	category	Number of people	percent
Relationship with students	parent	166	83.00%	Educational background	Below junior high school	98	49.00%
	grandparents	32	16.00%		High school, secondary school	85	42.50%
	Brothers and sisters	2	1.00%		Bachelor degree or above	17	8.50%
Visiting children	yes	199	99.50%	Contact frequency	At least once a week	98	49.00%
	no	1	0.50%		At least once a month	97	48.50%
Tuition source	Household income	145	72.50%		others	5	2.50%
	School aid, scholarships	26	13.00%	Income level	Monthly income < 2000 yuan	2	1.00%
	Part-time job	29	14.50%		Monthly income 2000-5000 yuan	158	79.00%
					Monthly income 5001-10000 yuan	32	16.00%
					Monthly income >10000 元	8	4.00%

4 Analysis of influencing factors of satisfaction

4.1 Reliability and validity test

SPSS 22.0 software was used for data processing, and A-MOS 21.0 software was used for structural equation analysis. As shown in table 4, the standard loads are all greater than 0.7, indicating that latent variables can be well explained by observed variables. SPSS 22.0 was used for reliability analysis, and the overallα of the questionnaire was 0.878, indicating a strong correlation between variables and suitable for factor analysis. Table 3 Reliability test related indicators Result Variable item Standard load Cronbach's combination reliability AVE degree validity test is used to test the validity of the questionnaire, which usually includes discriminated validity and aggregation validity. The discrimination validity was tested by comparing the square root of each variable's AVE value with the correlation coefficient, and the aggregation test was conducted by whether the AVE value of each variable was greater than 0.5. As shown in table4, AVE values of all variables are greater than 0.5, and as shown in Table 5, square roots of AVE of all variables are greater than correlation coefficients, indicating that the questionnaire has good validity.

Table3. Reliability and validity test results

			Estimate	alpha	CR	AVE
al1	<---	AL	0.743	0.878	0.882	0.654
al2	<---	AL	0.915			
al3	<---	AL	0.827			
al4	<---	AL	0.737			
cl1	<---	CL	0.842	0.905	0.908	0.767
cl2	<---	CL	0.939			

cl3	<---	CL	0.842			
nr1	<---	NR	0.853	0.868	0.869	0.768
nr2	<---	NR	0.899			
ps1	<---	PS	0.821	0.908	0.910	0.771
ps2	<---	PS	0.903			
ps3	<---	PS	0.908			
pt1	<---	PT	0.890	0.882	0.882	0.789
pt2	<---	PT	0.886			
pu1	<---	PU	0.852	0.837	0.837	0.719
pu2	<---	PU	0.844			
py1	<---	PY	0.789	0.929	0.932	0.775
py2	<---	PY	0.904			
py3	<---	PY	0.942			
py4	<---	PY	0.880			
ra1	<---	RA	0.864	0.867	0.867	0.766
ra2	<---	RA	0.886			
tr1	<---	TR	0.837	0.922	0.923	0.801
tr2	<---	TR	0.927			
tr3	<---	TR	0.918			

Table 4. Discriminated validity analysis

	py	cl	nr	tr	al	ra	pu	pt	ps
py	**0.881**								
cl	0.451	**0.876**							
nr	0.380	0.313	**0.876**						
tr	0.376	0.362	0.329	**0.895**					
al	0.420	0.475	0.369	0.431	**0.809**				
ra	0.372	0.345	0.444	0.361	0.338	**0.875**			
pu	0.506	0.509	0.546	0.480	0.562	0.479	**0.848**		
pt	0.504	0.502	0.522	0.484	0.543	0.475	0.578	**0.888**	
ps	0.458	0.492	0.479	0.492	0.549	0.484	0.538	0.529	**0.878**

Table5 .Model fitting indicator

Model fitting indicator						
index	χ^2/df	RMSEA	NFI	IFI	TLI	CFI
Judgment standard	<3	<0.10	>0.9	>0.9	>0.9	>0.9
value	2.05	0.062	0.907	0.95	0.94	0.95

4.2 The overall fit test of the model

By referring to the partitioning method of the overall model fit evaluation indexes of Hair et al., the overall model fit evaluation indexes are divided into three categories: absolute fit index, value-added fit index and reduced fit index. In terms of the overall fit test of the model,NFI, IFI ,CFI, TLI and are all greater than the fitting parameter standard 0.9, close to 1， As shown in table6 The fitting index RMSEA is 0.062, which are less than 0.1. The Chi-square degree of freedom ratio of students' parents' satisfaction model is 2.05, which is less than the fitting parameter standard 5. It can be seen that the conceptual model of parental satisfaction of depressed students has a good validity.

Table6.Model path coefficient table

			Estimate	S.E.	C.R.	P	Label
PU	<---	PY	.182	.043	4.217	***	significant
PU	<---	CL	.248	.044	5.613	***	significant
PU	<---	NR	.446	.061	7.362	***	significant
PT	<---	TR	.213	.057	3.748	***	significant
PT	<---	AL	.533	.081	6.552	***	significant
PT	<---	RA	.437	.078	5.629	***	significant
PS	<---	PT	.355	.061	5.778	***	significant
PS	<---	PU	.553	.084	6.605	***	significant

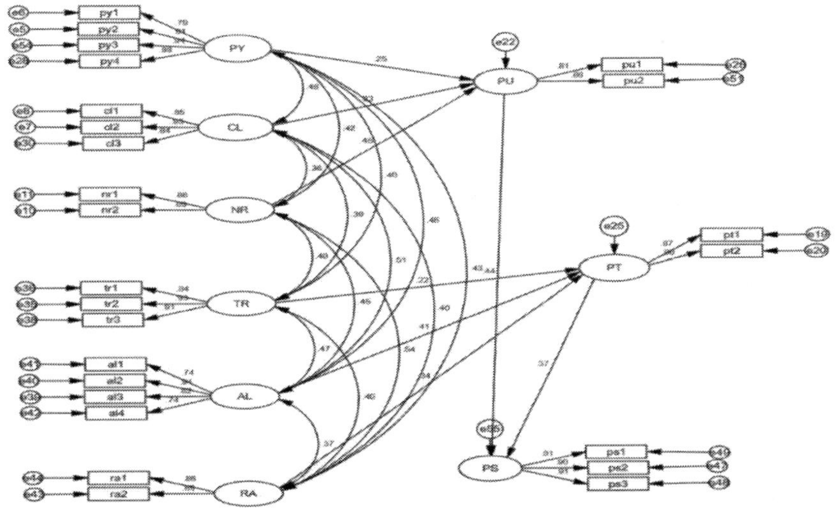

Figure 2. Model test result

5. Results andRecommendations

1.As shown in table6 and figure 2, Cognitive policy interpretation, cue mutuality and natural language factors have positive effects on parents' perceived usefulness. This shows that the school's transmission of information is very effective, and parents can feel it more clearly, thus improving individual cognition. The school's interactive communication can deliver more information needed by parents. The more similar cases of such information, the more abundant and easy to understand communication channels such as text and language exchange, and the easier parents will accept the school's suggestions.

2.Emotional authenticity, altruism and interaction have a positive impact on parents' perceived trust, indicating that emotional communication is very important in the home-school communication process of parents. The more objectively described information and the more successful cases conducive to students, the more parents can eliminate doubts and create a sense of trust. When the school responds to the parents' questions in a timely and consistent manner, the parents will be more sensitive to the school's decision and think that the school's decision is the best choice for the students.

3.Parents' perceived usefulness and perceived trust are bound to be satisfied with parents' handling of the school. In practice, it is not that parents think the school's decision is unreasonable, but that some schools handle things with simple explanations and rude answers. This will certainly not narrow the psychological distance between parents and schools. However, if there is advice from other parents in the previous case, it will undoubtedly have a greater significant influence on the decision of the subsequent parents. Parents of students will have a great sense of empathy and sureness, and will also positively improve their trust in the school, believing that the school will provide their children with perfect psychological counseling services, a good growth environment and an optimal decision-making structure. In turn, the trust of parents in the school will affect parents' judgment of the school, which is conducive to the school's enrollment publicity and student management.

4.Parents have different communication skills and communication quality due to different human and cultural capital. Parents' judgment on students is not due to their different economic ability, but due to their inherent cognitive ability. For some parents with problems, it is suggested that schools can do the ideological work of students first, and then do the work of parents, so that the management of students can get twice the result with half the effort. At the same times, for students who have violent behaviors and tend to hurt others, if their parents ignore them in school, it is necessary for the relevant government health agencies to intervene and resolve the problems in timeand could be working with the parents' community or parents' units.

5.In the telephone and face-to-face interviews at the later stage of this survey, the families and schools reached the following consensus: the current education focuses on the accumulation of scientific knowledge, and lacks the connection between students and students, and between teachers and students. The current education does notfocus on trust, respect and other emotional factors as the core elements of education. And the most essential and effective communication between people should be based on true feelings and trust. The research results show that the parent satisfaction model constructed in this paper has high credibility, but because the sample comes from families in Jiangsu province, its representativeness has certain limitations, and it is expected that other scholars can further expand and improve it.

Acknowledgment

The study was funded by"General Research Project of Philosophy and Social Sciences in universities in Jiangsu Province in 2023", Project Number: 2023SJSZ0844, A study on the management and governance system of university students based on the case analysis of forced suspension of depressed students abroad.

The study was also funded byZhenjiang City Social Science Applied Research Law Special Project in 2023", Project Number:2023FX86.

References

1. Wang Shaobo, A Case Study on Administrative Mediation of University Compulsory Suspension Decision, Legal System and Society [J] : 2020 • 12 (Part 1).
2. Hunter College Reaches Settlement With Depressed Student Who Was Barred From Dorm The Chronicle of Higher Education September 8, 2006 Frida STUDENTS; Pg. 34; Vol. 53; No. 3.
3. Jeffrey Benson(2018), When Teacher Self-Care Is Not Enough,Educational Leadership [J] : 75(4): 38-42.
4. McClain, S., & Cokley, K. (2017),Academic disidentification in Black college students: The role of teacher trust and gende, Cultural Diversity and Ethnic Minority Psychology[J]: 23(1), 125. - PubMed - DOI.
5. Liu Xinyan, Research on Customer Satisfaction index model [M], Beijing: China Finance and Economics Press.
6. Zheng Qing, The influence of teachers' caring behavior on depression in junior high School students: The mediatingrole of redundancy and forgiveness, Chinese Journal of Clinical Psychology, volume 25, No 1, 2017.
7. Liu Kai, Construction of a Satisfaction Index Model for Parents of college students -- A case study of parents of college students in Xizang Province, Journal of Changchun University of Technology (Higher Education Research Edition) :volume 35, Issue 2,2014.
8. Wei Haiying, Yang Guoliang, Analysis of the Impact of enterprise-Customer Interaction on Brand Trust -- From the perspective of crisis prevention [J], Finance and Trade Economics, 2011 (4): 79-84.

Research on the Training Path of Accounting Talents in the Integration of Industry and Finance under the Goal of High Quality Employment

Dongqi Han, Yan Xu*, Nanyun Xiao, Jinlong Wang

Department of Accounting,Harbin Finance University,Harbin Heilongjiang 150030,China

Abstract. Talents are an important symbol of a country's comprehensive strength. Accounting talents, as an important component of modern society, are an important support for promoting economic development. Especially with the rapid development of digitalization and the continuous emergence of new business formats, technologies, and business models in social development, cultivating comprehensive talents that adapt to social development has become an important topic and reform direction for accounting talent cultivation. However, under the significant impact of the epidemic, issues such as an increase in the number of graduates majoring in accounting, a lack of social practice, and a mismatch between abilities and positions have become prominent. This article is oriented towards high-quality employment and analyzes the training path of accounting talents from the perspective of integrating enterprise business and financial business. It attempts to address the challenges arising from changes in employment structure and environment, achieve efficient integration between accounting talents and society, and meet the special needs of society for accounting talents.

1 Introduction

With the continuous construction of digitization and digital intelligence, the Ministry of Education has issued the "Declaration on the Construction of New Liberal Arts", providing guidance for the construction of new liberal arts[1]. The rapid development of science and technology has put forward higher requirements for social employment,. The finance and accounting major has become the main carrier of the construction of the new humanities. Combining the characteristics of the discipline, the construction of the new business discipline is proposed. The concept of the new business discipline is based on the background of the fifth technological revolution, fully utilizing diversified information technology on the basis of the existing business discipline, and integrating more humanities and social content, forming an interdisciplinary discipline that includes economics, management, philosophy, law, computer science, and other disciplines, Committed to cultivating high-quality business talents that meet the needs of the times. It is precisely because of the comprehensiveness and progressiveness of the new business concept that new opportunities and challenges are brought to the training of new business talents in colleges and universities.

The current employment of accounting talents is mainly characterized by polarization. On the one hand, the market for ordinary accounting talents is becoming saturated, and on the other hand, there is a shortage of senior accounting talents, which is in contrast to the requirements of enterprises[2]. High quality employment has become an important indicator for cultivating talents, and high-end accounting talents need to meet the needs of highly integrated business and finance in enterprises, becoming a key target for cultivation. Therefore, it is necessary to increase the importance of accounting talent cultivation, and optimize and improve accounting talent cultivation with the guidance of high-quality employment, taking advantage of the requirements of the new business construction era, in order to comply with the requirements of the times.

2 Analysis of the Employment Quality of Accounting Talents in Current Higher Education Institutions

2.1 Mismatch between personal abilities of accounting talents and the needs of enterprises

After nearly three years of the impact of the epidemic, most accounting students have completed their studies through online learning, and there are few practical teaching methods. The learning effect is not as good as offline teaching, and the holding of financial and accounting professional qualification exams is limited, further resulting in students' personal abilities far from meeting the needs of accounting positions in enterprises. On the one hand, the number of accounting students

* Corresponding author: 724259289@qq.com

obtaining qualification certificates such as the Junior Accounting Professional Title Certificate and 1+X vocational skills is relatively small, which cannot meet the threshold for entering enterprises[3]; On the other hand, traditional accounting teaching still focuses on accounting processing and report analysis, with less involvement in digital and intelligent teaching, resulting in mismatched abilities[4]. In addition, there are few internship opportunities in the accounting major, which makes it difficult for students to truly perceive the talent needs of enterprises. Many students are unwilling to participate in practical learning organized by the school and have not obtained internship qualifications from enterprises, resulting in a lack of understanding of accounting positions among students.

2.2Accounting students have immature employment mentality

The accounting position is a necessary and very important position for every enterprise, belonging to the middle and high-level positions. However, accounting students who have just graduated cannot meet the requirements of this position, but their expectations for the position are too high, both in terms of salary and corporate environment. However, from the perspective of enterprises, it is believed that newly graduated students lack the ability to handle accounting, as well as analytical, predictive, and good execution skills. However, schools lack the cultivation of students' comprehensive abilities, and enterprises are unable to provide accounting students with higher salaries, resulting in many accounting students choosing to take postgraduate or public entrance exams to alleviate employment pressure[5]. Parents also provide more learning environments for students, However, the lack of better attention to the substantive issues of employment has led to the phenomenon of "slow employment" and "non employment" among accounting students.

2.3Insufficient prediction of future employment patterns by accounting professionals

With the continuous emergence of new technologies and formats, the business and finance of enterprises are undergoing digital transformation and information construction. More enterprises have moved towards intelligent and digital production processes, and accounting positions are gradually shifting from being single to diversified. The business and finance of enterprises are closely related, and accounting professionals need to master more financial information from the business of enterprises. It is not as simple as entering enterprises and using computerized accounting, but rather to master more information technology to help enterprises achieve rapid development. Obviously, the current accounting students have insufficient predictions about the changes in employment forms when they graduate. Even if the school offers courses such as big data, internet, and artificial intelligence, the students' course selection rate is not high. Coupled with the special reasons of the epidemic in recent years, online teaching is far less effective than offline teaching, further exacerbating the serious disconnect between accounting talents and practical needs.

3 Current situation of accounting talent cultivation in universities

3.1 Insufficient hardware conditions for cultivating digital accounting talents

The digital construction of teaching facilities is the foundation of talent cultivation in universities. In the continuous development of the economy, universities are the training bases for providing talents, and should develop at a faster pace in infrastructure construction. However, the construction of accounting laboratories in universities has not yet fully popularized financial functional platforms such as intelligence and financial sharing, and existing information technology equipment has not been updated. Obsolete equipment and systems have long been phased out by enterprises, and new financial processes cannot be applied under existing experimental conditions[6]. Due to various reasons, accounting students are unable to access practical training conditions that keep up with the needs of enterprises and the requirements of the times. Not to mention the learning and application of information technologies such as artificial intelligence and blockchain. Students lack practical teaching resources, making it difficult to adapt to the employment environment. Schools should pay more attention to the sense of teaching presence and increase opportunities for students to have cognitive internships and exercise before employment.

3.2Lack of integration between industry and education and failure to achieve collaborative education

Schools are the foundation for building students' theoretical foundations, and the practice of enterprises is the only way for students to move towards job positions. However, the accounting talents trained by schools cannot meet the basic requirements of enterprises, and enterprises are unwilling to spend too much cost and energy on cultivating students. Under the impact of functional finance, the transformation of accounting professionals has become particularly important, and the lack of integration between industry and education has further caused difficulties in the employment of accounting talents. The training of accounting professionals is a complex and systematic project, and some teaching reforms such as student training objectives, enterprise needs, teaching methods, and school characteristics cannot be achieved overnight. Despite facing huge challenges and obstacles, universities should attach importance to the reconstruction of the training methods for accounting talents, relying on the integration of industry and education, and school enterprise collaboration to cultivate talents is an effective path to quickly cultivate digital and intelligent accounting talents.

3.3The expansion of curriculum boundaries due to limitations on teacher resources

In the context of the new business discipline, there is a greater emphasis on the cross integration of relevant disciplines, and more cutting-edge courses related to new methods and technologies have been set up, which puts higher requirements on the professional literacy and disciplinary perspective of teaching teachers. The lack of high-level teaching staff has also become the biggest challenge for many universities to carry out the construction of new business disciplines[7]. Currently, accounting teachers in universities have received traditional accounting education in the past, especially experienced teachers who have worked for a long time. They have poor learning and acceptance abilities towards new technologies and have a certain degree of resistance. For example, when offering the course 'Python Data Analysis', teachers need to first self-study Python programming, then combine it with professional knowledge, and finally transmit it to students. This process requires a long and difficult preparation period, which is a significant challenge for teachers without a background in computer science, and the final teaching quality cannot be guaranteed.

4 Future Development Trends of Accounting Talent Training

In the future, high-quality employment oriented integration of industry and finance accounting talent cultivation will become an important trend. Here are some related trends and development directions:

4.1 Diversified skill development

Traditional accounting majors no longer meet the needs of enterprises for talent. In the future, accounting talents need to possess a wider range of skills, such as financial analysis, risk management, data analysis, information systems, etc. Therefore, while cultivating students in the field of accounting, they should also focus on expanding their knowledge and skills in other related fields[8].Here are some possible directions for diversified skill development:

Financial analysis ability: Accounting professionals need to have the ability to deeply understand financial data and conduct analysis. They need to be able to extract valuable information from financial statements and accurately evaluate the financial condition and performance of the enterprise.

Data analysis and processing capabilities: With the advent of the big data era, accounting professionals need to possess the ability to analyze and process data. They need to master relevant data analysis tools and technologies, be able to extract key information from massive data, and provide support for enterprise decision-making.

Information system application: Accounting professionals need to understand and master the application of financial information systems. They need to be familiar with the operation and application of financial software, and be able to efficiently manage and process financial data.

Tax knowledge: Tax policy is one of the important factors in business operations, and accounting professionals need to understand tax laws and policies to provide advice on tax planning and planning for enterprises.

Risk management ability: Accounting professionals need to possess the ability to manage risks, identify and evaluate various risks faced by enterprises, and provide corresponding control and response strategies.

Communication and collaboration skills: Accounting professionals need to have good communication and collaboration skills, be able to collaborate with different departments and teams, effectively transmit financial information and interpret financial data.

4.2 Digital transformation

With the continuous development of technology, accounting work is also undergoing a digital transformation. Future accounting talents need to be familiar with and master technical tools such as financial software, cloud computing, and big data analysis to improve work efficiency and accuracy[9]. Therefore, while cultivating accounting talents, digital skills should be incorporated into the curriculum system and practical teaching should be strengthened.

Accounting professionals need to continuously learn and master digital technology to meet the requirements of the digital era. Here are some key directions for digital transformation:

Mastering financial software and tools: Accounting professionals need to be familiar with and master various financial software and tools, such as financial management systems, financial analysis tools, data processing software, etc. These tools can help accountants process financial data, generate reports, and analyze financial conditions more efficiently.

Data analysis and big data application: Future accounting talents need to have the ability to analyze data, extract valuable information from massive financial data, and provide support for enterprise decision-making. At the same time, they also need to understand big data technology and applications, and be able to use big data analysis tools to process and analyze financial data.

Cloud computing and mobile technology: Cloud computing and mobile technology have become key driving forces for enterprise digital transformation. Accounting professionals need to understand the principles and applications of cloud computing, and be able to use cloud computing platforms to store and process financial data. In addition, mobile technology also provides accountants with more flexible and convenient working methods.

Automation and Intelligence: The development of automation and intelligence technology will have a profound impact on accounting work. Future accounting professionals need to understand and apply automated tools and intelligent systems, capable of automating repetitive and cumbersome tasks, thereby improving work efficiency and accuracy.

4.3 Business awareness and innovative thinking

Future accounting talents need to possess good business awareness and innovative thinking abilities. They need to understand the strategic goals of the enterprise, be able to extract value from financial data, and provide decision support. Therefore, cultivating students' business awareness and innovative thinking should become an important part of accounting talent cultivation.

Business awareness: Accounting professionals need to have a deep understanding of the company's business model, market competition environment, and industry trends. They should be able to identify business opportunities and risks from financial data, and provide corresponding advice and decision support for the enterprise. Having a keen business awareness can help accounting professionals better understand the meaning behind financial data and how to use financial information to drive the development of enterprises.

Innovative thinking: Traditional accounting work focuses on accuracy and compliance, but with the development of digitization and automation, many repetitive accounting work will be replaced by automation. Therefore, accounting talents need to have innovative thinking, be able to find new perspectives to handle financial problems, and propose improvement and optimization plans. They need to think about how to use new technologies, tools, and methods to improve the efficiency and quality of accounting work.

4.4 Cross disciplinary integration

Accounting is no longer an isolated field, but closely related to other disciplines. In the future, accounting talents need to integrate interdisciplinary knowledge with fields such as finance, economics, and management to cope with the complex and ever-changing business environment. Therefore, when cultivating accounting talents, cross disciplinary training with other disciplines should be strengthened to improve students' comprehensive qualities.

Data Science and Statistics: Accounting professionals need to possess basic knowledge in data science and statistics, and be able to process and analyze large-scale financial data. They should understand the methods of data collection, cleaning, and analysis, and be able to use statistical tools and models to extract valuable information to support decision-making and prediction[10].

Information technology and computer science: Modern accounting cannot do without the support of information technology and computer science. Accounting professionals need to understand the principles and applications of information systems, and be familiar with knowledge in database management, software development, and network security. They should be able to use and maintain financial software and tools, and use computer technology to improve the efficiency and accuracy of financial data processing.

Economics and Finance: Accounting is an important component of the economic and financial fields, and accounting professionals need to possess a certain level of knowledge in economics and finance. They should understand the basic theories and principles of economy and finance, and be able to analyze the impact of economic environment and financial markets on the financial condition of enterprises.

4.5 Practical ability cultivation

Traditional theoretical courses can no longer meet the needs of enterprises for accounting talents. Future accounting talents need to possess solid practical skills and be able to handle complex accounting problems in practical work[11]. Therefore, cultivating students' practical abilities should become an important goal of accounting education, and improving students' practical operation abilities through internships, practical training, and other means.

With the changes in the business environment and the development of technology, pure theoretical knowledge is no longer sufficient to cope with complex accounting work. The following are some important aspects of cultivating practical abilities:

Internship experience: Providing internship opportunities for students is an important way to cultivate practical abilities. By participating in accounting work in a real work environment, students can apply their knowledge to practical situations, understand practical problems and challenges, and cultivate problem-solving abilities.

Case analysis: Through case analysis, students are exposed to real financial data and corporate situations, cultivating their analytical and judgmental abilities. Students can analyze cases, extract core issues, develop solutions, and evaluate results.

Teamwork: Accounting work usually requires collaboration with other departments and teams. Cultivate students' teamwork skills, enabling them to learn to communicate, coordinate, and collaborate with others, and adapt to diverse work environments.

Simulation Practice: Utilize simulation practice to enable students to engage in accounting practice in a virtual environment. Accounting software and simulation systems can be used to enable students to simulate business processing, prepare financial statements, and other tasks, improving their practical operational abilities.

In summary, the future high-quality employment oriented integration of industry and finance accounting talent cultivation will focus on diversified skill cultivation, digital transformation, business awareness and innovative thinking, interdisciplinary integration, and practical ability cultivation. These trends will enable accounting professionals to better adapt to the rapidly changing business environment and provide high-quality financial management and decision-making support for enterprises.

5A path for cultivating high-quality employment-oriented accounting talents through the integration of industry and finance

5.1Strengthen the digital infrastructure of universities and provide a business finance integration environment

In the context of the construction of the new business science, the concept of "cultivating accounting talents required by enterprises" has become the main direction of training program reform. The more enterprises develop, the more important accounting becomes. With rapid economic development, accounting talent training programs should be updated and followed up in a timely manner. In order to help accounting students strengthen their financial skills and theoretical foundation, strengthen infrastructure construction, improve intelligent teaching methods, accelerate the construction of intelligent platforms such as financial sharing, update financial information systems, and achieve the integration of industry and finance. Through on-site training and off campus internships, students' understanding of accounting positions can be improved. Only by applying rich theoretical knowledge to practice can the high-quality goal of accounting talent cultivation be reflected. Accounting teachers need to constantly lead students to explore accounting principles, combine practice with multi-dimensional teaching, explore relevant financial knowledge in the context of industry finance integration, and help students strengthen their own skills learning. In addition, the construction of the new business course breaks through the barriers of traditional knowledge and focuses on cultivating students' diverse qualities. Guided by the construction of the new business course, the current training plan for financial accounting professionals has been revised. On the basis of mastering high-level accounting knowledge and combining industry characteristics, it achieves the integration of multidisciplinary knowledge, so that students can have a rich theoretical foundation and practical work experience after truly entering the job position.

5.2Seeking cooperation with digital enterprises to achieve deep integration of industry and education

Cultivating accounting talents is no longer just a task for universities. In order to better cultivate the talents that enterprises truly need, enterprises have also become an important way to cultivate accounting talents. Only through deep cooperation between universities and enterprises can we cultivate professional talents that meet social needs. In the digital era, the main direction for cultivating digital and intelligent accounting talents is to seek cooperation from accounting firms, financial software companies, commercial banks, and listed companies to establish an industry education integration base, and cultivate accounting professionals with high-quality employment as the guidance. The deep integration of industry and education has also become an important guiding basis for business integration. Accounting talents cultivated by universities need to have knowledge and skills in accounting, as well as the ability to analyze data in enterprise business and the comprehensive ability to integrate industry and finance. Therefore, the joint training of enterprises and universities is particularly important. Through students' feedback in enterprise practice, universities take corresponding measures and collaborate to cultivate high-quality accounting talents.

5.3Building a "dual teacher" teaching team to achieve interdisciplinary integration

The most basic requirement for teachers' ability and quality in the high-quality training process of accounting talents is that 'forging iron requires one's own hardness'. Accounting teachers should first have a more solid professional quality and richer practical experience. New business education needs to continuously stimulate students' creative thinking, so students' thinking in the classroom is more active, the boundaries of the curriculum are also wider, and the classroom content is no longer limited to books. Teachers need to have a broad professional perspective and deep professional knowledge to flexibly respond to professional problems that may be involved in the classroom. At the same time, in order to achieve the integration of accounting talent cultivation and employment, and to implement the quality and effectiveness of talent cultivation in undergraduate colleges, teachers must have a certain level of professional competence to integrate more practical links into the classroom. Teachers are not only the imparter of theoretical knowledge, but also the guide of professional practice. Universities should encourage teachers to participate in professional skill training and exams, and cultivate a "dual teacher" teaching team; And create opportunities for teachers to engage in practical work, promote teachers to go out of school and deep into the enterprise, experience the similarities and differences between financial theory and practical work firsthand, and better achieve the complementarity of theoretical teaching and practical teaching in teaching. On this basis, accounting teachers also need to have a certain level of informatization. In the era of big data and intelligence, both teaching content and methods are closely integrated with modern information technology. Teachers should be adept at using advanced teaching methods in the teaching process, combining traditional teaching content with modern information technology, and cultivating students' innovative thinking by using new technologies to solve old problems. In promoting interdisciplinary integration, it is necessary for teachers to step out of their professional comfort zone, actively engage in related disciplines outside of the accounting profession, actively participate in interdisciplinary research activities, broaden academic horizons, think about the intersection and correlation between disciplines, form a closed loop for the cultivation of composite accounting talents, and improve the quality of accounting talent cultivation.

6Conclusions

The cultivation of accounting talents is moving towards a higher level, and cannot only focus on mastering basic business knowledge proficiently, otherwise it will be replaced by artificial intelligence. Higher education institutions should deeply reflect on cultivating accounting talents guided by high-quality employment, and cultivate accounting talents with practical application abilities such as understanding accounting, management, and operation. To cultivate high-quality accounting talents for society, it is necessary to possess advanced financial accounting thinking, highly integrate finance and business, integrate theory and practice, and effectively use technological means to solve practical problems.

Project source

2023 Key Project of Education Science Planning in Heilongjiang Province: "Innovative Research on New Business Talent Training Model Guided by High Quality Employment" (No.GJB1423319);The 2021 Provincial Higher Education Teaching Reform General Project Approval Project "Research and Practice on Integrating Ideological and Political Elements into the Accounting Course under the Background of New Liberal Arts" (No.SJGY20210369);2023 Provincial Undergraduate University Basic Research Business Fee Project(No.2023-KYYWF-004);2023 the key project of provincial educational science planning, "Practical Research on Teaching Reform of Intelligent Finance Courses in Application-oriented Undergraduate Colleges Based on CDIO Concept", (No. GJB1423316);2020 College level Education and Teaching Reform Project "Research on the Practice of Teaching Reform and Innovation of Adaptive Education for Accounting Specialty under the Background of Artificial Intelligence " (No.JGB2020010);2023 Heilongjiang Provincial Universities Basic Scientific Research Business Expenses Scientific Research Project(No.2023-KYYWF-E009)

References

1. Wang Yujuan. Research on the Training Path of Higher Vocational Finance and Accounting Talents Based on the Integration of Industry and Finance [J]. Finance and Accounting Learning, 2022 (27): 140-142
2. Yu Meng Research on the Practical Teaching of Financial Management Professional Industry's Integration of Finance and Taxation under the Background of "Great Intelligence Moving to the Cloud" [J]. Accountant, 2021 (14): 74-75
3. Yu Shufan. Exploration of the Training Path for Higher Vocational Finance and Accounting Talents under the Background of Integration of Industry and Finance - Taking ERP Enterprise Sand Table Simulation Course as an Example [J]. Finance and Accounting Learning, 2021 (20): 155-159
4. Lv Le, He Chengli, Pan Yao Innovation and Practice Research on the Training System of Financial and Accounting Talents under the Background of "New Business" [J]. Modern Commerce and Industry, 2021, 42 (02): 77-78
5. Zhang Xiaocai. Construction of an applied finance and accounting talent training model guided by the integration of industry and finance [J]. Accountant, 2020 (23): 63-64
6. Fan H, Lu P, Li T. Reform and Challenge of Accounting Education in the Intelligent Era[J]. Advances in Educational Technology and Psychology, 2023, 7(1): 34-38.
7. Jamil N N, Seman J A. The impact of fintech on the sustainability of Islamic accounting and finance education in Malaysia[J]. Journal of Islamic, Social, Economics and Development, 2019.
8. Cohen J R, Dalton D W, Holder-Webb L L, et al. An analysis of glass ceiling perceptions in the accounting profession[J]. Journal of Business Ethics, 2020, 164: 17-38.
9. Rajgopal S. Integrating practice into accounting research[J]. Management Science, 2021, 67(9): 5430-5454.
10. Albert E T. AI in talent acquisition: a review of AI-applications used in recruitment and selection[J]. Strategic HR Review, 2019, 18(5): 215-221.
11. Jędrzejka D. Robotic process automation and its impact on accounting[J]. Zeszyty Teoretyczne Rachunkowości, 2019 (105): 137-166.

Mechanisms of Pet Engagement in the Formation and Strengthening of Urban Social Support Networks: A Sociological Investigation

Wei-tong Wang*

Shaanxi Normal Univercity School of Philosophy, Xi'an, Shaanxi, China.

Abstract: In today's urban landscape, an increasing number of residents are embracing pet ownership. The social support networks established through pets can effectively address issues of isolation and social disconnection within corresponding communities, bolstering individuals' sense of identity and belonging, and contributing to the strengthening of social support networks. This paper employs field research methods, conducting interviews with pet owners in a community in Chang'an District, Xi'an City, Shaanxi Province, as well as on various social media platforms. The aim is to explore the mechanisms through which pets function within urban social support networ

1. Introduction

Pets, as companions to humans, have a long history. With economic growth and improved living standards, the variety of pets has significantly increased. Statistics show that, in just four years until 2002, the number of pets in the United States increased by a staggering 366 million[1]. In 2004, there were 417,000 registered pet dogs in Beijing, a number that soared to over 1 million in 2006. The surge in pet ownership has made activities like walking dogs or cats a common sight in the corners of major cities, and interactions facilitated by pets as social mediators are becoming increasingly prevalent.

Pets are often seen as a direct source of companionship and support for their owners. Pet ownership has been shown to have significant positive effects on human health, including reducing the risk of heart disease, lowering blood pressure, increasing physical activity, providing emotional support, enhancing overall well-being, and strengthening psychological resilience in the face of adversity. Moreover, in the relatively impersonal realm of urban living, pets, acting as benign social catalysts, significantly facilitate human social interactions. Numerous studies have indicated that dog walkers, for instance, have more opportunities for social contact than those who walk alone, and pets can serve as conversation starters among strangers in environments like parks. So, how do pets, as benign catalysts, influence interpersonal interactions and the establishment and promotion of social support networks?

Therefore, this paper aims to describe and explore the current state of benign interactions between people mediated by pets and how pets, as social catalysts, can be used to build and strengthen social support networks, providing practical support for promoting the smooth functioning of society.

2. Literature Review

2.1. Research on Interpersonal Communication and Social Support Networks

2.1.1 Research Related to Theories of Interpersonal Communication

Social exchange theory, pioneered by Peter Blau, Emerson, and Homans, applies economic concepts to sociology. This theory focuses on mutual exchanges and reciprocal relationships in people's social lives. Peter Blau's social exchange theory asserts that the fundamental characteristic of interpersonal interactions is reciprocity. Human interaction is a process of satisfying self-needs within the framework of interdependence with others. Starting from the exchange theory of communication activities, one can fully explain the motivations behind interpersonal interactions and the phenomenon of interpersonal attraction.

Urban interaction theory originated in communication studies but can be traced back to the Chicago School of Sociology, represented by Georg Simmel. Simmel believed that urban interactions were filled with rational emotions, and complete societies could only be formed when media affected the relationships between individuals. Robert E. Park, on the other hand, believed that city dwellers lacked traditional emotions, and individuals had no opportunity to understand their positions within the entire city. This led to personal chaos and spiritual collapse. Therefore, Park pessimistically summarized rational interactions as interest-based interactions.

*1479156429@qq.com

Social support refers to the provision of material and spiritual assistance to vulnerable groups by social support networks without compensation. Social support theory generally believes that the larger an individual's social support network, the better they can cope with challenges posed by various pressures.

2.1.2 Research on Social Support Networks

Social support networks are a branch of social support theory. Scholars such as Li Qiang[7][9] and He Zhaoping[10] argue that social support should be defined as various forms of unpaid assistance and services provided to vulnerable groups by various social forms or the positive effects obtained by individuals through social connections, which can alleviate psychological stress, relieve tension, and enhance social adaptability, among other benefits. Social support includes both formal and informal support, often emphasizing reciprocity in social exchanges. Social support networks refer to the social networks through which individuals can obtain various forms of support (such as financial, emotional, and friendship support), which helps reduce dissatisfaction with social inequality, alleviate social conflicts, enhance individual risk resistance, and contribute to maintaining social stability. Current research on social support networks focuses on the flow and backflow of resources within these networks[5], the role of resources in receiving social support, and how the overall structure of social support networks influences social support, as well as the measurement of social support networks.

2.2. Research on Pets as Social Mediators

Studying benign interactions between people and the establishment of social support networks through pets holds significant research value. McConnell AR[2]suggests that the social support provided by pets complements the social support offered by human resources, as pet ownership offers robust social support for meeting social needs[3] . Wood L[4] has more comprehensive conclusions in this research area, considering pets as valuable channels for understanding others, tangible forms of social support, powerful protective factors for community friendships and personal well-being, and neutral, safe opportunities for dialogue in social interactions. Pets can effectively address issues of isolation and social disconnection, thus contributing to the strengthening of social support networks[6].

2.3. Literature Review Summary

Foreign scholars have made significant achievements in humanities and social science theories related to social interaction and interpersonal communication, forming relevant theories on interpersonal interaction and social support networks, providing a solid theoretical foundation for this paper. Domestic scholars have also conducted research on interpersonal interaction patterns, interaction patterns, and communication behaviors from various perspectives. However, most studies view interpersonal interactions as negative, suggesting that urban interactions are gradually becoming less warm. Moreover, most research in China focuses on social support from a theoretical perspective, with limited exploration from other theoretical angles on how to enhance the construction of social support networks.

In the field of research on pets as social mediators, foreign research on their relationship with sociology has progressed more rapidly than in China. Current research in China primarily focuses on epidemiology and medical aspects of pets, with more literature reviews and commentary articles than academic research. Sociological studies on the role of pets remain relatively scarce[8].

3. Research Content and Methods

This paper primarily focuses on a community in the southern suburbs of ChangAn District, Xi'an City, Shaanxi Province, and online pet exchange platforms. It examines pet owners as the primary subjects of study, with a focus on their current interpersonal interactions and social support status. The goal is to explore the social support networks formed through pet-mediated interpersonal interactions and their mechanisms.

This research predominantly employs interview methods and participant observation, supplemented by literature description methods, to analyze and summarize the interview results theoretically.

4.Conducting the Investigation

4.1. Selection of Survey Subjects

The main research location for this paper is a community in ChangAn District, Xi'an City, Shaanxi Province. The community has a complex composition of residents (with no obvious group characteristics), featuring a variety of professions, age groups, and social classes. Upon observation, it was noted that there are many pet owners in the community, with residents forming their own pet groups. The community has designated times for walking dogs and cats, which are from 6:00 AM to 9:00 AM and 8:00 PM to 10:00 PM, respectively. During these times, pet owners commonly gather for outings, primarily along the community's main roads and lawn areas. In addition to offline visits, online interviews were conducted with super-topic bloggers on pet-related topics. The content of these interviews was consistent with the offline interviews. (Super-topics are unofficial online organizations formed by netizens with similar interests, and people of different genders, ages, and professions can voluntarily join, breaking geographical limitations and expanding the scope of research.)

4.2. Commencement of the Investigation Phase

Due to the complex classification of research subjects in this paper, a flexible approach involving participant observation and interviews was adopted. For the first category of interviewees, detailed observations were made during the dog-walking hours in the community from August 2nd to August 5th, 2022. This identified the primary spaces where interviewees frequently gathered for pet dog activities. Formal interviews were then conducted from August 6th to August 10th, 2022, with a total of 10 offline interviews and 7 online interviews using the community's WeChat pet group.

For the second category of interviewees, while conducting offline research, investigators also read posts and asked questions on various social media platforms. After obtaining consent, online interviews were conducted with pet bloggers through private messages, totaling 13 online interviews.

5. Mechanisms by Which Pets Contribute to the Formation of Urban Social Support Networks

Social support encompasses formal and informal assistance and help. It is not merely a one-way form of care or assistance; in most cases, it involves social exchange. Social support networks, as a beneficial supplement to the social security system, help alleviate people's dissatisfaction with social injustices, mitigate various social conflicts, enhance individual risk resilience, and contribute to maintaining social stability. Pet owners have a broad range of social contacts, transcending occupational, blood, and geographical boundaries, forming a new type of interest-based network under the emerging situation. Among the 30 interviewees, 25 mentioned that they would strike up conversations with strangers because of their pets. Nearly 90% of the interviewees reported that they were not familiar with other community residents when they first moved into the community. However, they got to know many people through their pets, which facilitated the formation of friendships on both sides. Pets, as bonds, no longer exist only in face-to-face communication. Pet enthusiasts establish clubs, online gathering places, and organize various activities such as offline meetings. The similarity in pet interests attracts people to each other, and interpersonal interactions facilitated by pets involve the exchange of emotions, evaluations, information, and tangible support.

Based on literature reading and data research, this paper categorizes social support into four types:

Informational Support: Referring to beneficial or helpful information that individuals can obtain during interactions, such as location or product information.

Tangible Support: Involving practical actions or physical items of assistance, like purchasing items or providing pet care.

Emotional Support: Meeting emotional needs of individuals during interactions, such as sharing feelings, venting, and providing a listening ear.

Appraisal Support: A more abstract form of support than the previous three, which involves individuals feeling accepted, acknowledged, or appreciated during interactions, satisfying their self-esteem.

5.1. Informational Support Provided by Pets to Social Support Networks

In this survey, 50% of the interviewees mentioned that they had received or provided informational support during their interactions with other pet owners. Informational support, being the shallowest form of support, is relatively common in social interactions. For example, the owner of a pet dog named Cindy, interviewed offline, said, "When we greet a dog we don't know for the first time, besides exchanging the dog's name, we also ask where they usually play in the community. If the conversation goes smoothly, we might inquire about what snacks their small dog usually eats at home and where they buy them." Similar cases also exist in the communication between various pet bloggers on online platforms. In the pet forums observed in this paper, there is a dedicated section in the forum's sticky posts where pet owners from all over the country can share their pet-raising experiences. These posts include recommendations for good products, common diseases in pets, exotic pet hospitals in various provinces and cities, and honorary doctors, all of which constitute informational support provided by pets as connecting bonds.

5.2. Tangible Support Provided by Pets to Social Support Networks

Sociologist Lin Nan pointed out that in the operation of social support networks, people gain additional resources by using resources from direct and indirect relationships. Pet owners get to know many other pet owners through communities, networks, and other means, and they exchange resources through social interactions, thereby gaining support. Here, resources specifically refer to tangible, not spiritual or advisory, resources. Tangible support goes beyond informational support in terms of depth. For instance, the owner of the cat Tim stated, "Actually, when you get to know each other well, there are many places to interact. Adding each other on WeChat is very convenient. Sometimes, if someone's cat needs to buy cat supplies, a few of us will place an order together. When I go on a business trip, I'll directly leave my cat at their place. Sometimes, on weekends, we'll drive into the mountains together. We don't bring our cats because it's not very convenient." Cases like Tim's mom, who became close friends with someone because of their cats, are numerous. This shows that pets serve as positive bonds in interpersonal relationships, and the pet enthusiast community, which initially served as an entertainment group for the masses, is also constantly evolving.

5.3. Emotional Support Provided by Pets to Social Support Networks

Emotional support mainly involves understanding, confiding, and empathizing with each other during interactions. People often seek emotional support more than instrumental or tangible support, and pet bonds play a significant role in providing emotional support. During the survey, it was observed that, whether offline or online, when the interaction relationship reached a certain level of familiarity, if one party needed emotional support or advice, they would seek emotional support from others, and the other party would provide emotional support in return, such as being a good listener. The owner of Doggy said, "I remember that day when our Doggy couldn't make it through due to canine distemper and passed away at the hospital. At that time, my world really darkened. Once, I ran into the parent of Snow (a dog) at the supermarket, and to console me, they brought Snow to play with me and would often call me to chat and help ease my mood. Everyone in the group found out and would frequently send messages to comfort me. It was through this that I gradually emerged from the shadow of losing psychological dependence. I'm really thankful to those who provided me with emotional support."

Interactions between individuals in urban settings often take on a unique form of non-personal, emotionally uncharged interactions. However, within the emotional support provided by pets, we frequently observe that the communication between pet owners extends beyond mere discussions about pets. While offering emotional value, it helps individuals grappling with emotional turmoil to overcome their struggles, enhances mutual fondness, and fosters deeper friendships. This transformation allows this relationship, from being strangers to becoming familiar, to evolve into more profound and extensive dimensions.

5.4. Appraisal Support Provided by Pets to Social Support Networks

Appraisal support is the psychological journey that individuals seeking to integrate into a collective group undergo, feeling accepted and acknowledged by that group. It is only when a medium affects the mutual influence between people that a complete society is formed, thereby creating a complete social support network. The larger an individual's social support network is, the better they can cope with the challenges posed by various pressures. Therefore, as a crucial component in constructing social support networks, appraisal support allows inter-actors to feel affirmed and appreciated by other community members, leading to satisfaction in terms of self-esteem and promoting further positive integration.

In the comments section of the social media topic #Golden-bear#, investigators came across a post: "Today, I spent half a month's living expenses on buying a new habitat for my pet, and I spent an afternoon decorating it. But when I saw Bear (the pet) happily playing around in the sawdust, I instantly felt that it was all worth it." The post was accompanied by a heartwarming photo of the pet's new den. In the comments on Weibo, netizens responded with comments like "Wow! It's so beautiful; can you send me the link?" and "You've really transformed from a beginner to an expert!" The original poster then replied to all the comments, saying, "Thank you all for your praise. Friends who need links to good products can just message me directly!" This Weibo user often shares various experience posts in the topic, engaging in friendly conversations with other members.

6. Conclusion

Based on the interview results of this study, almost all interviewees indicated that they had formed some level of friendship through their pets. Pets serve as a beneficial channel for understanding others and are a tangible form of social support. As the bonds created by pets deepen, the levels of social support provided by both parties in their interactions continue to increase. Pets, as important sources of social support networks, can transcend blood, geographical, and occupational boundaries, forming subtle relationships that bridge different ages and personalities. This makes the bond-based pet social support network more solid than other types of support networks. The social support network established through pets effectively addresses issues of isolation and social disconnection within specific segments, enhances individuals' sense of identity and belonging to society, and contributes to the further strengthening of social support networks. A well-established social support network serves as a strong protective factor for community friendships and individual life satisfaction, while also providing a neutral and safe platform for dialogue in social interactions.

References

1. Zhang Di. (2008). Research on the Influence of Raising Pet Dogs on Urban Interpersonal Communication.Harbin Institute of Technology. DOI:CNKI:CDMD:2.2009.292020.
2. McConnell AR, Brown CM, Shoda TM, Stayton LE, Martin CE. (2011). Friends with benefits: on the positive consequences of pet ownership. Journal of Personality and Social Psychology, 101(6), 1239-1252. DOI:10.1037/a0024506. Epub 2011 Jul 4. PMID: 21728449
3. Aragunde-Kohl U, Gómez-Galán J, Lázaro-Pérez C, Martínez-López JÁ. (2020)Interaction and Emotional Connection with Pets: A Descriptive Analysis from Puerto Rico. Animals (Basel), 10(11), 2136.DOI:10.3390/ani10112136.PMID:33212986;PMCID: PMC7698559.
4. Wood L, Giles-Corti B, Bulsara M.(2005) The pet connection: pets as a conduit for social capital? Soc SciMed.61(6):1159-73. DOI: 10.1016/j.socscimed.2005.01.017. Epub 2005 Mar 3. PMID: 15970228.
5. Cai He, Ye Baoqiang, Kuang Ziwen, Zhuo Huixing. (1996) Comparative Study on Social Relationship Intention Between Urban Residents and Suburban Rural Residents Seeking Social Support. Sociological Research, 10-17. DOI:10.16059/j.cnki.cn43-1008/c.2001.04.011.
6. Bao Gongmin, Wang Jie. (2007) A Review of Research on Interpersonal Conflict Theory.

Technology Economy, 12-16. DOI:10.3969/j.issn.1002-980X.2007.11.003.

7. Li Qiang. (2009) A Study of Interpersonal Communication Patterns of College Students from the Perspective of Social Exchange Theory. Law andSociety,225. DOI:10.19387/j.cnki.1009-0592.2009.09.149.
8. Ding Peng, Wang Jing. (2012)New Characteristics of Interpersonal Communication from the Perspective of Social Exchange Theory. Science, Technology and Education Times (Mid-month Issue),159-160. DOI:10.3969/j.issn.1672-7894.2012.07.092.
9. Li Qiang. (1998)Social Support and Individual Psychological Health. Tianjin Social Sciences, 1998(01),66-69. DOI:10.16240/j.cnki.1002-3976.1998.01.013.
10. He Zhaiping. (2002)Socioeconomic Status, Social Support Networks, and the Physical and Mental Health of Rural Elderly People." Chinese Social Sciences,135-148+207. DOI:CNKI:SUN:ZSHK.0.2002-03-011.

Analysis of the breakthrough of gender equality advertisement on gender stereotypes – Take PROYA's 2021 advertisement film "It's gender,not border" as an example

Li Yue*

Wuhan institute of technology, Wuhan, Hubei, China

Abstract. This paper adopts text analysis to interpret gender equality advertisements, and selects PROYA 2021 "It's gender, not border" advertisement film as a research case to analyze its breakthrough in gender stereotypes. From the perspective of advertisement creation, the breakthrough of gender stereotypes in this advertisement film can be clearly divided into three stages including insight, questioning and denial, which presents, criticizes and breakthroughs the gender stereotypes layer by layer, which provides the significance of creating advertisements from women's perspectives on the practical level.

1 Introduction

In recent years, women's marketing has become a hot trend. In the advertising industry, many brands have created advertisements from women's perspectives to convey brand attitudes and propositions, such as PROYA's "It's gender, not border", NEIWAI's "No body is nobody", YOUDAO's "Interests are not about gender", and Durex's "Celebration starts from this moment", and so on. The creation and communication of these advertisements break the traditional gender stereotypes to a certain extent and convey the concept of gender equality, which can be defined as gender equality advertisements. The popularity of gender equality advertisements provides a model for the creation of advertisements from a female perspective, as well as case support for the study of gender roles in advertisements.

2 Research question

Gender roles are a set of behavioral norms corresponding to one's gender that an individual acquires through imitative learning during the socialization process[1]. Gender stereotypes refer to the specific expectations, norms, and evaluations that society places on individuals based on their gender, limiting their ability to develop diverse personalities, behaviors, and roles[2]. Gender stereotypes are often manifested in the social division of labor between men and women, the power relationship between men and women, the character traits of men who are strong and women who are soft, the abilities and intelligence of men who are superior and women who are inferior, the consumer relationship between men who earn and women who spend, and the emotional traits of men who are sensible and women who are romantic or fragile, and so on.

Numerous studies have shown that gender stereotypes are reflected in advertisements in the form of stereotyped images of women and men. Women in early advertisements were fixed as gentle and virtuous housewives in the family, male supporters in society, beautiful vases or sex objects, while men were fixed as mature and stable successful men, breadwinners, drinkers and sex addicts, which is the reproduction and reinforcement of gender stereotypes in advertisements, reflecting the concept of gender discrimination. With the awakening of women's consciousness and the improvement of their economic status, gender stereotypes in advertisements have been transmuted, showing the development of gender diversification, which pushes scholars to re-examine gender roles in advertisements out of the traditional critical perspective. Han S and Han Y[3] found that there has been a significant weakening of gender stereotypes in TV advertisements over the past decade of the market economy in China, and the gender roles of both sexes have begun to change, in which the principle of consumerism and the masking of gender identities by the new middle class cannot be denied. Ding L[4] studied the performance, reasons and positive significance of the softening of the male image in TV advertisements, and at the same time pointed out that the excessive softening of the male image should be prevented. The studies of Xu J[5] and Liao J[6] showed that the image of women in China's TV advertisements is being redefined, and the traditional stereotypes of women are beginning to be broken through and disintegrated, and the image of women is becoming more diversified.

Most of the above studies discuss the performance and reasons for the changes produced by the traditional

* Corresponding author: wit_ly296@163.com

gender stereotypes in advertisements, and few scholars explore how to break through the gender stereotypes from the perspective of advertisement creation. Meanwhile, the gender equality advertisements of PROYA, NEIWAI and other brands out of the loop in feminist marketing provide case support for exploring the breakthrough of gender stereotypes in advertisements.

The above clever encounter between academia and industry finds an entry point for this study, which aims to interpret the gender equality advertisement and analyze its breakthrough of gender stereotypes. Specifically, the following questions are examined: What traditional gender stereotypes does the advertisement break through? How is it a breakthrough? What is the inspiration for contemporary advertisement creation from a female perspective?

3 Research method

Text analysis is a research method used by researchers to describe and interpret media messages, focusing on describing the content, structure, and function of the text, explaining the deep underlying meaning, and applying it to the deep reading of case texts thus unearthing the theoretical meaning[7]. The text in this method includes all meaningful symbols that can be recorded, such as words, drawings, videos, and so on. This paper adopts text analysis to interpret gender stereotypes in gender equality advertisements and analyze how they break through traditional gender stereotypes.

This paper selects PROYA's "It's Gender, Not Border" advertisement which was released on Women's Day 2021 as the subject for analysis. There are two reasons why it has been selected. Firstly, unlike many advertisements that are created solely from a female perspective, this advertisement also considers the male perspective, which is typical and inspiring in the creation of gender equality advertisements. Secondly, the advertisement became a benchmark case for communication effectiveness as it was the most forwarded and discussed brand campaign on Women's Day that year, simultaneously sweeping the public welfare, academic, and advertisement circles[8]. Both of these reasons justify the choice of the "It's gender, not border" advertisement film as a case study of gender equality advertisements.

4 Research finding

The breakthrough of gender stereotypes in this advertisement film can be clearly divided into three stages including insight, questioning and denial, which presents, criticizes and breakthroughs the gender stereotypes layer by layer.

4.1 Insight into gender stereotypes goes beyond women

The advertisement is introduced with the question "Under what circumstances would you evaluate a thing by gender", which quickly attracts the audience's attention and triggers them to think about gender stereotypes simultaneously. Afterward, the main characters use several "we say" to state a series of objective gender stereotypes in life and adopt the first and third-person perspectives to present the corresponding images alternately.

Analysis shows that the advertisement has gained insight into gender stereotypes at five levels: ability and intelligence, personality traits, family roles, career choices, and hobbies. In terms of ability and intelligence, "traffic accidents", "female drivers" and "girls can't learn science well" reflect the gender stereotypes of women's poor driving skills and weak logic and abstract thinking; In terms of personality traits, "bold" character is considered "not like a girl", "quiet" and "crying" are considered "not manly", reflecting the gender stereotypes that women are quiet and fragile while men are brave and strong; In terms of family roles, "good women" are considered to "balance family and career" and "emphasize family", and do not need to "fight for a career" and "read so many books", while "real men" are considered to be "successful in their career" and should not "squat at home" or "act like a woman", reflecting the gender stereotypes of men's roles outside the home and women's roles inside the home. In terms of career choices, women are considered to be "not good at science" and men are considered not to "study nursing", reflecting gender stereotypes of women engaging in careers such as nursing, education, or social work and men engaging in careers such as technology, business or manufacturing; In terms of hobbies, women who "practice boxing" are considered "unmarriageable" and men are considered not to "dance ballet" or "like pink". This reflects the gender stereotypes that women love art, cultural activities, and pink, while men love powerful and intense activities. These are summarized and summarized in Table 1.

Table 1. Gender stereotypes at five levels.

gender stereotypes	women	men
ability and intelligence	poor driving skills, weak logic, and abstract thinking	better than women
personality traits	quiet and fragile	brave and strong
family roles	inside the home	outside the home
career choices	nursing, education, or social work	technology, business, or manufacturing
hobbies	art, cultural activities, and pink	powerful and intense activities

The first stage of the advertisement presents the above gender stereotypes to trigger the audience to think about the fact that gender bias exists for both women and men, limiting the development of both genders in the same way.

4.2 Questioning about gender stereotypes through the vocalization of gender groups

Based on the insight of gender stereotypes, the advertisement shifts from objective statements to tit-for-tat questions and rhetorical questions, ending the one-sided output of the main character. Next, as shown in Fig 1, it uses the expression of group images and voices, focusing on the close-ups of different people's mouths and frequently using question words such as "why" and "could it be said that " to further guide the audience to think about the logical confusion and irrationality behind gender stereotypes. "Why is it necessary to evaluate by gender?" , "Why is it required for a powerful woman to act like a man?", and "Why don't men have the right to be vulnerable?". The joint voices of women and men are layered with the background music that gradually increases in tempo, causing the audience's emotions to build up and pushing the advertisement to a climax.

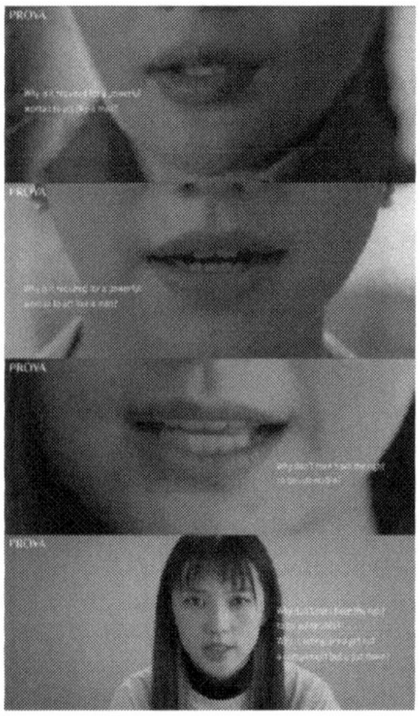

Fig. 1. Advertisements with group images and voices.

The second phase of the advertisement adopts the expression of group images and group voices to ask questions about gender stereotypes, to trigger the audience's resonance, and building up the emotion of criticizing gender prejudice to the peak.

4.3 Denying gender stereotypes and calling for diversity

Different from the traditional gender images in advertisements, this film features several new images,as shown in Fig 2 and 3. The women who skydive, scuba dive, fly a flying machine, practice boxing, and study hard in science, and the men who read a book in a bedroom alone, take care of a sick person, dance ballet, and wear a pink shirt, all these new images show the courageous, independent, strong, powerful, and rational side of women, and the quiet, gentle, fragile, and sensual side of men at the same time. This is an indirect negation of the gender stereotypes and sets the stage for the direct negation that follows.

Fig. 2. Images of women in the advertisement film.

Fig. 3. Images of men in the advertisement film.

The third phase of that advertisement, first breaks and then builds. First of all, it is a direct negation of gender stereotypes in the form of self-answering questions: "Women should be gentle", "How can women balance family and career", "Men should be strong", and "Men should not cry" is "wrong". Before discussing what women or men should do, it should be recognized that both men and women are first and foremost "human beings". The traits and behaviors such as "gentle", "brave", "romantic", "calm", "strong" and "crying" are possessed by humans, rather than unique to women or men. Human beings should be equal and diverse. In the end, the advertisement once again calls for "Let's be gentle, brave, romantic and ......" in a group voice, which leads to the core idea of the advertisement, "It's gender, not border".

The third phase of this advertisement breaks away from traditional gender stereotypes through the double negation of images and language, intending to convey the idea of gender equality and calling on everyone to live diversely.

5 Conclusions

Through the above analysis, the breakthrough of gender stereotypes in the advertisement "It's gender, not border" is divided into three stages: insight, questioning and denial. Firstly, based on the insight into the gender stereotypes in life, it presents the audience with five levels of gender stereotypes: ability and intelligence, personality traits, family roles, career choices, and interests, which triggers the audience to think about them. Secondly, through the expression of the group images and voices, it asks questions about gender stereotypes and accumulates the audience's critical emotions to the peak. Finally, by presenting new images and the double negation of audiovisuals, it breaks through the traditional gender stereotypes and conveys the concept of gender equality.

The advertisement's breakthrough of gender stereotypes brings many inspirations for creating advertisements from women's perspectives. First, in the face of increasingly sensitive gender issues, advertisement creators should enhance gender awareness, maintain gender sensitivity, and be able to recognize gender inequality in gender stereotypes, which include both women and men. Secondly, they should have the courage to break through gender stereotypes and adopt expressive creative techniques to show new images of both genders, which is the way to make advertisement creativity innovative and eye-catching to the audience. Finally, the mass media is also an important source of gender stereotypes, in other words, the mass media has the function of constructing gender meanings or patterns[9]. We should take the initiative to utilize the social value of advertisements, transcending the binary position of men and women to convey the concept of gender equality, contributing to the discussion and development of the issue of gender equality, and thus promoting the formation and maintenance of harmonious gender relations.

References

1. Pleck, J. (1984). The Theory of Male Sex-Role Identity: Its Rise and Fall,1936 to the Present.In: Lewn ed. In the shadow of the Past: Psychology Portryals the sexes.NewYork. pp. 205–225.
2. Liu, B. (2015). Achieving gender equality requires breaking through traditional gender stereotypes. Chinese Women's Movement, 5, 17–18. https://doi.org/CNKI:SUN:ZFYZ.0.2015-05-011
3. Han, S., & Han, Y. (2004). An Analysis of Changes in the Gender Image of Television Advertisements over the Decade of Market Economy. Journalism & Communication, 2, 39-46+96. https://doi.org/CNKI:SUN:YANJ.0.2004-02-006
4. Ding, L. (2012). A Brief Discussion on the Softening Phenomenon of Male Image in China's TV Advertisements. Journalism and Mass Communication, 1, 65–68. https://doi.org/10.15897/j.cnki.cn51-1046/g2.2012.01.010
5. Xu, J. (2015). Stereotypes of Women in China's TV Advertisements and Breakthroughs. Youth Journalist, 2, 44–45. https://doi.org/10.15997/j.cnki.qnjz.2015.02.030
6. Liao, J. (2018). The Use and Change of Female Images in Chinese TV Advertisements. *Youth Journalist*, *14*, 97–98. https://doi.org/10.15997/j.cnki.qnjz.2015.02.030
7. Chen, Y. (2007). *An Introduction to Communication Research Methods*. China Renmin University Press, Beijing.
8. PROYA. (2021). *What did PROYA do right on International Women's Day?* Adquan.Com. https://creative.adquan.com/show/302526
9. DeFleur, M., & Ball-Rokeach, S. (1989). *Theories of Mass Communication*. Xinhua Publishing House, Beijing.

Rethinking Ethos: A Comparative Analysis of Persuasive Character Building in Classical Antiquity through the Lenses of Aristotle, Cicero, and Quintilian

Ganlin Zhuang*

Xianda College of Economics and Humanities, Shanghai International Studies University, Shanghai, China

Abstract. This text aims to reexamine the role of ethos in persuasion through a comparative analysis of the viewpoints of the great classical rhetoricians Aristotle, Cicero, and Quintilian. By comparing these viewpoints, it seeks to demonstrate their similarities and differences. The study reveals that Aristotle emphasizes the moral aspect of character, Cicero focuses on the artistry of rhetorical techniques, and Quintilian places greater importance on adapting to the practical context of the audience. The perspectives of Aristotle, Cicero, and Quintilian provide us with profound insights that contribute to a reconsideration of the role of ethos in persuasion.

1. Introduction

The realm of ethical persuasion, rooted in the art of rhetoric, has held a central place in human communication for millennia. Among the luminaries who shaped and expounded upon the concept of ethos, or personal appeal, during the ancient times were philosophers and orators from Greece and Rome. Aristotle, Cicero, and Quintilian, whose profound contributions continue to influence modern rhetoric and communication theory, explored the multifaceted dimensions of ethos and its pivotal role in effective persuasion.

In this article, we delve into the rich philosophical and rhetorical heritage of these three distinguished figures, each hailing from distinct periods within the ancient world. We embark on a journey through time and thought to examine their unique perspectives on ethos and its implications for the art of persuasion. By shedding light on their profound insights, we aim to gain a deeper understanding of how ethos has evolved over centuries and how these enduring philosophies continue to resonate in contemporary discourse.

2. Definition of ethos

The concept of ethos, deeply rooted in the art of persuasion, has evolved over time while maintaining its central role in effective communication. To understand the essence of ethos as conceived by Aristotle, Cicero, and Quintilian, let's embark on a historical journey that uncovers the nuanced layers of this persuasive tool.

Ethos, which concerns the character and credibility of the speaker, has always been pivotal in influencing the beliefs and decisions of an audience. In ancient times, its meaning had subtle differences. For instance, in the *Iliad*, ethos referred not to character but to a habitual place of activity.[1] It was a public matter, connecting the speaker and audience, as noted by Tindale. [2]Ethos personalized statements, giving the audience a voice to connect with, not just a series of claims.

Aristotle also emphasized the importance of ethos, stating that we become virtuous by practicing virtuous actions. [3]This echoes the idea that ethos is built through actions and character.

As we delve into the philosophies of these ancient thinkers, we'll explore their unique interpretations of ethos, recognizing the subtle variations and resonances in their definitions. This exploration will help us comprehend the multifaceted nature of ethos and its enduring significance in persuasion. In the subsequent sections, we will unravel how Aristotle, Cicero, and Quintilian applied their respective ethos definitions to rhetoric, shedding light on the nuances of their persuasive techniques and their lasting impact on contemporary discourse.

3. Insights on Ethos from Aristotle

Aristotle, often regarded as the progenitor of rhetorical theory, laid the foundational definition of ethos within his seminal work, *Rhetoric*. For Aristotle, ethos extended beyond mere character; it encompassed the perception of trustworthiness, competence, and moral standing projected by an orator to their audience. It was, in essence, the bedrock upon which successful persuasion was built, as Aristotle believed that an audience was more likely to accept arguments from a speaker they found credible and virtuous.

* Corresponding author: 13611768898@163.com

3.1 The Three Modes of Persuasion

Aristotle identifies three key elements that inspire confidence in the orator's character: good sense, good moral character, and goodwill. According to Aristotle, these elements are instrumental in persuading the audience to believe in the orator's message even without requiring explicit proof. As stated in *Rhetoric*, Aristotle writes, "There are three things which inspire confidence in the orator's own character—the three, namely, that induce us to believe a thing apart from any proof of it: good sense, good moral character, and goodwill".[3]

To cultivate good sense, the orator must demonstrate a comprehensive understanding of the subject matter. Aristotle mentioned in *Rhetoric*: "The way to make ourselves thought to be sensible and morally good must be gathered from the analysis of goodness already given"[3]. Through careful analysis and well-reasoned arguments, the speaker can project wisdom and intellect, fostering credibility and trust. As Aristotle points out, discussing topics related to virtue, vice, nobility, and baseness in persuasive oratory can align arguments with the audience's ethical values, further enhancing the orator's persuasive power. Referring to Aristotle's words, "We have now to consider Virtue and Vice, the Noble and the Base, since these are the objects of praise and blame" [3], it becomes clear that understanding and discussing moral concepts are fundamental in building a persuasive case.

Moreover, Aristotle emphasizes the significance of narration in portraying character and moral purpose, influencing the audience's perception of the speaker's credibility. In *Rhetoric*, he states, "The narration should depict character; to which end you must know what makes it do so. One such thing is the indication of moral purpose; the quality of purpose indicated determines the quality of character depicted and is itself determined by the end pursued" [3]. By using narration to reveal moral intent behind actions and choices, the orator can effectively shape the audience's view of their character.

3.2 Maxims or common sayings

Aristotle further highlights the use of maxims or common sayings to persuade the audience, particularly if they align with moral goodness. Quoting his words from *Rhetoric*, "And one should even use trite and common maxims if they are applicable; for because they are common, they seem true, as though everyone agreed. if the maxims are morally good, they make the speaker seem to have a good character"[3], it becomes evident that employing morally sound maxims can contribute to enhancing the speaker's character perception.

3.3 Language and style

Language and style play pivotal roles in persuasion. The orator should adapt their speech to match the audience's character, using appropriate language that reflects emotion and character while remaining in harmony with the subject matter. Aristotle stressed that "People always think well of speeches adapted to, and reflecting, their own character"[3], highlighting the importance of tailoring the message to resonate with the audience. This approach fosters rapport and trust.

Moreover, Aristotle cautioned against artificiality in language and style, as it can diminish persuasiveness. He advised, "Authors should compose without being noticed and should seem to speak not artificially but naturally. Naturalness is persuasive, artificiality is the contrary".[3]When the audience detects insincerity or contrived language, it can lead to suspicion and reduced persuasiveness.

In line with this perspective, the aptness of language is a crucial element that influences whether people believe in the truth of a story. As cited in *Rhetoric*, "This aptness of language is one thing that makes people believe in the truth of your story: their minds draw the false conclusion that you are to be trusted from the fact that others behave as you do when things are as you describe them; and therefore they take your story to be true, whether it is so or not"[3]. In essence, a natural and sincere approach to language and style is more likely to be persuasive, as it aligns with the audience's expectations and fosters trust.

In conclusion, Aristotle's principles of persuasion, including good sense, good moral character, goodwill through effective narration, use of maxims, and appropriate language and style, all contribute to building the orator's credibility and character in persuasive discourse. By understanding and skillfully employing these elements, the orator can gain the audience's trust and belief, making the persuasive message all the more compelling.

4.Cicero's Oratory Philosophy

Cicero, the eminent Roman orator and philosopher, expanded upon Aristotle's groundwork. In *De Inventione*, Cicero delineated a triad of virtues that constituted ethos: wisdom, justice, and benevolence. He emphasized that ethos was not confined to the speaker's character alone but was intrinsically tied to the ethical virtues they embodied. Thus, an orator's ethos was not merely a matter of credibility but a reflection of their moral character and ethical principles. This part explores Cicero's insights on these concepts through direct quotations from his works, highlighting their significance in crafting persuasive discourse.

4.1 The Three Modes of Persuasion

Cicero emphasized the critical role of extensive knowledge in oratory. He stated that "a knowledge of various subjects must be grasped to avoid oratory becoming empty and ridiculous verbiage" [4]. In essence, he argued that without a broad knowledge base, oratory lacks substance and depth, appearing superficial and immature. Cicero went on to stress that true oratory excellence requires the orator to acquire knowledge across various subjects and arts. He asserted, "No one can be a complete orator without attaining knowledge of all important subjects and arts. Oratory derives its beauty and fullness from knowledge, and without it, speeches lack depth and appear almost childish" [4].

Moreover, Cicero pointed out that achieving excellence in both wisdom and eloquence demands unwavering dedication, zeal, and rigorous study. He maintained, "No one can attain high distinction in wisdom and eloquence without a great amount of zeal, industry, and dedicated study" [4].

Cicero's doctrine differs significantly from Aristotle's perspective. While Aristotle believed that ethical persuasion primarily relies on the actual content of the speech, Cicero places a significant emphasis on the speaker's character and the prior impression they create[5]. According to Cicero, achieving the goals of oratory requires a deep understanding of human character, the vast spectrum of human nature, and the underlying motivations driving individuals. In his own words, Cicero explained, "A speaker cannot hope to accomplish their objectives through words alone unless they have acquired profound insights into the intricacies of human character, the entire panorama of human nature, and the underlying motives that either propel our souls forward or hold them back"[4].

Cicero extended his perspective to propose a profound connection between eloquence and virtue, suggesting that eloquence itself represents a form of virtue[4].Cicero advocated for qualities such as kindness, loyalty, and a disposition that resonates with the audience, as these traits hold the power to garner favor and resonance. He noted, "Displaying good-nature, kindness, calmness, loyalty, and a pleasing disposition is very helpful. Qualities like uprightness, unassumingness, and avoiding haste, stubbornness, strife, or harshness are powerful in winning goodwill" [4].

Additionally, Cicero's ethos is intricately associated with style[6]. Cicero highlighted the connection between the speaker's amiability and the style of expression. He suggested that sometimes, a speaker may adopt a slightly different style to magnify the virtues of another person or downplay their own qualities. This strategic adjustment, when motivated by goodwill rather than insincerity, reflects the speaker's ability to connect with the audience through artful manipulation of their speech. In essence, Cicero's approach places greater emphasis on elements such as style[5].

In summary, Cicero's oratory philosophy integrates knowledge, character, and style into a holistic approach to persuasion. It underscores the importance of understanding human nature, adopting virtuous qualities, and skillfully employing style to connect with the audience and achieve persuasive success.

4.2 Fostering Persuasion Through Humor

Cicero's exploration of oratory extends to the artful use of humor as a persuasive tool and its intimate connection to an orator's character. He recognizes that humor can be a powerful means to engage emotions, establish rapport, and reveal aspects of the speaker's personality. Cicero's insights into the strategic deployment of humor underscore its role in fostering connection and making arguments more relatable: "what excites laughter is disappointing expectations and ridiculing other people's characters and imitating a baser person and dissembling and saying things that are rather silly and criticizing points that are foolish, and consequently a person who wants to speak humorously must be equipped with a disposition and character that is suited to artifices of this kind, so that even his expression of countenance may be adapted to each kind of variety of the ridiculous." [5]

"For my own part, I vow I am also much amused by those pettish and rather ill-tempered jests—but not when they are spoken by an ill-tempered person, for then it is not his wit but his character that we laugh." [5] Cicero's insights highlight that humor is not only a skillful technique but also a reflection of the orator's disposition and character. "More specifically,verbal jesting and irony are seen as appropriate because those forms of humorare not used primarily to entertain the audience, but to demonstrate the speaker's clevemess."[7].When humor is thoughtfully integrated into the fabric of a speech, it can effectively break down barriers, create emotional resonance, and offer glimpses into the speaker's personality. This interplay between humor and character enhances the speaker's ability to persuade, making their arguments more relatable and their presence more engaging.

4.3 Eliciting Emotions and Demonstrating Character in Oratory

Cicero's perspective on oratory places virtue and emotion at its core, highlighting their pivotal roles in effective persuasion. He eloquently states, "Who indeed does not know that the orator's virtue is pre-eminently manifested either in rousing men's hearts to anger, hatred, or indignation, or in recalling them from these same passions to mildness and mercy"[4]. This reflects Cicero's belief in the transformative power of oratory, where an orator can influence and guide the emotions of their audience.

In comparison to Aristotle's more rational concept of ethos, Cicero's ethos is broader and encompasses a range of emotions, particularly milder ones closely related to pathos[8]. Cicero's ethos is intertwined with the speaker's character and their ability to evoke and manage emotions effectively.

Cicero also underscores the significance of style in persuasive speaking, emphasizing the importance of surprise and emotion as key elements in effective oratory. He argues that a speaker's character plays a pivotal role, whether by expressing their own judgment, humanity, and liberality of mind, or by skillfully modifying their style to enhance others while downplaying their own views, all driven by goodwill rather than insincerity [4]. This aligns with Sattler's analysis, which suggests that Cicero primarily associates ethical proof with the mild emotions conveyed by the speaker. Thus, Cicero places great importance on the speaker's character and emotional impact in the realm of persuasion.

In conclusion, Cicero's exploration of emotion and character in oratory underscores the intricate interplay between language and personality. By strategically employing unexpected elements, evoking emotions, and aligning their style with their own character, an orator can craft speeches that resonate deeply with their audience. This not only showcases the speaker's skill but also establishes a genuine connection that enhances the impact of their words. Cicero's oratory philosophy is a testament to the enduring power of ethos and emotion in the art of persuasion.

5.Quintilian's Institutio Oratoria: A Holistic Approach to Oratory Mastery

Quintilian, another luminary of Roman rhetoric, contributed a nuanced perspective. In *Institutio Oratoria*, he argued that ethos emerged from consistent ethical behavior and virtuous living. Quintilian posited that a persuasive speaker must not merely simulate ethical virtues but genuinely possess them. He underscored the moral and ethical dimensions of ethos, emphasizing the significance of sincerity and authenticity in persuasion.Throughout the text, Quintilian's insights offer a holistic approach to oratory that resonates with Cicero's emphasis on virtue and authenticity.

Quintilian firmly believed that a good man makes for a good orator, emphasizing that the perfect orator should possess not only exceptional speaking skills but also virtues that exemplify integrity, ethics, and good conduct. He stated, "My aim, then, is the education of the perfect orator. The first essential for such a one is that he should be a good man, and consequently we demand of him not merely the possession of exceptional gifts of speech, but of all the excellences of character" [9].

Quintilian's definition of ethos aligns with this perspective. He defines ethos as character and courtesy (decorum, the becoming) and repudiates the notion that it is limited to mild emotions. Quintilian sees ethos as an end in itself, divorced from the character of the speaker, giving new emphasis to the strictly classical concept of rhetorical ethos [5].

Moreover, Quintilian places considerable emphasis on the speaker's authority and character, noting that an orator's credibility and influence depend on their perceived wisdom and moral character. He asserts that a speaker's advice should align with their moral character, resonating with Cicero's contention that an orator's character should harmonize with their message. Both philosophers emphasize that genuine trust and influence stem from the speaker's wisdom and virtue. Quintilian stated, "For he, who would have all men trust his judgment as to what is expedient and honorable, should both possess and be regarded as possessing genuine wisdom and excellence of character" [9].

Quintilian's views align with Cicero's belief that eloquence and virtue are intertwined. He emphasized that the teacher of oratory should be distinguished for both eloquence and good character. In both Cicero and Quintilian's perspectives, the perfect orator is not only a proficient speaker but also a virtuous individual. They view goodness and skill as inseparable components of an orator's character. Quintilian stressed that a skilled orator not only presents compelling arguments but also upholds a virtuous character, ensuring authenticity and resonance with the audience.

Furthermore, Quintilian's perspective on the relationship between knowledge and virtue underlines his emphasis on well-roundedness. He argued that a truly effective orator should possess both virtuous character and a comprehensive knowledge base, acknowledging that knowledge extends beyond rhetoric and encompasses a broad understanding of various subjects and arts.

In addition to virtue and knowledge, Quintilian also highlights the importance of goodwill in oratory. He advises orators to include words of praise for their audience to secure their goodwill. Quintilian's work represents a comprehensive approach to oratory that integrates character, knowledge, adaptability, and eloquence. This alignment between their philosophies reveals a holistic paradigm for oratory mastery. Quintilian's practical guidance and incorporation of various dimensions of effective communication enrich Cicero's foundations with nuanced techniques, making his work an invaluable resource for aspiring orators.

6. Conclusion

In our exploration of the views on ethos by ancient Greek and Roman philosophers Aristotle, Cicero, and Quintilian, we have gained profound insights into this concept's pivotal role in the art of persuasion. Each philosopher contributed unique perspectives that illuminate the multifaceted nature and enduring significance of ethos in communication and rhetoric.

Aristotle, often regarded as the father of rhetoric, underscored that ethos encompasses the persuasive power of an individual's character and credibility. He emphasized the importance of an orator's trustworthiness, competence, and moral standing as essential elements for effective persuasion. Aristotle's principles revolved around cultivating good sense, demonstrating wisdom, and employing appropriate language and style to establish ethos.

Cicero, the Roman orator and philosopher, built upon Aristotle's foundation by introducing a triad of virtues that constitute ethos: wisdom, justice, and benevolence. Cicero expanded the concept by highlighting that ethos is not solely dependent on one's character but also reflects the ethical virtues they embody. He emphasized the significance of goodwill, kindness, and a disposition that resonates with the audience in establishing a persuasive ethos.

Quintilian, another influential Roman rhetorician, offered a nuanced perspective, stressing that ethos emerges from consistent ethical behavior and virtuous living. He believed that persuasive speakers must genuinely possess ethical virtues rather than merely simulating them. Quintilian placed strong emphasis on sincerity and authenticity as fundamental elements of persuasion.

In summary, the philosophies of Aristotle, Cicero, and Quintilian collectively provide a comprehensive understanding of ethos, emphasizing its deep connection to character, knowledge, and virtue. Their enduring ideas continue to shape contemporary discourse and the art of persuasion. These ancient perspectives serve as a foundational guide for modern communicators seeking to effectively employ ethos in their rhetoric.

References

1. Frobish T S. An Origin of a Theory: A Comparison of Ethos in the Homeric Iliad with that Found in Aristotle's *Rhetoric*[J]. *Rhetoric Review*, 2003, 22(1): 16-30.
2. Tindale C W. Character and knowledge: Learning from the speech of experts[J]. *Argumentation*, 2011, 25: 341-353.
3. Aristotle. *Rhetoric*. Translated by W. Rhys Roberts,[M]. Dover Publications, 2004.
4. Cicero. *De Oratore*. Translated by E. W. Sutton and H. Rackham, [M].Harvard University Press, 1942.
5. Sattler W M. Conceptions of ethos in ancient rhetoric[J]. *Communications Monographs*, 1947, 14(1-2): 55-65.
6. Mc Tavish J. The Ethos of the Praltice of Rhetoric[J]. *Philippiniana sacra*, 2010, 45(133): 66-78.
7. Perks L G. The ancient roots of humor theory[J]. *Humor*, 2012, 25(2): 119-132.
8. May J M. *Trials of character: The eloquence of Ciceronian ethos*[M]. UNC Press Books, 1988.
9. Quintilian. *Institutio Oratoria*.Translated by Donald A. Russell, [M].Harvard University Press, 2001.

Research on the legal regulation of Generative Artificial intelligence—— Take ChatGPT as an example

Luodongni Yang *

Faculty of law, Qingdao University, Qingdao, Shandong, China

Abstract. Generative artificial intelligence is an emerging hot topic in the field of artificial intelligence, with ChatGPT being the most representative worldwide. This article focuses on ChatGPT,.In the first and second parts, we dissect the micro operational structure of ChatGPT and categorize the legal risks/issues it may bring, in order to reduce the risks and solve the current problems brought by generative artificial intelligence, improve user safety during use, encourage the development of generative artificial intelligence industry and protect technological innovation and cultivate entrepreneurial spirit. In the third part, this paper comprehensively examines the legal means and rule systems adopted by the EU, the UK and China to regulate generative AI, and evaluates the respective characteristics of the current Western model and Chinese model. Subsequently, in the fourth part, the macro principles for regulating generative artificial intelligence are proposed..Also, micro rules about regulations are constructed to address specific legal issues. It constructs a reasonable and feasible systematic legal regulatory framework. Finally, we will return to mainland China and provide some forward-looking suggestions for regulating the legal issues related to China that arise during its use.

1. Introduction

Artificial intelligence plays an important role in the development of science and technology in the world. In recent years, the generative artificial intelligence represented by ChatGPT has become the latest form of the development of artificial intelligence. When it comes to generative artificial intelligence, we have to discuss the concept of supremacy, namely artificial intelligence. The concept of artificial intelligence was first introduced at the Dartmouth Conference in the United States in 1956.[1] With the development of science and technology, artificial intelligence technology has made new breakthroughs, and the concept system has become increasingly diversified, but it has never formed a formal and unified interpretation. For example, Professor Nils John Nilsson believes that artificial intelligence is the science of how to represent knowledge and how to acquire and use knowledge,[2] while Professor Patrick Winston (Patrick H. Winston) believes that it is artificial intelligence studying how to enable computers to do intelligent work that only people can do in the past.[3] The most popular, Wikipedia defines artificial intelligence as intelligence expressed by artificially made systems. It also refers to the scientific field of studying whether and how such an intelligent system can be implemented.[4] Generally speaking, artificial intelligence can be derived from the appropriate meaning of its text composition, and the artificial intelligence consists of two parts: artificial and intelligence. Among them, artificial refers to the full use of their own subjective initiative, through the way of labor, reflecting human wisdom and strength; intelligence refers to through the perception of the outside world and their internal calculation, so as to make the corresponding behavior, is the expression of human will externalization. Therefore, a safer general concept can be concluded that artificial intelligence refers to the theory, method, technology and application system that uses computer or computer-controlled machines to simulate, extend and expand human intelligence, perceive the environment, acquire knowledge and use knowledge to obtain the best results.[5]

Compared with artificial intelligence, the meaning of generative artificial intelligence (AI Generated Content / AI Generated Media, AIGC) is determined by its directions of usage and ways of utilisation —— generation. That is, compared with other forms of AI, generative AI calculates and processes the learned data, and then produces new output results. Generative artificial intelligence, as a universal concept, does not have a unified definition in various countries. Wikipedia interprets it as short for the production, manipulation and modification of data or media through artificial intelligence algorithms. From the common understanding of current Chinese academic researchers and business practitioners, AIGC refers to a new production mode that automatically generates content with artificial intelligence technology after professional generated content and user-generated content.[6] At the same time, Chinese regulators further define it from the perspective of generating source as generating text technology based on algorithms, models,

* Corresponding author:yangluodongni@163.com

rules, pictures, sound, videos, code and other content.[7] Although the generative artificial intelligence represented by ChatGPT has not been officially recognized at the official level in China, Chinese local products such as Chat intelligent AI, Sheniu oral English, AI creation Max, AIbot and AI Xiaoice have shown a trend of all flowers bloom. To this end, it has sparked a heated discussion in China. Researchers have done a lot of research and proposed a series of relevant theories on generative artificial intelligence and its legal problems. Meanwhile, although there is no effective legal document regulating the generative AI, and there is no section related to generative AI in the current law, the Cyberspace Administration of China has released notices that some provinces and cities issued a series of local pilot legislation with the role of administrative regulators to regulate the legal risks of generative AI technology developments, commercial promotion and usage of users. On the whole, we can see the attitudes, methods, ideas and wisdom of Chinese regulators on regulating generative artificial intelligence.

Therefore, this paper will take the most representative form of generative artificial intelligence in the world, ChatGPT, as the analysis object, dissect its micro-operation structure, and categorize the legal risks / problems it may bring. It is not to find problems than the purpose. It is the proper goal of this paper to reduce and solve the risks brought by generative AI, improve the safety of users and encourage the development of generative AI industry, protect scientific and technological innovation and cultivate the spirit of entrepreneurship. In the end, the third part of this paper will comprehensively examine the legal means adopted by the European Union, the UK and China, and evaluate the characteristics of the current Western model and the China model. Then, in the fourth part, this paper puts forward the macro principle of regulatory generative artificial intelligence, and gives the micro rational rule structure for specific legal issues, so as to construct a reasonable and feasible systematically legal regulation framework. Finally this paper will return to China to provide some forward-looking suggestions on the legal issues related to China in the process of regulating its use.

2. Risks and hidden dangers of generative artificial intelligence: taken ChatGPT As a typical example

2.1 The microscopic mechanism and super function of ChatGPT

The full name of ChatGPT is Chat Generative Pre-trained Transformer. As an AI chatbot program developed by Open AI, an artificial intelligence research institute in America, was launched in November 2022. This program uses a large language model based on the GPT-4 architecture and trained with reinforcement learning. Currently ChatGPT interacts with users in a written way. As a typical example of generative artificial intelligence, ChatGPT has super capabilities in many aspects.

First, the super-strong ability to process language. With robust text generation capabilities and strong scalability, it can automatically generate text that adheres to grammatical specifications and logical coherence. It can also finish different tasks such as text generation, summary and translation. At the same time, ChatGPT is a multilingual expert, supporting multiple natural languages, and can process translation and communication between different languages.

Second, the strong plasticity of user experience. ChatGPT can be adjusted according to the different dialogue scenes, so that users can form deja vu. ChatGPT has a high immediacy, responding rapidly. It can immediately answer the user's inquiry. ChatGPT is also highly integrated so that developers can arbitrarily integrate it into their own developed software.

Third, super-adaptability in learning and training. ChatGPT has powerful language processing and generating functions, which are mainly derived from the reinforcement learning and training of large-scale text generated and fed back by human's data. Specifically, learning and training of ChatGPT can be divided into two stages. In the pre-training stage, ChatGPT grasps the basic structure and semantic rules of the language by learning a large amounts of unlabeled text data. These data mainly come from the Internet, including news articles, blogs, forums, books and so on. During training, the model randomly masks some words using a method called Mask Language Model , and the model needs to predict these masked words based on the context information. In this way, ChatGPT learns to capture the semantic and syntactic relations in the text. Then during the fine-tuning phase, the model is trained using task-specific label data to adapt to different application scenarios better. These labeled data often include high-quality conversations generated by humans and question-and-answer conversations relevant to specific tasks. During the process of the fine-tuning, the model learns how to generate more accurate and relevant responses based on the input.

2.2 Legal risks and hidden dangers of ChatGPT

As a new generative artificial intelligence product, ChatGPT not only brings convenience to people, but also inevitably issues some risks and hidden dangers, while the legal risks and hidden dangers in the process of usage and development mainly focus on rights protection and supervision mode.

2.2.1 Rights protection

First, the production process of generative AI software and the product may infringe on the legal rights of others. As mentioned above, ChatGPT relies on massive information of human language databases, including a large amount of information entered by Internet users themselves. Therefore, when users input information such as personal data or trade secrets, ChatGPT may incorporate it into its own corpus then cause the risk of leakage. Although the developers of ChatGPT promise to delete all personally identifiable information, it does not explain the standard and methods of deletion. In the case that it cannot evaluate effectively and safely delete the information and data sources, these personal information still has the risk of disclosure, which may pose a threat to the individual privacy right. In addition, when using data in existing databases for ChatGPT training, it is possible to use some data involving improper behaviors and materials, including but not limited to data about personal attacks, gender discrimination and racial discrimination and so on. When ChatGPT generation directly or indirectly reflects these data, it is likely to violate the human rights and human dignity of specific individuals or groups. More seriously, the authenticity of the information presented by the ChatGPT generation is questionable. Take China as an example, in February 2023, a news that the Hangzhou municipal government of China will cancel the restriction policy on the final number of motor vehicles on March 1 went viral on the Internet. But Hangzhou police later confirmed that it was a fake news written by ChatGPT. [8]It can be seen that as ChatGPT will make serious nonsense, users need to have certain professional knowledge to distinguish the authenticity, otherwise they will believe the false information easily.

Second, the infringement process of generative AI software is relatively hidden, so it is difficult to allocate the tort liability and the principle of attribution is difficult to be clear, leading to the difficulty of accountability and the high protection cost. Due to the ability to learn of generative artificial intelligence, it is difficult to judge whether the infringement results are implemented by AI after autonomous learning. For example, in the AIGC application scenario, if the code developer is responsible for creating algorithm, which changes through self-learning, and then in a specific application scenario, the generated content ultimately leads to damage. Should the developer bear tort liability? Obviously, the developer did not directly commit the infringement, and had no subjective intention of negligence. However, due to the limitation of technical standard, most of the existing machine learning of artificial intelligence generation tools is not high, and its generated content is mostly subject to the input of operators. On this basis, the corresponding text, image, audio and other content are generated (for example, the ChatGPT application is an AI generation tool based on input of users). In this kind of AI generation scene, if the generated content objectively causes damage results, how to confirm the responsibility subject of it? After all, it is also said that the infringement is completely and independently implemented by AI through autonomous learning. Therefore, this means that the constituents of the tort liability of the generative artificial intelligence are different from those of the general tort liability, but the specific differences need to be further discussed.

Third, the intellectual property rights of generative artificial intelligence products have become the main legal risk and problem about ChatGPT, including (1) the copyright ownership of the products (2) whether the generation infringes the intellectual property rights of others. The two issues related to intellectual property rights will be systematically introduced respectively.

The first question is the copyright of ai-generated content belonging. To answer this question, first we should discuss its upper concept —— work. According to Article 3 of the Copyright Law of the People's Republic of China, the works mentioned in this legislation refer to intellectual achievements that are original and can be expressed in a certain form in the field of literature, art and science.[9] Article 2 of the Regulations for the Implementation of the Copyright Law of the People's Republic of China stipulates: The works mentioned in the Copyright Law refer to the intellectual achievements with originality in the field of literature, art and science and that can be reproduced in some tangible forms.[10] It can be concluded from the definition of work in two official legal documents that the work is the achievement of the author's independent intellectual and creative activities. But when it is applied to unscramble AI products, it seems impossible to give a proper interpretation. After all, the process of AI generating content cannot be interpreted as intellectual activities of human. Artificial intelligence generating content is essentially a process of computing for computers based on big data and algorithms. In contrast, human intellectual activities are more complex. It includes not only rational thinking activities (such as the mastery and application of writing rules), but also perceptual thinking activities (such as thoughts, emotion and inspirations). Even if the content generated by artificial intelligence has the superficial features of the achievements of human intellectual creations, it is difficult to be included in the category of works in the meaning of human intellectual creation activities. Therefore, whether there is copyright in the content generated by AI (that is, whether it can be called works) and if there is copyright, where is the copyright belonging? Current law in China cannot answer these questions.

The second question is whether the ChatGPT product may involve infringement of intellectual property of others. The learning and training process of ChatGPT is described in detail above. Obviously, ChatGPT will definitely use the existing human language data in the network database during the training process, and some language data may be written works where others enjoy intellectual property rights. Although ChatGPT officials have constantly claimed that the materials used for its training come from Open Data Bases（ODB）to avoid risks of violating others' intellectual property rights. After all, open databases are often seen as an updatable platform for technicians (developers, users) to share and reuse code among themselves. Anyone can modify it to meet the needs of their own specific application scenario. In addition to providing software code, open databases can also provide documents, such as requirements analysis documents, design documents, test documents and so on. In addition, open databases include data structures, device drivers, sample projects, and language supplement application components and toolkits. Besides, many open databases also include examples to help users with specific tasks to adopt a certain technology in real applications. However, it is hard to say that the data of ChatGPT users comes entirely from open source databases. At the same time, even if the open source database comes from the open source database, ways of utilizing data of the open source database is generally stipulated in advance, so the production process of using the data beyond the permission undoubtedly infringes the intellectual property rights of others, because it goes beyond the scope of reasonable use. Therefore, the process of AI learning and its products do indeed have the possibility of infringing intellectual property rights.

2.2.2 Risks and hidden dangers of absent supervision

First of all, the generative artificial intelligence represented by ChatGPT has been applied in some important industries. If its application is not reasonably and effectively regulated, it may hinder the development of the industry under the new trend of widespread application of artificial intelligence. Taking the legal industry as an example, generative artificial intelligence is mainly used to help legal workers to deal with a large amount of information and effectively complete various tasks. It has replaced a part of the judicial auxiliary personnel who do repetitive work. However, the existing generative artificial intelligence software still lacks the ability to create, understand emotion and think comprehensively. Whether it can give professional legal advice like lawyers or make fair judicial decisions like judges is still uncertain, but this possibility cannot be ruled out in the future. It can be predicted that with the deep development and popularization of artificial intelligence, the legal profession will be subversively shaped by the power of science and technology. The career function and development direction of the legal industry under the new trend still need to be further discussed.

Secondly, generative artificial intelligence presents great challenges to transaction security. Generative artificial intelligence technology products may be used to mining transaction data on the network because it has ability to capture data and generate text according to the scene of usage. If the network was attacked by hackers, or received Trojan virus, the information stored in the database of the program would be stolen without awareness of users and developers. For example, in application scenario of marketing, interest, which is closely related to the finance, if users use such an artificial intelligence technology, the content of the process of chatting with ChatGPT may be disclosed in other unsafe scenarios, in which the content is collected and become learning materials of algorithm training. It may cause financial data leakage accidents and threaten the financial security, data security and regulatory security.

Finally, some of the actions implemented by using generative AI, which potentially do harm to the public interest and subvert social ethics. In 2022, China issued the Internet Information Service Algorithm Recommendation and Management Regulations. It has declared: Algorithms should adhere to the mainstream value orientation, actively spread positive energy, and shall not use algorithmic recommendation services to endanger national security and social and public interests, disrupt economic order and social order, and infringe on the legitimate rights and interests of others.[11] As an artificial intelligence technology, algorithmic architecture of ChatGPT is planned and programmed by human engineers. It is still a black box model, and it is unknown whether its algorithm can sufficiently disclose information. It cannot be excluded that some illegal elements take advantage of the technical loopholes of their algorithm architecture or database review loopholes to carry out illegal activities, disturbing public order and endangering public interests. For example, some criminals convey false information to the public; deliberately fabricate inflammatory speech; intentionally influence political ecology; maliciously incite violence and crime. Some people will make use of this kind of generative artificial intelligence program technology vulnerabilities, collect different types of user's privacy data, and then suit the remedy to the case by providing phishing site links to directional fraud. These illegal and criminal behaviors by using generative artificial intelligence have greatly challenged the social public interest and social ethical standards.

3.Comparative observations of generative AI regulation: the European Union, the UK and China

3.1 The UK has a regulatory system centering on the Artificial Intelligence Regulatory Methods to Promote Innovation

The UK has established a regulatory framework for the regulatory needs of specific high-risk scenarios in the use of AI to address the problems of how to regulate the usage of AI applications in such scenarios. Earlier, there was no special law in the UK to regulate AI products and services, and the relevant content of the regulation was scattered in various departmental laws. Its current laws related to AI include the Equality Act 2010, the Product Safety Act and the UK Consumer Protection Act. However, some official documents have set forth requirements for the application and development of AI in related fields, such as the Ethics Principles and Policies of AI released by the Ministry of Defence. In April 2023, for specifically high-risk sences of applications, the British department for science, innovation and technology has issued an official document —— A pro-innovation approach to AI regulation white paper (hereinafter referred to as The Approach). [12]This document has put forward its artificial intelligence management methods based on the five principles. The purpose is to provide enterprises and the masses with confidence of using artificial intelligence, provide industries with certain and consistent regulatory methods. At the same time, the white paper noted that the AI industry will not be strictly regulated, so as to encourage innovation and ensure that the response to challenges is in a timely manner. In addition, the white paper particularly emphasizes the importance of collaborative governance by the government, industry, enterprises and other entities, for the purpose of strengthening global cooperation and interoperability in AI governance, so as to achieve the UK's global position as a leader in AI as soon as possible. It should be noted that the Approach emphasizes the pilot implement, observing the results of the proposed regulatory methods, and then further legalized the regulatory principles of regulatory agencies, such as the Office of the Information Commissioner (ICO), the regulatory Agency (the Equality and Human Rights Commission (EHRC), and the Financial Conduct Regulatory Authority (FCA)).

In terms of the level of regulatory power, The Approach lays more emphasis on the decentralized regulatory structure and methods, emphasizing the participation of existing regulatory agencies and the functions of regulatory agencies. Firstly, the document stresses that there is no need for a dedicated AI regulator, and the responsibility for regulating AI should be borne by existing regulators. Secondly, The Approach emphasizes the need for coordination of regulation. While abiding by their profession and fulfilling their duties, they should be consistent with law enforcement. At the same time, they should not abuse power of regulation. Thirdly, if the regulators need supports, the central government will provide these regulators with multifaceted support. But in cases where regulators are unable to handle more than two conflicts of principles, the government wants regulators to work with the government in a timely manner to identify potential actions. Finally, the government also hopes that regulators can effectively coordinate with multiple departments to jointly formulate regulatory guidelines to clarify their own understandings of regulatory principles. We also want to develop a cross-economic and whole-social assessment framework of risk, so as to ensure that risks fall in the responsibility range of diverse regulators.

In terms of regulatory subjects, The Approach proposes to establish a special department to develop a centrally reasonable monitoring and evaluation framework (M & E); develop a reasonable and multi-channel feedback framework, and extensively solicit opinions and suggestions from government, regulators, AI industry, civil sector, and society; provide central support to regulatory agencies; supervise and evaluate the regulatory framework mentioned above.

In terms of regulatory professionalism, The Approach proposes that all regulatory agencies should try their best to learn professional knowledge of artificial intelligence to narrow the gap among diverse agencies in professional knowledge and ensure that regulatory behaviors are scientific and correct. For example, the government invites regulators to pay attention to the risks arising from the application of new basic models (such as LLM large-scale language models); the government wants to work with existing regulators to develop sandbox models to assist innovators in the AI industry better. The government also wants regulators to develop generic technical standards. The Approach also mentions that the above measures need to be piloted.

In terms of regulatory coverage, The Approach specifically mentions the need for effective linkage and response to the international regulatory framework. While formulating the domestic regulatory standards and technical standards, we should also pay attention to the international standards, ensuring that these principles of risk management are consistent with international standards. By confirming opportunities for regulatory interoperability, supporting cross-border coordination and collaboration, so as to achieve comprehensive coverage of regulation inside and outside the region and prevent problems caused by the absence of cross-border regulation.

3.2 The regulatory system formed by the EU cetreing the Artificial Intelligence Act

The EU's regulation of generative AI is mainly reflected in the issuance of the Artificial Intelligence Act (hereinafter referred to as the Act), so as to ensure that development of AI in Europe is people-oriented and moral. Specifically, the Act, which was amended and passed in May 2023, responds to AI infrastructure models, responsibilities of value chain and the recent popularity of generative AI. At the same time, it studies exemptions for businesses, especially small or medium-sized enterprises and start-ups. The Act mainly puts forward regulatory requirements for generative AI from the following aspects, the specific contents of which can be seen in the following table:

Table 1. The Construction and essence of the Artificial Intelligence Act[13]

Regulatory aspects	Law serial number	Core essence
Definition of the regulated objects	Section 3 of Part I of the Act	The AI system and its related usage, developments, regulatory subjects and their behaviors are clearly defined.
Introduction of the regulatory principles	Section 4a of Part I of the Act	It proposes six general principles which are applicable to all AI systems: human subject and supervision, and technology robustness and security, Privacy and data governance, and transparency , diversity, non-discrimination and fairness, and social and environmental well-being
Setting-up of the regulatory authorities	Section 3 (42) of Part I of the Act	It proposes to set up a centrally regulatory agency (the European Committee on Artificial Intelligence) and a nationally regulatory authority in each member state
AI accomplishment	Section 4d of Part I of the Act	It are requirements for regulatory authorities, providers and deployers of AI systems to strengthen the training and education of AI expertise, requiring the above subjects to ensure that their staff and other personnel who handle the operation and use of AI systems on their behalf have sufficient AI knowledge level

risk management	Section 6,7,8, and 9 of Part III of the Act	The AI system is divided into three risk levels: unacceptable risk; high risk; non-high risk. According to different risk levels, the Act regulates AI systems at that level differently, especially high-risk AI systems, and the Act hopes that evaluate high-risk AI systems in terms of both rights impact and technical standards
The behaviors of the supervised subjects	Chapter III, full chapter of Part III of the Act	Detailed guidance and requirements are put forward for the behaviors that are prohibited and encouraged to implement. It also introduced a new penalty of 6% or 30 million euros (whichever ever)

3.3 Summary of the regulatory model in the UK and Europe

The UK and the European Union belong to Europe, and Europe is leading the world in the development of artificial intelligence. Comparing the UK and the legal regulation of the AI system with the EU, we can find something meaningful that the regulatory direction chosen by the UK is clearly different from the EU, and shows the differences in the following five aspects:

First, a broader definition. The UK White Paper defines AI as any system with a function of combining adaptability and autonomy. The broad definition ensures that regulations can be flexibly applied and modified to accommodate evolving technologies. In contrast, the EU's AI regulations use a fixed and narrow definition of AI.

Second, it concentrates on specific situations. The UK White Paper avoids a classification of specific products, instead favors outcome-based regulation. The application of AI will be analyzed on a case-by-case basis, and considered risk Regulators which will balance the benefits and risks. In contrast, the Act proposes a formal risk classification (from low to high) and bans the use of AI technologies that are deemed to pose an unacceptable risk.

Third, there is no new regulator. A decentralized UK system will not create new regulators; instead, it will rely on existing regulators. By contrast, the EU proposes a central regulator (the European Council on Artificial Intelligence) and national regulators in each member state.

Fourth, there is no new punishment. The white paper did not introduce new penalties for violations. Instead, UK regulators will use their existing enforcement powers to punish relevant offences. By contrast, the EU system introduced new fines of up to 6% of the global annual turnover or 30 million euros (whichever highest).

Fifth, the role of the government in the regulation is different. The government will provide central support functions to UK regulators, including monitoring effectiveness and consistency, assessing economic impact, and promoting interoperability with foreign institutions. The white paper recommends that the government will work with regulators to implement its recommendations. Its roles will include identifying risks to AI applications, working with regulators to identify central risks across sectors, and updating laws and regulations to address risks marked by regulators.

It can be concluded that while both are generally designed to promote the adoption of people-oriented and trustworthy AI and to ensure a high degree of protection of health, safety, fundamental rights, democracy, rule of law and the environment from the harmful effects of AI systems within the effectiveness of the law domain, supporting innovation in the meanwhile. However, the two main regulatory concepts are not the same. In terms of the overall concept of legal regulation, the UK prefers the open-regulatory concepts, without imposing too many restrictive provisions on AI systems and their related subjects. While the EU prefers the regulatory concepts which are relatively rigorous, putting forward a large number of prohibitive provisions on AI systems and their related subjects.

3.4 The legal means adopted by China and the rule system formed

In recent years, China has mainly issued administrative regulations and local regulations to supplement the current laws for the lack of AI systems and product sectors. Meanwhile, China puts forward new requirements for the development of the AI industry through releasing policy documents. There are two policy documents concerning the regulatory and normative Generative AI systems or products issued by Chinese regulators: one is the Measures for the Administration of generative ARTIFICIAL Intelligence Services (Draft) (hereinafter referred to as the Measures) and the Regulations for the Administration of Internet Information Services jointly issued by the Cyberspace Administration of China (hereinafter referred to as the Regulation) , the Ministry of Industry and Information Technology and the Ministry of Public Security. Specifically, the Measures set certain obligations for users and developers of generative AI technology, including but not limited to not to release illegal content, no discrimination; respect the rights and privacy of others. In view of the practical harm and potential risks brought by deep synthesis technology, the Regulation puts forward scientific, refined and systematic governance requirements, delimit the bottom line and red line of deep synthesis services, and promote the effective use of deep synthesis technology in accordance with the law and regulations. Generating a pre-trained transformation model (GPT) is a kind of deep synthesis technology, so the ChatGPT developed based on the GPT model is also within the regulatory scope of this regulation. In addition, the local legislation issued by Chinese regulators includes the Regulations of the Shenzhen Special Economic Zone on Promoting the AI Industry and the Regulations of Shanghai Municipality on Promoting the Development of the AI Industry. The two local regulations are actually a manifestation of pilot studies on

AI regulation by Chinese regulators in some developed regions of China.

Overall, Chinese regulators have a relatively positive regulatory awareness. Through the above official documents, we can see that on the one hand, the regulators actively promote the continuous development of the AI industry in China, and on the other hand, they also conduct preventive regulation on the scientific ethics, public interests and other issues that may be involved in the development process. Compared with the UK and the European Union, China covers more comprehensive industries in the legal means and the rule system of regulating generative AI. However, there are also some deficiencies. For example, for regulatory generative AI technology, the regulatory suggestions put forward by Chinese regulators are normally relatively vague, which cannot be implemented directly.

4. The future improvement of the legal regulation of generative artificial intelligence

As a new thing, the supervision of generative artificial intelligence must appropriately learn from the advanced legislative experience of various countries. On the premise of promoting the development of generative AI that is more conducive to people's work and lives and developments, we should establish and gradually improve the legal regulatory framework of generative artificial intelligence. In general, a scientific, complete regulatory system should be able to clarify respective rights, obligations and relationships of the developers, users, the third party and related subjects, effectively protecting the legal rights of the relevant subject. Meanwhile, some special behaviors implemented by related subjects of generative AI should be effectively regulated.

4.1 Regulatory measures aimed at protecting the rights of relevant subjects

Regulators must clarify the legal status of generative AI. For example, does generative AI have an independent personality at this stage? Should there be a certain capacity of rights and behaviors? In other words, in some situations, it has partially independent rights and (or) capacity. Only when their legal status is clarified can regulators discuss whether the current laws can be applied to restrain, regulate and even punish them under this premise. In addition, regulators should type the infringement issues in specific scenarios of generative AI, and then issue guiding official documents such as legal applications, legal interpretations and guiding opinions for these issues. In particular, the concept of the intellectual property infringement should be clarified, otherwise it will lead to the overlap and confusion of the regulatory system. For instance, regulators should set rules for using code, documents, images and other resources in the open source database about the entry and range. To further illustrate, if resources from a specific open source database are commercially developed, it may be involved in infringing copyrights of others, so these resources can only be used for personal studies and researches, instead of being used for developing commercial systems, products, etc.

4.2 Regulatory measures aimed at the reform of regulatory methods and regulatory standards

First of all, regulators should clarify the risk level of the target. Regulators should assess the risk of the generative artificial intelligence existing at the present stage, and then classify these AI. They should adopt different measures with different priorities to regulate the generative artificial intelligence with different risk levels. But they should always pay attention to the aforementioned premise, otherwise it will hinder the development of generative AI. At the same time, regulators should promote the joint participation of multiple departments in special legislative activities and the formulation of cross-departmental technical standards. The different functions of the regulatory authorities lead to their different regulatory priorities, and it is also different that targeted professional knowledge related to generative AI mastered by these regulatory authorities. If multiple regulatory authorities make different nature judgments on the same behavior and adopt different regulatory behaviors, it will cause the regulated objects themselves and the innovators within the artificial intelligence industry to be confused, which may weaken the innovation power of the industry. Therefore, at present, the more urgent goal is to establish a cross-departmental regulatory system and specific technical standards as well as to improve the professional knowledge level of various regulatory departments. Finally, the regulators should work hard to promote the implementation of the regulatory measures. The current official regulatory documents and standards are more designed and regulated in the general concept, lacking of implementation rules that can be implemented. For some phenomena of generation in generative artificial intelligence in actual production and life, grassroots regulatory authorities may know they need to regulate, but do not know what regulations should be followed. Therefore, the grass-roots regulatory departments should conduct more coordination with the superior regulatory departments to further standardize the regulatory behaviors in actual operations.

4.3 To refine the constructions of specific rules as the core of the regulatory measures

At present, the regulatory suggestions put forward by regulators are often relatively vague, so these suggestions are difficult to be implemented by the regulatory enforcement agencies. For this reason, it is necessary to further refine regulatory objects more clear, the regulatory scope is more comprehensive, and the regulatory basis is clearer. An ples will be given below.

Example 1: For the using permission and scope of the code, documents, pictures and other resources in the open source database.

Article 7 of China's Measures for the Management of Generated Artificial Intelligence Services (Consultation

https://doi.org/10.1051/shsconf/202317802017

Paper) stipulates the following provisions: Providers shall be responsible for the legitimacy of the pre-training data of generative AI products and the optimization of the source of training data. The pre-training and optimization training data used for generative AI products shall meet the following requirements: (1) meeting the requirements of the Network Security Law of the People's Republic of China and other laws and regulations; (2) not containing the infringement of intellectual property rights; (3) obtaining the consent of the personal information subject for the personal information or other circumstances complying with the laws and administrative regulations; (4) ensuring the authenticity, accuracy, objectivity and diversity of the data; (5) meeting the other regulatory requirements of the national cyberspace department on generative AI services. [14]But there are some problems in this article alone, such as the fourth article, which is stricter. Large models are inseparable from the training of using existing data. If too strict regulations are made for the training data, it means that the development speed of large models will be seriously hindered. The scale of data training will determine the quality of the content generated by AI, and the quantity and quality of data are more important than algorithms and models. Therefore, if the relevant training data cannot be used due to the authenticity, objectivity and other problems, it will have a great negative impact on the development of artificial intelligence technology. At the same time, the first model also suggested adding but not including synthetic data sources for AI-generated content. That is, the provider should be responsible for the legitimacy of the pre-training data of the generative AI products and the optimization of the training data sources, but not including the synthetic data sources of the AI-generated content. It is required to explain that pre-training and optimized training data for generative AI products are divided into natural data and synthetic data. The natural data includes personal information in the physical world; public data acquired directly and data purchased through data transactions. According to the five requirements stipulated in this article, it can be understood that the pre-training data and optimized training data stipulated in this article exclude the source of synthetic data, and only refer to the natural data of the three sources. Otherwise, requiring the provider to take responsibility for the authenticity, accuracy and objectivity of the synthetic data, which obviously aggravates the obligation of the provider. It is difficult to ask the provider still take responsibilities to the legitimacy of data resources of the generated content reproduced by generative AI from the previous content that as the origin of synthetic data. Technically, it is also impossible to achieve.

Example 2: the risk level classification for the regulatory objects.

Article 15 of the Measures for Generated Artificial Intelligence Services (Consultation Paper) stipulates: For the generated content found in operations and reported by users that does not meet the requirements of the Measures, besides taking measures like content filtering, it shall be prevented from recurring within 3 months through model optimization training and other methods.[14] ChatGPT has been on line for several months, which obscure problems still cannot be correctly understood and not correctly answered. At present, China's online AI is still in the private beta stage, the time of implementation is not long enough, the obscure problems expose insufficiently. Therefore, it is required to prevent recurrence through model optimization and training within 3 months, the mentioned period is too short. In the large-scale model of artificial intelligence, it is difficult to accurately predict and control most of the generated content. Some experts believe that it is difficult to prevent the regeneration through model optimization training in 3 months technically. It is advisable to firstly stipulate the remedial measures that should be taken for violations, and a reasonable time limit should be determined after a lot of practices after the stable development of artificial intelligence. At the same time, the article does not specify the legal liability of the same problem still occurring three months later. The legislation is not clear whether three months is a period of exemption or increased liability.

Example 3: The generative artificial intelligence is classified according to its functions and uses.

For example, according to the European Union's AI Liability Directive (AILD), only high-risk AI applications can bear AI strict liability and assume different responsibilities according to the level of risk. In view of the fact that generative AI products and services are still in the development stage, the relevant regulations should not be too strict to hinder the development of artificial intelligence. The production should be encouraged and too much restrictions also should not be imposed on it, so as to improve the international competitiveness of AI industries. In addition, the scope of the regenerated content can be clarified, such as preventing the recurrence of the same kind content or the content with the same characteristics. If the model optimization fails, or it is difficult to avoid regenerating, it can be further clarified how to regulate.

Example 4: For the joint participation of multiple departments in special legislative activities and the formulation of cross-departmental technical standards.

Article 20 of the Measures for Generated Artificial Intelligence Services (Consultation Paper) has regulated: the provider in violation of the provisions of these measures shall be punished by relevant departments in accordance with this law, or Cybersecurity Law of the People's Republic of China, or the Data Security Law of the People's Republic of China, or the Personal Information Protection Law of the People's Republic of China or other revelant laws and administrative regulations. In the absence of laws and administrative regulations, the cyberspace department and relevant competent authorities shall, according to their duties, give a warning and circulate criticism, and order them to correct within a time limit; if they refuse to correct or if the circumstances are serious, they shall be ordered to suspend or terminate their services provided using generated artificial intelligence and impose a fine of not less than 10,000 Yuan but not more than 100,000 Yuan. If the case constitutes a violation of public security administration, it shall be given administrative punishment for public security according to law; if the case constitutes a crime, criminal responsibility shall be investigated according to law.[14] However, in the second paragraph of this Article, the responsibility of providers is too strict, and even for the current laws and regulations that have not been clearly stipulated, the practice of ordering suspension does not comply with the provisions of the Administrative Punishment Law. Article 14 of China's Administrative Punishment Law stipulates that if no laws and administrative regulations have been formulated, the departmental rules of The State Council may impose administrative penalties of warning, notice of criticism or a certain amount of fine for acts violating the administrative order. Visibly, with regard to unspecified violations of the management order in laws and administrative regulations, the highest penalty limit is just a warning, criticism, and fines.[15] However, the measures also stipulates that certain subjects shall be ordered to suspend or terminate the utility of generative artificial intelligence services. This belongs to some subjects that should be restricted in production and business and be ordered to suspend production and shut down. They may also be restricted from becoming practitioners. The penalty limit is significantly higher than that stated in the law on administrative punishment. This is not friendly to the subsequent development of the emerging generative artificial intelligence service industry. It is suggested that the limit of responsibility should be consistent with the Administrative Punishment Law. The suspension or termination of its services using generative artificial intelligence should be deleted, and the fine should be retained.

5. Conclusion

The use and development of generative artificial intelligence has facilitated to the user's work and life, greatly promoting the development of the AI industry. However, the technical complexity of generative artificial intelligence technology leads to legal risk of some process of use, strengthening the complexity of the relevant legal issues, such as generated content infringement of intellectual property rights and the difficulty of proof, damage of the network transaction security owing to data leakage. These things can be concluded as algorithm infringements. Their occurrence also helps the legal regulation of generative artificial intelligence technology which becomes an important issue and a common problem in the field of researching artificial intelligence. Overall, if Chinese leaders want to develop generative AI in an orderly, rapid and vigorous manner in China, it is necessary to refine the specific rules of regulatory measures related to generative AI. In the process of detailed rules, we can find many hidden, controversial legal issues and form a more comprehensive legal regulation advice in the collision of thought. At the same time, we can refer to foreign legislation, regulatory ideas and measures, helping Chinese regulators with the development of generative artificial intelligence constantly adjusting and modifying regulations, so as to better promote the development of Chinese native generative artificial intelligence related industries.

Reference

1. Yongyu,2021.Overview of the Development History of Artificial Intelligence.https://www.zhihu.com/column/c_1637861790871711744
2. Nilsson, N. (1983) Principles of Artificial Intelligence. Science Press, Beijing.
3. Winston, P. (2005) Artificial Intelligence. Science Press, Beijing.
4. Anonymity,2023.Artifical Intelligence.https://encyclopedia.thefreedictionary.com/Artifical+Intelligence
5. Standing Committee of Shanghai Municipal People's Congress,2022.Regulations of Shanghai Municipality on Promoting the Development of Artificial Intelligence Industry.https://law.sfj.sh.gov.cn/#/detail?id=63ac062e60b224bab986e200
6. Qingfeng Gan,2023.On the Legal Regulation of Artificial Intelligence Jurisdiction.https://mp.weixin.qq.com/s/VriLWFYzzMaNJk-prx0XhQ
7. National Internet Information Office,2023.Management Measures for Generative Artificial Intelligence Services (Draft for Soliciting Opinions).http://www.cac.gov.cn/2023-04/11/c_1682854275475410.htm
8. JIMU NEWS,2023.Hangzhou will lift travel restrictions on March 1st? The media claims it was written by ChatGPT, and the police will release the investigation results.https://baijiahao.baidu.com/s?id=1758049381599427211&wfr=spider&for=pc
9. the National People's Congress (NPC),2020.Copyright Law of the People's Republic of China.https://www.gov.cn/guoqing/2021-10/29/content_5647633.htm
10. the State Council,2013.Regulations for the Implementation of the Copyright Law of the People's Republic of China.https://www.ncac.gov.cn/chinacopyright/contents/12230/346377.shtml
11. Ministry of Industry and Information Technology of the People's Republic of China,National Internet Information Office,Ministry of Public Security of the People's Republic of China,2023.Regulations on the Management of Deep Synthesis of Internet Information Services.https://www.gov.cn/gongbao/content/2023/content_5741257.htm?eqid=fb80e86900094abc00000004645e095c
12. the Secretary of State for Science, Innovation and Technology ,2023.A pro-innovation approach to AI regulation.https://www.gov.uk/government/publications/ai-regulation-a-pro-innovation-approach/white-paper
13. European Union,2023.EU Artificial Intelligence Act.https://techonomy.com/the-eus-artificial-

intelligence-act-a-pragmatic-approach/

14. Cyberspace Administration of China, 2023. Measures for Generative Artificial Intelligence Services (Consultation Paper). http://www.cac.gov.cn/2023-04/11/c_1682854275475410.htm
15. the National People's Congress (NPC),2021.Law of the People's Republic of China on Administrative Penalties.https://www.gov.cn/xinwen/2021-01/23/content_5582030.htm

https://doi.org/10.1051/shsconf/202317802018

Analysis of the Challenge in Fake News and Misinformation Regulation Comparative in Global Media Landscape

Eugene Chan*

Head, Department of Management and Hospitality, Technological and Higher Education Institution of Hong Kong, Hong Kong SAR, China

Abstract. In the age of online social platforms, increasing interactions beyond the wildest imagination, dealing with fake news has been the most challenging. Perhaps the most visible victims of fake news today are political individuals and institutions. The delicate balance between regulation and respect of individual rights and maintenance of constitutionalism in various countries today creates a unique regulatory dilemma that allows for fake news and the use of it to continue thriving. At the same time, politicians and other institutions occupy a critical role in creating legislation towards combatting fake news. Such moves could significantly undermine the ability to observe some of the fundamental rights of individuals and the overall rule of democracy [1]. Misinformation and other cases of fake news have a relatively significant impact on societies. However, resolving fake news has proved not to be easy because of numerous issues, among them, being the need to preserve fundamental rights [2]. Desktop-based research has been conducted to achieve the objectives of this research. Critical to note that most of the information vital in completing this research is electronically availed. Hence, secondary sources of information have been utilized to reach conclusions on the existence of regulations and its effectiveness toward combating fake news in the three jurisdictions.

1.Introduction

Falsified information has always existed, and it is unlikely that one can point out where and when false information. However, in the advent of the 2016 United States presidential election, the term 'Fake News' received enormous attention. The presidential candidates at that juncture, the media, and the public went frenzy over the existence of fake news in the United States and the potential damage that it could cause in the country's social and political fabric. Later into the presidency of the 45th president of the United States, a probe began into the ways that foreign agents might have sought to interfere in the United States' most sanctified democratic process [5]. Despite the inconclusive ending of the report on the probe, it became apparent that fake news was more than anyone could imagine. Since then, it has also dawned on other democracies and societies that fake news is indeed existent, and it can bear detrimental effects on societies, groups, individuals, and even organizations (Park & Youm, 2019) [11].

The practice of publishing fake news has been around for so long. As early as the 18th century, the United States Congress had addressed the issue of fake news, placing a legal narrative on it. The Alien and Sedition Act of 1978 was one of the earliest formal legal measures to combat fake news. The Act aimed at deterring the individual from publishing fake news by punishing those guilty of malicious and falsehood that aimed at the government. Since then, numerous developments have been made to resolve the spread of false information that may be targeted at individuals.

2.Significant Problems

2.1. Theoretical framework

This study adopts a conceptual framework from Wardle & Derakhshan (2017) that has been used to explore the problem of fake news from a social point of view [14]. This conceptual framework has been adopted because it serves as a common denominator in exploring fake news in different legal jurisdictions and societies that differ in values but yet share the most aspect of constitutionalism and the rule of law.

* Eugene Chan: eugeneyhchan@thei.edu.hk

Figure 1. Conceptual Framework breaking into 3 forms of information disorders by Council of Europe [14].

	Forms
Types	Misinformation; Disinformation; and Malformations
Phases	The agent; The message; and The receiver
Elements	Misinformation; Disinformation; and Malformations

2.2. Defining "Fake News"

As the term fake news becomes more and more popular, it has become challenging to define. In the development of any set rules or regulations to combat a particular vice in society, it is in the interest of the law and the law enforcers to determine the definition of the vice and other underlying concepts; similarly, the significance of adopting a standard definition of fake news is vital to understanding the role of the law and what the current legislatures on global, national and local levels toward combating its existence in society today [3].

In the United States, where the term has recently gained attention and shined a light on it in the global stage, there lacks a standard definition to describe the word and the phenomenon of fake news. For instance, one of the Journalists in The Washington Post defines fake news as 'deliberately constructed lies that bear the intention of deceiving the public' (Engesser et al. 2017) [7]. However, many of the legal scholars warn that defining fake news does not only overlap with the existing and communication laws but also interfere with the First Amendment's legal protection of the freedom of speech. Numerous other societies have found themselves in this kind of legal ambiguity due to the precise definition and description of the term fake news.

In attempts to offer a working definition for the term fake news, Egelhofer and Lecheler (2019) show that there two dimensions to the usage and perceived importance of the word 'fake news.' They include; fake news as a genre and fake news as a label. There are three main characteristics of fake news as a genre; low in is facticity, journalistic, and bears the intention to deceive [6]. Researchers and authors agree that fake news is made up of false information. For instance, in line with the adopted conceptual framework, Butler (2018) lists a variety of fake news; misinformation, disinformation, and manipulated content [4]. Finkel et al. (2017), on the other hand, describes fake news to be either false in its totality or intentionally contains deliberately misleading aspects from the information provided [9]. Hence, the use or presence of facts does not mean that a message cannot be fake news.

As a genre, it has also been observed that fake news is often seeking to appeal to potential consumers, hence mimics real news or media content. From this approach, fake news is not only false but also tends to imitate real news. Hence, articles of fake news can have a structure that includes a heading, body, and even media for illustrations. Mimicking real news is a creation or production antique aimed at achieving the pretence of being real. Waldman (2014) notes that, in such cases, fake news can be better described as pseudo-journalistic [15].

2.3. Research Methodology

Desktop-based research has been conducted to achieve the objectives of this research. Critical to note that most of the information vital in completing this research is electronically availed. Hence, secondary sources of information have been utilized to reach conclusions on the existence of regulations and its effectiveness toward combating fake news in the three jurisdictions. The desktop research approach has also been utilized in completing the literature review section of this research. The scope of this research is guided through the use of a theoretical framework.

3 Evidence on The Fight Against Fake News

The law against defamation is perhaps one of the pre-existing legal frameworks that have been used by some countries to protect individuals and the general public from the malicious intents of fake news. A good example is South Korea. The constitution of South Korea protects individuals against defamation and tarnishing of reputation in the exercise of one's free speech. However, this is in contrast with other countries like the United States whose constitutional weight is behind the exercise of free speech. In South Korea, fake news is perceived to be both a crime and civil wrong. In November 2017, under the Public Official Election Act, an individual that posted false information on the election was convicted of conducting an election crime (Park and Youm, 2019) [11]. The ruling court ordered that the defendant pay a fine for posting misleading information that had potential effects on the general public. In France, the 1881 Law on Freedom of the Press and Article L.97 of the Legal code is still enforced (Rampersad 2020) [12].

The existing legislature dictates that causing disturbance to public peace through the production, reproduction, or dissemination of fake news in bad faith is punishable by a fine. Numerous other state legislatures have been enforced.

Despite the apparent challenges of the current state of flooded information, numerous measures have been put in place by state regulators and other players to combat information disorders and the raging wave of fake news. Legislative measures appear to be the most consequential in some of the countries that have waged war against fake news [4]. There are broadly two categories of approaches that have indicated success in such country's enforcement of the existing legislature and creation of new legislature to overcome legal loopholes.

3.1. New Legislation

On the other hand, nations throughout the world have realized the need to exert more control by the creation of a new legislature. Some of the legislatures are general in manner while others target specific activities and platforms within the process of informational disorders [8]. In Germany, the 'Act to Improve the Enforcement of Law in Social Network' began to take full effect in 2018. The Act is specific to social media platforms (Malda and Milan, 2018). The Act requires that malicious information shall be removed from the internet within a specified period in which it has been determined by a law of court to be malicious [10].

Japan has been struggling with the fake news issue and has established several laws in an attempt to curb the spread of fake news. Among the laws is the broadcasting act that regulates broadcasters by establishing a system that prohibits them from distorting facts. The penal code focuses on news aimed at defamation. The penal code provides the punishment for fake news mongers who defame an individual or obstruct a business which is three years and above or not less than 500,000 Yen in fines. The election law prohibits individuals from spreading defamatory news during election campaigns, and lastly, the internet provider law that exempts internet providers from liabilities for preventing the distribution of infringing information about other individuals. Out of all the given laws, Japan does not have one designed to curb the spread of fake and defamatory news through social media [10].

3.2 Service Providers' action

With over 160 million Indian citizens using WhatsApp, the Indian government has moved swiftly to protect its citizens and the general public from the potential effects of fake news. Platform regulation and collaboration with service providers to enforce jurisdictional laws are some of the measures that countries have taken against the spread of the threat of fake news.

Facebook and other social media websites are the primary points of targets for regulations in Germany focused on curbing the problem of fake news. As noted in the literature review, Germany developed 'The Act to Improve the Enforcement of Law in Social Networks.' The new regulation aims at reducing the spread of disinformation and other forms of information among millions of Germans users of such social medial platforms. The federal government, under this law, provides that all any forms of information that may undermine the wellbeing of the public or individuals are pulled down with 24 hours of the court's determination of their illegality. The Act did provoke concerns about the infringement of personal freedoms of expression and speech. Nevertheless, Germany authorities hold the stand that such rights are only exercised within the provision of the country's law to protect the rights of other members of the society that have spelled out in the country's criminal law [14].

3.3 State exercises over the dissemination of information

Unlike other countries, China's fake news problems tend to be countered on issues of healthcare, food safety, and other social issues. Interestingly, China has managed to curb issues centered around politics. This has been made possible with the use of old and new regulations surrounding the increased use of social media among members of its society.

The case of citizens of China being subject to anyone involved in the spread of fake news that holds the potential of disturbing public peace in an information network or media is culpable of punishment by imprisonment. For instance, the Manufacturing and Spread of Fake News Law prohibit most of the procedures that are associated with the manufacturing and distribution of misinformation and disinformation in the country (The Law Library of Congress, Global Legal Research Directorate, 2019) [13].

Some of the laws include the Ninth Amendment to the Criminal Law of the People's Republic of China. The law provides that anyone involved in the spread of fake news that holds the potential of disturbing public peace in an information network or media is culpable of punishment by imprisonment. There also exists the Real-Name Registration legal provision. Under this law, members of the public are legally required to use their official names on social or official websites that may be used to disseminate information that may have a potential impact on the broader population.

3.4 Finding

The individual case studies from above highlight some of the legislative measures that countries across the globe have sought to put in place in order to deal with fake news. The study establishes that there is a minimal reflection of the term 'fake news' in new and old legislation. However, as outlined from the theoretical framework, fake news has a genre that has been tackled through the identities of misinformation, disinformation, and mal-information. Hence, the issue of fake news has label largely remains unresolved with the existing legislature. The essence of addressing fake news in the form of propaganda and opinionated expression appears to be alienated. The probable victims in this scenario include journalists that stand the threat of a poor reputation.

Curbing the problem of fake news presents new challenges to individual countries. Some of the challenges are unique to the country's political and social background. Most of the countries are still in the process of developing legislative measures that reflect the new problem of fake news. Enforcement issues are even more troubling. For instance, Germany and Kenya face resistance to how some of the legislative measures are exercise. Lastly, collaboration with service providers like Facebook and Twitter are primary issues that may require more than legislative measures to deal with them [13].

4 Conclusion

Fake news is an issue many countries are struggling with and have been unable to put in control. The spread of fake news has evolved with technological developments and is now easy, with almost everyone having smartphones and the internet. The manufacturing and dissemination of false information have become easier. Spreading fake information impacts communication, one of the essential aspects of society negatively. Proper communication is vital in informing society on essential issues occurring, for example, with the current coronavirus pandemic.

Misinformation and fake news impact society negatively, but most jurisdictions have been unable to effectively curb the spread of fake news, especially through venues such as social media without infringing the rights of its citizens of free speech and their privacy and confidentiality. This research aimed at determining how regulation contributes to the stopping of the adversity of fake news. The different measures different nations have taken in the attempt to curb the spread of fake news depend on the political, social, and regulative uniqueness of the different societies.

References

1. Allcot, Gentzkow. *Social media and fake news in the 2016 election*. JEP, **211–236**. (2017)
2. Allen, J., Howland, B., Mobius, M., Rothschild, D., Watts, D. J. *Evaluating the fake news problem at the scale of the information ecosystem*. S.A, **1-18**. (2020).
3. Bhaskaran, Mishra, Nair., *Contextualizing fake news in the post-truth era*: JE. AP M.E, **41–50**. (2017)
4. Butler, A., *Protecting the Democratic Role of the Press: A Legal Solution to Fake News*. WULR. (2018)
5. Cheruiyot. *Popular criticism that matters: Journalists' perspectives of "quality" media critique*. JP, **1008–1018**. (2018)
6. Egelhofer, J. L., Lecheler, S. *Fake news as a two-dimensional phenomenon: a framework and research agenda*. ICA, **97-116**. (2019)
7. Engesser, Ernst, Esser, Büchel. *Populism and social media: How politicians spread a fragmented ideology. Information, Communication & Society*, **1109–1126**. (2017)
8. Figueir, Á., Oliveirab, L. *The current state of fake news: challenges and opportunities*. PCS, **817-825**. (2017)
9. Finkel, J., Luo, M., Metaxa-Kakavouli, D., Peeples, C. *Fake News & Misinformation Policy Practicum*. SLS. (2017)
10. Malda, V., Milan, S. *Wisdom of the Crowd; Multistakeholder Perspective on Fake News*. UofP. (2018)
11. Park, A., & Youm, K. H. *Fake News from a Legal Standpoint; The United States and South Korea Compared*. SJIL, **100-110**. (2019)
12. Rampersad, G. *Fake news: Acceptance by demographics and culture on social media*. JITP, **1-11**. (2020)
13. The Law Library of Congress, *Global Legal Research Directorate. Initiatives to Counter Fake News in Selected Countries*. USC. (2019)
14. Wardle, C., Derakhshan, H. *Information Disorder*. Council of Europe, pp.20-27. (2017)
15. Waldman, A. E. *The Market Place of Fake News*. J.C.L. (2014)

An exploration of selective justice in the International Criminal Court

Zhimin Geng*

School of International Law,University of World Economics and Diplomacy,Tashkent 100174,Uzbekistan
School of Jining Normal Unicersity,Wulanchabu 012000,China

Abstract: With the development of history, human society has experienced the development stage from disorder to law, from domestic law to international law. International law came into being and developed in the context of increasing exchanges between states. International criminal law has emerged and developed in modern times on the basis of the continuous development of international law. The International Criminal Court is also a form of international legal system that has emerged when human society has developed to a certain stage.To punish international criminal criminals more effectively through the trial of the International Criminal Court, to stop, prevent or deter the occurrence of international criminal crimes and potential international criminal criminals is an inevitable historical product of the development of human society to a certain stage, and it is a milestone progress. The International Criminal Court has played a positive role in the protection of human rights in all countries, especially in developing countries. Based on the principle of complementary jurisdiction, the jurisdiction of the International Criminal Court encompasses a set of jurisdictional principles, conditions and procedures to ensure the legality and impartiality of its exercise. These principles, conditions and procedures of jurisdiction have certain particularity. Since the establishment of the International Criminal Court, there has been a problem of selective justice, so this paper tries to discuss the meaning of selective justice in the International Criminal Court, the development of selective justice in the International Criminal Court and the influencing factors of selective justice in the International Criminal Court.

1.INTRODUCTION

The entry into force of the Rome Statute of the International Criminal Court (hereinafter referred to as the Statute) and the establishment of the International Criminal Court are important symbols of the improvement and development of the international criminal law system. But in the first decade of the ICC's existence in 2002, one of the main controversies that emerged was the issue of selective justice, as the court did not investigate or prosecute outside Africa. The international community has criticized the "selective justice" of the International Criminal Court, especially African countries, which have expressed strong dissatisfaction. Therefore, the international community has been highly concerned about the judicial selectivity of the International Criminal Court, and domestic and foreign scholars have conducted in-depth discussions on its connotation from both broad and narrow perspectives, providing a solid theoretical basis for the study of this issue. Due to a combination of factors, the International Criminal Court often adopts a selective approach to justice in practice, which may jeopardize national sovereignty, call into question national judicial systems and have a negative impact on non-States parties.

2.THE CONCEPT OF SELECTIVE JUSTICE IN THE INTERNATIONAL CRIMINAL COURT

In recent years selective justice has been directed at African leaders, so enforcement of international criminal law has been more selective than enforcement by domestic courts. Due to the higher selectivity of international criminal law enforcement, international criminal law is more likely to be accused of injustice, and this selectivity controversy is triggered by the choice of international criminal law and its institutions. Selective justice refers to a system in which when a crime is committed in a country or region, the judicial authorities will make different treatment according to its specific circumstances.[1] To the extent that judicial selectivity exists, limitations are necessary but acceptable. Under the current legal framework, selective justice should be appropriately restricted. Given the realities of the time constraints and limited resources that prosecutors face in their work, a comprehensive assessment of the information and evidence available, as well as the environment in which investigations and prosecutions are

* Corresponding author: 515467315@qq.com

carried out, must be made in order to prioritize the selection of cases for investigation and prosecution.[2]

According to Chinese and foreign scholars, the selective justice of the International Criminal Court covers a series of selective acts, including but not limited to treaty making, discretion, the UN Security Council and state contracting, etc., which have a certain impact on the operation of the International Criminal Court system; It refers to a set of rules and mechanisms based on "jurisdiction" in the field of international law.[3] In the International Criminal Court, the discretion of judges and prosecutors leads to selective judicial actions.

3. THE INTERNATIONAL CRIMINAL COURT CHOOSES THE DEVELOPMENT OF JUSTICE

Before the birth of the International Criminal Court, the establishment of some international criminal tribunals or international military tribunals to punish international crimes marked the progress of the international criminal system, but the emergence of the statute proved the codification of international criminal law. But with that comes selective justice.

3.1 Embryonic stage

After the end of World War II, the establishment of the Nuremberg, Tokyo and ICTY Tribunals, as well as the Rwanda Tribunals, injected targeted judicial shoots into the field of international criminal justice. With the development of the practice of selective justice, various countries have enacted or revised relevant laws to restrict selective justice. Prior to the Rome Statute and the subsequent establishment of the International Criminal Court, advocates of international criminal justice were increasingly concerned about the judicial selectivity involved in the arbitrary creation of single conflict tribunals. With the gradual maturity of the selective justice system, selective justice has become a special legal department independent of traditional litigation. Since these international criminal judicial bodies were established in the context of selective justice, selective justice was seen in its infancy as a criticism, whether well-intentioned or ill-intentioned, and has been a difficult problem for the ICC from its inception to the present day. The main feature of the sprout of selective justice is that it does not conform to the principle of neutrality and non-discrimination, which has become the biggest obstacle to its development.[4] The credibility of the International Criminal Court has suffered direct damage in the embryonic stage of selective justice, which is one of the most typical negative acts.

3.2 Formation stage

In 1994, the draft Rome Statute of the International Criminal Court was introduced; In 1995, the United Nations established the Preparatory Committee for the International Criminal Court; On July 1, 2002, 60 countries ratified the Rome Statute, making the International Criminal Court the world's first permanent international criminal institution responsible for enforcing jurisdiction over international crimes in accordance with the Rome Statute. The establishment of the International Criminal Court is a wise move, but it cannot ignore the fact that while reducing selectivity towards the accused, it also adopts a selective stance towards the law.[5] Throughout the process of compiling the Rome Statute, legislators and compilers constantly compromised and screened to ensure the quality and accuracy of the final version. From the cautious consideration in the initial formulation of the Convention to hasty accession and modification later, and finally to becoming a fully effective text, there have been multiple iterations, ultimately forming the current massive code. In the process of compiling the provisions of the Rome Statute, the power of judges and prosecutors has caused the greatest controversy. Therefore, the compilers of the Rome Statute aim to limit the discretion of law enforcers to the maximum extent possible, which also gives the International Criminal Court more ability to transcend boundaries than the former Yugoslavia Tribunal. It is undeniable that the compilation of the Rome Statute provides selective judicial guarantees for legal practitioners, which is beyond doubt. In the relevant provisions of the Rome Statute, judges and prosecutors of the International Criminal Court enjoy extensive discretionary power, and the legalization of these discretionary powers is a symbol of the selective judiciary formation of the International Criminal Court.

3.3 Practical stage

On January 27, 2016, the Pre-Trial Chamber, at the request of the Prosecutor, authorized it to investigate the situation in Georgia. However, as the situations investigated by the Prosecutor were all African countries, there were voices criticizing the Prosecutor for selectively administering justice to African countries. Although this claim was supported, it also raised some doubts.The disappointment of the African Union with the Court and the United Nations Security Council mainly stems from its belief in the politicization of international criminal justice. As a result, the United Nations Security Council was authorized by the Rome Statute to submit cases of suspected international crimes within the jurisdiction of the International Criminal Court to the Court, while the transfer of the Security Council extended the Court's jurisdiction to non contracting parties, including referring the situations in Darfur, Sudan, and Libya to the Court. African countries account for the vast majority of investigations and prosecutions by the International Criminal Court, regardless of the contracting situation. During this period, the United States launched wars in Afghanistan and Iraq, and the crimes committed by them were widespread within the jurisdiction of the International Criminal Court.[6] Until 2020, Judge Hofmansky authorized prosecutors to conduct a comprehensive investigation into the alleged war crimes and crimes against humanity committed by the United States in Afghanistan in 2003, determining that the International Criminal Court had jurisdiction over these crimes. Regardless of whether the United States is a party to the Rome Statute or not, the International Criminal Court had investigative capabilities and powers during this period, but ultimately failed to investigate and try the United States.

3.4 Regulatory Stage

In the selective jurisdiction of the International Criminal Court, the discretion of prosecutors has caused the greatest controversy. When the Pre-Trial Chamber authorizes the Prosecutor to investigate a particular situation, or when the State Party or the Security Council submits the situation to the Prosecutor, and the Prosecutor's Office begins a formal investigation of the specific situation, some people believe that once the jurisdiction of the court is triggered, the Prosecutor will have considerable discretion to choose cases in a specific situation. This viewpoint is not entirely in line with the actual situation, as prosecutors' judgments on different situations are often influenced by the judge's personal stance and other factors, and in the actual operation process, prosecutors may also make appropriate adjustments as needed. The principle of fairness is the primary principle for regulating judicial selectivity. The prosecutor's office screens and analyzes potential criminal offenders involved in multiple objects to ensure that they are not subject to partisan bias.

4.FACTORS INFLUENCING THE SELECTION OF JUSTICE BY THE INTERNATIONAL CRIMINAL COURT

The biggest problem facing the establishment of the International Criminal Court is the existence of different legal systems and different criminal trial systems, that is, the court must be able to adapt to civil law countries that adopt interrogative trial methods, but also be satisfied with common law countries that adopt procedures that focus on discovery and presentation of evidence. States must also reach agreement on all relevant judicial issues. This different legal system directly affects the judicial selectivity of the International Court of Criminal Law. The selective justice of the International Criminal Court has active selective justice, which mainly includes the choice of legislation, the choice of law enforcement and the choice of the state.

4.1The choice of legislation

The establishment of the International Criminal Court is based on the Rome Statute, which is a universally adopted form of international treaty-style legislation of international law, among which the most important legislative principle is the "principle of legality of crime", which for the first time explicitly stipulates the principle of legality of crime in the form of international criminal legal documents.[7] In addition, the Statute also selectively identifies four types of crimes, the selection of which is closely related to the specific historical background, the scale of the crime, armed conflict and other factors, so that the most serious types of crimes in the international community are diverse and need to be investigated by the International Criminal Court. Given the dual nature of the ICC as a treaty-based international organization and a court for the trial of criminals, it must apply the basic legal principles of criminal law in accordance with the rules of treaty interpretation in international law. Therefore, the complexity and prominence of the problem of judicial selectivity are more significant.

The main body is the International Criminal Court itself, composed of judges and prosecutors; There is also passive selective justice, including political selectivity, the selectivity of States parties, including the UN Security Council and relevant countries. The characteristics of these two types of selective justice have inherent theoretical and logical relations. The subject of selective justice is not only the dominant player but also the participant, influencing and penetrating each other.

4.2 The choice of law enforcement

In the statute, the ICC has adopted a unique organizational structure that organically combines judges and prosecutors to form a unified law enforcement system, making the relationship between judges and prosecutors both independent and interrelated.[8] The judge's discretion is a relatively independent judicial power. The discretion of the International Criminal Court is a dual structure composed of two different levels of discretion of the judge and the prosecutor, which is characterized by the interweaving of the discretion of the judge and the prosecutor, forming a unique legal system. Many States were inclined to accept an independent international criminal prosecutor to enable preliminary investigations into allegations of core international crimes, but this did not meet their expectations because the Prosecutor's investigative powers were too strong. As a result, many States have decided to authorize the Pre-Trial Chamber to act as the authority for the Prosecutor to initiate investigations, in order to represent the inherent constitutional constraints on the Prosecutor, while some States require the Pre-Trial Chamber to accept the autonomous authority to initiate investigations. Although the Pre-Trial Chamber has the dual functions of prosecution and trial, it has not been as effective as it should be, mainly because it lacks a self-regulating mechanism, that is, internal checks and balances.[9] Although the Statute is intended to impose constraints on the Prosecutor through the Pre-Trial Tribunal, there are still doubts about the trust in the impartiality of the judge's supervision due to the delicate cooperation and game relationship between the judge and the prosecutor, as well as the lack of detachment in their status. At the same time, the procurator's power is too concentrated and its own lack of effective checks and balances makes it difficult to play a role.

4.3 Choice of country

The political choice to accede to the Statute is different from the political choice of the Security Council, because the Council influences the selective justice of the International Criminal Court through legal means, while States can only rely on diplomatic means, which is the selectivity of national interests based on national sovereignty, and thus affects the selective justice of the International Criminal Court.[10] Due to different national conditions, the scope and degree of application of international law are also different. It punishes criminals in large countries and produces selective justice compared to countries with less political 'influence'." The influence of powerful countries in the international criminal law system is needless to say, and how should weak countries choose, especially weak African countries. But the ICC, which was set up with the support of African countries, now wants to "break the hearts of our African brothers". At a time when African countries have little influence in some important international institutions, they are pushing for the creation of another body in the global governance system, independent of the UN Security Council. They want the ICC to be an institution that transcends the old centers of diplomatic and economic power, which they fear will turn the court into a political organ of powerful countries and suffer injustice. Since its inception, the ICC has touted its "principle of complementary jurisdiction" to ensure the integrity of its jurisdiction. It can not only safeguard its own sovereignty by interpreting international law, but also provide protection and assistance to other States by exercising jurisdiction. However, the jurisdiction principle of the International Criminal Court, as a sovereign state, is constantly disputed, and this dispute constantly touches the bottom line of national sovereignty.

Since its inception, the International Criminal Court has been controversial, but it is undeniable that its judicial activities have played a certain role in maintaining international peace and security. Selective justice has existed in the International Criminal Court since its establishment, and has gone through four stages: germination, formation, practice and regulation, and has shown different characteristics of selective justice. The International Criminal Court is mainly affected by the factors of legislation, law enforcement, politics, state contracting, etc., resulting in two aspects of active and passive selective justice theory. As a result of these factors, selective justice in the International Criminal Court has often led in practice to situations that undermine State sovereignty, question national judicial systems and target non-States parties. Therefore, it is necessary to deal with it from the aspects of foreign policy, national criminal law system, international judicial assistance, etc., which can not only protect national interests, but also effectively combat international crimes and maintain their justice.

REFERENCES

1. Wang Xiaojing. Mechanisms to check the Prosecutor's own investigative power in the International Criminal Court: Judicial review by the Pre-Trial Chamber [J]. Journal of International Law, 2023(1):20.
2. Zeng Delu. Study on Trial in Absentia System in International Criminal Court [D]. Southwest University of Political Science and Law,2021.
3. Feng Mingchao, Li Fantang, Wang Shengsheng, et al. Discussion on the jurisdiction of the International Criminal Court [J]. Legal Review, 2017-02-04.
4. Wang Yi. Study and Response on Selective Justice in International Criminal Court [J]. Journal of Shenyang Normal University, 2022-04-01.
5. Zhang Guiling. Research on the International Criminal Court [M]. Beijing: People's Publishing House, 2013.
6. Yu Nan. A Study on the discretion of Judges of the International Criminal Court [M]. Beijing: Law Press, 2014.
7. JIA Yu. International Criminal Law [M]. Beijing: Law Press, 2019.
8. Wang Xiumei. Research on the International Criminal Court [M]. Beijing: China Renmin University Press, 2002.
9. Eckart Klein, "Parallel Activities of the Security Council and the International Court of Justice in Dealing with Disputes Endangering Peace," in Essays on Contemporary Federal German International Law, Beijing University of Aeronautics and Astronautics Press, 1992.
10. Bedjiaoui Mohammed.The New World Order and the Security Council :Testing the Legality of its Acts[M]:Doedrencht:Martinus Nijhoff Publisher,1994.

Analysis of Spatio-temporal Change Characteristics of Water Environment in Nansi Lake Based on Satellite Remote Sensing Monitoring

Xingkai Wang, Wenlong Yu, Rui Wang, Bin Li, and Xiangyang Cao*

Shandong Jiaotong University, School of Civil Engineering, 5001 Haitang Road, Jinan, China

Abstract. Lakes play a crucial role in the preservation and distribution of surface water resources, making them integral components of environmental protection. This study utilizes remote sensing data obtained from Landsat-8 to investigate the temporal changes in water area, land use patterns, and distribution of cyanobacteria in Nansi Lake between 2015 and 2023. Furthermore, it conducts an extensive analysis of the environmental condition of Nansi Lake, considering the escalating environmental crises resulting from industrial and agricultural development in recent decades. By adopting a multidimensional approach, this article offers a comprehensive examination of the environmental situation in the Nansi Lake region. Notably, the findings of this study exhibit enhanced accuracy and practicality when compared to conventional environmental analysis methods. Consequently, this research contributes novel ideas and methodologies for the monitoring and safeguarding of the ecological environment of lakes through the application of remote sensing techniques.

1 Introduction

Aquatic ecosystems are an important foundation for the existence of life on Earth. In recent decades, human activities have had some impacts on freshwater lakes and their surrounding environment. Compared to traditional sampling analysis methods, conducting a series of environmental analysis on a certain research area based on satellite image data is a fast, effective, and cost-effective method.

In the past decade or so, due to the rapid development of domestic industry and agriculture, the phenomenon of water pollution in China has become increasingly apparent. This article conducted a series of environmental analyses on Nansi Lake from 2015 to 2023 based on Landsat-8 data. This study conducted a two-year cycle to monitor changes in water area, analyse land use types, and extract and analyse cyanobacteria from the selected five time periods in the Nansi Lake area.

Compared to previous research on remote sensing of lakes, this article conducts a more objective and rigorous analysis of the environmental situation of Nansi Lake from multiple dimensions. It can accurately reflect the current environmental situation of Nansi Lake, providing new ideas and theoretical basis for using aerial remote sensing technology to monitor changes in the lake environment on a large scale.

2 Area and methods

1.1 Research area

Nansi Lake is the largest freshwater lake in northern China, located in Weishan County, Jining, Shandong Province. It plays a crucial role in the East Route of the South to North Water Diversion Project [1][2]. Since the beginning of the 21st century, due to the development of heavy industry along the lake coast and other reasons, the water pollution in this area has been severe. The study area of Nansi Lake is shown in the Figure 1.

1.2 Data and used methods

This article uses remote sensing images provided by Landsat-8 satellite as the data source and analyses the environmental situation of Nansi Lake through a combination of three methods [3-6].

This study utilized a New Normalized Difference Water Index *(NNDWI=ρ(RED+NIR)/ρ(GREEN*C)))* to extract the water body of Nansi Lake. Support vector machine classification [7][8] and cluster analysis were used to study the land use types in the Nansi Lake area. The Normalized Difference Vegetation Index (NDVI=ρ(NIR-RED)/ ρ (NIR+RED)), which eliminated outliers, and parameter assignment were used to extract and analyse cyanobacteria in the water body of Nansi Lake. And it is shown in Table 1.

* Corresponding author: caoxiangyang@sdjtu.edu.cn

Table 1. Parameter table for assignment

Year	Parameters corresponding to cumulative area ratio less than 5%	Parameters corresponding to cumulative area ratio greater than 90%
2015	-0.220927	0.444555
2017	-0.429156	0.485164
2019	-0.462526	-0.095810
2021	-0.729019	0.108627
2023	-0.346604	-0.197162

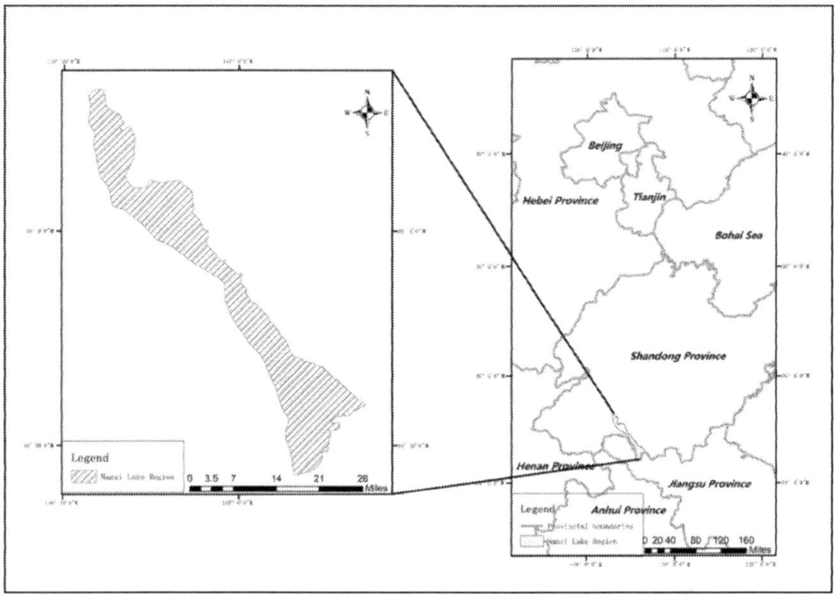

Fig. 1. Research area of Nansi Lake.

3 Results & discussion

This article uses Landsat-8 image data from 2015 to 2023 as the data source to conduct three different studies on the Nansi Lake area.

The current research on extracting water based on remote sensing technology is generally divided into two directions: enhancing water information and suppressing non-water information. This article selects the method of enhancing water information. Compared with other water indices, NNDWI has a lower probability of classification errors and higher accuracy in extracting results [9][10]. In order to accurately extract the boundary of the Nansi Lake water area, this paper further refined the above results using threshold method based on the NNDWI results. The threshold range is 0.0000001 to 0.8. The final water area changes in the Nansi Lake area from 2015 to 2023 are shown in Figure 2.

Due to the high classification accuracy of the support vector machine classification method compared to the minimum distance classification method and the maximum likelihood classification method in this research area. So, this article uses SVM to divide the Nansi Lake area into four categories: water area, arable land, residential land, and other land. The classification results of land use types in Nansi Lake based on SVM are shown in Table 2.

In order to further optimize the extraction results, this study excluded the interference of suspended impurities in water through threshold method based on NDVI, and combined with real research data in the Nansi Lake area, outliers in the results were removed through parameter assignment, which can more intuitively reflect the distribution of Cyanobacteria in the water. Figure 3-7 show the distribution of Cyanobacteria in Nansi Lake.

Table 2. Land use classification based on SVM

	2015	2017	2019	2021	2023
Water area	542.962km^2	569.941km^2	626.818km^2	632.908km^2	487.053km^2
Arable land	397.062km^2	358.315km^2	259.044km^2	393.999km^2	492.419km^2
Residential land	14.657 km^2	15.543 km^2	13.318 km^2	14.318 km^2	14.155 km^2
Other land	158.282km^2	169.163km^2	213.161km^2	70.985km^2	118.692km^2

* Corresponding author: caoxiangyang@sdjtu.edu.cn

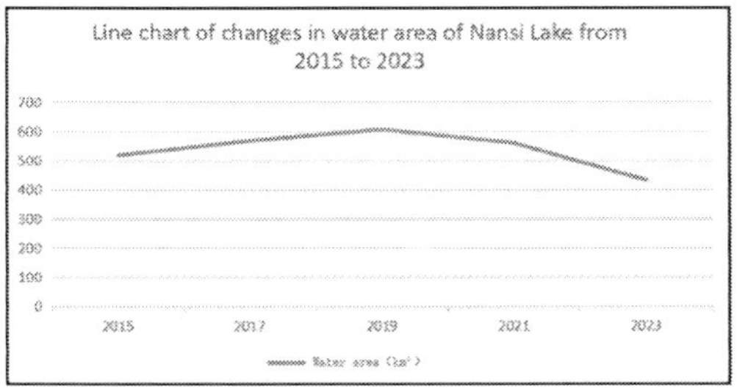

Fig. 2. Changes in water area of Nansi Lake from 2015 to 2023.

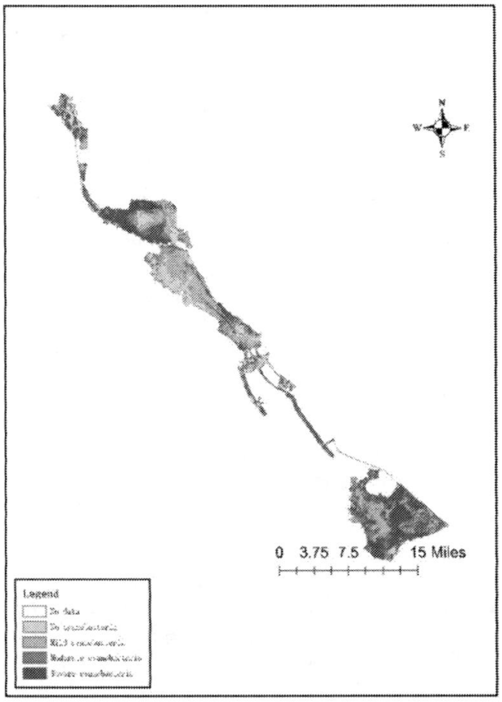

Fig. 3. Distribution map of cyanobacteria in Nansi Lake in 2015.

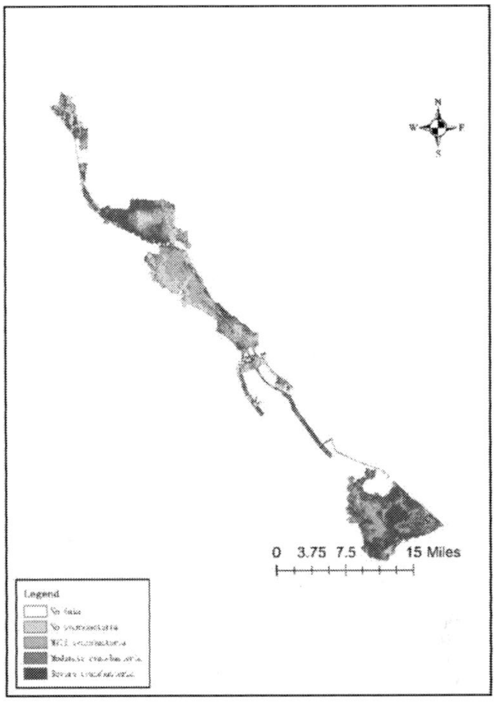

Fig. 4. Distribution map of cyanobacteria in Nansi Lake in 2017.

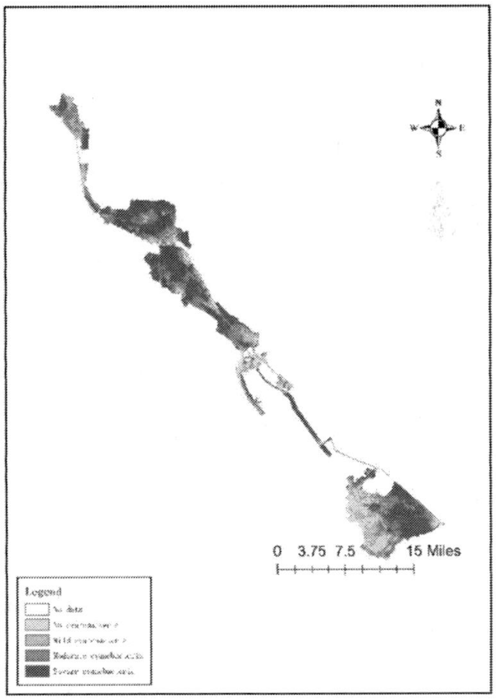

Fig. 5. Distribution map of cyanobacteria in Nansi Lake in 2019.

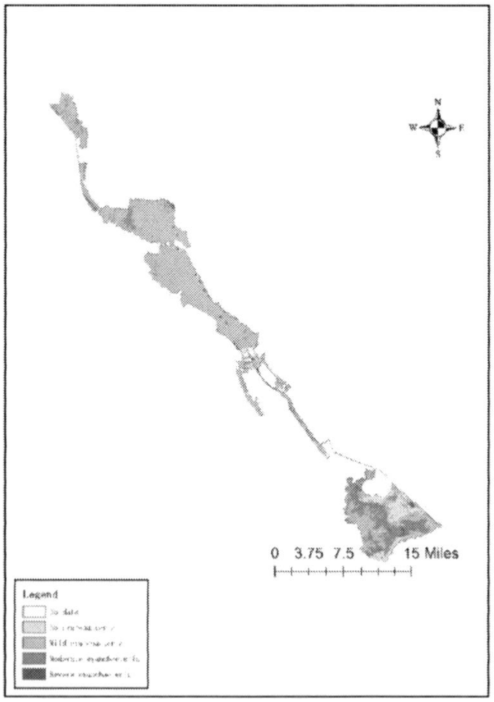

Fig. 6. Distribution map of cyanobacteria in Nansi Lake in 2021.

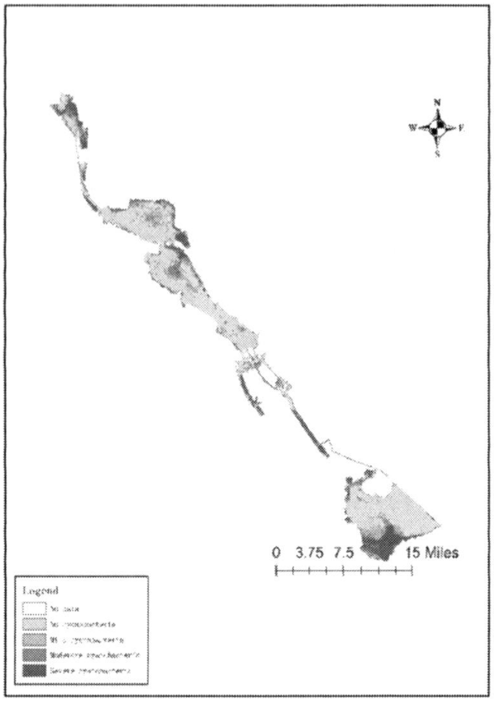

Fig. 7. Distribution map of cyanobacteria in Nansi Lake in 2023.

4 Conclusion

Analysing the changes in the water area of Nansi Lake from 2015 to 2023, the water area in this area first increased and then decreased. As of 2023, the total decreased area was 85.428 square kilometres. Through the study of land use types in the Nansi Lake region, this article found that during this period, the proportion of water area in the Nansi Lake region decreased from 49% in 2015 to 44% in 2023. With the decrease in the proportion of water bodies, the proportion of arable land in the region has increased, and the changes in the two are extremely similar. Through the extraction and analysis of cyanobacteria in this area, it can be concluded that the distribution of cyanobacteria in Nansi Lake is random, and the frequency of occurrence is periodic. The situation of cyanobacteria is most severe in the southeast region.

Through a comprehensive analysis of the Nansi Lake area, the following conclusions can be drawn: between 2015 and 2023, the water area of Nansi Lake had some fluctuations, but overall it was still within the self-regulation range of the lake; The proportion of arable land in the region has increased, while the proportion of water bodies has decreased, and there is a correlation between the two; The problem of cyanobacteria in the Nansi Lake area has been weakening over time, and the data from 2023 can reflect that the vast majority of water bodies are not covered by cyanobacteria. In summary, during this period, environmental protection in the Nansi Lake region achieved phased results, and environmental issues were greatly reduced. The local government has made effective efforts in protecting the environment.

In the field of environmental research in Nansi Lake, this article uses remote sensing technology to combine water area, land use status, and cyanobacteria analysis. By comparing and analysing multiple dimensions, the analysis of the regional environment is more intuitive, accurate, and efficient. This article uses NNDWI for the first time to extract the water bodies of Nansi Lake. Compared to other water body indices, this index has significant advantages in accuracy and reliability. In terms of cyanobacteria extraction and analysis, this article uses a combination of NDVI and assignment parameter method to accurately study the distribution of cyanobacteria in Nansi Lake. In the future, if the relationship between lake water quality parameters and cyanobacteria reflectance spectra can be identified, and combined with on-site sampling data, it will be more efficient to complete large-scale quantitative analysis of water pollution.

Acknowledgements

This research was supported by the Scientific Research Foundation(Z202106) and the Teaching Reform Research Foundation(2021XJYB45) of Shandong Jiaotong University.

References

1. L. Meng, J. Dong., LUCC and Ecosystem Service Value Assessment for Wetlands: A Case Study in Nansi Lake, China. Water. **11,** 1597-1597 (2019)
2. Y. Li, Z. Yu, S. Ji et al., Diverse drivers of phytoplankton dynamics in different phyla across the annual cycle in a freshwater lake. Journal of Freshwater Ecology. **36**, 13-29 (2021)
3. X. Ma, L. Yang, E. Liu et al., Evaluating the release risk of potentially toxic elements from sediments in the New Zhuzhao River Estuary of Nansi Lake, using high-resolution technology and sequential extraction. Environmental monitoring and assessment. **195**, 353-353 (2023)
4. K. Vemuri, P. Reddy, B. Kumar et al., Deep learning based remote sensing technique for environmental parameter retrieval and data fusion from physical models. Arabian Journal of Geosciences. **14**, 1-10 (2021)
5. Y. Wang, L. Fan, R. Tao et al., Research on cropping intensity mapping of the Huai River Basin (China) based on multi-source remote sensing data fusion. Environmental science and pollution research international. **29**, 12661-12679 (2021)
6. W. Yu, Y. Wen, R. Wang., Water pollution environment information extraction based on Landsat images. International Conference on Geographic Information and Remote Sensing Technology. **2023**, 12552 (2022)
7. M. Danish, M. Talib, M. Mohammad., Support vector machine (SVM) approach to develop the discharge prediction model for triangular labyrinth weir. Water Supply. **22**: 8942-8956 (2022)
8. R. Andry, J.B. Martijn., Evaluation of MODIS-Landsat and AVHRR-Landsat NDVI data fusion using a single pair base reference image: a case study in a tropical upstream catchment on Java, Indonesia. International Journal of Digital Earth, **15**: 164-197 (2022)
9. P. Yutthana, K. Ekasit., Revision of Vajiralongkorn Dam's Reservoir Characteristic Curves Using NDWI Derived from Landsat 8 Data. The Environment and Natural Resources Journal. **18**: 134-145 (2020)
10. W. Arief, W. Pramaditya., Geometric Accuracy Assessment for Shoreline Derived from NDWI, MNDWI, and AWEI Transformation on Various Coastal Physical Typology in Jepara Regency using Landsat 8 OLI Imagery in 2018. Geoplanning: Journal of Geomatics and Planning. **6**: 55-55 (2019)

A Study of Strategies for the Legitimation of News Discourse Based on the Perspective of Convergence Theory ——Take the Washington Post's news coverage of the Federal Reserve's bailout policies during the subprime crisis as an example

Tingxuan Li *

Affiliation: Nanjing tech University, Nanjing, China

Abstract: In news reporting, legitimising discourse is usually present in subtle and implicit ways, potentially influencing people's social cognition and decision making. Based on Cap's convergence theory, this paper examines the legitimising discourse strategies in the Washington Post's coverage of the Federal Reserve's bailout policy. Through the integration and analysis of the three dimensions of the corpus, it is found that temporal convergence aims to link future presuppositions with present events, highlighting the Federal Reserve's large role in solving the subprime mortgage crisis, in order to construct the legitimacy of the act. Spatial convergence, on the other hand, highlights the negative impact of the external entity represented by the NFIs on the internal entity represented by the US and the positive impact of the measures taken by the internal entity, thus paving the way for arguing the legitimacy of one's own discourses and behaviours. Value convergence, on the other hand, demonstrates the potential for ideological conflict to be transformed into actual conflict by showing the clash of values and ideologies between external and internal entities in the discursive space. The study of convergence theory in this paper contributes to the construction and interpretation of the legitimation of political discourse and, at the same time, to some extent provides a reference for subsequent research on convergence and legitimation theory.

1 Introduction

In 2007, the US financial system was hit by an unprecedented shock with the collapse of the real estate market and the outbreak of the subprime mortgage crisis. In early 2008, the US Federal Reserve (hereafter referred to as the Fed) stabilised the financial market through a variety of means, including interest rate cuts and asset purchases, and also took some unconventional measures, such as quantitative easing, i.e. increasing the money supply to stimulate economic growth and ease pressure on financial markets. While the US has had many negative effects in reversing the financial crisis, its construction of legitimacy in political discourse has attracted widespread attention. In its coverage of the event, the Washington Post used various discursive strategies to legitimise its goals, actions and policies, and to find grand excuses for illegitimate US actions.

At the beginning of the 21st century, Piotr Cap proposed convergence theory, which is cognitively oriented and looks at the process of interpreting discourses and critical discourse analysis through multiple channels. [1][2] Based on convergence theory and the STA analysis model, Wang Yuxia studied the strategic reports issued by the US government against China, and the analysis found that they promoted the construction of political legitimacy of the US government at that time through the use of convergence strategies. [3]

This study finds that there is already a theoretical framework for foreign research in this discipline, but domestic research in this area started later than abroad, focusing mainly on the specific core discourse strategies used in specific discourse pieces, but lacking the study of the specific social background, international dynamics, value concepts and realpolitik contexts embedded behind them. With reference to these research directions and shortcomings, this study takes the news coverage of the Fed's bailout policy in the Washington Post as an example corpus and conducts innovative research on this topic.

2 Overview of the research design

2.1 Research Questions

This study analyses numerous news reports in the Washington Post about the interpretation of the Federal Reserve's bailout policy. It explores the construction of the discourse legitimation system of media political

*Email address: 806495488@qq.com

discourse strategies from Cap's discursive convergence perspective and examines how target media organisations utilise these different forms of discourse strategies to interpret the politically undesirable effects resulting from the introduction of bailout policy measures by the Federal Reserve. The analysis examines three aspects related to the legitimisation of the Fed's bailout policy by the Washington Post.

(1) How the Washington Post constructs the legitimisation of the Fed's bailout policy through temporal convergence;

(2) How the Washington Post constructs the legitimation of the Fed's bailout policy through spatial convergence;

(3) How the Washington Post constructs the legitimisation of the Fed's bailout policy through value convergence.

2.2 Research corpus

The corpus of the study was obtained from the Lexis Nexis database, from which the news reports of the Washington Post containing "Fed" or "Federal Reserve" in the title from 1 December 2007 to 31 May 2008 were retrieved. By removing descriptions of the macroeconomic situation, descriptions of the economic behaviours of specific companies or economic conditions, and reports that were not related to the economic theme, a total of 41 news reports with the highest correlation with the Federal Reserve bailout policy were selected.

3 Analysis and Discussion

3.1 Temporal Convergence

On 8 March 2008, the Washington Post commented on the Federal Reserve's new bailout policy, and its discourse is linked to the temporal convergence theory.

Case 1： The world's financial plumbing is so clogged that the central bank sees a need for new steps to clean it out to prevent severe damage. [...]"The Fed has been running around putting fingers in dikes," said Diane Swonk, chief economist of Mesirow Financial. "Without that, the dike would have imploded, and water would have been spilling in." (8 March 2008)

Case 2： Since the central bank was created in 1913, it has served as a lender of last resort for ordinary banks, allowing them to post high-quality loans at a "discount window" in exchange for cash.[...]Last night, it announced a new provision that will in effect do the same for major investment firms.(19 March 2008)

Temporal convergence is an important element of news discourse legitimisation strategy, which refers to the narration of past, present and future stories or situations for the purpose of legitimisation, and various types of narration can be used together. [4] In case 1, the Washington Post predicts the future direction of the market, citing the allusion to the boy who avoided flooding by plugging his finger against the dike, and rendering the economic crisis through descriptive terms such as "implode" and "spilling in", so that readers can visualise the challenges facing the national economy. The case looks to the future and shows readers the challenges facing the country's economy. This forms a "present-future" linkage structure, which builds legitimacy for the Federal Reserve's bailout initiative in the news discourse, leading readers to speculate on the direction of the economy and to understand and support the country's policies, thus enhancing the credibility of the U.S. government.

In case 2, the Washington Post's discourse constructs a "past-present" structure by using non-finite imitations such as "in 1913" and "last night", and the use of nominalised words such as "high-quality loans" and "in effect" is intended to elaborate that the Federal Reserve, as the central bank of the United States, has provided many loan facilities to other banks since its establishment. The media use this to make the interviewees consciously associate the US central bank with the long period of time it has been established and the many facilities it has provided during this period, which inadvertently gives the interviewees the psychological implication that the new policy will continue to bring facilities. In addition, this narrative easily reminds readers of the Federal Reserve's contribution to the relief of the Great Economic Crisis of 1929, the Second World War, and other similar economic difficulties, thus believing that it will play a key role in the subprime crisis and constructing a favourable value orientation towards the Federal Reserve and the United States government, which ultimately stabilises the market situation and corporate psychology.

3.2 Spatial convergence

In discourse analysis, the concept of spatial convergence is used to decipher the encroachment of external entities on internal entities, allowing entities in different spatial locations to converge. Internal entities mainly refer to the sender of the discourse and his allies, and the main difference between them and external entities is reflected in their ideology. Internal entities generally perceive their own values as dominant and positive, while the values of external entities are negative and can have a detrimental effect.

Case 3： Geithner a senior official in the Treasury Department told a congressional hearing April 3 that the risks of the Fed's actions "are modest in comparison to the substantial damage to the economy and economic well-being that potentially would have accompanied Bear's insolvency."(29 April 2008)

Analysed in terms of the construction of internal versus external entities. In case 3, the Washington Post compares the risks generated by the measures taken by the Federal Reserve with the impact generated by the bankruptcy of Bear Stearns, setting such non-state financial institutions as external entities and highlighting their negative impacts on the US economy, in order to achieve the effect of reinforcing the legitimisation of the bailout policy.

Analyse from the perspective of real and expected impacts. The case uses abstract noun phrases such as "substantive damage" to describe the expected impact of financial institutions on the US economy. Verb phrases such as "damage to the economy" are used to mark the displacement and direction of movement from external to internal entities, emphasising the potential threats faced by internal entities, and thus expressing the urgency and legitimacy of policy implementation.

3.3 Value convergence

Cap sees value convergence as a discursive marker that establishes a potential ideological conflict between the 'native' values of the speaker and his or her audience and the values of the antagonist. Similar to the other two convergences, value convergence is a formalised abstraction, an enforced recognition of the growing ideological conflict that instructs both internal and external entities, and Cap argues that this potential clash of values may eventually lead to physical conflict (e.g. armed conflict) in real life. [5]

Case 4: "There's been a lot of focus on Bear Stearns," Paulson said. "But the Fed took, I believe, dramatic and very important, powerful action to make liquidity available to . . . the investment banks." (18 March 2008)

Case 5: These regulations seek to "strive to protect borrowers from practices that are unfair or deceptive," said Randall S. Kroszner, a Fed governor who oversaw the drafting of the proposed rules. He added that the agency wanted "to do so without unintentionally causing responsible lending to shrink or unduly limiting consumer choice." (19 December 2007)

In case 4, the report quotes Treasury Secretary Henry M. Paulson Jr.'s comments on the Federal Reserve's economic behaviour, using a lexico-grammatical resource common in value convergence. A series of positively oriented words such as "dramatic and very important, powerful action" are used to fully demonstrate the objective correctness of the internal entity, i.e. the Fed's external bailout policy itself, emphasising the positively oriented and correct values towards the Fed's interest rate hike. This external bailout policy should be fully in line with the psychological condition of the social audience and market expectations.

Case 5 uses three lexico-grammatical resources to constitute value convergence. Firstly, positive words such as "strive" and "protect" are used to reflect the positive values and ideology of the internal entity. Meanwhile, the use of negativity words such as "unfair" and "deceptive", which juxtapose the positive values of IDCs and the negative values of ODCs, is meant to highlight the threats and negative impacts of the external entity on the internal entity, so that the internal entities to legitimise their strategic discourse. In addition to this, 'shrink or unduly limiting consumer choice' materialises the negative consequences of external entities on internal entities, leading the audience to associate these consequences with external entities and creating a value dichotomy. The case emphasises that the Federal Reserve's bailout policy can efficiently solve the economic problems and injustices faced by the general public, highlighting its emphasis on positive values and ideologies such as human rights and democracy, and thus arguing for the legitimacy of the bailout policy. The consolidation of ideal beliefs shared by the addressee is the basis of value convergence, on which the sender subsequently imposes on the addressee messages that deviate from his or her presuppositions. [6]

4 Revelations

Through the above in-depth analysis of the legitimising construction of the Washington Post's news report on the Federal Reserve's bailout policy, we can explore and understand the deeper American political and social psychology and its cognitive factors that are hidden behind its economic policy. It is these social and cognitive factors that further contribute to the legitimisation of the news discourse.

4.1 Socio-cultural factors

The social culture of a country influences the domestic public's judgement of the international political landscape. Nationalism is one of the most prominent features of social and cultural factors.

The so-called nationalism refers to the belief in American values that the United States is God's chosen people with the gift and mission of saving all mankind, and that it is right to contribute to the development of the world. According to this cognitive feature, the Washington Post related reports link the subprime mortgage crisis in the United States and the collapse of the world economy, and is committed to constructing an internal entity that stresses or exaggerates the positive role of the Federal Reserve's bailout policy, which is not only conducive to its own development, but also conducive to the economic recovery of the world's major economies.

Case 6: After the cut was announced, stock pricesmoderated in Europe and Asia.(23 January 2008)

The phrase "after the cut was announced" emphasises the improvement of economic conditions after the Federal Reserve's interest rate cut, linking the interest rate cut to the improvement of economic conditions. Reporters often use the word after to induce a causal relationship between the two, in order to emphasise the positive aspects of the policy and to construct a legitimisation of the Fed's policy. "in Europe and Asia" extends the impact of the Fed's bailout policy to Europe and Asia, conforming to and fulfilling the nationalistic perceptions and expectations of the American public.

4.2 Ideological factors

Ideology generally refers to the systematic ideas and concepts formed on the basis of a certain socio-economic foundation and representing the interests of a certain class or social group. [7] National ideology profoundly affects the value judgement and value choice of the American people, influencing their perception and understanding of things and information. This facilitates the legitimisation of news discourse constructs, which are often constructed with the help of utilitarian thinking to bring about changes in people's perceptions.

People's cognitive judgements about things are largely influenced by two factors, the first being personal interests and the second being the extent to which the thing coincides with their own interests. [8] The principle of safeguarding national interests is embodied in most of the policies formulated by the U.S. Under the influence of the ideology of utilitarianism, the U.S. public holds a positive attitude towards such policy behaviours of safeguarding national interests, and thus will consciously safeguard the implementation of these policies. In the news discourse, it tends to emphasise that the internal entity, i.e. the United States, is suffering or has suffered major threats and damages to its interests, so as to construct spatial convergence and value convergence, and to win the public's support for its so-called legitimate policies. The word "severe damage" in case 1 is a good example of this.

4.3 Psychological Factors of the Audience

The biggest audience of the news discourse is the American people, therefore, in order to achieve the purpose of the discourse writing must consider the psychological factors of the audience. Audience psychological factors include patriotism and individualism.

The patriotism and individualism of the American public is reflected in the close connection between national interests and personal interests, which makes it unacceptable for the public to accept that the national interests are damaged or threatened, and also makes them try to defend the national interests. News discourse always takes advantage of this psychology in order to gain the public's legitimised approval of its policies, and tries its best to show the audience that the interests of the United States and the American people are being greatly undermined, or to reinforce the positive impacts and positive effects of its policies in favour of defending the interests of the public.

Case 7: The rate cuts, coupled with proposed taxcuts and spending increases, could shore up confidence among companies and consumers.(23 January 2008)

In this case, "shore up confidence" and the word "could" constitute legitimation by describing the positive effects of the Federal Reserve's policy on the economy in a predictive tone, while "companies and consumers" reinforce the positive effects of the policy on the population.

5 Conclusion

This study analyses the legitimation construction of news discourse from three dimensions of time, space and value based on the convergence theory. It is found that the application of convergence theory not only promotes the deconstruction and in-depth interpretation of news discourse, but also helps to strengthen the legitimation construction of its own policy. At the same time, the study of discourse legitimisation also contributes to the further development of convergence theory. In addition, this study analyses the social cognitive factors affecting the construction of discourse legitimation from the audience's perspective (i.e., the American public), which provides a new way of thinking about the use of convergence theory and the construction of legitimation. The original materials of this research work has certain limitations. Subsequent research can take this paper as a basis to expand the corpus and corpus resources, conduct more in-depth and systematic discussion and research, and promote the development of convergence theory.

References

1. Cap, P. Towards the Proximization Model of the Analysis of Legitimization in Political Discourse [J] . Journal of Pragmatics, 2008(4) .
2. Cap , P . Axiological Aspects of Proximization [J] . Journal of Pragmatics, 2010(2) .
3. Wang Yuxia. Analysis of the legitimation of political discourse under the perspective of convergence theory[D]. Wuhan University of Science and Technology,2021.
4. VAN LEEUWEN, T. Discourse and Practice: New tools for critical discourse analysis [M]. Oxford: Oxford University Press, 2008 : 105,6,6.
5. Zhao Lin,Wang Xiaozhong. The Fed's bailout action in the subprime crisis and the revelation to China[J]. Inner Mongolia Financial Research, 2008(11):42-44.
6. Lv Wan. Study on the legitimisation of US diplomatic press conference discourse under the perspective of convergence theory [D]. Heilongjiang University,2022.
7. Piotr Cap. Towards the proximization model of the analysis of legitimization in political discourse[J]. Journal of Pragmatics,2007,40(1).
8. Robert Jervis, Perception and Misperception in International Politics (translated by Qin Yaqing), Beijing: World Knowledge Publications, 2003:201.

https://doi.org/10.1051/shsconf/202317803001

On the Significance and Measures for Effectively Aligning the Consolidation of the Achievements in Poverty Alleviation with Rural Revitalization
—a case study of Halu Village in Lüchun County, Yunnan Province

Xuhua Liu[1,a*], Qiu Jin[2,b], Qiong Lei[3,c]

[1] School of Marxism, Yunnan Land and Resources Vocational College Kunming, Yunnan Province, China
[2] School of Marxism, Yunnan Land and Resources Vocational College Kunming, Yunnan Province, China
[3] Office of Party and Government Affairs, Yunnan Land and Resources Vocational College Kunming, Yunnan Province, China

Abstract The Central Committee of the Communist Party of China (CPC) and the State Council have assessed the situation and made sound decisions to further consolidate and expand the achievements of poverty alleviation and to successively promote the development of poverty-alleviated areas and comprehensive rural revitalization. In order to ensure that the "relay baton" of poverty alleviation and rural revitalization is handed over successfully, the No. 1 Central Document has clearly proposed to set up a 5-year transition period from the date of poverty alleviation, maintain the overall structure of the existing major supportive policies, give poverty-alleviated counties a leg up to get them going. The government will continue to consolidate poverty alleviation's achievements, promote comprehensive rural vitalization, strengthen assistance for low-income rural residents on a regular basis, and gradually realize a smooth transition from concentrated support for poverty alleviation to the comprehensive promotion of rural revitalization. [1]The purpose of effectively aligning the consolidation of the achievements of poverty alleviation with rural revitalization is to help rural areas to secure the achievements of poverty alleviation, prevent the reoccurrence of poverty for those who had been lifted out and diminish the effectiveness of poverty alleviation, and further transfer the battle of poverty alleviation to the battlefield of rural revitalization and achieve a good alignment.

1. preface

Since the 18th CPC National Congress, the Central Committee, with Comrade Xi Jinping at its core, has put poverty alleviation as the top priority of governance, comprehensively planned, scientific deployed, with the whole Party and the whole country, won a huge battle against poverty, resolved overall regional poverty, completed the arduous task of eradicating extreme poverty, and created yet another miracle that will go down in the annals of history! Poverty alleviation is not the end, but the starting point of new life and new endeavor. After winning the battle against poverty and building a moderately prosperous society in all respects, the country will consolidate and expand the achievements of poverty alleviation, effectively carry out rural revitalization and promote the development of poverty-alleviated areas and the improvement of people's lives. [2]At the same time, aligning the consolidation and expansion of poverty alleviation's achievements with rural revitalization are also the key tasks of the work concerning "agriculture, rural areas and farmers" during the "14th Five-Year Plan (2021-25)" period, especially the work of "agriculture, rural areas and farmers" in the poverty-alleviated areas. It is of great significance to the whole Party and the people of all ethnic groups.

2. The Significance of Aligning the Consolidation of Poverty Alleviation's Achievements with Rural Revitalization

2.1It is the requirement for continuously promoting the development of poverty-alleviated areas and the improvement of people's lives.

To effectively align the consolidation of poverty alleviation's achievements with rural revitalization is necessary to continuously promote the development of poverty-alleviated areas and improve people's lives. Although the goal of poverty alleviation is, by 2020, to steadily realize assurances of adequate food and clothing, and guarantees of access to compulsory education, basic medical services and safe housing for impoverished rural residents, i.e. "The Two Assurances and Three Guarantees". This goal only helps the poor to get out of

* Corresponding author: [a]727685298@qq.com
[b]E-mail:798446616@qq.com
[c]E-mail:694369175@qq.com

extreme poverty under the current standard, which still has a long distance from rural revitalization and common prosperity. In addition, the measures to alleviate poverty are not once and for all. Over time, the people who have been lifted out of poverty may return to poverty again for various reasons, and the poverty-alleviated areas may fall into poverty again. Therefore, it is necessary to set a five-year transition period between poverty alleviation and rural revitalization so as to consolidate the achievements of poverty alleviation in the early stage and prevent the return to poverty. In the meantime, this period can help realize the effective alignment of poverty alleviation and rural revitalization so as to lay the foundation for the following rural revitalization. [8]

2.2It relates to the establishment of the new development paradigm featuring dual circulation, in which domestic and overseas markets reinforce each other, with the domestic market as the mainstay.

The domestic market as the mainstay means focusing on getting through all stages of domestic production, distribution, circulation and consumption and taking full use of the advantages of China's super large-scale market. Meeting domestic demand is the starting point and the ultimate goal of economic development. China's domestic market includes not only the urban market but also the massive rural market, where farmers, rural areas and agriculture have vast development space and market potential. Therefore, consolidating the achievements of poverty alleviation and rural revitalization will be a matter of whether China can establish a new development paradigm.

First, establishing a new development paradigm requires expanding domestic demand. The latest data released by the National Bureau of Statistics regarding the number of China's population shows that as of the end of 2019, the number of the country's rural population was 551.62 million, accounting for nearly 40% of the total population. It is evident that farmers are the major driving force of the domestic demand market. The reason for the current lack of domestic demand is the lack of demand from farmers, and farmers' demand depends on the level of farmers' sustainable income. Therefore, expanding domestic demand must increase farmers' sustainable income and boost farmers' demand. Consolidating the achievements of poverty alleviation and rural revitalization are the main strategy to increase farmers' sustainable income and boost their continuous demand.

Secondly, speeding up the establishment of a new development paradigm of agriculture and rural areas must improve the quality and competitiveness of agriculture, continuously promote the supply-side structural reform of agriculture, optimize the layout of the agricultural industry and create new types of agriculture. Poverty alleviation and rural revitalization strategy are important strategic initiatives to promote rural development and policy guarantee to build a new development paradigm of agriculture and rural areas. Rural areas are important markets and carriers of the new development paradigm of dual circulation. Agriculture is the promising industry and pillar industry of the new development paradigm for dual circulation. Rural areas and agriculture have a vast field for development and great potential. At present, family farms, farmers' cooperatives, specialized agricultural service organizations and other new types of agricultural operating entities have gradually become important forces to ensure continuously increased income of farmers, effective supply of agricultural products and agricultural transformation and upgrading. Therefore, it is of great significance to effectively align the consolidation of the achievements of poverty alleviation with rural revitalization to achieve the next-stage goal of rural revitalization and establish a new development paradigm.

2.3It relates to the overall framework of building a modern socialist country in all respects and realizing the Second Centenary Goal.

The effective alignment of consolidating the achievements of poverty alleviation and rural revitalization is related to whether the goal of building a modern socialist country can be achieved, and whether the CPC's Second Centenary Goal can be achieved on schedule. The victory in the battle of poverty alleviation marks the realization of a moderately prosperous society in all respects, and also means that the First Centenary Goal of the CPC has finally been achieved. The next strategic arrangement is to build on the foundation of a moderately prosperous society with 15 more years of hard work to see that socialist modernization is basically achieved in 2035. From 2035 to the middle of the 21st century, having achieved basic modernization, to work hard for 15 more years and develop China into a great modern socialist country that is prosperous, strong, democratic, culturally advanced, harmonious, and beautiful, which is the Second Centenary Goal of the CPC.[3]

Chinese socialist modernization is unique and different modernization from capitalist countries. One of the most important features is that Chinese modernization is the modernization of common prosperity for all people, and common prosperity is an essential requirement. This requires that the government must consciously and proactively solve the problems of regional disparity, urban-rural disparity and income distribution disparity in the process of realizing modernization. Consolidating the achievements of poverty alleviation and continuing rural revitalization are the fundamental measures to narrow the urban-rural disparity. Only the modernization of farmers, rural areas and agriculture will lead to the modernization of the whole people and realize the strategic overall framework of building a comprehensive socialist modern country.[4]

3.Emphasis on Strengthening the Effective Alignment of Poverty Alleviation and Rural Revitalization

2021 is the transition year of realizing the effective alignment of the consolidation and expansion of poverty alleviation and rural revitalization. In order to realize the effective alignment of the consolidation and expansion of poverty alleviation and rural revitalization, and to successively promote the consolidation of poverty alleviation and comprehensive rural revitalization, Lüchun County, according to the spirit of the state government and province government on the effective alignment of the consolidation and expansion of poverty alleviation and rural revitalization, has formulated the Key Points for Consolidating and Expanding Poverty Alleviation Work in Lüchun County in 2021. It emphasizes strengthening the effective alignment of poverty alleviation and rural revitalization. Halu Village, under the guidance of the key points for consolidating poverty alleviation in Lüchun County, will make efforts in the following eight aspects.[5]

3.1To effectively align the consolidation of poverty alleviation achievements with the goal of rural revitalization

On the journey to achieve the great rejuvenation of the Chinese nation, poverty alleviation is the first step, consolidating and expanding the results of poverty alleviation is the successive power, and promoting rural revitalization is the ultimate goal. At present, it is in a critical period of effectively aligning the consolidation and expansion of poverty alleviation with rural revitalization, and the primary focus of the alignment is to connect the achievements of poverty alleviation with the goal of rural revitalization. First of all, leading groups should coordinate well with each other. The rural revitalization's leading group, office and work team cooperate with the command office of poverty alleviation so that the leading group and office can be seamlessly connected to ensure that the tasks of poverty alleviation and rural revitalization are not separated, measures are not interrupted and persons in charge remain at their posts. Second, the specific target tasks should be connected. In the aspects of rural industry integration, infrastructure construction, ecological construction, tourism development, cultural construction, health construction, management and autonomy, resource revitalization, talent cultivation, etc., the village should further enhance its competence, from industry poverty alleviation to industry prosperity, from infrastructure construction to infrastructure improvement, from good ecology to ecological livability, from tourism development to tourism revitalization, from cultural construction to rural civilization, from health construction to healthy China, from management and autonomy to effective governance, from resource revitalization to resource utilization, from talent cultivation to talent revitalization. These specific tasks of each goal should be improved and enhanced to lay a solid foundation for rural revitalization and create a beautiful countryside and a beautiful China.

3.2To effectively align industry poverty alleviation with industry revitalization.

Industry poverty alleviation is an important method and endogenous driving force to win the battle against poverty. For the next step of rural revitalization, industry development is still the main channel and driving force. However, the goal of industry development has undergone a change, requiring a transition from industry poverty alleviation to industry revitalization. The Key Points for Consolidating and Expanding Poverty Alleviation Work in Lüchun County in 2021 pointed out that Lüchun County should, in the following work, highlight the cultivation of traditional leading industries such as tea and rubber, and moderately cultivate new industries such as macadamia nuts and regional medicinal herbs, etc. Halu Village has already expanded new economic crops such as new bananas, sugar cane, lemon, passion fruit and red rice based on local conditions. However, the disadvantage is that the planting is relatively scattered, weak and simple, and has not formed large-scale plantings and regional industries. Therefore, the key work of Halu Village in industry revitalization should be carried out from the following three aspects: the first is to firmly retain the red line of arable land. Agriculture is still the foundation of current development in Halu Village, which requires consolidating and improving the food production capacity on the existing basis and exploring high-standard farmland. The second point is to develop and expand new industries and new businesses. Halu Village should select the regional economic crops from the crops already planted, regularly carry out technical training, increase the technical training of planting and breeding, accelerate large-scale planting, build special brands, drive with leading enterprises and promote the increase of farmers' income. In terms of sales of agricultural products, it should use modern digital means, provide villagers with training in e-commerce business sales skills, broaden the sales channels of agricultural products, and integrate the development of agricultural planting, agricultural products processing and agricultural products sales. The third point is to promote the deep integration of primary, secondary and tertiary industries. It should vigorously develop regional planting of agricultural products, promote the development of agricultural products processing industry and support the integrated development of agricultural production, processing and circulation. It should also expand the function of agriculture, actively developed "agritainment", leisure agriculture and countryside tourism, organize villagers who have knowledge and skills to go out learning the experience of other villages in the development of modern agriculture, promote the deep integration of agriculture with tourism, education, culture and other industries, innovate the benefit linking mechanism, lead farmers to increase their income and become prosperous.[5]

3.3To effectively align ecological livability and beautiful countryside construction.

Based on the preliminary achievements of poverty alleviation, Halu Village has a certain foundation in the construction of beautiful countryside, and the appearance of the countryside has been greatly improved. However, there is still a distance from the standard of ecological livability and beautiful countryside. According to Key Points for Consolidating and Expanding Poverty Alleviation Work in Lüchun County in 2021, it should adhere to the vigorous implementation of the upgraded version of "beautiful countryside" well-off action, further improve the standard of rural roads, reconstruct rural roads, improve the level of infrastructure construction and break the bottleneck that restricts regional development. In terms of road and traffic facilities, it should further raise the standard of rural roads, improve and maintain them, and provide good conditions for villagers to travel and business development. In terms of logistics services, it should increase the number of express logistics stations and improve the efficiency of logistics delivery, so as to enhance villagers' well-being. In terms of the construction of farmland water conservancy, water and power supply facilities, it should set up special funds to maintain water and power facilities, so as to secure villagers' needs in water irrigation, water and power supply. In terms of village appearance, it should combine the requirements of the patriotic health campaigns to promote the improvement of the rural living environment and keep the village clean and tidy. In terms of the village's public environment, it should put the concept of green ecology into practice, create an "ecological and livable" environment and foster rural tourism. In terms of education, culture and health facilities, it should focus on improving the quality and level of basic education for the village, increasing teachers' training, improving the construction standard of the village's cultural sites, enriching villagers' spiritual life, improving the medical level of the village health center, improve the medical environment for the villagers and meet villagers' basic needs for medical treatment. The primary goal of village construction is achieving ecological livability, and the ultimate goal is to build beautiful countryside [5]

3.4To effectively align the boosting of motivation with good governance in the rural area.

For both villagers' autonomy and rural revitalization, the principal part is the villagers. It is necessary to make the most efforts to effectively align the boosting of villagers' endogenous development motivation with good governance of the villagers and countryside. The Key Points for Consolidating and Expanding Poverty Alleviation Work in Lüchun County in 2021 emphasize the need to strengthen the combination of poverty alleviation with improving the will and wisdom, enhance the positive incentive mechanism of more pay for more work, and guide the impoverished people to shake off poverty and get rich gradually through their own hard work. First of all, it is needed to guide the villagers to raise their sense of participation in rural revitalization. The main battlefield of rural revitalization is the rural area, the main force is the villagers, and the beneficiaries are also the villagers. Thus, the villagers are the leading power. It is necessary to guide villagers to be the master of the revitalization and development of their hometowns, cultivate their overall and long-term views of rural revitalization and development, and temporarily work hard for a better life in the future. Only when families unite together and pool the efforts of the whole village can they make their hometown develop better. Second, rural skills training should be carried out on all fronts to improve villagers' income-earning skills. Some families cannot increase their income, which is not because they are lazy, dependent, or take for granted, but because they lack the skills to increase income through work. Therefore, the village committee should try its best to adopt assistance methods such as work for food, production subsidies, labor subsidies, creating public welfare jobs, etc., strengthen the training of poor villagers in practical skills, market management, employment and entrepreneurship, encourage poor villagers to be self-reliant, work hard and develop diligently, and strive to cultivate new types of farmers who can adapt to the development of the new era. To cultivate a new type of farmer who can adapt to the development of the new era. Finally, it should also give full play to the role of Party building in poverty alleviation and revitalization. Primary-level Party organizations and Party members are the key role of the party working at the first line, and can give play to the exemplary and vanguard role for villagers. The village can select and advertise a number of poverty-alleviated and wealthy models, play the exemplary and leading role of people and deeds around the village, let the primary-level Party organizations take the leading work, adhere to the integration of villagers' autonomy, rule of law and moral governance, build a modern social governance system in rural areas, and realize "effective governance and social etiquette and civility,".

3.5To effectively align financial input policies.

Financial support and assistance can provide a strong financial guarantee for the development of rural revitalization. In the stage of effectively aligning the consolidation of poverty alleviation with rural revitalization, it is crucial to effectively align financial input policies. According to the requirements of Key Points for Consolidating and Expanding Poverty Alleviation Work in Lüchun County in 2021, first of all, government finance at all levels should reasonably arrange the amount of financial input. According to the needs and financial situation of the effective alignment of consolidation and expansion of poverty alleviation achievements and rural revitalization, it should retain, adjust and optimize the original financial funds specialized for poverty alleviation, and focus on consolidation and expansion of poverty alleviation achievements and rural revitalization. To increase financial input, Halu Village should make a detailed plan for the financial needs of the village and seek more financial support from the government at each level for the construction of public infrastructure such as roads, water conservancy, water and electricity, education, sports, culture and health, and logistics services in the village. Second, in terms of financial expenditure in Halu Village, it should be based on support from the higher level of the financial department and social support, and optimize the structure of expenditure. It should gradually increase the proportion used for business development, slowly provide a preference of the financial funds to business development, especially to increase support for the development of the village collective business, improve the collective income of the village and stimulate the villagers to work to become rich. Finally, it should adjust the focus on financial support. For industries affected by epidemics, disasters and economic downturns, more financial support should be given. For areas where drought, floods, earthquakes and other disasters arouse "Three Guarantees" and drinking water safety problems, they also should have a preference for financial support.

3.6To effectively align land support policies.

In terms of the effective alignment of land support policies, Halu Village should strictly follow the requirements of the national policy on the protection of arable land, strengthen the responsibility of arable land protection, strictly control the occupation of arable land by non-agricultural construction and resolutely retain the red line of arable land. Generations in Halu Village have relied on arable land to produce food and make a living, which is indeed villagers' primary source of livelihood. Whether it is for poverty alleviation or rural revitalization, the existing arable land cannot be damaged for development. Arable land cannot be occupied for business development and tourism development, etc. In terms of territorial space planning, according to the principle of protecting as much as possible, new construction land plan indexes will give priority to guarantee the needs of consolidating and expanding the achievements of poverty alleviation, rural revitalization and infrastructure construction. According to the policy of the national natural resources department on the construction and utilization of land resources, Halu Village should also carry out comprehensive land improvement in the whole village and provide systematic support for rural revitalization through projects such as agricultural land consolidation, construction land consolidation, rural ecosystem protection and restoration, rural history and culture preservation and characteristic promotion [6]

3.7To effectively align intellectual and talent support policies.

Poverty alleviation and rural revitalization cannot be realized without the contribution of various types of talents. In terms of the effective alignment of intellectual and talent support policies, the Village should continue to extend the various talent and intellectual support policies that have been adopted during the poverty alleviation period. According to the requirements in the Key Points for Consolidating and Expanding Poverty Alleviation Work in Lüchun County in 2021, first of all, it should continue to implement the special-post program for teachers in rural compulsory education, the national training program for teachers in kindergarten, primary, and secondary school, the priority programs and preferential policies for rural teachers, such as the living subsidy. These will attract more talents to the countryside and primary-level areas, make more talents willing to stay and retain talents for the countryside. Second, it should continue to implement the program "taking community-level posts in education, agriculture, health care and poverty relief" for college graduates, continue to implement the special-post for general practitioners and the free training program for directed medical students in rural areas. In addition, it should explore a new mechanism for talents promoting agricultural science and technology, namely, "managed by the county and work in the countryside", This can attract more college students and skilled talents to go to the countryside, which can solve the problem of shortage of skilled talents for rural revitalization and reduce graduates' employment pressure. Finally, it is suggested to appropriately relax the recruitment conditions for primary-level civil servants and institution staff, and according to the actual need, to offer a certain number of positions for people with household registration (or student record) or retired soldiers in the county or surrounding counties. This policy will undoubtedly be a preferential policy for those talents who want to apply for local civil servants and public institution positions to contribute to the construction of their hometown.

3.8To effectively align East-West poverty alleviation collaboration with various types of support policies.

First, the government of Lüchun County should continue to improve the East-West collaboration mechanism, take the initiative to strengthen the coordination with the paired assistance departments, such as Shanghai Changning District, Chongqing University and Ocean University of China, continue to take advantage of new ideas, new technologies, new promotion, etc. in the eastern coastal region, coordinate and plan the alignment work between the county and village with paired assistance departments and make sure each village obtaining guidance from paired assistance departments. Halu Village committee should take the initiative to liaise with the assistance provider, communicate the specific supportive policy for both sides to concentrate and discuss the program and methods according to local conditions and local talents. Second, for the specific content of East-West poverty alleviation collaboration, it should seek more opportunities for industry cooperation, labor collaboration, and talent support. In terms of industry cooperation, commitment is made to seek guidance and help in industrial planting and industrial sales, learn from the eastern coastal region of its new ideas, new technologies and new marketing, and seek cooperation with enterprises in the eastern coastal region. if possible, it makes sense to attract investment and strive for the campaign of "10,000 enterprises assisting 10,000 villages". In terms of labor collaboration, it should strengthen labor export to the eastern coastal region. In terms of talent support, it should send village committee officials and active members of industry development to study and train in the eastern coastal region, and seek support or remote technical guidance in the fields of education, medical care and agriculture. Finally, for the specific form of East-West poverty alleviation collaboration, more diversified and flexible guidance, assistance and cooperation should be adopted, such as remote video for technical guidance, information-sharing assistance, etc., to address the issues that experts and scholars cannot come to the field for guidance.[7]

4. conclusion

In short, consolidating the achievements of poverty alleviation and effectively connecting poverty alleviation with rural revitalization is the top priority of the next three rural work, and also an important part of the national "14th Five Year Plan". The No. 1 central document of the Central Committee clearly proposed to set up a five-year transition period from the date of poverty alleviation to maintain the overall stability of the existing main assistance policies, and to help counties out of poverty, gradually realize the smooth transition from concentrated support to poverty alleviation to comprehensive promotion of rural revitalization.[9]

Acknowledgments

"Study on the effective alignment of consolidating and expanding the achievements of poverty alleviation and rural revitalization in Halu Village, Lüchun County, Yunnan Province" Philosophy and Social Science Planning Project of Yunnan Province (Provincial Social Science Research Base) (PY202103)

References

1. Xinhua Network ,Opinions of the Central Committee of the Communist Party of China and the State Council on Comprehensively Promoting Rural Revitalization and Accelerating the Modernization of Agriculture and Rural Areas,1-2page, (February 21, 2021)
2. Xinhua News Agency ,Opinions of the Central Committee of the Communist Party of China and the State Council on Comprehensively Promoting the Key Work of Rural Revitalization in 2022, 5-6page, (January 4, 2022)
3. State Council Gazette ,Opinions of the Central Committee of the Communist Party of China and the State Council on Comprehensively Promoting the Key Work of Rural Revitalization in 2023, 1-4page, (January 2, 2023)
4. Xinhua News Agency ,Opinions of the Central Committee of the Communist Party of China and the State Council on Realizing the Effective Alignment of Consolidating and Expanding the Achievements of Alleviation with Rural Revitalization ,1-5page,(December 16, 2020)
5. Lvchun County People's Government ,Key Points for Consolidating and Expanding Poverty Alleviation Work in Lüchun County in 2021, 2-19page, (March 12, 2021)
6. Halu Village Committee , Report of the Committee of Halu Village, Gekui Township, Lüchun County on the Work in the Period of Poverty Alleviation , 1-6page, (November 16, 2020)
7. Halu Village Committee, Report of the Committee of Halu Village, Gekui Township on Consolidating the Achievements of Poverty Alleviation in 2022, 1-8page,(January 10, 2023)
8. James Galbraith, "The Essence of Poverty," pp.30-38, Eastern Publishing House (October 2014)
9. Martin Ravalle, "Learning from Success: Interpreting China's Progress in Anti Poverty", 3-5page (October 2004)

South Korea Releases Indo-Pacific Strategy: Analysis of Shifting Attitudes and Motivations

Ren Qing*

Korean Studies, Graduate School of International Studies Department, Hanyang University, Seoul, South Korea

Abstract. Since Trump's "Indo-Pacific Strategy" was formally unveiled in 2017, the South Korean government's perspective has undergone a dramatic shift. From being "very cautious" in the early years of the Moon Jae-in administration to "gradually approaching" in the latter years, to the Korean Indo-Pacific Strategy's release immediately following Yoon Seok-yeol's assumption of office to benchmark with the United States, which has raised concerns on all sides, this essay explores the domestic and external factors that have affected South Korea's perception of the Indo-Pacific Strategy. On the domestic front, the primary drivers are rooted in economic interests and diplomatic considerations, with political party competition exerting a significant influence. On the international level, the continuous change in the situation in Northeast Asia and the pressure from the US have also compelled South Korea to seek more influence through the Indo-Pacific policy.

1 Introduction

The Indo-Pacific Strategyis a U.S.national security strategy designed to contain China's rise, which was officially announced by the Trump administration in 2017 and officially released in June 2019. Subsequently, the Biden administration also continued this strategy, emphasizing U.S.-Japan-South Korea trilateral cooperation and the "U.S.-Japan-India-Australia" quadrilateral consultation mechanism, with the intention of jointly suppressing China.

For South Korea, the ROK-US alliance has been a central tenet of its long-term diplomatic and security strategy since the Cold War ended. The Indo-Pacific Strategy, on the other hand, is directed at China, which is South Korea's major trade partner. Due to its complex diplomatic dynamics with Japan and neighboring nations, South Korea initially harbored reservations regarding the Indo-Pacific Strategy. However, given the fast-shifting internal and worldwide environment, South Korea is becoming more and more aligned with the U.S. Indo-Pacific Strategy as a result of both national and international considerations.

After President Yoon Seok-yeol took office in May 2022, South Korea upgraded the South Korea-U.S. alliance to a "comprehensive global strategic alliance" and formally unveiled South Korea's version of the "Indo-Pacific Strategy" on December 28. This attracted the attention of the international community.

The goal of this paper is by analyzing the shift in South Korea's stance toward the Indo-Pacific strategy of the United States, examining the causes of the shift in stance in light of the policy changes in both the South Korea and the United States, gaining a deeper understanding of the foundation of South Korea's fundamental interests, and getting to grips with the complex situation in Northeast Asia at the current stage of ROK-US relations, ROK-China relations, and other regional relations.

2 A shift in South Korea's attitude toward the Indo-Pacific Strategy

Following Japanese Prime Minister Shinzo Abe's suggestion, Trump formally announced the Indo-Pacific Strategy in 2017. Led by the United States, Australia, Canada, and other countries have also joined the Chinese restriction campaign. Although South Korea was initially included in the U.S. Indo-Pacific strategy, it was cautious and evasive because of its relationship with China, Japan and its neighbors. However, South Korea has grown more active in pursuing its own interests and collaborating under this strategy as the situation in Northeast Asia has evolved. And there has been a significant shift since President Yun Seok-yeol took office in 2022.

2.1. Negative attitude toward the Indo-Pacific Strategy

Initially, upon the release of the official announcement of the Indo-Pacific Strategy by the United States, South Korea exhibited a negative disposition towards it. Korea's initial reservations pertaining to the Indo-Pacific Strategy centered around three distinct points. First, it is highly alert to the U.S. intention to advance the strategy. South Korea generally believes that behind the Strategy

* Corresponding author: cn4419@naver.com

lies the U.S. ambition to confront China and achieve hegemony and that South Korea is only a pawn in its geostrategic plan1. Second, South Korea worries that supporting the US may worsen relations between China and Korea2. Since their diplomatic ties were established 30 years ago, South Korea and China have grown to be significant trading partners. In order not to damage economic cooperation with China, South Korea must also distance itself from the U.S. "Indo-Pacific Strategy" and not be involved in the U.S. "containment network"3. Thirdly, South Korea has doubts about the United States' ability to implement the Strategy and its prospects for growth4. The Indo-Pacific Strategy, which is still in its early stages, is not yet fully developed in terms of how it is operated and structured. Additionally, it emphasizes the United States' desire to safeguard its own interests in matters of international commerce and security, suggesting that the connected nations may not be overly eager to take part. As a result, there are still too many unknowns for South Korea, and it would be better for both national emotion and practical considerations to maintain a distance from them.

With this assessment, South Korea was very circumspect when the Indo-Pacific Strategy was first put forth. Given the delicate situation, President Moon refrained from making any public pronouncements, but many of his policy choices mirrored South Korea's negative attitude. For example, the U.S. side has repeatedly made official statements like "the U.S.-ROK alliance promotes a free and open Indo-Pacific region," and Moon has sent Cheong Wa Dae officials several times to explain that South Korea does not agree or has not agreed to join. South Korea has also been absent from QUAD, a key component of the Indo-Pacific strategy, which consists of the United States, Japan, Australia, and India.

2.2. A shift in attitude toward the Indo-Pacific Strategy

Starting in 2019, during the latter part of the former President Moon Jae-in's administration, South Korea's attitude toward the Indo-Pacific Strategy showed positive signs, and South Korea began to participate in the Indo-Pacific Strategy in a cautious and limited manner. On November 2, 2019, while attending an ASEAN meeting, South Korea and the U.S. jointly released a briefing paper entitled "South Korea-U.S. Efforts to Enhance the Indo-Pacific Strategy. On November 2, 2019, while attending an ASEAN meeting, South Korea and the United States jointly released a briefing entitled "South Korea-U.S. Efforts to Enhance Cooperation between the New South Policy and the Indo-Pacific Strategy5," which was the first time South Korea explicitly stated its position on its participation in the Indo-Pacific Strategy. In order to strengthen its economic links with the nations of the Indo-Pacific, South Korea has also joined the Regional Comprehensive Economic Partnership (RCEP), a framework agreement for free trade that brings together 15 nations from the region, including China, Japan, Australia, and New Zealand.

After President Yoon Seok-yeol took office in May 2022, South Korea changed its previous cautious attitude toward the Indo-Pacific Strategy and joined the U.S.-led Indo-Pacific Cooperation System in a more active manner. Yoon Seok-yeon is pursuing value-based diplomacy, repairing the Korea-U.S. alliance, and pushing for the transformation of diplomatic security strategy. He has upgraded the Korea-U.S. alliance to a "comprehensive global strategic alliance" and takes an active part in the Indo-Pacific Economic Framework, as well as expressing strong interest in the Quadripartite Security Dialogue (QUAD) and the Indo-Pacific Partnership (IPP). On December 28, 2022, the President's Office of South Korea officially released the final version of the Indo-Pacific Strategy for Freedom, Peace, and Prosperity, which is in line with the U.S. Indo-Pacific Strategy. Since then, South Korea has been more active in improving Korea-Japan relations and promoting Korea-U.S. cooperation.

3 Motivations for South Korea's shift in attitude toward the Strategys

Equations should be centred and should be numbered with the number on the right-hand sideThere has been a shift in South Korea's mindset as a consequence of both domestic and external causes. On the domestic front, economic interests and political and diplomatic needs are the fundamental motivations, and domestic regime rivalry plays a direct role. Internationally, the shifting circumstances in Northeast Asia and the increasing pressure from the US have compelled South Korea to shift its focus to the Indo-Pacific strategy in an effort to gain more clout.

3.1. Domestic Factors

Economic interests and diplomatic needs are the fundamental reasons for South Korea's shift in posture toward the Indo-Pacific Strategy. In recent years, South Korea's economy has been sluggish due to a limited domestic market and a lack of resources, and it relies on external markets to a large extent. Since the Moon Jae-in government took charge, many efforts have been made to improve the external market environment and raise the level of foreign economic cooperation, but with little success. South Korea's trade surplus in the U.S. and Chinese markets has been decreasing, and the long-standing policy of "economic dependence on China" is no longer effective6. Coupled with the escalating trade dispute between South Korea and Japan, South Korea's economy has been further aggravated. Considering the above, South Korea has started to consider the Indo-Pacific Strategy to promote its economic growth and competitiveness, as many of its major trading partners are located in the Indo-Pacific region, which is also a major market for South Korean exports.

In terms of diplomatic needs, South Korea would prefer to seek more international discourse through the

Indo-Pacific Strategy7. In the previous relatively non-reciprocal South Korea-U.S. alliance, South Korea was passive on many bilateral and regional issues. Therefore, from the perspective of diplomatic independence, South Korea urgently needs to gain more leverage through the Indo-Pacific Strategy to improve its voice. After the final version of the Indo-Pacific Strategy was released, South Korean Foreign Minister Park Jin immediately said that South Korea will now plan its strategic future beyond the Korean Peninsula, ushering in a new era of proactive diplomacy on regional and global issues. It was also interpreted as a sign that South Korea's diplomatic focus will no longer be limited to solving the Korean Peninsula issue but will instead shift to a broader regional strategy, hoping to gain more diplomatic autonomy in the international arena.

Meanwhile, South Korea's domestic regime rivalry and conservatism provided the domestic foundation8 for the launch of the South Korean version of the Indo-Pacific Strategy and directly promoted the launch of South Korea's Indo-Pacific Strategy. The former Moon Jae-in government focused on restoring North-South relations but achieved limited results, coupled with escalating trade disputes between South Korea and Japan and cooling relations between South Korea and the U.S., which led to the Moon Jae-in government and its ruling party coming under heavy criticism from conservatives and opposition parties as well as a decline in support. In order to break the passive situation, the Moon Jae-in government started to approach the "Indo-Pacific Strategy." In last year's presidential election, Yoon Seok-yeol, the candidate of the National Power Party, defeated Lee Jae-myung, the candidate of the Joint Democratic Party, to become president. After taking office, the Yoon Seok-yul government changed its previous negative attitude and vigorously promoted pro-U.S. diplomacy, proposing the values of "liberal democracy and market economy." This coincided with the Biden administration and directly contributed to the official release of the Indo-Pacific Strategy.

3.2. International Factors

The alteration in South Korea's stance toward the Indo-Pacific Strategy is significantly influenced by the evolving situation in Northeast Asia9. In recent years, China's rise to power in Northeast Asia has had a tremendous impact on the power structure of the region, and its ranking as the world's second-largest economic entity and its increasing influence in the international political arena in Northeast Asia have made the United States and Japan uneasy. To offset China's influence, many nations in the area, notably South Korea, have worked to fortify their connections with the US and other Indo-Pacific nations. The launch of the Indo-Pacific Strategy by the U.S. and its major allies has provided an international environment for the emergence of the Korean version of the Indo-Pacific Strategy.

At the same time, tensions on the Korean Peninsula are one of the important factors to consider. The former President Moon Jae-in has been prioritizing the improvement of relations with North Korea in the early years of his administration. South Korea has been making active efforts to ease tensions on the peninsula and promote the process of denuclearizing the peninsula as a solution. However, since the post-Hanoi summit dialogue between North Korea and the U.S. has been difficult and sanctions against North Korea have not been eased, the highly anticipated cooperation between Russia and North Korea has remained on paper. The "New North Policy" is unable to make substantial progress, so at this time, South Korea also needs to come up with another solution to the North Korean problem10.

There is also the U.S. factor. The U.S. government has contributed to the shift in South Korea's attitude in two ways. On the one hand, the U.S. has been very aggressive in putting pressure on South Korea and showing sincerity to South Korea, such as by including South Korea in the Indo-Pacific Strategy in several official statements and making Seoul the first stop on Biden's first trip to the Indo-Pacific to show its recognition of South Korea's importance, hoping that South Korea will be more closely integrated into the U.S.-centered system and strategic arrangements. On the other hand, the renewal of the Indo-Pacific Strategy itself has also given South Korea more room to participate in it. Over time, the Indo-Pacific Strategy has increased the options for countries to cooperate with the U.S. in infrastructure, finance, high technology, social governance, and other economic and social areas within the framework of the Indo-Pacific Strategy from its initial concentrate on military and security areas. This allows South Korea to see the possibility of strengthening the U.S.-Korea alliance while avoiding stimulating China too much.

4 Conclusion and prospects

In conclusion, the South Korean government preferred to avoid discussing it because of its ties with China, Japan, and other nearby nations at the commencement of the formal release of the Indo-Pacific Strategy by the Trump administration. The change in South Korea's attitude is the result of a combination of domestic and international factors. From a domestic point of view, South Korea's own economic interests and diplomatic needs were the root causes of the change, and the rivalry between the Conservative Party and the Progressive Party played a direct role. At the international level, the changing situation in Northeast Asia, and the tensions on the Korean Peninsula, as well as pressure from the U.S., have forced South Korea to turn to the Indo-Pacific strategy to gain more voice.

Regarding the prospects for the future, South Korea's participation in the Indo-Pacific strategy has both benefits and disadvantages,contributing to the protection of its own interests and security but also having the potential to exacerbate regional tensions and hostilities. At the economic level, the revival of trade and investment with Japan could alleviate the enormous pressure exerted by the epidemic and the semiconductor crisis. Cooperation with ASEAN in the Indo-Pacific

Economic Framework (IPEF) to find new economic growth points could effectively improve Korea's economic situation. From a diplomatic point of view, the integration of the Indo-Pacific Strategy and the promotion of "value diplomacy" to win the recognition of the international community, such as Japan and the United States, will contribute to raising the voice of Korea in multilateral platforms such as the G7 and the U.S.-Japan-India-Australia quadrilateral mechanism and strengthening the position of Korea in the regional order.

However, the United States and South Korea have collectively exerted great pressure on North Korea, which has led to North Korea's retaliation measures and caused the situation on the Korean Peninsula to deteriorate in an unprecedented way. South Korea has made excessive concessions on the issue of forced labor, and relations between South Korea and Japan have improved, but there is unprecedented resistance in South Korea, so there is no guarantee that relations between South Korea and Japan will improve in the long term11. Although the Korean Indo-Pacific Strategy does not contain any content aimed at China, the strategy itself has the attribute of the United States forming a small circle with several countries to resist China's strategy. Therefore, South Korea should be very cautious as the slightest mistake would make it a victim of confrontation between the U.S. and China.

References

1. Park, H. (2020). An Analysis on the "US Indo-Pacific Strategy" and Implications for South Korea: Risks of "Tucydides' Trap.". Journal of International Relations, 23(2), 105-129. https://doi.org/10.15235/jir.2020.06.23.2.105
2. Andrew Yeo. (2022). Past, Present, and Future of the ROK-U.S. Alliance in the Indo-Pacific Era. KRINS QUARTERLY, 7(2), 7-27. https://doi.org/10.46322/KRINSQ.7.2.1
3. Park, H., & Yang, C. H. (2021). South Korea's Strategic Options amid the US–China Strategic Competition Under the Biden Administration. East Asian Policy, 13(02), 49-66.
   https://doi.org/10.1142/S179393052100012X
4. Chung, K. (2020). Korean Peninsula and the evolving Sino-US strategic stability in the Indo-Pacific. In Panda, J. P. (Ed.), The Korean Peninsula and Indo-Pacific Power Politics (1st ed., pp. 15). https://doi.org/10.4324/9780429345852
5. U.S. Embassy Seoul. (2019, November 2). Joint Fact Sheet by the United States and the Republic of Korea on Cooperation between the New Southern Policy and the Indo-Pacific Strategy. Retrieved from https://kr.usembassy.gov/110219-joint-fact-sheet-by-the-united-states-and-the-republic-of-korea-on-cooperation-between-the-new-southern-policy-and-the-indo-pacific-strategy/
6. Choo, J. (2023). China's strategic distastes for Korea in the Indo-Pacific strategy: Major concerns and countermeasures. Asian Politics and Policy, 15(1), 128– 140. https://doi.org/10.1111/aspp.12674
7. Huynh, T. S. (2023). Bolstering middle power standing: South Korea's response to US Indo-Pacific strategy from Trump to Biden. The Pacific Review, 36(1), 32-60.
   https://doi.org/10.1080/09512748.2021.1928737
8. Lee, J. Y. (2020). The Geopolitics of South Korea-China Relations Implications for US Policy in the Indo-Pacific. Korea Foundation Seoul Korea, South. https://www.jstor.org/stable/resrep27748
9. Park, J. J. (2023). The US-led security network in the Indo-Pacific in international order transition: a South Korean perspective. The Pacific Review, 36(2), 329-350.
   https://doi.org/10.1080/09512748.2022.2160790
10. Chung, K. (2023). Recalibrating South Korea's role and regional network in the Indo-Pacific: An analysis from a network-centered approach. Asian Politics and Policy, 15(1), 21– 36. https://doi.org/10.1111/aspp.12670
11. Koga, K. (2023). Struggle for coalition-building: Japan, South Korea, and the Indo-Pacific. Asian Politics and Policy, 15(1), 63– 82. https://doi.org/10.1111/aspp.12679

https://doi.org/10.1051/shsconf/202317803003

A study on the "excessive" emergency behavior of local governments in the context of the covid-19 epidemic – Based on the perspective of phenomenological sociology

Yang Shen*

School of Sociology and Population Studies, Renmin University of China, 100872, China

Abstract. The "excessive" emergency behavior of local governments at all levels in China has caused adverse social impacts during the covid-19 epidemic prevention and control period. From the perspective of phenomenological sociology, this paper uses the method of case study to analyze the motivation of "excessive" emergency behavior in the process of local government epidemic prevention, and understand the "real cause motivation" and "purpose motivation" behind such behaviors. The study found that the "excessive" emergency behavior of local governments is more caused by irrational factors from personal, social risks and local fiscal capacity.

1 Introduction

As of March 2023, under the leadership of the Party Central Committee with Comrade Xi Jinping at the core, China has achieved a major and decisive victory in the prevention and control of the covid-19 epidemic. However, in the process of epidemic prevention and control in the past three years, various "excessive" emergency behaviors such as "one-size-fits-all" and "layer-by-layer" adopted by local governments at all levels, hospitals, communities and other epidemic prevention entities have hindered the implementation of the decisions of the Party Central Committee, resulting in adverse social impacts, and have repeatedly attracted public attention.

Among these epidemic prevention subjects, hospitals, neighborhood offices, communities, villages and towns and other grassroots subjects are greatly affected by irrational factors such as public sentiment, staff social relations, special environment, and individual emotions due to direct contact with the general public, and behavioral choices are highly flexible, flexible and uncertain, and research is difficult and conclusions are difficult to generalize. In contrast, the behavior of local governments at all levels causes a wider range of effects and affects a larger number of people, so their behaviors are inherently rational, impersonal, and stylized, and are easier to study. Therefore, this paper identifies the research object as the "excessive" emergency response behavior of local governments at all levels.

When analyzing the "excessive" emergency behavior of local governments at all levels, the researchers found that under the same laws, political systems, and government agencies, only some local governments had "excessive" emergency behaviors in the process of implementing prevention and control policies. And in areas with different levels of development, material resources and infrastructure conditions, there are "excessive" emergency incidents of some local governments. From these facts, it can be concluded that the reasons why local governments choose "excessive" emergency behavior are complex and may affect a variety of factors. Therefore, it is not appropriate to choose a theoretical model analysis with a single indicator and a single variable.

Therefore, from the perspective of phenomenological sociology, this paper adopts the method of case study to understand the meaning and motivation of the "excessive" emergency behavior of local governments at all levels during the epidemic prevention and control period.

2 Literature Review

Max Weber, in his Sociology of Domination, pointed out the "tangible" and "impersonal" characteristics of modern bureaucracy - that is, "according to the rules that can be calculated, affairs are handled "regardless of who the object is" [1] (48 pages). The result of this characteristic was that all "status and honor" from feudal society were leveled, and officials in bureaucratic organizations had only "official loyalty" and no obligations to any specific individual. This theory gives the premise of "excessive" emergency behavior of local governments in China, and "excessive" implementation behavior in the normal field, that is, the bureaucratic organization on which modern administration relies is difficult to pay enough attention to individual differences and complex specific situations, but can only provide rational and computable policies and rules.

* Corresponding author: 1961608558@qq.com

Current domestic research on the "excessive" implementation of policies by local governments under normal conditions suggests that the institutional roots of local governments' "excessive" implementation of policies lie in the fact that China's high-level governments have centralized and strong personnel and financial control over local governments at all levels without effective administrative supervision, which on the one hand relies heavily on higher-level policy directives, personnel appointments and dismissals, and financial support, but on the other hand has sufficient autonomy. [2] (Kou Haoning and Li Pingju, 2017). Under this "pressure-type system", in order to complete the tasks of their superiors, local governments will adopt a working mechanism derived from "increasing the quantity, reducing the time limit, and raising the standard" around "layer by layer" [3](Zhou Zhenchao and Zhang Jincheng, 2018). At this level, the choice of local governments to "excessively" implement policies is actually a helpless move to complete the tasks of their superiors.

On the other hand, local governments have their own interests due to their own interests, which mainly include the interests of local governments, government departments and the personal interests of government officials. For example, according to Li Weiquan and Huang Yang in their research on stereotypes in policy implementation, although the government uses discretion in the process of policy implementation to simplify or even "one-size-fits-all" treatment of the problems encountered, resulting in policy implementation deviations and reduced service quality. However, this treatment is the most economical and efficient way to accelerate the progress of public policy [4](2019). However, if local governments pursue narrow departmental interests and individual officials' interests, and implement such interests in the implementation of policies, they are likely to harm the public interest and the overall interest (He Xiuling and Kong Xiangli, 2011). [5]From this perspective, the "excessive" implementation of policies by local governments may be a manifestation of their lazy and neglectful administration and even opportunistic mentality [6](He Yongsong, 2021).

In the study of the "excessive" emergency behavior of local governments in the context of public health emergencies, Ding Huang and Zhang Shaofei pointed out from the perspective of structural functionalism that local governments are subjected to the "triple squeeze" of higher level governments, governments at the same level and external subjects in the implementation of prevention and control policies, and choose to adopt "excessive" emergency behaviors such as "one-size-fits-all" and "layer by layer"[7] under the combination. Zhang Weijing and Zhou Mi started from China's emergency management system, pointing out that the shortcomings of the incentive mechanism in China's current emergency management system make local government officials tend to take responsibility avoidance as the starting point of policy implementation, resulting in two strategies of "procedural performance" and "layer by layer"[8].In addition, Sun Yufei and Yang Xuedong pointed out that whether local governments choose "excessive" emergency response is related to their "capacity space"[9].The stronger the governance capacity, the larger the "capacity space", and the less likely it is to choose "layer by layer".

As mentioned above, at present, sufficient attention has been paid to the research on "excessive" executive behavior in the field of normalization and "excessive" emergency behavior in the scene of public health emergencies represented by the covid-19 pneumonia epidemic, and fruitful results have been achieved.

First, in terms of research subjects, the above studies lack a more precise division of "excessive" performance or "excessive" emergency behaviors, and do not note that there are many types of these behaviors, such as "one-size-fits-all" and "layer-by-layer", and the causes of these different types may be different and must be analyzed separately.

Based on the research perspective and theory, the above research basically follows the logic of structural functionalism to understand "excessive" emergency behavior, treating local government officials at all levels as absolute rational actors, and their behavior choices are based on a set of behavioral norms provided by their "status-role". In other words, these studies ultimately equate the "motivations" derived from the elements of the social system directly to the real motives of local governments at all levels in each specific situation, so that their behavior cannot be truly understood. However, Li Mingxia and Chen Lin pointed out in their research on the "persuasion" announcements issued by local governments at all levels during the epidemic that these "excessive" emergency announcements have loopholes in the legality of the subject of the persuasion announcement, the legality of the procedure, and the legality of the content. In other words, local government officials are influenced by their personal education level and awareness of the rule of law, and do not have a clear understanding of the systems and rules of the social system in which they live, so their behavior choices are more likely to come from various irrational motives, rather than highly calupculable and predictable motives such as "weak responsibility" motives, "profit-seeking" motives and "stability maintenance" motives.

In terms of research methodology, the above studies also have their flaws. They all conduct motivation analysis from the institutional framework and authoritative theoretical model of China's reality, and only use empirical materials as proof of theoretical models, and lack the understanding and motivation analysis of individual cases of "local governments" as "actors" in the process of policy implementation at specific times and places. So these studies eventually return to theoretical speculation, reasoning and deduction. The disregard for empirical material has led these studies to provide only theoretical possibilities for the occurrence of "excessive" emergency behavior by local governments, but not to fully explain any specific phenomenon that actually occurs.

Therefore, this study chooses to start from individual cases, that is, from the specific events of "excessive" emergency behavior of local governments at all levels during the prevention and control of the covid-19 pneumonia epidemic to carry out meaning understanding and motivation analysis, so as to make up for the shortcomings of previous research.

3 Theoretical perspective and conceptual definition

3.1 Theoretical perspective: Schutz's phenomenological sociology

The German sociologist Schütz continued the tradition of Max Weber's sociology in his major work "The Composition of Meaning in the Social World", arguing that the purpose of sociology is to provide an interpretive understanding, or meaning understanding, of various social actions. To compensate for Weber's ambiguity in his definition of "meaning", Schütz drew on Bergson's "philosophy of life" and Husserl's phenomenological theory to define "meaning" as "intentional achievement"—"the 'meaning meaning' of an experience is nothing more than a self-interpretation of that experience starting from a new experience[10]." (96 pages). And a distinction is made between two types of motivation that influence the behavioral choices of social actors – the "real cause motivation" and the "purpose motive". The former mainly refers to the personal and social factors that cause the actor's behavioral choices but are not fully aware of the actor, such as family background, class status, etc., while the latter is a clear goal set by the actor in the form of "future completion". These two concepts are also the two that will be mainly utilized in this article.

When discussing the problem of "others", Schütz analyzes two tools for "understanding" others—"indicators" without informing function and "marks" with informing functions. The former mainly includes actions, sounds, and words with fixed meaning, while the latter refers to those words that have an informative function and must be understood in specific situations. After a detailed analysis of these two tools, Schütz distinguishes between "objective meaning" and "subjective meaning", thus laying the foundation for the discussion of "ideal type" later.

3.2 Conceptual definition

From the perspective of "management is decision-making" in public policy theory, Ding Huang and Zhang Shaofei define "excessive" emergency behavior as "expanded" implementation behavior of emergency policy, that is, emergency policy implementation behavior that exceeds the original policy requirements in terms of policy interpretation and means selection, and deviates from the original policy objectives. The flaw of this definition is that it is defined from structural functionalism and public policy theory, limiting "excessive" emergency behavior to "wrong decisions" and ignoring the holistic and multifaceted nature of "behavior".

Schütz, on the other hand, defines "behavior" through a phenomenological perspective as "the conscious experience that gives meaning through spontaneous initiative" (p. 66), and it is "marked by the original intentionality of spontaneous activity preserved in all intentional variations within the 'general experience'" (p. 67). In other words, "act" is not a rational planning towards a particular goal and the decisions made therefrom, but refers to an experience that can be considered subordinate, served, and therefore "meaningful" when a completed act is reviewed.

According to this theory of Schütz, I define "excessive" emergency behavior as all words and deeds of the policy implementation subject that deviate from the original policy in the process of implementing the emergency policy, such as policy issuance, implementation means, relevant remarks of leading cadres responsible for implementation, and some body actions. This definition provides a more comprehensive examination of the behaviour of policy implementers and an analysis of their motivations.

"Excessive" emergency behavior can be further divided into three types: "restriction of return", "excessive control", and "charging in isolation places". Among them, "restricting the return to their hometowns" refers to taking other compulsory measures such as unreasonable persuasion to return to their hometowns; "excessive control" refers to the epidemic prevention behavior of local governments exceeding the necessary limits and interfering with the normal production and life of the masses; and "charging fees for quarantine places" refers to local governments collecting fees from people who have been forcibly quarantined in the mainland. It should be noted that local government actions related to "restricting return" are often accompanied by "over-control" behavior, but are listed as a separate type because of their focus on "return". The policy of "charging for quarantine places" targets persons who are forcibly quarantined from the mainland, and those who take the initiative to undergo self-payment quarantine and those who return from abroad are not counted. These three types will serve as the "ideal types" for this study.

4 Case selection and data collection

4.1 Cases of restriction of return to home

In the cases of "excessive" emergency response behavior of local governments exposed by the press conference of the joint prevention and control mechanism of the State Council on September 8, 2022, Suzhou City, Jiangsu Province and Sanmenxia City, Henan Province were both reported that there was "non-approval and non-response to the reporting of persons coming (returning) from low-risk areas." Blocking and persuading people to return"[11]. By further comparison of the cities of Suzhou and Sanmenxia. First of all, Suzhou and Sanmenxia have very different levels of economic development and social governance capabilities, but they have both chosen to "restrict return to their hometowns", so the "capacity space" of local governments is not a decisive factor for such behavior. Second, as prefecture-level cities, Suzhou and Sanmenxia are consistent with the "vertical squeeze" from the provincial government in their respective provinces, but only they and a few other prefecture-level cities have adopted "excessive" emergency actions. And we have shown that this response to orders from higher levels of government does not depend entirely on the material resources that local governments are able to deploy. Third, the "horizontal squeeze" of Suzhou and Sanmenxia by the same level of government is not obvious, and they are almost the only cities in their provinces that have adopted "excessive" emergency actions. Fourth, the "supplementary squeeze" from external subjects suffered by Suzhou and Sanmenxia is also not obvious, because on the one hand, prefecture-level cities do not need to face the masses directly and respond immediately to the opinions of the masses like grassroots governments and autonomous organizations. On the other hand, although Suzhou City and Sanmenxia took "excessive" emergency actions, they did not cause serious adverse consequences and did not attract widespread attention from public opinion, but were widely known through mass reports and comprehensive reports of the national joint prevention and control mechanism. Therefore, the "triple squeeze" analyzed by Ding Huang and Zhang Shaofei is not obvious and does not fully explain the "excessive" emergency behavior that occurs in the covid-19 epidemic.

From the above analysis, we can conclude that the "real cause motives" that cause "excessive" emergency behavior should be more complex and diverse, and cannot be explained only by the "capacity" space of local governments and the squeeze they are subjected to in the bureaucratic system. Therefore, we must go deep into those real "actors" in the epidemic prevention process, and analyze the "real cause motivation" and "purpose motivation" of their choice of "excessive" emergency behavior through some of their specific measures, words and deeds. To this end, the author chose the "malicious return" event that attracted public attention and official criticism for analysis.

On January 20, 2022, the county magistrate of Yuncheng County, Henan Province, claimed at a meeting: "All medium- and high-risk areas, trying to return, do not talk about whether you have a vaccination certificate, do not mention whether you have a 48-hour nucleic acid test, you just need to return, first isolate and then detain!" The county governor also said that he made such remarks at the time because he encountered many hard-line returnees in his work, which caused great difficulties for the government.[12]

This case meets the requirements of phenomenological sociological research, because the speeches of public servants at meetings and in front of the media are necessarily not "meaningless" unconscious reactions, but have a "informing" function, from which the conscious experience of meaning-giving consciousness of the actor's thoughts and motivations can be understood.

From this case, we can find that if the county governor still does not realize his statement in the process of clarification after the fact, the biggest problem is that the statement of "isolation before detention" seriously violates the principle of "governing the country according to law" and is ultra vires. If the county governor were really a rational actor with a full understanding of our country's legal and political system, he would never have chosen to make such inappropriate statements in public, whether it was to deter returnees or if he really intended to take the unreasonable measure of "isolation before detention", and remain unaware of it afterwards, because such statements would do more harm to his personal future and to the image of the government. Therefore, we can conclude that the prefect is a "limited rational actor" who lacks sufficient knowledge of the legal and political system of our country, but has made wrong decisions based on limited experience and knowledge in order to implement the epidemic prevention policy. Then the "real motive" that caused him to choose "excessive" emergency behavior was his limited knowledge and the difficulties he encountered in his work. The "purpose motive" was to solve the local problem at that time, that is, a large number of field workers returned home during the Spring Festival.

Through the analysis of "restricted return" cases, we can conclude that policy implementers are often "limited rational actors" whose limited literacy and knowledge will cause him to engage in "excessive" emergency behavior, regardless of whether his purpose is for the public good or personal interest. And the quality and knowledge of these individuals are actually irrational factors in the bureaucratic system, because it varies from person to person, from time to time and place, and it is difficult to strictly control and regulate it.

4.2 "Over-control" cases

On December 3, 2022, at the press conference on the prevention and control of the covid-19 pneumonia epidemic held in Jinan, the deputy secretary of the party group of the Jinan Urban Administration introduced to the public the newly formulated "Guidelines for the Prevention and Control of the Epidemic in Public Toilets in Jinan" (hereinafter referred to as the "Guidelines") formulated by Jinan City. Its main content is the upcoming QR code toilet system in Jinan, and all public toilets strictly implement the "venue code" scanning code toilet system, and check the nucleic acid test certificate[13]. As soon as this policy was launched, it caused criticism, and public opinion accused it of adding unnecessary burdens to people's normal lives and being "excessive" epidemic prevention.

This case is a typical case of "over-regulation" and is suitable for phenomenological sociological analysis. Because it extends the scope of control to public toilets, which are widely distributed, numerous, have a large flow of people, and are extremely closely related to daily life, the control cost is extremely high. And the "scan code to go to the toilet" system is not controlled by a group with specific attributes, but everyone who uses the public toilet.

By analyzing the content of the Guidelines, we can see that the introduction of the QR code toilet system is not a local government's arbitrary and disregard of the actual "pat on the head" decision. In fact, the Guidelines provide detailed provisions on many details of the toilet scanning system, such as the daily sanitize of toilets, staff allocation, queuing standards for toilet personnel, and the difficulties of special groups such as the elderly and children. And the original intention of this policy is to prevent cross-infection caused by contact with public toilet personnel. In early December 2022, when the policy was released, Guangzhou, Chongqing and other places have begun to relax epidemic prevention and control, except for special public places such as hospitals, other public places will no longer check for negative nucleic acid test certificates, and Chongqing advised citizens to "do not do nucleic acid unless necessary" at a press conference on the prevention and control of the covid-19 pneumonia epidemic on December 3. By comparing the epidemic prevention and control policies of the two places with the "scan code toilet" system in Jinan, we can conclude that the "purpose motive" of the Jinan municipal government in choosing "excessive control" cannot be to evade responsibility or seek personal interests, but to bear too much responsibility for epidemic prevention and control and prevention of transmission risks, that is, to extend the supply boundary of public services to public toilets. However, this kind of public service requires extremely high human and material costs, and it is difficult to achieve good results, and even interferes with the normal life of the masses, so it is condemned by public opinion as "layer by layer" and "excessive control".

The "risk society" theory has a strong explanatory power for this case. Wang Jiao's research on government public services in the context of a risky society (2023) mentioned that the interaction between people in modern society is extremely frequent, and the world is becoming more and more closely connected, resulting in risks caused by human activities that are far greater than those caused by nature itself, and "there are many uncertain factors, which may cause public risks in any field" [14]. The role of the government in such a risky society is to inject certainty into social life through the provision of public services. The reason why governments often adopt "over-control" behavior is precisely because the supply boundary of public services is extremely difficult to clear. Because in the context of the "risk society", individual risks can easily be transformed into public risks, and in the process of the covid-19 pandemic, the activities of a single infected person in public places may trigger cluster epidemics, causing immeasurable consequences. Therefore, if the government wants to completely mitigate the risk of transmission, it must monitor the health status of each individual in real time, which is almost impossible to achieve in a highly mobile and highly concentrated modern society. Therefore, "excessive control", that is, ignoring the differences between individuals, classifying individuals into several groups divided according to common attributes, and carrying out "one-size-fits-all" form of control, so as to save administrative costs, have operability, and prevent and resolve the risk of epidemic transmission.

Therefore, we can conclude that the harmfulness and uncertainty of social risks and the resulting ambiguity of the government's public service supply boundary are the "real motives" that cause local governments to choose the "excessive" emergency behavior of "excessive control". The "purpose motive" of local governments is to use government power to strongly cut off the risk of epidemic transmission. "Social risk" is also an irrational factor that is accidental and difficult to predict and control.

4.3 "Segregated Charges" Cases

Before 2022, local governments in China rarely had "quarantine charges" in the process of epidemic prevention. However, after May 2022, the phenomenon of "isolation charging" appeared in many places, such as Zhenxiong County in Yunnan Province, Changshou District and Shapingba District in Chongqing, Haishu District in Ningbo in Zhejiang, Jinxian County in Nanchang, Jiangxi Province, and Sansui County in Qiandongnan Miao and Dong Autonomous Prefecture in Guizhou Province. Through observation, we can find that most of the county-level and district-level governments have adopted "quarantine charging" measures.

The administrative payment for the quarantine expenses of administratively quarantined persons during the epidemic mainly tests the financial capacity of local governments, rather than the decision-making and judgment ability in the face of public health crises, which is different from the other two types of "excessive" emergency behaviors. Chen Wangye's research on the financial security capacity of district-level governments under the impact of the covid-19 epidemic pointed out that China's district-level administrative regions are subject to the influence of geography, talents, resources and the financial system of districted cities, and lack stable and sustainable sources of fiscal revenue. The sudden outbreak of the covid-19 epidemic has affected the normal operation of the economy and increased the financial burden [15]. Bi Guoqiang's research on the impact of the covid-19 epidemic on the county-level fiscal economy also came to a similar conclusion - the covid-19 epidemic has led to a significant decline in the fiscal revenue of county-level local governments, but at the same time has increased the pressure on fiscal expenditure [16]. Therefore, we can infer that the excessive emergency behavior of local governments choosing to "quarantine fees" is mainly due to the financial burden caused by the epidemic. In the following, I will further analyze the motivations of local governments through case studies.

On September 22, 2022, the WeChat public account "Wei Zhenxiong" of the Publicity Department of the Zhenxiong County Party Committee of Yunnan and the WeChat public account "Zhenxiong Chengguan" of the Zhenxiong County Urban Management Comprehensive Law Enforcement Bureau released news that a few days ago, a reporter from the Zhenxiong County Rong Media Center learned from the Zhenxiong County Covid-19 Pneumonia Epidemic Prevention and Control Headquarters that from 0:00 on September 21, 2022, Zhenxiong decided to implement fee management for all new personnel staying in centralized isolation places. The quarantine fee ranges from 100 yuan to 150 yuan per person per day depending on the conditions of the quarantine place. Zhenxiong County has a large population, but its economic development is poor, and while facing greater pressure to prevent the epidemic, it only has relatively weak financial strength. Then the "real reason" motive that caused them to choose "quarantine fees" should be that the pressure of epidemic prevention is too great, and local finances are difficult to support. It is also worth noting that in the news released by the official public account of Zhenxiong County, there is a statement that "if the quarantined person refuses to pay fees, submits false certification materials, or refuses to cooperate with the relevant regulations of the centralized isolation place, it will be handed over to the public security department to pursue legal responsibility in accordance with laws and regulations". The "segregated fee" itself is contrary to the law, and this statement is even more ultra vires. Therefore, the "real reason motive" of the Zhenxiong County government's action included the same factor as when we analyzed the "excessive return" case, that is, the lack of understanding of relevant laws and regulations by local government officials. The "purpose motive" of the Zhenxiong county government is to reduce the burden on local governments through fees.

5 Conclusion and discussion

Through the phenomenological sociological analysis of different types of "excessive" emergency behaviors of local governments, this study concludes that there are three main "real motives" that lead local governments to make "excessive" emergency behaviors: limited personal literacy and knowledge level of local government officials, unpredictable social risks, and insufficient financial capacity of local governments. If local government officials have limited personal literacy and knowledge, they may act rashly, make wrong judgments and wrong decisions; and unpredictable social risks will make the supply boundary of government public services unclear, so that local governments choose "excessive control" in order to cut off risks most efficiently, interfering with the normal production and life of the masses; the financial capacity of local governments is the material conditions that must be relied on to take any active epidemic prevention measures, and if there are financial difficulties, local governments will be forced to adopt "isolation fees" and other unreasonable behavior.

These three factors are "irrational factors" that vary from time to place and person to person, and it is difficult to estimate in advance, which is why "excessive" emergency behavior is an occasional event during the covid-19 epidemic prevention and control period. In view of these three elements, in order to reduce the occurrence of such behaviors as much as possible, it is necessary to do regular work, improve the personal quality of local government officials, strengthen the rule of law and political education; explore the boundaries of public service provision in the daily policy implementation process, so as to satisfy as many people as possible and live without interference; be prepared for danger in times of peace, enhance the financial support ability to deal with various risks, and ensure the stability and sustainability of local financial supply capacity.

Local governments' "purpose motives" for choosing "excessive" emergency response are varied, because local governments have different specific circumstances when addressing any specific problem in the epidemic prevention process. However, local governments, whether they regard the public interest, local interests, or the personal interests of officials as "end motives" in various situations, tend to choose "excessive" emergency response because of the three factors mentioned in the previous paragraph. Because on the one hand, the prevention and control measures taken against a specific group are not only related to the safety of people's lives and property, but also to the normal operation of local production and life, and also to the responsibility of local officials. On the other hand, the three factors of limited personal literacy and knowledge, difficulty in predicting social risks and financial difficulties, once present, will permeate the entire decision-making and implementation process of local governments, prompting them to choose "excessive" emergency response.

There were also certain flaws in this study. This study lacks direct dialogue with local government related staff, but only inferences based on the performance of local governments in different cases, which may be wrong. However, this is mainly due to the fact that if the interview method is used, direct dialogue with local government staff may interfere with their normal work, and some details may not be convenient to disclose. Therefore, the use of interviews violates the academic ethics of "value neutrality" and does not interfere with the normal life of the research subjects. Future improvements to this study will mainly focus on expanding more cases, and not sticking to the context of the covid-19 epidemic, but studying the "excessive" emergency behavior of local governments in a broader field, so as to enrich and improve this research.

References

1. Max Weber. (2020)Sociology of Domination. Shanghai Sanlian Bookstore. Shanghai
2. Haoning Kou and Pingju Li. (2017)"Excessive Implementation": The Implementation Logic of Grassroots Governments and Rural Dibao Policies". Journal of Shenzhen University (Humanities & Social Sciences).34:135-141.
3. Zhenchao Zhou and Jincheng Zhang. (2018)"Layer-by-layer" under Duty Isomorphism: An Explanatory Framework for the Long-term Existence of

Formalism".THEORETICAL INVESTIGATION.203:28-33.

4. Weiquan Li and Yang Huang, (2019)"Stereotypes in Policy Implementation: An Analytical Framework of "Activation-Application": A Case Study of Family Doctor Policy Implementation in a Street Social Health Center".Journal of Public Management.1-18
5. Xiuling He and Xiangli Kong, (2011)"On the Government's Profit-Seeking Behavior and Its Norms". THEORETICAL INVESTIGATION.162:36-39.
6. Yongsong He. (2021)"An Investigation on the Occurrence of the "One-size-fits-All" Phenomenon in Grassroots Social Governance". Administrative Science Forum. 8:50-55
7. Huang Ding and Shaofei Zhang. (2022)"Why Local Governments Constantly Choose "Excessive" Emergency Behaviors". 398:127-139
8. Weijing Zhang and Mi Zhou. (2022)"Emergency Management Research on Public Health Emergencies——— Based on Comparative Analysis of Central and Local Policies". Comparative Economic & Social Systems. 219:127-138.
9. Yufei Sun and Xuedong Yang. (2023)The "Capacity Space" of Local Governance and the Transmission Variation of Epidemic Prevention and Control Pressure———An Empirical Study Based on Local Epidemic Prevention and Control Policies".2023,22(02):71-81.
10. Schütz. (2012)The Composition of Meaning in the Social World,.The Commercial Press. Beijing.
11. Upstream news. (2022)excessive epidemic prevention layer by layer has been exposed by the State Council in many places. https://www.163.com/news/article/HGTNHR720001899O.html
12. Red Culture Network. (2022)What is homecoming coming? https://www.hswh.org.cn/wzzx/llyd/zz/2022-01-23/73495.html
13. Xinming Network. (2022)Central media angrily criticized: compulsory fees for epidemic prevention and quarantine personnel are suspected of being illegal. https://mp.weixin.qq.com/s?__biz=MzAwOTYyMTMzMA==&mid=2650026897&idx=7&sn=0c42385a4dda83571373bdcb63cf626d&chksm=835c49acb42bc0ba96de6e0fc392dbdfd90415199d736942f6b1466883b50fb7118d88034c10&scene=27.
14. Jiao Wang. (2023)"Innovation of Public Service Theory for Risk Society: Rediscussion Based on the Hypothesis of "Public Risk". FISCAL SCIENCE. 86:31-40.
15. Wangye Chen. (2023)"Research on the Path of Improving District-level Financial Security Capacity under the Impact of the Covid-19 Epidemic: A Case Study of District A of J City, Zhejiang Province". 2023,(19):76-79.
16. Guoqiang Bi. (2020)"Analysis and Policy Suggestions on the Impact of the Covid-19 Pneumonia Epidemic on County-level Financial and Economic Operations: A Case Study of Ruzhou City, Henan Province". 2020,(04):46-48.

Public Participation in China Urban Planning Development

Han Li*

Luskin School of Public Affairs, UCLA, Los Angeles, CA, United States

Abstract. Public participation usually acting as an important role in urban planning of developing countries. In the process of developing, the government need to build new cities and fix the old areas. However, questions must come with the revolution. Citizens want a new, clean, and orderly city to live that has a safe community with some culture characteristic. During this process, citizens would use the tools to make their requirements to the government. This paper will analyze two examples in China of people trying to make demands on local government. The research question is how do social movements differ from the third-party institutions in the public participation of urban planning.

1. Introduction

As one of greatest work for developing countries, urban planning has been studies by all different scholars around the world. It is an important topic for the governments of developing countries to make a fitful plan for the urban development. A local government would like to make a new city for citizens with enough public needs that also good for city development. During this time, the public participation is playing a great role so that the local government are not building whatever they want.

However, it is not that easy for citizens to make the government hear the voices as the social and political system are also in the process of developing. More public participation is not only asking for the openness of governments, but also the consciousness of people in society. This paper will compare two examples in the same city of China on different building program to analyze how citizens get involved when they are unsatisfied with government's plan. The analyze would show out how public get involved in urban planning of Chinese cities. Finally, this paper will argue the difference between social movements and the third-party institutions in public participation.

The two cases used in this paper are from Guangzhou city of Guangdong province in south of China. Guangzhou is a city coastal city with 18870.6 thousand population and 86.46% of urbanization rate[1]. Guangzhou is one of the National Historical Culture City and first group of cities that open to the international trade after China open up to the world economy. Thus, Guangzhou develops faster than other province in China, which is also one of the first few cities that shows out developing problems. Thus, the cases from this city could be particularly classic to the statement of China.

The two cases happened in 2006 and 2007, which is the year of developing time of the city. For the first case, Guangzhou city started a new urban planning program to rebuild up the old part of the city in 2007. The government notices citizens that they are going to take down all old traditional buildings on Enning Road. By doing this, the Enning Road will be build as a new shopping road that might become a new economic center of the city[2]. The old Enning Road was full of illegal construction, old building, and massive wires. The government believes that they should take down all constructions so a new modern road could be built up. Politicians also want a new economic center can carry up the economic development of the city. However, the civil society organizations and 220 civilians live on Enning Road disagree with the plan and suggest the government to protect historical relic. The local NGOs recognizes the requirement of citizen and intervene into the problem. The government changes plan for few times and finally makes a new plan to keep all historical building in 2011.

For the second case, the local government of Guangzhou city Fanyu area decides to build up a waste treatment plant between two villages. The waste treatment plant should burn away the refuse from the whole area and generate electricity with fire. However, the place they choose to build the factory surrounded by a community lived with more than 300000 civilians, which might threat the health condition of citizens. People then start a social movement to state their requirements. Finally, the central government send officers to learn about the situation and move the plant to a new place with less environmental effect.

2. Problems

In case 1, the government announces a new plan of taking down the old building after the decision has made. Some of old buildings are seen as an important part of the city culture. A new modern shopping center might make more opportunities of economic development. However, economy is not the only thing people wants. In this case, the main conflict of urban planning is a decision between

* Corresponding author: hl8030@g.ucla.edu

modern life and old culture. It is also an important topic for the cities with long history to consider when they have to rebuild some old buildings. The is always some hidden dangerous or problems in the old streets because the buildings or structures could not hold for so many years. Thus, the government must take care of the public safety when doing the urban planning. Usually the easiest way is to take down the old buildings so they can build up new buildings, which is prettier, safer, and fit more in the new modern city. However, as it shows in the case that Chinese people are sensitive to traditional and cultural life by living in a long, multi-culture country[3]. It is normal to see that in a multi-culture country that people prefer old styles than new things. The longer history a country has, the more aggressive people are on the cultural topics[3]. People would have stronger emotion of having old lives rather than modern styles. Sometimes the most aggressive groups would think the modern stuffs are ruining their lives. Once people feel that they need to resist to the popular trends, they would be much more difficult to be convinced on accepting new buildings in the area around their living place[3,4]. The local government needs to balance the relationship between development and history.

Unlike case 1, case 2 is more about new development but not old culture protection. The site selection of waste treatment plant has a close relationship with daily life of citizens. People understand that waste treatment is not the only public needs, a health environment is also important to the community[4]. Usually, the waste treatment plant would built in the industrial area because the smoke of burning waste will raise the risk of lung cancer or asthma of people living around[5]. Once the plant successfully build, more than 300000 people will living under the affect of the smoke. On the other hand, the place is also one of the important transportation hubs of the city. The plant site is only 3 miles far from Asian biggest train station in Guangzhou. There is also a wildlife reserve 1 mile away from the plant. The location also have more than three main subway trail going through. Thus, everyone in this city will be impact by the smoke pollution of burning waste because it is almost the central of the neighborhood. According to the possibility of great dangerous to the environment, citizens suspect that the local government are not taking the responsibility of public health. Thus, citizens in case 2 are acting more aggressive than case 1 to make the public participation available.

3. Demands

Citizens in case 1 on Enning Road program are easy to communicate. In this case, there are more elder citizens involved because they have stronger emotion bonds with the old buildings[6]. The citizens would also prefer a better environment to live in. This wish includes a clean street, enough public services, convenient public transportation, and so on. Thus, citizens in case 1 accept the local government wish of rebuilding up Enning Road for a new economic center to make city more pretty and cleaner. However, Enning Road is an old neighborhood that full of elders[6]. It is more difficult for elders to move around for a new place to live. From this point of view, the local government has more elements to consider while they are making a decision. In case 1, the government is only thinking of the normal needs of citizens as they are planning, which underestimate the compilation of the issue. Usually, the local government would choose to pay citizens some subsidy for moving so they can rebuild the area easier[2,3]. However, the subsidy is not as important for elders as it for young people. The elders would have more emotion with the place they live in because they have been stay around for a long time. It is also more difficult for elders to move than young people. The elders in case 1 would prefer to stay in old place rather than looking for a new house. The old neighbors are also important to them that elders would have less interests on making new friends in a new neighborhoods[5,6]. In this case, the first conflict is that original neighbors of the area disagreed with moving out for new streets building up. On the other hand, the old buildings that local government wants to take down has been there for more than hundreds years. There are lots of historical decoration and cultural subject in the building. Compare to a new modern shopping center, citizens prefer an old building with cultural relic being protected[3,7]. The grievances of the public is that local government is only focusing on the economic income but not paying enough attention to the historical culture or real life of citizens.

Moreover, citizens in case 2 is asking more than only moving the plant away. As the local government resist to change the plan for the first time and makes a fake poll result, people start to think of accountability to the government[7,8,9]. The problem in case 2 has changed from urban planning issue into an acceptability problem of the local government. The unusual action of the local government has raised the suspicion up from citizens. The problem became worse when it might related with some corruptions in the local government. Thus, the citizens in case 2 got angry for the attitudes and actions of the local government. The public participation became imminent because people in the case believes that some hidden corruptions must be found. The citizens are asking the government give a new plan of waste treatment plant and also be honest with the reason why the government wants to site the plant at the beginning.

4. Allies and Resources

Since there are more elder people in case 1 participating with the problem, they find limited resources to help with the issue. Compare to citizens, the third party play an important role in this case. There is a NGO called China Biodiversity Conservation and Green Development Foundation (CBCGDF) hear about the situation and the rebuilding plan of Guangzhou government. CBCGDF is a non-government organization that focus on the green life, environment protect, and culture protect[8]. The organization introduce them as an organization that work for the public benefits of green and sustainable urban living. Thus, CBCGDF recognize the requirements of citizens on Enning Road about cultural relic protection. As an organization focus on living environment of the public, they want to join and show the public how they can help on the public benefits so more people could hear of them and give some supports. The organization first contact with newspaper called Guangzhou Daily to make a poll for the public to get a result of citizens' opinion because there are voices that local government never asked before they announce the decision[9]. After the poll, the CBCGDF have a meeting with the Guangzhou Municipal Institute of Cultural Heritage and Archaeology, which is an institution responsible with the protection of cultural heritage and archaeology. The meeting talk about the Enning Road with diverse culture for hundreds of years and how to protect them[9,10]. The cultural department of Guangzhou government then publish a new law about protecting cultural heritage and rebuilding of old road. The document is suggesting all local government focus on fixing old building and recover intangible cultural heritage but not taking down to rebuild a city into modern style[3,9]. The document stops the whole plan of rebuilding Enning Road. However, the department of urban planning only says the plan would be paused but not stopped. This announcement enrage the citizens again because they think the government wants to prolong the process of making decision until the public forget about this topic[1,10]. CBCGDF then sues the whole department to the supreme people 's court for against the national law of protecting cultural relic[5,10]. The central government finally recognize about the issue and send a investigation team to talk with local government, CBCGDF, and citizen deputy.

Citizens in case 2 are more aggressive with more allies and resources. First, the citizens post problem on the social media so people in other cities could see and support them[4,9]. The topic quickly drags attention of people from the whole nation. The media are also reporting the newest requirement and situation of the case. More than ten daily newspapers around nation are posting related information. However, the media posts have no impact to the plan. The local government announces in 2009 that the waste treatment plant will be built in 2010 and everyone who against or trying to stop the plan will be sued as hinder the national work[6,8,9]. This document suddenly enrage the citizens so they work as communities to make litigation. Citizens unite as few neighborhood coalitions to sign on a letter for declaring their requirements[5]. After receiving the letter, the local government suggest to have a meeting between mayor and 30 deputies from the community that selected by citizens. However, citizens state that they do not want to be represent so no one show up on the meeting hold by the government[6]. The mayor wait for two hours but never get a chance to start talking. Two month later, an anonymous netizen post on the social media to report that more than one politicians in the local government is taking benefits from the contractor of waste treatment program[4]. The true reason that local government persist to build waste treatment plant become an important "public event" to the central government because there are more than one citizen reporting the evidences of corruption after this post. A litigation start again in 2011 by the neighborhood community of citizens. However, this time the public are not reporting to the local government but the central government. Citizens skip the local government and send a letter with more than 100000 signatures to the central government that ask for investigating the local government for corruptions and redo the site selection of the plant[2,10]. As the central government learn about the case, the National People's Congress Deputy (NPC Deputy) from Guangdong make a public speech that the local government should stand out and claim their position of this issue[10]. The pressure in case 2 is from the central government to the local government, which is from top to bottom.

5. Outcome

The outcome of both cases are successful that the public meet their requirements. As a result, the government in case 1 agree to keep the old historical building on Enning Road. The government also spend more money to protect more than 100 traditional related building on the road[1]. On the other hand, the government makes a meed for the citizens. The meeting allows citizens choose their representative join the discussion of rebuilding Enning Road[1,10]. The community could tell the professions which part they want to keep so the government could make a city with both urbanization and cultural characteristics. The public also meet their requirement in case 2 that the local government choose a new place for the plant. The central government send officers to check on the politician who are responsible to the problem.

6. Conclusion

As it shows above, case 1 and 2 were happened in the same city and both of them spent a long time to achieve the goals. It is clear that the public is choosing different strategies in case 1 and 2 based on the different structures they have. Compare to the social movement, the public organization are acting more orderly and risky.

The process of social movement in case 2 was a mess and always repeating the same steps. The neighborhood communities start litigation for two times and there are more than one communities doing litigation, which cause a confusion to the citizens. Some of citizens are too tired to support the litigation because they have a feeling of doing one thing over and over again without any responses. One of the most important part of public participation is letting the citizens see the feedback or any other progress of their work. Some times the citizens are only asking for responses no whether positive or not. In another word, no information could kill the enthusiasm of public participation. Too much work with less efficiency reduces the willing of the public on fighting for their needs, which also raise the difficulty of social movements. The participatory institutions are more orderly and efficient on making demands. An well-developed institution would have a stable structure with clear work division of everyone. There is no issue for organizations on unite people for the public benefits because the institution will decide the path to go.

Furthermore, the participatory institutions could act more risky than the social movement. The organizations tend to have a chance to talk with different government departments about the problems. It is also less risky for the institutions if they act against the government because they usually have great impact in society. The local government does not want to mess up with the participatory institutions since they are presenting the public benefits with social power. Unlike the NGOs, people in social movement are acting more carefully to make demands. Citizens are worried about the vengeance from the politicians if they posts the negative information but could not bring the politicians down. As we can see in the case 2, people started to report tons of evidence of corruption only after one person post anonymously online first. A leader that start the social movement might encourage others to join, but the first step is always a great risk to citizens[11].

However, the social movements have wider net work in society than the participatory institutions. As the example in case 2, citizens use the social medias to get support from other cities. One of the main strategies of social movements is changing the issue of one city into an important public event so the central government could take it serious and pay attention to the hidden problem of local governments[11]. In conclusion, the public organization are acting more orderly and risky while the strategies of social movements are relying more on few great leaders and wide net work in society.

Although the methods and outcomes of two cases are different, the citizens in both cases are facing the dilemma of participating into urban planning. The culture diversity with globalization and the conflict between history and new technology are raising up the requirements of public participation in urban planning. In this process, the government might choose to build up easier application path or other choices for the public. This will lower the cost of participation of citizens with easier process, which could make citizens more willing to participate with less work requirements. The local governments might also want to advertise the importance of public participation so more citizens could have an idea of participating into the urban planning. Same as education or advocacy of public participation, the local governments could also make a public survey before doing a new arguable urban project. However, the most important is that urban planners need to change opinions so they could fit into the new social environment. As Davidoff says that pluralism urban designers need to see themselves as technician during the process of urban planning[12]. Traditional city value is no longer the only element to consider for urban planning. The urban planners need consider more about social culture and public interests when they are designing a city. The suggestion for designers is to have a wide knowledge of philosophy, social work, law, the social sciences, and civic design.

Reference

1. The People's Government of Guangzhou Municipality, https://www.baidu.com/link?url=0ihDkFgWJB73Vvk0IimyNaCXvuLWa_sRHK-KbwimHce&wd=&eqid=c99de1fd000117eb0000000663f2ce60
2. Lin, D., Zhou, K., &Wang, S., "Consider Old City Rebuilding and Public Participation in Guangzhou -- A case study on Enning Road Problem." https://wendang.xuehi.cn/doc/2i8r0mh9p6223s9ejzsg- 11.html
3. "All News Reports History of Fanyu Problem", Wangyi News, China Biodiversity Conservation and Green Development Foundation, http://news.163.com/special/00013VUU/panyuroll.html
4. Huang, D., "How Do People Get Engaged in Civic Participation: A Case Study of the Citizen Activism in Rebuilding Enning Road, Guangzhou."CJS (33), 131- 158 (2013). https://wenku.baidu.com/view/18bed7e95bf5f61fb7360b4c2e3f5727a5e924d3.html?_wkts=1675039599788&bdQuery=%E5%B9%BF%E5%B7%9E%E5%B8%82%E6%81%A9%E5%AE%81%E8%B7%AF%E4%BA%8B%E4%BB%B6
5. Wangyi News"Site Selection Case of Fanyu Waste Treatment Plant: The Experts of Government are Not Clean" (2010), Housing and Real Estate. https://xueshu.baidu.com/usercenter/paper/show?paperid=c228a25d2e430a7a81952081544cd757&site=xueshu se
6. Hou, L. & Liu, Y. , "A Study on Adjacency Effect on Site Selection of Public Building and Public Participation -- Guangzhou Fanyu Area Waste Treatment Event as An Example", Urban Planning Journal (5), 7 (2014).
7. Huang, D., "City Public Participation and Social Accountability -- Guangzhou Enning Road Rebuilding as An Example", Wuhan University Newspaper: Philosophy Social Science (1), 8 (2013).
8. Cui, J. "The Promoting Action of Social Power on Local Rule by Law -- A Case Study on Guangzhou Fanyu Waste Treatment Plant Site Selection." News Amateurs (24), 31-32 (2012). doi:10.16017/j.cnki.xwahz.2012.24.005.
9. Shen, C. , "A Study of Historical Area on Guangzhou Enning Road Buildings", Gaungzhou University (2007).
10. Wu, Z. , "Analysis the Role of Third Party in Urban Planning Public Participation -- A Case Study to Guangzhou Enning Road.", Urban Planning (2), 8 (2014).
11. Malgorzata Hanzl, "Information technology as a tool for public participation in urban planning: a review of experiments and potentials", Design Studies (28), 3 (2007). Pages 289-307, ISSN 0142-694X, https://doi.org/10.1016/j.destud.2007.02.003.

https://doi.org/10.1051/shsconf/202317803004

12. Davidoff, P., "ADVOCACY AND PLURALISM IN PLANNING" Journal of the American Institute of Planners (31), 331-338 (1965)

Cooperation: a Systematic Review of how to Enable Agent to Circumvent the Prisoner's Dilemma

Jiateng Pan[1,a] [0000-0001-6187-9459], Atsushi Yoshikawa[1, 2], Masayuki Yamamura[1]

[1]School of Computing, Tokyo Institute of Technology, Japan
[2]College of Science and Engineering, Kanto Gakuin University, Japan

Abstract. It is widely accepted that rational individuals are unable to create cooperation in a prisoner's dilemma. However, in everyday life, cooperation, for example, during a fishing moratorium, can be observed frequently. Additionally, the appearance of cooperation in the prisoner's dilemma can be seen in numerous simulation studies.

This paper reviews 31 simulation studies published between January 2017 and January 2023 in which agents can be observed in the results to improve cooperation in a prisoner's dilemma.

The proposed methodologies were sorted into seven categories, including Bounded Rationality, Memory, Adaptive Strategy, Mood Model, Intrinsic Reward, Network Dynamics, and Altruistic Attribute. Based on their impacts, the effectiveness of these seven approaches was classified into three categories: generating cooperation, maintaining cooperation, and spreading cooperation.

This review is expected to be helpful for scholars conducting future research on multi-agent cooperation and irrational agent modeling.

1 Introduction

1.1 Background and Rationale

The Prisoner's Dilemma is a game that has been analyzed in game theory (Tucker & Straffin, 1983; Felkins, 2001; Chong, et al., 2007)[4, 9, 35]. It presents a dilemma for two entirely rational agents, where they must decide to cooperate with their partner for the common good or to betray their partner for a personal reward (defect).

Table 1. Payoff table for the prisoner's dilemma

	Cooperate	**Defect**
Cooperate	**R, R**	**S, T**
Defect	**T, S**	**P, P**

The classic prisoner's dilemma depicts a scenario in which two rational prisoners, A and B, are held in solitary confinement and are each faced with an either/or situation: either they choose to be silent (Cooperate), or they choose to betray (Defect). This leads to 4 different outcomes:

1. If A and B both remain silent, they will each serve the lesser charge of 2 years in prison.
2. If A betrays B but B remains silent, A will be set free while B serves 10 years in prison.
3. If A remains silent but B betrays A, A will serve 10 years in prison and B will be set free.
4. If A and B both betray the other, they share the sentence and serve 5 years.

Since both Prisoner A and Prisoner B would avoid the worst of 10 years of incarceration by choosing to "Defect", while having a chance of being released outright, they both end up selecting to "Defect" and receive 5 years of imprisonment. However, overall the outcome would have been more favorable if they had chosen to cooperate. A total of 4 years in prison would have been shorter than the 5 years for both.

The definition of the prisoner's dilemma (Hofstadter, 1983) [16] can be standardized, as shown in Table 1, by defining the gain of cooperating with each other as Reward (R), the gain of betraying each other as Punishment (P), the gain of betraying alone as Temptation (T), and the gain of being betrayed alone as Suckers (S). When the four gains satisfy the inequality:

$$T > R > P > S$$
$$2R > T + S$$

the game is a prisoner's dilemma game.

Many studies related to the prisoner's dilemma have utilized agent-based simulations (Gotts, et al., 2003) [13], which use the ABS model to examine possible social problems in the prisoner's dilemma and to suggest possible solutions.

In these studies, researchers have proposed several ways to prevent agents from falling into a prisoner's dilemma, the most famous of which is the tit-for-tat

[a] pan.j.aa@m.titech.ac.jp

strategy (Rapoport, 1989) [28], which allows agents to mimic the choices their opponents made in the previous round. It is widely believed that only repeated prisoner's dilemma games can cause participants to shift from focusing on $T > R > P > S$ to focusing on $2R > T + S$ (Hofstadter, 1983) [16], which leads to cooperation.

Repeated prisoner's dilemma games come in various forms, and the ontology mainly considers the "game → elimination → reproduction → game" model. In this model, all participants play the game, then the participants with the lowest scores are eliminated and the remaining participants are proportionally reproduced to the original number and the game continues. In this way, the strategies and overall gains of the last remaining participants can be observed to determine whether cooperation is generated.

Firstly, the two most basic policies, AllD and AllC, need to be defined.

Where AllD refers to a participant's unconditional choice of "defect" and AllC refers to a participant's unconditional choice of " cooperate".

It is easy to see that AllD can gain a lot by exploiting AllC. In the "game → elimination → reproduction → game" model, AllC will be eliminated quickly. Since the TFT strategy suppresses the return of AllD and protects the return of AllC, AllC will be retained and not eliminated in studies where the TFT strategy is used.

However TFT still has many drawbacks such as no fault tolerance mechanism (Kopelman, 2020) [19], so more methods or better strategies need to be added to achieve more cooperation.

This review will summarize and conclude these methods.

Also with the development of artificial intelligence, reinforcement learning has been introduced into the study of the prisoner's dilemma (Sandholm & Crites, 1996; Fujimoto & Kaneko, 2019; Gill & Rosokha, 2020) [11,12,29].

The agent that introduces learning can learn through its own choices and gains, and constantly change its strategy. However, it is impossible to escape the prisoner's dilemma by learning only the immediate gains. In recent years, a plethora of methods exist that enable agent to learn group gains.

This review also summarizes such studies.

1.2 Purpose of the Review

Research on cooperation between multiple agents has focused on how to encourage rational agents to cooperate (Jiang & Lu, 2018; Xuan, et al., 2001, May; Lazaridou, et al., 2016) [18, 20, 44]. Simulations of the prisoner's dilemma show that cooperation can be observed under certain approaches.

However, the absence of a unified framework to consolidate these methods has resulted in numerous studies repeating the same work.

The purpose of this paper is to summarize the methods used in recent years for how to get agents out of the prisoner's dilemma and to categorize the methods used so that subsequent researchers can more easily begin their work. Furthermore, it is anticipated that this review will contribute to further research regarding the modeling of irrational behavior.

The objective of this paper is to reach a general conclusion on how agents can evade the prisoner's dilemma by answering the following research questions:

1. What methods are used in these studies to circumvent the prisoner's dilemma?
2. Which environments are these methods applicable to?
3. How effective are these methods?

2 Methods

2.1 Search Strategy

The literature for this review was searched for the period January 2017 to January 2023,

where the subject terms need to include either variable-sum, non-constant-sum or Prisoner's Dilemma,

and in the abstract, it needs to mention either agents, Artificial Intelligence or AI,

also in the abstract needs to mention one of Cooperation, Cooperate or Win-Win.

The search statement is as follows:

("VARIABLE-SUM" OR "PRISONER'S DILEMMA" OR "NON-CONSTANT-SUM" IN SUBJECT TERMS) AND ("AGENTS" OR "ARTIFICIAL INTELLIGENCE" OR AI IN ABSTRACT) AND ("COOPERATION" OR "COOPERATE" OR "WIN-WIN" IN ABSTRACT)

The databases used in this paper are Web of Science and IEEE Xplore, respectively, and the search results are shown in Table 2.

Table 2. Database, search statements and search results

Source	Search string	Result
Web of Science	TS=("variable-sum" OR "Prisoner's Dilemma" OR "non-constant-sum") AND AB=("agents" OR "Artificial Intelligence" OR AI) AND AB=("Cooperation" OR "cooperate" OR "Win-Win") Filter used:2017.01.01-2023.01.31	124
IEEE Xplore	("All Metadata":"variable-sum" OR "All Metadata":"Prisoner's Dilemma" OR "All Metadata":"non-constant-sum") AND ("Abstract":"agents" OR "Abstract":"Artificial Intelligence" OR "Abstract":AI) AND ("Abstract":"avoid" OR "Abstract":"circumvent" OR "Abstract":"solve") AND ("Abstract":"Cooperation" OR "Abstract":"cooperate" OR "Abstract":"Win-Win") Filters Applied: 2017 - 2023	23

In total, there were 124 Web of Science search results and 23 IEEE Xplore search results, making a total of 147 articles.

2.2 Inclusion/Exclusion Criteria

There were five inclusion criteria for the corresponding studies in this review, see Table 3.

First, it must be research related to the prisoner's dilemma; other research is not included whenever it does not meet the definition of the prisoner's dilemma.

Also, the study must be for surrogate research, and primary work in examining humans or other organisms should be excluded.

Other studies, such as those analyzing the effects of the prisoner's dilemma and the social problems that may result from particular strategies, need to be removed.

Finally, all review articles and articles written in languages other than English were also excluded from inclusion.

Table 3. Inclusion/Exclusion Criteria

Inclusion Criteria	Exclusion Criteria
Studies on Prisoner's Dilemma	Not studies on Prisoner's Dilemma
Studies on Artificial Intelligence	Studies on humans or other organisms
An increase in cooperation can be observed in the results	Not increased level of cooperation in the results.
Empirical studies	Literature reviews, commentaries or meta-analysis
Written in English	Written in other languages

2.3 Quality Assessment

Based on the Inclusion/Exclusion Criteria of 2.2, 31 studies were finally screened in this paper. The specific screening process is shown in Figure 1.

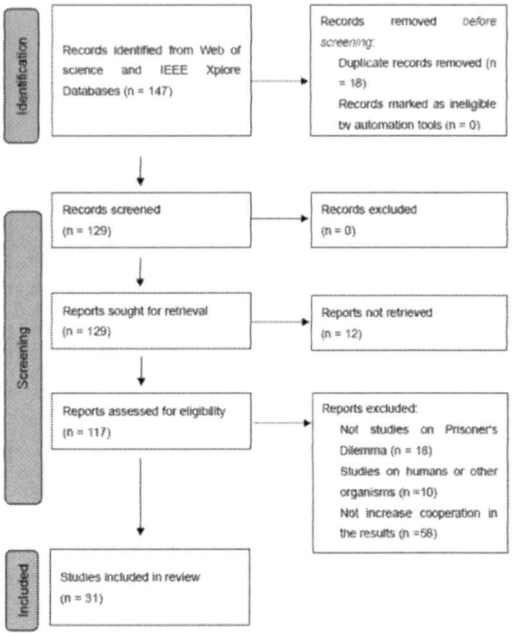

Fig. 1. Screening flow chart

3 Results

3.1 What methods are used in these studies to circumvent the prisoner's dilemma?

3.1.1 Bounded rationality

Among the 31 studies, 5 (Xu, et al., 2018, May; Moriyama, et al., 2017, July; Wang & Jiang, 2019; Otsuka & Sugawara, 2017, August; Otsuka & Sugawara, 2018) [25, 26, 27, 37, 43] used the Bounded rationality method.

If rationality is the cause that drives agents into a prisoner's dilemma, then limited rationality can help to alleviate it.

A low rationality level implies that it is possible to choose a random strategy instead of the strategy with the highest payoff (Xu, et al., 2018, May) [43].

Based on this definition, Bounded rationality includes allowing the agent to make mistakes (Moriyama, et al., 2017, July) [25], as well as all artificially set strategies of the game, such as Tit for Tat (TFT) (Wang & Jiang, 2019) [37], Extended Expectation-of-Cooperation (EEoC) (Otsuka & Sugawara, August 2017; Otsuka & Sugawara, 2018) [26,27] and others.

The Tit for Tat (TFT) strategy is a classic approach that enables an agent to imitate the opponent's decision from the previous round, after the initial round of cooperation. On the other hand, the Extended Expectation-of-Cooperation (EEoC) strategy involves cooperating multiple times when agents encounter cooperation, regardless of whether the object changes.

3.1.2 Memory

Among the 31 studies, 6 (Wang, et al., 2016, August; Fernández-Domingos, et al., 2017; Wang, et al., 2017; Heller & Mohlin, 2018; Lotfi & Rodrigues, 2022; Tao, et al., 2022, February) [10, 15, 23, 34, 38, 39] used the Memory method.

With memory, agents can be less concerned with immediate interests. Agents in the population update their strategies not by all the previous interactions but decided by the length of history records (Wang, et al., 2017) [39].

Alternatively, the agent can be allowed to remember the results of previous games between its opponent and itself to choose this game choice (Tao, et al., 2022, February) [34], or to choose whether to interrupt the link with an agent that is less cooperative (Fernández-Domingos, et al., 2017) [10].

Of course, memory here also refers to the fact that the agent can know the choices made by other agents in the past few rounds of the game (whether or not the game was played with itself) and decide its own strategy based on these (Lotfi & Rodrigues, 2022; Wang, et al., 2016, August; Heller & Mohlin, 2018) [15, 23, 38].

3.1.3 Adaptive strategy

Among the 31 studies, 6 (Xue, et al., 2017; Shang, et al., 2021, July; Wang, et al., 2021; Seredyński & Gąsior, 2019; Liu, et al., 2021, August; Xu & Hui, 2019) [22, 30, 31, 36, 42, 45] used the Adaptive strategy method.

In contrast to the humanly set strategy in Bounded rationality, agents can learn and adapt their strategies through reinforcement learning method (Xue, et al., 2017) [45], or adjust their strategy update rates according to the environment they are in (Shang, et al., 2021, July) [31]. For example, if an agent is surrounded by many cooperators, it will lower its strategy update rate to maintain a cooperative state.

In some cases, such as when payoffs are lower than expected (Wang, et al., 2021) [36]; payoffs are lower than those of neighbors (Xu & Hui, 2019; Seredyński & Gąsior, 2019) [30, 42] or communities (Liu, et al., 2021, August) [22], agents learn the strategy of neighbors or game objects that are better than themselves (Wang, et al., 2021; Xu & Hui, 2019; Seredyński & Gąsior, 2019) [30, 36, 42] or the strategy of the best individual in the community (e.g., neighbors of neighbors) (Multi-hop learning) (Liu, et al., 2021, August) [22] to update its strategy.

3.1.4 Mood model

Among the 31 studies, 5 studies (Collenette, et al., 2017, September; Feehan & Fatima, 2022; Collenette, et al., 2019, July; Zeng, et al., 2017; Zeng & Li, 2020) [5, 6, 8, 47, 48] added for agent Mood model.

Unlike Bounded rationality, the degree of rationality of an agent will be determined by an emotion value under the mood model.

Agent has an expectation of payoff, and if the actual payoff is lower than that expectation, its mood value decreases, and a lower mood value means more rationality (Collenette, et al., 2017, September) [5] and more risk seeking (Zeng, et al., 2017; Zeng & Li, 2020) [47, 48], and vice versa.

When the mood is high the agent cooperates with new agents, and when the mood is very high, the agent will always cooperate (Collenette, et al., 2019, July) [6].

The methods used to compare payoffs can be divided into two categories (Zeng, et al., 2017; Zeng & Li, 2020) [47, 48], historical-comparison and social-comparison.

For historical comparison, the expectation of payoffs depends on the agent's past payoffs (Zeng, et al., 2017; Zeng & Li, 2020; Collenette, et al., 2017, September; Feehan & Fatima, 2022) [5, 8, 47, 48].

For social comparison, the expectation of benefits depends on the average of the benefits of the agent's neighbors or the community (Zeng, et al., 2017; Zeng & Li, 2020; Collenette, et al., 2019, July) [5, 6, 47, 48].

3.1.5 Intrinsic reward

Among the 31 studies, 3 different Intrinsic reward approaches (Yuan, et al., 2022; Wu, et al., 2017; Fan, et al., 2022) [7, 40, 46] were proposed to induce cooperation.

An intrinsic reward is an additional reward that is independent of the game, but this reward is involved in the judgment of the game choice. While it seems to be no different than directly modifying the payoff table of the game, they tended to be based on some sociological research that conferred some known psychological need for satisfaction on agents.

Using adherence as an intrinsic reward enables agents to consider the collective, thus promoting cooperation (Yuan, et al., 2022) [46].

Assume that each agent has an Internal-standard, a behavior that they consider to be the right. If their behavior is consistent with the Internal-standard, they receive an additional reward; if it is not, they deduct the reward (Wu, et al., 2017) [40].

Using social payoff as an intrinsic reward enables agents to consider the collective (Fan, et al., 2022) [7].

3.1.6 Network dynamics

Among the 31 studies, 5 (Takesue, 2018; Takesue, 2021; Ichinose, et al., 2018; Li, et al., 2020; Guo, et al., 2022) [14, 17, 21, 32, 33] used the Network dynamics method.

In the Network dynamics approach, agents are allowed to change their game objects in different ways, so this approach can also be regarded as "partner selection".

Agents can choose to interrupt the link with an uncooperative object (i.e., refuse to game) and create a new link (Takesue, 2018; Takesue, 2021) [32, 33], or agents can choose to move to an empty node in the network to game with a new neighbor (Ichinose, et al., 2018) [17].

There are also some more specific partner selection mechanisms:

The agents will send a signal with strength to each other, and they will play if both sides receive each other's signal, they will not play as long as one side's signal is not delivered to the other side. The weaker the signal strength is, the harder it will be delivered to the other side. At this point, allowing agents to adjust the strength of the signal based on their own payoffs (Li, et al., 2020) [21] is also one of the approaches of Network dynamics.

A concept of active or inactive can also be introduced to the agents, so that if the agents are inactive, they will stop playing with any neighbor. In each round, agents select the action of being active or inactive via iterative Q-table (Guo, et al., 2022) [14].

3.1.7 Altruistic attribute

One (Wu, et al., 2018) [41] of the 31 studies mentioned Altruistic attribute.

Several altruistic agents are designed for the network. When an altruistic agent cooperates, its neighbors, regardless of their strategies, can gain additional benefits (Wu, et al., 2018) [40].

3.2 Which environments are these methods applicable to?

For narrative convenience, all environments in this paper are divided into only 2 cases: IPD and Network.

IPD stands for Iterative Prisoner's Dilemma, although it usually includes the game of iterative prisoner's dilemma among multiple agents, in this paper, it refers specifically to the game between 2 agents.

Network refers to all prisoner's dilemma games between more than 2 agents, which includes various network forms such as square lattices network, scale-free network, etc., and includes all game models with partner selection. NIPD (The agent plays with all its neighbors one at a time and has the choice of cooperating with all or none) is also among them.

Among the 31 studies that used Bounded rationality, there were 26 applied to Network and 5 applied to IPD.

Details as follows:

3.2.1 Bounded rationality

Among the 5 studies that used Bounded rationality, there were 4 (Xu, et al., 2018, May; Otsuka & Sugawara, 2017, August; Wang & Jiang, 2019; Otsuka & Sugawara, 2018) [26, 27, 37, 43] applied to Network and 1 (Moriyama, et al.,2017, July) [25] applied to IPD.

3.2.2 Memory

Among the 6 studies that used Memory, there were 5 (Tao, et al., 2022, February; Fernández-Domingos, et al., 2017; Lotfi & Rodrigues, 2022; Wang, et al.,2016, August; Wang, et al., 2017) [10, 34, 38, 39] applied to Network and 1 (Heller & Mohlin, 2018) [15,23] applied to IPD.

3.2.3 Adaptive strategy

All of the 6 studies (Xue, et al., 2017; Shang, et al., 2021, July; Wang, et al., 2021; Seredyński & Gąsior, 2019; Liu, et al., 2021, August; Xu & Hui, 2019) [22, 30, 31, 36, 42, 45] that used Adaptive strategy were applied to Network.

3.2.4 Mood model

Among the 5 studies that used Mood model, there were 2 (Feehan & Fatima, 2022; Collenette, et al., 2019, July) [6,8] applied to Network and 3 (Collenette, et al., 2017, September; Zeng, et al., 2017; Zeng & Li, 2020) [5, 47, 48] applied to IPD.

3.2.5 Intrinsic reward

All of the 3 studies (Yuan, et al., 2022; Wu, et al., 2017; Fan, et al., 2022) [7, 40, 46] that used Intrinsic reward were applied to Network.

3.2.6 Network dynamics

All of the 5 studies (Takesue, 2018; Takesue, 2021; Ichinose, et al., 2018; Li, et al., 2020; Guo, et al., 2022) [14, 17, 21, 32, 33] that used Network dynamics were applied to Network.

3.2.7 Altruistic attribute

The only study (Wu, et al., 2018) [41] that used Altruistic attribute was applied to Network.

3.3 How effective are these methods?

3.3.1 Bounded rationality

Bounded rationality only makes it possible that the cooperators are not all destroyed (Xu, et al., 2018, May) [43]. Agent will try to cooperate several times in a short period of time (Moriyama, et al.,2017, July) [25].

EEoC make it possible to spread cooperation rather than generate it (Otsuka, et al.,2017, August) [26] and if the number of ProbD (A strategy of always choosing defects) agents is not large, EEoC agents can spread and maintain mutual cooperation (Otsuka & Sugawara, 2018) [27].

TFT can promote the cooperative behavior of agents in a multi-agent system and improve the overall benefit of the system (Wang & Jiang, 2019) [37].

3.3.2 Memory

Memory can increase cooperation frequency substantially (Tao, et al., 2022, February; Wang, et al., 2016, August) [34, 38], but requires the network to be as dense as possible (Fernández-Domingos, et al., 2017) [10], because if game partners are randomly matched and difficult to match again with the same agent, cooperation cannot be induced (Heller & Mohlin, 2018) [15].

For memory length (the number of innings that can be remembered), the longer the memory length, the more it promotes cooperation (Lotfi & Rodrigues, 2022) [23], while there exists an optimal memory length to develop cooperation (Wang, et al., 2017) [39].

3.3.3 Adaptive strategy

By updating the strategy with Adaptive strategy, agents were able to cooperate with their opponents without losing competitiveness (Xue, et al.,2017) [45] and, in some cases, can even make it the only stable state (Shang, et al.,2021, July) [31].

If it is limited to imitating the strategies of other superior agents, Adaptive strategy can promote the formation and maintenance of cooperation, especially when there is a significant payoff difference between cooperators and defectors (Seredyński & Gąsior, 2019; Wang, et al., 2021) [30, 36]. Multi-hop learning can enhance cooperation, and there is an optimal hop number (Liu, et al., 2021, August) [22].

With the extreme tendency to imitate the superior agents, cooperators can dominate a limited population, and the level of cooperation increases with population size (Xu & Hui, 2019) [42].

3.3.4 Mood model

Positive moods can facilitate and achieve a stable cooperation (Collenette, et al., 2017, September; Collenette, et al., 2019, July) [5,6], and even could reach high levels of cooperation (Zeng & Li, 2020) [47]. Negative emotions are detrimental to cooperation, but exploitation can be avoided in extreme environments (Collenette, et al., 2017, September).

The import of mood models can facilitate cooperation (Collenette, et al., 2017, September; Collenette, et al., 2019, July; Zeng & Li, 2020; Feehan & Fatima, 2022; Zeng, et al., 2017) [5, 6, 8, 47, 48], however, highly connected networks will reduce effectiveness (Feehan & Fatima, 2022) [8].

With Historical-comparison a high level of cooperation can be achieved and with Social-comparison the high level of cooperation can only be sustained in a portion of the population (Zeng, et al., 2017) [48].

3.3.5 Intrinsic reward

Intrinsic reward can enhance cooperation (Yuan, et al., 2022) [46], significantly affect the outcome of cooperation evolution (Wu, et al., 2017) [40], and ensure cooperators coexist with defectors even at high temptations to defection (Fan, et al., 2022) [7].

3.3.6 Network dynamics

Network dynamics enhance the cooperation in the prisoner's dilemma (Takesue, 2021; Li, et al., 2020) [21, 33], networks with medium density (a higher number of links between agents playing the game) can increase cooperation (Ichinose, et al., 2018) [17], and cooperation can be the best strategy when the density of the network increases (Takesue, 2018) [32].

For games that introduce the concept of signal strength (Li, et al., 2020) [21], Network dynamics can not only help cooperators escape from the risk of extinction but also can greatly contribute to the population size of cooperators.

For games that introduce active and inactive states (Guo, et al., 2022) [14], Network dynamics can make Inactive agent form a belt to separate cooperative and defective clusters.

3.3.7 Altruistic attribute

Altruistic attribute promoted cooperation (Wu, et al., 2018) [41].

4 Discussion

4.1 Summary

This paper summarizes 31 studies from January 2017 to January 2023 on how to make AGENT cooperate in a prisoner's dilemma and summarizes seven methods: Bounded rationality; Memory; Adaptive strategy; Mood model; Intrinsic reward; Network dynamics; and Altruistic attribute.

The Boundary rationality approach enables agents to engage in certain "prescribed" behaviors, often of a perceptual nature. Numerous strategies have been proposed based on this idea, including the TFT strategy. This perceptual strategy disregards the "benefits of Temptation" that may not provide maximum returns in the short term but can enhance overall benefits in the long run.

In general, not all agents use the same strategy; they are often placed into populations with a certain proportion of ProbD and ProbC, and the proportion of mutual cooperation in the whole population is observed after several rounds of "game → elimination → reproduction → game". Therefore, this approach is more akin to "spreading" cooperation than "generating" cooperation.

The exception to this is allowing agents to make mistakes, which does generate cooperation. If agents are allowed to learn, they will reach cooperation on occasional mutual mistakes and as a result learn to cooperate for a short period of time, somewhat like the operant conditioning chamber (McLeod, 2015) [24], where cooperation is more rewarding and becomes an incentive due to the long-term environment of mutual betrayal.

Unlike the fixed strategy of Bounded rationality, both Adaptive strategies and Memory allow agents to change their strategies through learning. They differ only in the agent's ability to observe the outcome of past rounds of the game.

When an agent holds Memory, it can learn from past gains and learn to cooperate without being tied to immediate benefits, since multiple Reward gains are more than Punishment gains. If agents can also observe other agents' past game choices, they can learn richer strategies to gain more while avoiding exploitation; a simple example is the TFT strategy that can be learned when Memory length is 1 (Anastassacos, et al., 2020, April)[1]. Memory is not only the agent's own, but also can be the whole society's. In fact, in real life, criminal records are an example of social memory means, and employers can judge whether to make employment (cooperation) based on criminal records.

Learning from immediate gains alone cannot produce cooperation (Axelrod, 1980) if Memory is not included[2]. Thus, the best approach is to learn the strategies of "better" performing agents, but this requires that agents observe the total gains of other agents, and agents with the highest gains tend to be able to exploit or cooperate with others over time, and exploitation is not sustainable when agents are able to change their strategies to avoid being exploited. When agents are able to change their strategies to avoid being exploited, the exploitation is not sustainable, and agents end up learning from those who cooperate in the long-term.

However, this method needs to ensure that cooperators do not disappear prematurely. If agents can observe the highest global gainers and all turn to exploitation, then cooperators will die out at an early stage and finally become unavailable for learning, which is one of the reasons why all six studies using Adaptive strategies only learn from their neighbors.

The Mood model is actually a kind of bounded rationality, but it is classified as an independent approach due to its dynamic adjustment. It is commonly believed that rationality makes agents choose the optimal solution in a prisoner's dilemma (Campbell & Sowden 1985) [3]-Defection. And the Mood model is set up in such a way that mood becomes low when agents feel their gains decrease and high when agents feel their gains increase. When sentiment is high, agents will be less "calculating" and ignore the optimal solution in favor of cooperation. At the same time, when they are exploited to reduce their gains, they "calm down" and return to rational thinking to avoid further exploitation.

If it is in a two-player IPD, when both parties are in a high mood state, both parties' gains will increase, which in turn will further increase the mood value. In fact, the TFT strategy can also be seen as an extreme Mood model, i.e., if the opponent cooperates, whether they choose to cooperate or defect, the increase in gain will lead to a rapid increase in the mood value and thus the choice to cooperate, and similarly, if they encounter a defection, they will immediately feel frustrated because of the decrease in gain. Thus, the disadvantage of the mood model is very obvious, when the number of participants increases, once someone chooses to defect when the mood value is low, it will quickly break the mutual cooperation and cannot stop the propagation of defections.

This is why sentiment models are asked to be performed in a less connected network. social-comparison performs better than historical-comparison for this same reason, because defections may be passed through the network, so the average gain for the society as a whole is reduced, which makes it less likely to compare with past gains, compared to see a decrease in gains to maintain good mood.

The final method related to the rationality level is Intrinsic reward, which allows agents to use some psychological satisfaction as a reward for participating in the prisoner's dilemma. From a global perspective, these agents act as if they have many "obsessions" in general, such as setting agents to be inclined to cooperate, so that when agents choose to cooperate, they receive an additional reward for whatever their opponent chooses, which will undoubtedly increase cooperation. Although at first glance this appears to be a "cheating" strategy because it directly modifies the payoff table and undermines the definition of the prisoner's dilemma, in reality, all human-set strategies are the same, such as the TFT strategy, which is a strategy that ignores the payoff table. And usually these intrinsic rewards are based on psychological findings or guesses, so they are also considered as a method.

Network dynamics is also a strategy, but it is fundamentally different from the previously mentioned strategies because the strategy is not a decision to cooperate or defect, but a decision to play or not to play with another agent, namely partner selection. Similar to Memory, agents can observe other agents' past choices or changes in their own past gains and choose whether to continue playing with a particular agent.

If the model is "game → elimination → reproduction → game", ProbD will end up being eliminated because no partner is willing to play with him and he cannot gain benefits as a result; if the model is with learning, the agent will know that he can gain more benefits only if he is allowed to play and learn to cooperate. Network dynamics disguises itself as a "self-protection" means, which is also a means of punishment. The example of "criminal record" is mentioned in the discussion of memory, but it is closer to the case of Network dynamics, because the employer does not choose to defect, but refuses to play together.

The last method is Altruistic property, i.e., adding some special agents to the game, and when playing with these special agents, if the special agent chooses to cooperate, the opponent will get an additional reward regardless of what the opponent chooses. Such agents are easy targets for exploitation, so it is necessary for agents to learn self-preservation strategies. Once such a special individual does not choose to cooperate, then his opponent does not receive an additional reward.

When the extra reward is large enough, Reward gains will exceed Suckers gains, so other agents will choose to cooperate. Similar to Intrinsic rewards, it alters the payoff table, but unlike Intrinsic rewards, the Altruism property alters the opponent's payoff, not his own. Even though the additional payoff is not large, other agents may choose to cooperate in order to receive additional payoffs in the long run and avoid special agents turning to defection. Because the method requires the observation of long-term outcomes, it requires a memory function.

In summary, after reviewing the seven methods for generating cooperation among agents in the 31 studies, it is easy to see that there is some interoperability among the seven methods, some of which are even variants of another method (e.g., Intrinsic rewards are complex manifestations of Bounded rationality). However, this paper distinguishes them as different methods for ease of description and also because they are all set up from different human perspectives. The various strategies in Bounded rationality are similar to the various "must follow" rules we were taught as children, Memory is like the records we acquire (criminal records, awards, etc.), Adaptive strategies are learning, Mood models are, as the name implies, emotional, Intrinsic rewards are psychological satisfaction, Network dynamics are the right to choose partners, and Altruistic attributes are like social welfare policies.

Some of them can generate cooperation (Memory, Adaptive strategy, Intrinsic reward), some can maintain it (Network dynamics, Altruistic attribute), and some can spread it (Bounded rationality, Mood model).

Returning to the three questions that this paper seeks to answer:

1. What methods are used in these studies to circumvent the prisoner's dilemma?

In total, there are seven methods to get agents to produce cooperation from the prisoner's dilemma. They are: Bounded rationality; Memory; Adaptive strategy; Mood model; Intrinsic reward; Network dynamics; Altruistic attribute.

2. Which environments are these methods applicable to?

All seven methods have been used in multi-agent networks.

Where Bounded rationality; Memory and Mood model were also used in IPD with only 2 agents.

3. How effective are these methods?

The study showed that Memory, Adaptive strategy, and Intrinsic reward performed well in generating cooperation, Network dynamics, Altruistic attribute was suitable for maintaining the generated cooperation, and Bounded rationality, Mood model could spread the cooperation out.

4.2 Future work

In future work, further research will be conducted to unify the seven methods summarized in this paper, in order to model the generation, maintenance, and propagation of cooperation and simulate real-life human-generated cooperation - such as the development of concepts like fishing moratoriums.

The strengths of these seven approaches will be integrated and explored, with a particular focus on extending the mood modeling component. This will enable agents to spontaneously generate emotions based on their environment and develop corresponding strategies, thereby creating a more realistic social environment.

Additionally, this review will be regularly updated to incorporate any new methods that emerge.

Acknowledgement

This research was conducted with the guidance of several members of the Yoshikawa Research Office, and I have greatly benefited from their valuable advice and opinions.

Additionally, this study received support from both the Shoji Kawashima Memorial Scholarship Fund and the Tokyo Tech Tsubame Scholarship for Doctoral Students.

Reference

1. Anastassacos, N., Hailes, S., & Musolesi, M. (2020, April). Partner selection for the emergence of cooperation in multi-agent systems using reinforcement learning. In Proceedings of the AAAI Conference on Artificial Intelligence (Vol. 34, No. 05, pp. 7047-7054).
2. Axelrod, R. (1980). Effective choice in the prisoner's dilemma. Journal of conflict resolution, 24(1), 3-25.
3. Campbell, R., & Sowden, L. (Eds.). (1985). Paradoxes of rationality and cooperation: prisoner's dilemma and Newcomb's problem. UBC Press.
4. Chong, S. Y., Humble, J., Kendall, G., Li, J., & Yao, X. (2007). Iterated prisoner's dilemma and evolutionary game theory. In The Iterated Prisoners' Dilemma: 20 Years On (pp. 23-62).
5. Collenette, J., Atkinson, K., Bloembergen, D., & Tuyls, K. (2017, September). Mood modelling within reinforcement learning. In ECAL 2017, the Fourteenth European Conference on Artificial Life (pp. 106-113). MIT Press.
6. Collenette, J., Atkinson, K., Bloembergen, D., & Tuyls, K. (2019, July). Stability of cooperation in societies of emotional and moody agents. In Artificial Life Conference Proceedings (pp. 467-474). One Rogers Street, Cambridge, MA 02142-1209, USA journals-info@ mit. edu: MIT Press.
7. Fan, L., Song, Z., Wang, L., Liu, Y., & Wang, Z. (2022). Incorporating social payoff into reinforcement learning promotes cooperation. Chaos: An Interdisciplinary Journal of Nonlinear Science, 32(12), 123140.
8. Feehan, G., & Fatima, S. (2022). Augmenting Reinforcement Learning to Enhance Cooperation in the Iterated Prisoner's Dilemma. In ICAART (3) (pp. 146-157).
9. Felkins, L. (2001). The Prisoner's Dilemma.
10. Fernández-Domingos, E., Loureiro, M., Alvarez-López, T., Burguillo, J. C., Covelo, J., Peleteiro, A., & Byrski, A. (2017). Emerging Cooperation in N-Person Iterated Prisoner's Dilemma over Dynamic Complex Networks. Computing & Informatics, 36(3).
11. Fujimoto, Y., & Kaneko, K. (2019). Emergence of exploitation as symmetry breaking in iterated prisoner's dilemma. Physical Review Research, 1(3), 033077.
12. Gill, D., & Rosokha, Y. (2020). Beliefs, learning, and personality in the indefinitely repeated prisoner's dilemma. Available at SSRN 3652318.
13. Gotts, N. M., Polhill, J. G., & Law, A. N. R. (2003). Agent-based simulation in the study of social dilemmas. Artificial Intelligence Review, 19, 3-92.
14. Guo, H., Wang, Z., Song, Z., Yuan, Y., Deng, X., & Li, X. (2022). Effect of state transition triggered by reinforcement learning in evolutionary prisoner's dilemma game. Neurocomputing, 511, 187-197.
15. Heller, Y., & Mohlin, E. (2018). Observations on cooperation. The Review of Economic Studies, 85(4), 2253-2282.
16. Hofstadter, D. R. (1983). Metamagical themas. Scientific American, 248(5), 16-E18.

17. Ichinose, G., Satotani, Y., & Nagatani, T. (2018). Network flow of mobile agents enhances the evolution of cooperation. Europhysics Letters, 121(2), 28001.
18. Jiang, J., & Lu, Z. (2018). Learning attentional communication for multi-agent cooperation. Advances in neural information processing systems, 31.
19. Kopelman, S. (2020). Tit for tat and beyond: The legendary work of Anatol Rapoport. Negotiation and Conflict Management Research, 13(1), 60-84.
20. Lazaridou, A., Peysakhovich, A., & Baroni, M. (2016). Multi-agent cooperation and the emergence of (natural) language. arXiv preprint arXiv:1612.07182.
21. Li, J., Park, J. H., Zhang, J., Chen, Z., & Dehmer, M. (2020). The networked cooperative dynamics of adjusting signal strength based on information quantity. Nonlinear Dynamics, 100(1), 831-847.
22. Liu, X., Guan, R., Wang, T., Han, L., Qin, Y., & Wang, Y. (2021, August). Multi-hop Learning Promote Cooperation in Multi-agent Systems. In Knowledge Science, Engineering and Management: 14th International Conference, KSEM 2021, Tokyo, Japan, August 14–16, 2021, Proceedings, Part I (pp. 66-77). Cham: Springer International Publishing.
23. Lotfi, N., & Rodrigues, F. A. (2022). On the effect of memory on the Prisoner's Dilemma game in correlated networks. Physica A: Statistical Mechanics and its Applications, 607, 128162.
24. McLeod, S. (2015). Operant Conditioning: What It Is, How It Works, and Examples.
25. Moriyama, K., Nakase, K., Mutoh, A., & Inuzuka, N. (2017, July). The resilience of cooperation in a Dilemma game played by reinforcement learning agents. In 2017 IEEE International Conference on Agents (ICA) (pp. 33-39). IEEE.
26. Otsuka, T., & Sugawara, T. (2017, August). Robust spread of cooperation by expectation-of-cooperation strategy with simple labeling method. In Proceedings of the International Conference on Web Intelligence (pp. 483-490).
27. Otsuka, T., & Sugawara, T. (2018). Promotion of robust cooperation among agents in complex networks by enhanced expectation-of-cooperation strategy. In Complex Networks & Their Applications VI: Proceedings of Complex Networks 2017 (The Sixth International Conference on Complex Networks and Their Applications) (pp. 815-828). Springer International Publishing.
28. Rapoport, A. (1989). Prisoner's dilemma. Game theory, 199-204.
29. Sandholm, T. W., & Crites, R. H. (1996). Multiagent reinforcement learning in the iterated prisoner's dilemma. Biosystems, 37(
30. Seredyński, F., & Gąsior, J. (2019). Emergence of collective behavior in large cellular automata-based multi-agent systems. In Artificial Intelligence and Soft Computing: 18th International Conference, ICAISC 2019, Zakopane, Poland, June 16–20, 2019, Proceedings, Part II 18 (pp. 676-688). Springer International Publishing.
31. Shang, L., & Luo, H. (2021, July). Environmental adaptability promotes cooperation in the evolutionary game. In 2021 40th Chinese Control Conference (CCC) (pp. 7486-7491). IEEE.
32. Takesue, H. (2018). Evolutionary prisoner's dilemma games on the network with punishment and opportunistic partner switching. Europhysics Letters, 121(4), 48005.
33. Takesue, H. (2021). Symmetry breaking in the Prisoner's Dilemma on two-layer dynamic multiplex networks. Applied Mathematics and Computation, 388, 125543.
34. Tao, W., Wei, W., Xin, Y., & Meiqi, H. (2022, February). Strategies to Promote Cooperation in Mobile Networks. In 2022 8th International Conference on Automation, Robotics and Applications (ICARA) (pp. 140-145). IEEE.
35. Tucker, A. W., & Straffin Jr, P. D. (1983). The mathematics of Tucker: A sampler. The Two-Year College Mathematics Journal, 14(3), 228-232.
36. Wang, S. Y., Liu, Y. P., Zhang, F., & Wang, R. W. (2021). Super-rational aspiration induced strategy updating promotes cooperation in the asymmetric prisoner's dilemma game. Applied Mathematics and Computation, 403, 126180.
37. Wang, S., & Jiang, L. (2019). Study of Agent Cooperation Incentive Strategy Based on Game Theory in Multi-Agent System. In Communications, Signal Processing, and Systems: Proceedings of the 2017 International Conference on Communications, Signal Processing, and Systems (pp. 1871-1878). Springer Singapore.
38. Wang, T., Li, L., Zhang, S., Peng, H., Yu, L., & Chen, Z. (2016, August). Memory mechanism enhances cooperation in mobile multi-agent system. In 2016 8th International Conference on Intelligent Human-Machine Systems and Cybernetics (IHMSC) (Vol. 2, pp. 476-479). IEEE.
39. Wang, X., Zhang, L., Du, X., & Sun, Y. (2017). Evolving cooperation in spatial population with punishment by using PSO algorithm. Natural Computing, 16, 99-117.
40. Wu, Y. E., Zhang, Z., & Chang, S. (2017). Effect of self-interaction on the evolution of cooperation in complex topologies. Physica A: Statistical Mechanics and its Applications, 481, 191-197.
41. Wu, Y. E., Zhang, Z., & Chang, S. (2018). Heterogeneous indirect reciprocity promotes the evolution of cooperation in structured populations. Chaos: An Interdisciplinary Journal of Nonlinear Science, 28(12), 123108.
42. Xu, C., & Hui, P. M. (2019). Emergence of cooperation in finite populations under biased selection. Physica A: Statistical Mechanics and its Applications, 535, 122371.

43. Xu, X., Rong, Z., & Tse, C. K. (2018, May). Bounded rationality optimizes the performance of networked systems in prisoner's dilemma game. In 2018 IEEE International Symposium on Circuits and Systems (ISCAS) (pp. 1-5). IEEE.
44. Xuan, P., Lesser, V., & Zilberstein, S. (2001, May). Communication decisions in multi-agent cooperation: Model and experiments. In Proceedings of the fifth international conference on Autonomous agents (pp. 616-623).
45. Xue, L., Sun, C., Wunsch, D., Zhou, Y., & Yu, F. (2017). An adaptive strategy via reinforcement learning for the prisoner's dilemma game. IEEE/CAA Journal of Automatica Sinica, 5(1), 301-310.
46. Yuan, Y., Guo, T., Zhao, P., & Jiang, H. (2022). Adherence Improves Cooperation in Sequential Social Dilemmas. Applied Sciences, 12(16), 8004.
47. Zeng, W., & Li, M. (2020). Selective attention to historical comparison or social comparison in the evolutionary iterated prisoner's dilemma game. Artificial Intelligence Review, 53, 6043-6078.
48. Zeng, W., Li, M., & Feng, N. (2017). The effects of heterogeneous interaction and risk attitude adaptation on the evolution of cooperation. Journal of Evolutionary Economics, 27, 435-459.

A survey on the current state of innovation and entrepreneurship knowledge and experience among university students in Jilin Province

Jie Yao *

Northeast Electric Power University, School of Economics and Management, 132012 Jilin, China

Abstract. A questionnaire was used to investigate the knowledge and experience of university students in Jilin Province regarding innovation and entrepreneurship. A non-parametric test was used to analyze whether there were significant differences in knowledge perception, and experience of innovation and entrepreneurship between paired samples and independent samples. The results showed significant disparities in students' knowledge perception, and experience of innovation and entrepreneurship. No significant differences in knowledge perception, or experience of innovation and entrepreneurship were found between the male and female samples. Significant differences in both aspects were found between samples at different stages. All this suggests that as students grow in age and work experience, their knowledge perception, and experience of innovation and entrepreneurship are continuously enriched. To ease the pressure on employment, university students should be actively encouraged to innovate and start businesses.

1 Introduction

College students are the future of the motherland and a reserve force for the country's economic and social development. How to further enrich the knowledge of university students about innovation and entrepreneurship, increase their experience in innovation and entrepreneurship, and adequately mobilize their enthusiasm for innovation and entrepreneurship based on professional knowledge is also an essential goal of university personnel training. To this end, numerous scholars have conducted related studies, which are mainly reflected in the following two aspects.

Some experts and scholars have conducted related studies on the status quo of innovation and entrepreneurship education for college students, the difficulties they face, and the education system for innovation and entrepreneurship. To improve the innovation and entrepreneurship education system in colleges and universities, Xu, T., Qi, L., and Tian, F. (2023) proposed countermeasures. These included strengthening the innovation and entrepreneurship teachers in universities as well as enhancing the operation of innovation and entrepreneurship incubation bases [1]. Zhang, X., and Yang, X. (2022) believe that application-oriented talents with professional vision can be cultivated by furthering the reform of innovation and entrepreneurship education in universities, enhancing the quality assurance system, focusing on the process control of practical links, and encouraging graduates to innovate and start business [2]. Jin, Q. (2022) proposed measures and recommendations for college students' innovation and entrepreneurship education in the digital economy, including the impact of the innovation and entrepreneurship environment, innovation and entrepreneurship curriculum development, innovation and entrepreneurship practice results transformation, and the establishment of an innovation and entrepreneurship evaluation system [3]. Sun, A. (2022) elaborated on the main ways to boost college students' innovation and entrepreneurship education and proposed a "five-in-one" training mechanism of building a sound mechanism, improving the system, building a platform, building a team, and integrating resources [4]. Huang, W. and Li, W. (2021) pointed out that the innovation and entrepreneurship education system of college students is facing different challenges in the post-epidemic era, such as current formats of the digital economy, a conservative view of career choice, online teaching modes, etc. [5].

Based on the questionnaire survey data of undergraduate students at Southwest University of Political Science and Law, Wu, M. (2021) specifically analyzed the development status of college students' innovation and entrepreneurship education under the background of "Internet +" [6]. Rivera, Y. et al. (2021) describe the level of student's knowledge and experience in entrepreneurship and innovation along with their participation in the related activities in University of Puerto Rico at Mayagüez [7]. Portuguez Castro, M., and Gómez Zermeño, M. (2021) seek to identify the entrepreneurial profile of university students through their experiences, interests, attitudes, and entrepreneurial skills at a university in Mexico[8].

* Corresponding author: leeyooab@163.com

Several scholars have studied the implementation and management of innovative and entrepreneurial training programs for university students. Zhao. Z. et al. (2022) believe that the innovation and entrepreneurship training program for college students should be results-oriented, and the conclusion and outstanding standards of the project should be defined for university level, provincial level, and national college students [9]. Lin, F. (2022) pointed out the main problems existing in the innovation and entrepreneurship processes of university students and proposed measures to improve the implementation mechanism and process of the project [10]. Dong, W. (2022) analyzed the current situation of the innovation and entrepreneurship training program for university students in Shanxi Province in 2020 and proposed that universities should attach importance to the process, management departments should implement dynamic management, and the government should establish a resource integration and sharing mechanism [11].

In summary, some scholars have studied the current state of innovation and entrepreneurship education for university students, the difficulties they face, and the construction of the innovation and entrepreneurship education system. Others have come up with ideas on how to give full play to innovation and entrepreneurship training programs for university students in terms of improving their practical ability to innovate and start businesses. There are relatively few surveys and studies on the current state of knowledge perception, and experience of innovation and entrepreneurship among university students. Therefore, this paper sets out to investigate the current state of knowledge and experience in innovation and entrepreneurship among university students in Jilin Province.

2 Research methods

2.1 Questionnaire survey

Respondents to the questionnaire were university students in Jilin Province, including current students and graduates. To enhance the validity of the survey results, the survey adopted the "Wenjuanxing" survey platform, mainly through the purchase of sample service, to accurately collect questionnaires from the survey objects. A total of 315 questionnaires were collected, of which 295 were valid and 20 were invalid.

2.2 Wilcoxon signed-rank test and Kruskal-Wallis test

A Wilcoxon signed-rank test is a nonparametric test. This test aims to infer whether there is a difference between the two population medians from the two correlated samples of the paired sample. It is also commonly used in nonparametric T-tests of paired samples.

A Kruskal-Wallis test is also a nonparametric test for comparing groups of two or more continuous or discrete variables. It does not assume a specific distribution of the data, similar to a one-way analysis of variance (ANOVA).

3. Difference analysis of innovation and entrepreneurship knowledge and experience among university students in Jilin

3.1 Normality test

To investigate the differences between student knowledge and experience of innovation and entrepreneurship, this paper first conducted descriptive statistics and normality tests on two sets of survey data, the results of which are presented in Table 1.

Table 1. Descriptive statistics and normality test.

Item	Mean±SD	Kolmogorov-Smirnov		
		Statistics	Degrees of Freedom	Salience
Perception of innovation and entrepreneurial knowledge	6.89±1.400	.165	295	.000
Experience in innovation and entrepreneurship	6.25±1.807	.131	295	.000

a. Rielly's significance correction

The valid sample size of this survey is 295, so the results of the Kolmogonov-Smirnoff (K-S) test are selected. The results show a significance of 0.000 for the normal test and less than 0.05 for both sets of data. Thus, the null hypothesis is rejected, and the data for both student knowledge perception and experience in innovation and entrepreneurship are assumed not to follow a normal distribution.

3.2 Wilcoxon signed-rank test

In this paper, we also investigate whether there is a significant difference between the two sets of data by using a non-parametric test, the Wilcoxon signed-rank test. The results of the tests are shown in Table 2.

Table 2. Wilcoxon signed-rank test.

Item	Z	P
Experience in innovation and entrepreneurship – Perception of innovation and entrepreneurial knowledge	7.283	0.000 **

As can be seen from Table 2, the P-value of the Wilcoxon signed-rank test is 0.000, which is less than 0.01. Therefore, the null hypothesis is rejected, and it is argued that there is a significant difference between the perception of knowledge and the experience of innovation and entrepreneurship among students.

As can be seen from the descriptive statistical results in Table 2, the average score of students' perception of innovation and entrepreneurship knowledge is 6.89, which is significantly higher than the average score of their experience of innovation and entrepreneurship of 6.25, indicating that students have mastered the relevant knowledge of innovation and entrepreneurship in various ways, but their practical experience is relatively weak.

4 Analysis of knowledge perception and experience of innovation and entrepreneurship among different groups of students

In this paper, the Kruskal-Wallis test is used to investigate whether there are significant differences in student knowledge perceptions, and experiences of innovation and entrepreneurship across different groups.

4.1 Gender

The survey data obtained from "Perception of innovation and entrepreneurial knowledge" and "Experience in innovation and entrepreneurship" in the questionnaire are classified according to gender, see Table 3 for details.

Table 3. Descriptive statistics (by gender).

Item	Gender (Mean ± SD)	
	Male (n=127)	Female (n=168)
Perception of innovation and entrepreneurial knowledge	6.86±1.435	6.92±1.377
Experience in innovation and entrepreneurship	6.31±1.726	6.21±1.870

As can be seen from Table 3, the score of the female sample population in understanding and perception of innovation and entrepreneurship knowledge is slightly higher than that of the male sample population, and the score of the male sample population in experience of innovation and entrepreneurship is slightly higher than that of the female sample population. In addition, both men and women have a higher perception of innovation and entrepreneurial knowledge than experience. To investigate whether there are significant differences between men and women in their perceptions of innovation and entrepreneurial knowledge and experience, the Kruskal-Wallis test was used for additional research in this paper, and the results are presented in Table 4.

Table 4. Kruskal-Wallis test (by gender).

Item	Perception of innovation and entrepreneurial knowledge	Experience in innovation and entrepreneurship
Inspection Statistics	0.114	0.047
Progressive significance (two-sided test)	0.736	0.829

As can be seen from the data in Table 4, the significance of the Kruskal-Wallis test is 0.736 and 0.829, both greater than 0.05, for the male and female sample populations in terms of their perception of knowledge about innovation and entrepreneurship and their experience with innovation and entrepreneurship. Thus, the null hypothesis is accepted. This suggests that there is no significant difference between the male and female sample populations in their perception of innovation and entrepreneurial knowledge or their experience with innovation and entrepreneurship.

4.2 Current stage: in school or graduated

The data obtained from the survey were classified and statistically analyzed according to the different current stages, and the results are presented in Table 5.

Table 5. Descriptive statistics (according to the current stage).

Item	Current stage (Mean ± SD)	
	In school (n=101)	Graduated (n=194)
Perception of innovation and entrepreneurial knowledge	6.42±1.485	7.14±1.290
Experience in innovation and entrepreneurship	5.73±1.821	6.53±1.743

As can be seen from Table 5, graduates have higher knowledge perception, and experience than current students. At the same time, the average score of knowledge perception on innovation and entrepreneurship is also higher than the average score of experience. To investigate whether there are significant differences in knowledge perception, and experience of innovation and entrepreneurship among the sample population at different stages, the Kruskal-Wallis test was performed separately on both sets of survey data, and the results are presented in Table 6.

Table 6. Kruskal-Wallis test (according to the current stage).

Item	Perception of innovation and entrepreneurial knowledge	Experience in innovation and entrepreneurship
Inspection Statistics	17.558	13.582
Progressive significance (two-sided test)	0.000	0.000

As can be seen from Table 6, the significance of the Kruskal-Wallis test on the perception of innovation and entrepreneurial knowledge and the experience of students at different stages is 0.000, which is less than 0.05. Thus, the null hypothesis is rejected, suggesting a significant difference in perceptions of innovation and entrepreneurial knowledge and experience between current students and graduates. Current students have significantly less knowledge and experience in innovation and entrepreneurship than graduates. As a result, people's knowledge and experience of innovation and entrepreneurship are continuously enriched as they grow in age and experience.

5 Conclusions

A difference analysis of student knowledge perception, and experience of innovation and entrepreneurship shows that, overall, students have significantly different perceptions of innovation and entrepreneurial knowledge and richness of innovation and entrepreneurial experience. Analysis of the different groups showed that the male and female sample populations did not show significant differences in either respect. Significant differences in these two aspects were found between the sample groups at different stages. All this suggests that the knowledge perception and experience of innovation and entrepreneurship are being continuously enriched by students as they grow in age and work experience. To ease the pressure on employment, university students should be actively encouraged to innovate and start businesses.

Acknowledgments.

Give special thanks to the Social Science Research Planning Project of the Education Department of Jilin Province, "Research on Cultivating and Improving Innovation and Entrepreneurship Ability of University Students in Jilin Province" (JJKH20220125JY).

References

1. Xu, T., Qi, L.i, Tian, F.: Innovation in colleges and universities entrepreneurship education problems and countermeasures research. J. Financ. Theor. Instr. **1**, 110 (2023)
2. Zhang, X., Yang, X.: University students' innovative undertaking of education research and thinking. Sci. Technol. Innovation. **19**, 95 (2022)
3. Jin, Q.: Digital economy from the perspective of college students' innovative undertaking of education path research. J. Liaoning Econ. **2**, 52 (2022)
4. Sun, A. Research on college students' innovation and entrepreneurship education under the background of Mass Entrepreneurship and Innovation. J. Higher Educ. **8**,34 (2022)
5. Huang, R., Li, W.: After the outbreak of college students' innovative entrepreneurship education new mode. J. Chin. Mod. Educ. Equip. **15**, 141 (2021)
6. Wu, M. Investigation of the development status of college students' innovation and entrepreneurship education under the background of "Internet+". Mod. Trade Ind. **20**,53 (2021)
7. Rivera, Y., Rodríguez, G., Espada, K., Medina, L., García, A., Hoyos, M.: A study of student knowledge and participation in entrepreneurship and innovation ecosystem. IISE Annu. Conf. Expo **2021**, 1130 (2021)
8. Portuguez Castro, M., Gómez Zermeño, M.: Identifying Entrepreneurial Interest and Skills among University Students. Sustainability **13**, 6995 (2021)
9. Zhao, Z., Luo, Z., Xia, W., et al. Project management of college students' Innovation and Entrepreneurship training based on OBE Theory. Theor. Res. Pract. Innovation Entrepreneurship. **5**,188 (2022)
10. Lin, F. Research on the project practice of College students' Innovation and Entrepreneurship training plan under the background of Mass Entrepreneurship and Innovation. Ind. Innovation Res. **16**, 196 (2022)
11. Dong, W. Innovation and entrepreneurship training program for college students in Shaanxi Province: status, problems, and suggestions. Shanxi Educ. (Higher Educ.). **4**, 69 (2022)

Ethicality of Business Leadership

Sirui Xiang*

Birmingham Business School, University of Birmingham, Edgbaston, Birmingham, B15 2TT, United Kingdom

Abstract: Ethical leadership plays an essential role in the success of an organization. Against this background, this study aims to investigate whether leadership is ethical through the case of Volkswagen and the concept of ethical leadership. The studies I have performed showed that leaders can follow both ethical and unethical styles of leadership. Therefore, at the end of the essay, it can be said that leadership can be both ethical and unethical.

1. Introduction

The essay aims to know better about whether leadership can be ethical or not. In the case study, Volkswagen is a German multinational automobile brand that manufactures and sells cars and other vehicles like motorcycles, commercial autos, passenger vehicles, etc. In 2012, Volkswagen became the second-largest automaker in the biosphere, behind Japanese Toyota and ahead of General Motors [12]. In 2013, Volkswagen was considered the ninth-largest company globally, producing nearly 10 million vehicles. The organization was affected by its unethical leader in 2015 when the US Environmental Protection Agency found many secret devices fitted with all the cars that the organization sold in the US market [8]. This unethical activity by the leader of the organization led the organization to a share price drop and replaced the previous leader, Matthias Mueller.

The essay will talk about how Volkswagen can be led by Matthias Muller with his ethical leadership and the impacts that the organization faced from the previous chairman's unethical activities. At the end of the essay, it will be identified whether the leadership can be ethical or unethical in Volkswagen's case.

2. The Concept of Ethical Leadership

Ethical leadership can be referred to as leadership in which the leader tries to follow normatively appropriate conduct in his actions, communication, and collaboration. Mitonga-Monga et al. [10] clarify that by following ethical leadership, the leaders become someone who will be an example of appreciation, motivation, and ethical conduct. Kim and Vandenberghe [8] explain that ethics is a concept that is being promoted in today's business. The idea of ethical leadership is also becoming popular day by day as it has so many right sides. Business leadership ethics is a crucial matter to understand, develop, and apply in professional life.

Figure 1. Characteristics of Ethical Leadership that the Leader of Volkswagen Can Follow
Source: Van et al. [13]

According to Figure 1, some specific characteristics are there in ethical leadership the current leader of Volkswagen can follow to build ethical leadership in the organization. All these characteristics are also interconnected [10]. The first characteristic of ethical leadership is to respect others, and the new chairman of Volkswagen, Matthias Mueller, respects all the employees, customers, and suppliers of the organization. He also appreciates others' laws and values, which is why he initiated to remove the secret that showed false emissions of the cars. It is a great virtue to respect others in this way. De and Farooq [6] argue that justice is the perfect measurement of behavior and treatment. By following this thing, one person can make everyone happy. It is a part of ethical leadership that Matthias follows to be a moral leader, whereas the previous leader did not follow this thing [13]. The chairman of Volkswagen has no favorites and treats everyone in the same way to maintain justice. Benevene et al. [2] elucidate that no worker has any aim to fear discriminatory action based on masculinity, ethnicity, population, or other factors under this leader. With justice in mind, this person quickly understands that

* Corresponding author: siruixiang22@outlook.com

he needs to win the customers of the United States of America.

It is called that honesty is the best policy and being an honest leader creates considerable value to the followers. In this leadership, employees are satisfied with coming up with innovative thoughts and fortified to recover how they do belongings as they know the leader will be honest and value them [6]. Employees are commended for taking the primary step in place of waiting for someone else. The antecedent chairman of Volkswagen did not follow the honest path to produce cars for the US market. That is why he was forced to resign when he got caught. The new chairman is honest, and he confessed to the scandal and was committed to remaking the wrong that the previous chairman did. Blome et al. [3] illustrate that building community is another aspect of ethical leadership. To be a moral leader, one should try to build a strong community. Matthias Mueller is determined to deliberate the high standards and expectations they place on themselves to make a strong community with the organization and its employees. He is also determined to make the community of the customers and the organization very strong.

3. Can business leadership Be ethical?

Yes, leadership can be ethical, but it can be unethical, too. Following ethical leadership can easily attract the minds of the followers. On the other hand, unethical leadership may push down the admirable relationship with the followers. Being an ethical leader can have many values, even if it can be stronger than values [2]. On the other hand, unethical leadership can significantly impact the followers' activities, but it does not have good deals. For example, it can be said from Volkswagen that the leader. Volkswagen's new leader encourages their employees to act ethically by enforcing appropriate actions and policies that allow employees to adhere to high standards and be held accountable for their actions. In the USA, 482,000 VW and Audi vehicles were equipped with a defeat device [13]. However, when the investigation went global, Volkswagen had to admit that around 11 million cars worldwide were equipped with such equipment. Under pressure from the world's media and various regulatory agencies, Volkswagen launched an internal investigation to clarify the situation and find the culprits. This case study clearly shows that the organization's previous leader forced its employees to do unethical activities [10]. With the help of that, he became a successful chairman until the scandal came true. It also shows that the next chairman is an ethical leader who tried to win back the customers' trust and took good initiatives to remove those devices at a considerable cost.

4. Importance of Ethical Leadership in the Perspective of Volkswagen

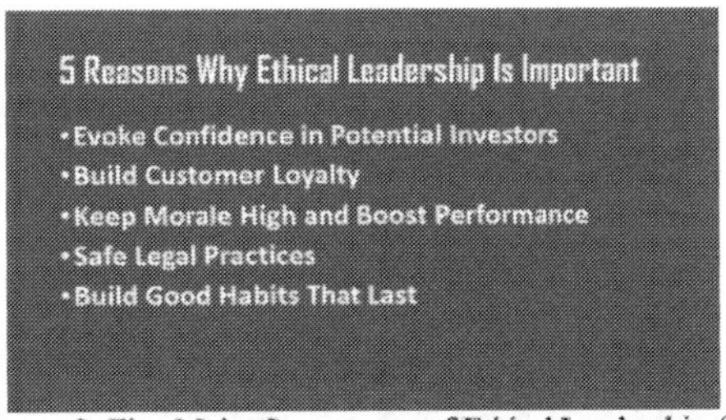

Figure 2. Five Major Importance of Ethical Leadership for which the Leader of VW follows Ethical Leadership
Source: De and Farooq [6]

A lot of importance of ethical leadership is found from a different perspective. It is seen from Figure 2 that there are five significant important ethical leadership that motivated Matthias Mueller to be an ethical leader. Van et al. [13] say that the first importance is that ethical leadership increases the potential investors' confidence to invest in the organization. Benevene et al. [2] elucidate that the antecedent chairman's unethical activities led the organization to a share price drop. It also resulted in a low level of confidence among potential investors. The ethical actions of Matthias Mueller can evoke the spirit of the investors to invest in Volkswagen.

Customer loyalty is essential to survive in today's competitive global market. The unethical activity of the antecedent chairman of Volkswagen resulted in the loss of many loyal customers. Then, it is the foremost duty of Matthias Mueller to retain them by taking the necessary steps. That is why he confessed to the previous chairman's unethical practice and took the initiative to remove those secret devices that showed false emissions of the cars of the organization [6]. Then, the new chairman also has to make cars that adjust to the laws of the US. Ethical leadership practices also help him to keep morale high and boost performance. It is evident because when the employees notice that their leader is righteous, they become more entrusted to work for this ethical leader. Blome et al. [3] illustrate that ethical leadership motivates leaders and other related persons not to undertake those activities that go against a country's laws. Undertaking activities against the laws of a country makes a leader unethical [2]. That is why it can be said that the former leader was unethical. Matthias Mueller, the new chairman of Volkswagen, is ethical as he makes sure the organization is not doing anything that is not legal. And all these things make a good organizational practice culture that positively impacts a long time.

5. Demerits of Unethical Leadership That the Antecedent Chairman Faced

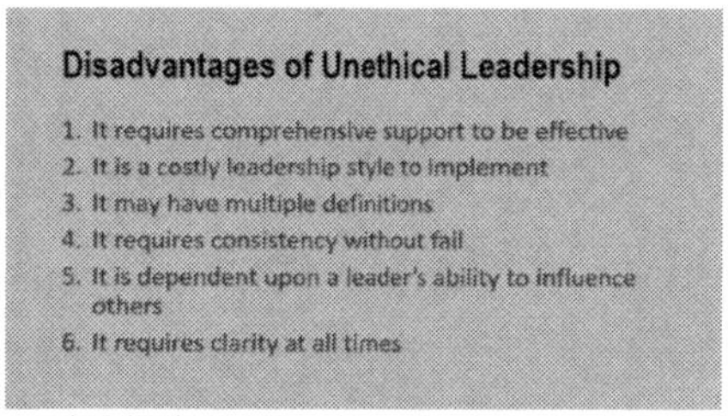

Figure 3. Disadvantages of Unethical Leadership That the Antecedent Chairman of Volkswagen Suffered
Source: Neubert et al. [11]

Several disadvantages are shown in Figure 3 that impacted the previous chairman of Volkswagen. All these things can be avoided if a leader tries to avoid unethical practices in his leadership. The first disadvantage of unethical leadership is that it needs support from others to be established [3]. In Volkswagen's case, the previous chairman did not get help when his unethical practice got caught. But before that, he took support from the employees and other related people with his power. But this did not last so long; he had to resign from his position for his unethical activities, and the organization, along with him, suffered hugely. Avey et al. [1] simplify that unethical leadership is costly as it requires different incentives for different kinds of people to get support for unethical activities. For example, when the antecedent chairman installed the secret device in Volkswagen's vehicles, the person who knew about this fact wanted more money from the chairman. He had to pay for that as it was required to keep it secret.

Ethical and unethical leadership differs from person to person and from situation to situation [11]. For example, installing a secret device in the US market perspective is unethical, but it may be ethical if any security or investigation department wants something like that. Another disadvantage of unethical leadership is that it seems successful until it gets failed. For example, Volkswagen's unethical leader was a successful one until his unethical activities were published in 2015. According to Hassan [7], To be successful in undertaking unethical leadership activities, he needed not to get caught by the US Environmental Protection Agency (EPA). It will also be a costly way for the new chairman to turn the situation right and need considerable time.

Unethical leadership is also dependent on the ability of the leader to influence others. It is a general ability required to succeed even if he follows ethical leadership [1]. The power of the previous chairman of the organization made him successful until 2015. According to Copeland [5], unethical leadership also seeks clarity every time. The leader needs to make sure he can make the other people understand through false words, fear, treaty, or something like that. These are the disadvantages of unethical leadership in the perspective of the case of Volkswagen.

6. Things to Be Considered in Ethical Leadership

Leaders give their people challenging and exciting work and trust them to do it. They inspire others and set an example by respecting their work ethic, integrity, and both employees and customers [7]. The performance of Volkswagen employees is the key to the achievement or disappointment of the business. Mihelic et al. [9] illustrate that being an ethical leader is not such an easy task in the competitive global market. The chairman of the Volkswagen Board of Directors was forced to resign for his unethical activity and was replaced by Matthias Mueller, Porsche's former head [5]. The company has recovered millions of vehicles and is earmarking nearly $ 5 billion minimum to cover restoration costs. It was to be done by the new chairman as he got what the organization needed to get the trust back from the customers. Otherwise, the organization may suffer massive damage in the future. So, there are a lot of things that are to be considered in ethical leadership, and a few of them are clarified here.

Brown et al. [4] say that ethical leaders are to be responsible for all the things under their control, like the organization, the employees, the relation with the customers, suppliers, and all the organization's stakeholders. Matthias Mueller's activities make it clear that he is very much responsible for the organization's goodwill, customers, employees, and so on. Ethical leadership is defined as leadership that respects moral principles and standards and others' self-respect and rights [9]. It is mainly about ethical expansion and virtuous behavior. Even if it damages, doing the true thing is the mantra of a moral leader [4]. The new leader of Volkswagen influences its employees and other related parties positively through ethical leadership.

Emotional stability or low-level neuroticism is related to moral leaders because they expect the leaders to be safe and secure. Shapiro et al. [12] say that an ethical leader must be emotional with his followers to ensure an adequate flow of activities throughout the organization. Agreeableness is another to be considered in ethical leadership. Ethical leadership needs to be more agreeable to all of his fellows [4]. For example, if Matthias Mueller does not become friendly with his followers in the organization, the organization will not succeed in the future. Tolerance is such kind of thing that must be followed to be an ethical leader. An ethical leader will get a lot of lousy offers, fear, opportunities, and so on not to follow the ethics in his leadership. But the ethical leader needs to tolerate all these things to be a successful leader [9]. For example, as an unethical leader, the previous chairman of Volkswagen installed a secret device in the vehicles so that these cars could be sold in the market of the United States. But the next chairman did not do this kind of thing again. After replacing the previous chairman, Matthias Muller wanted to get the trust and support of the customers back in favor of Volkswagen. he removed the secret device from 482,000 vehicles that were sold in the

US market at a very high cost as an ethical leader.

7. Conclusion

The essay shows that leadership style can be both ethical and unethical. It is the choice of the leader whether he will follow ethical or unethical leadership. The essay aimed to identify the things how leadership can be ethical. Volkswagen is a sizeable global car manufacturing organization that has its business activities all over the world. The essay is done by having examples of the case study of Volkswagen. It is shown in the essay that leaders can follow both ethical and unethical styles of leadership. The previous leader of Volkswagen followed an unethical means of leadership and was successful before 2015. But he did not hold this success for a long time. On the other hand, Matthias Mueller, the new chairman of Volkswagen, follows ethical leadership and is trying to get the organization's position back as soon as possible. However, due to the destructive activities of the antecedent chairman, he needs a long time to get things right. Therefore, at the end of the essay, it can be said that leadership can be both ethical and unethical.

References

1. Avey, J.B., Palanski, M.E. and Walumbwa, F.O. (2011) When leadership goes unnoticed: The moderating role of follower self-esteem on the relationship between ethical leadership and follower behavior. *Journal of business ethics*, *98*(4), pp.573-582.
2. Benevene, P., Dal Corso, L., De Carlo, A., Falco, A., Carluccio, F., and Vecina, M.L. (2018) Ethical leadership as the antecedent of job satisfaction, affective organizational commitment, and intention to stay among volunteers of non-profit organizations. *Frontiers in Psychology*, *9*, p.2069.
3. Blome, C., Foerstl, K. and Schleper, M.C. (2017) Antecedents of green supplier championing and greenwashing: An empirical study on leadership and ethical incentives. *Journal of Cleaner Production*, *152*, pp.339-350.
4. Brown, M.E. and Treviño, L.K. (2006) Ethical leadership: A review and future directions. *The Leadership Quarterly*, *17*(6), pp.595-616.
5. Copeland, M.K. (2015) The importance of ethics and ethical leadership in the accounting profession. In *Research on Professional Responsibility and Ethics in Accounting*. Emerald Group Publishing Limited.
6. De Roeck, K. and Farooq, O. (2018) Corporate social responsibility and ethical leadership: Investigating their interactive effect on employees' socially responsible behaviors. *Journal of Business Ethics*, *151*(4), pp.923-939.
7. Hassan, S. (2015) The importance of ethical leadership and personal control in promoting improvement-centered voice among government employees. *Journal of Public Administration Research and Theory*, *25*(3), pp.697-719.
8. Kim, D. and Vandenberghe, C. (2020) Ethical leadership and team ethical voice and citizenship behavior in the military: The roles of team morale efficacy and ethical climate. *Group & Organization Management*, *45*(4), pp.514-555.
9. Mihelic, K.K., Lipicnik, B., and Tekavcic, M. (2010) Ethical leadership. *International Journal of Management & Information Systems (IJMIS)*, *14*(5).
10. Mitonga-Monga, J., Flotman, A.P. and Cilliers, F. (2016) Workplace ethics culture and work engagement: The mediating effect of ethical leadership in a developing world context. *Journal of Psychology in Africa*, *26*(4), pp.326-333.
11. Neubert, M.J., Carlson, D.S., Kacmar, K.M., Roberts, J.A. and Chonko, L.B. (2009) The virtuous influence of ethical leadership behavior: Evidence from the field. *Journal of Business Ethics*, *90*(2), pp.157-170.
12. Shapiro, J.P., and Stefkovich, J.A. (2016) *Ethical leadership and decision making in education: Applying theoretical perspectives to complex dilemmas*. Routledge.
13. Van Quaquebeke, N., Becker, J.U., Goretzki, N. and Barrot, C. (2019) Perceived ethical leadership affects customer purchasing intentions beyond ethical marketing in advertising to moral identity self-congruence concerns. *Journal of Business Ethics*, *156*(2), pp.357-376.

Applications and Responses to RCEP Rules on Geographical Indications Protection in China

Ruolin Liao*

Guilin University of Electronic Technology, Law School/Intellectual Property School, 541004, China

Abstract. Regional Comprehensive Economic Partnership (RCEP) uses the World Trade Organization (WTO) rules as parent rules. Yet proble ms exist in the interpretation and application of intellectual property terms, which can easily cause controversy. The protection rules of the geographical indications (GI) are used as a magic weapon to promote Chinese high-quality agriculturalproducts to the world. But due to the left and right swaying of the legislators, they are chaotic and thus need to be improved urgently. In this case, China's GI protection should focus on disputed solutions, balancing legislators' initiatives and autonomy generated by geographical indication, and solving the historical problems of many departments overlapping management.

1 Introduction

The economic rationale for protecting GI derives mainly from the fact that the place of origin may be used as a quality signal, or alternatively, that the resources of the region may be captured as quality attributes.[1] However, China's GI of agricultural product brand competitiveness is still relatively weak, compared to its Asian neighbors and some European countries. Thus, its effect to increase agricultural exports, and improve the role of farmers' incomes is still very limited. China should seize this opportunity of RCEP, improve the existing protection system of geographical indications based on the Trademark Law, the Provisions on the Protection of GI, and the Measures for the Administration of GI of Agricultural Products, and improve the decentralized and fragmented features of China's legislation on geographical indications. [2]

2 Some problems of dispute rules on GI of RCEP

RCEP is based on Article 24 of the General Agreement on Tariffs and Trade(GATT) and Article 5 of General Agreement on Trade in Services(GATS), incorporates several WTO rules and, in some respects, develops as a modification of the rules by some WTO members. RCEP incorporates a number of WTO rules and develops some aspects of them and represents a modification of the rules by some WTO members. Chapter 6 lists a number of provisions that are incorporated into the TBT Agreement, but clearly states: 'In the event of inconsistency between the two, this Agreement shall prevail'. Confusingly, Chapter 11 of the RCEP (Intellectual Property Rights) states: 'In the event of inconsistency between the contents of this Chapter and TRIPS, TRIPS shall prevail'. Although RCEP members are free to establish 'conflicting norms', this provision is inconsistent with the principle of *lex posterior* and contrary to the TBT, which creates confusion. Moreover, the notion of 'TRIPS precedence' is puzzling: why should the old prevail over the new? 'TRIPS overrides' lead to two consequences: (1) it is contrary to the TBT provisions, resulting in the emergence of conflicting norms in the same treaty; (2) it is not in line with the intent of the FTA, as RCEP IPR protection should be more extensive than TRIPS.

While RCEP does not have a general conflict norm, it does have a treaty conflict arrangement in Article 20(2)(2), whereby if a Member considers that the provisions of this Agreement are inconsistent with the provisions of other agreements (including the WTO), it shall consult with the other Members with a view to arriving at a mutually satisfactory solution. RCEP does not have a general conflict norm, but it does provide for a conflict of treaties arrangement in its Article 20(2)(2). The use of case-by-case consultations rather than conflicting norms to resolve RCEP-WTO inconsistencies, while solving specific problems, can have institutional implications. For example, can two members disregard the principle of priority of subsequent law and arbitrarily adopt a solution similar to the priority of TRIPS, or even create a solution that is neither RCEP nor WTO? What would be the impact of two members' proposals on other members? Is it possible to reach a different formula with other members? Such a complex situation may not arise, but logically there are various possibilities. In short, there is a great deal more uncertainty in case-by-case consultations than in conflict regulation. [3]

Here, article 22 (1) of TRIPS defines GIs as 'indications which identify a good as originating in the territory of a WTO Member, or a region or locality in that territory, where a given quality, reputation or other characteristic of the good is essentially attributable to its geographical origin'. Articles 22, 23 and 24 of TRIPS can be enacted through specific laws

* Corresponding author: 2257916516@qq.com

https://doi.org/10.1051/shsconf/202317803008

such as China's Regulation on Protection of Products of Geographical Indications (2005), India's Geographical Indication of Goods (Registration and Protection) Act (1999) or Thailand's Act on Protection of Geographical Indications (2003). [4] Though China is privileged as a developing country, the huge blank on dispute rule leaves China with no roads but to prepare for future better interpretations or any other solutions.

3 Differences between RCEP and China's rules on GI protection

With the internationalization of GI worldwide, there is growing debate about the governance of GIs at a national level. How can this public good be reconciled with the logic of competition in a free market in which standard procedures and normalization of the contents are increasing? How should a state interact with private stakeholders in the regulation of such a public good? [5]

GI, in summary, refer to the relationship between products and their place of origin. According to different international legislative approaches, this correlation can currently be understood as 'subjective correlation' and 'objective correlation'. The essence of subjective correlation is to protect goodwill, which actually reflects the connection between products and specific places of origin in consumers' subjective cognition. Objective correlation, also known as the terroir element (translated by some scholars as 'soil'), refers to the objective nurturing relationship between a land and the people who have relied on it for generations to grow. This nurturing relationship is a de facto input-output relationship that is unrelated to the subjective perception of the public.

From the perspective of legal philosophy, regardless of the nature, type, and form of property, 'labor' is the basic way of obtaining property and one of the important sources of property rights. The intellectual property value of geographical indications first comes from the trademark use of relevant place names by relevant market entities on specific goods, which fully conforms to Locke's 'property labor theory', that is, 'labor should be rewarded'.

In fact, the lack of geographical indication concepts among ordinary people is not unique to China, which is a common situation. Even in Europe, where geographical indication protection is at the forefront of the international community, research on 600 Irish consumers shows that only 7% of respondents are aware of Ireland's geographical indication situation[6]. Another issue is that with the internationalization of global geographical indications, there is an increasing debate on national level governance of geographical indications. How does this public interest align with the competitive logic of the free market? In the free market, the standardization of standard procedures and content is increasing. How should the state interact with private stakeholders when regulating such a public product[7]?

In some aspects, China chose its own way while RCEP choose the other way. In terms of the standard of review for objections, the RCEP emphasizes 'the customary term in the common language as the common name for the goods in question in the territory of the Contracting Party' and 'the understanding of the term by consumers in the *territory* of the Contracting Party, whereas China's *Measures for the Protection of Foreign GI* (MPFGI) emphasize the technical quality requirements, quality requirements, popularity, and relevance of the product. RCEP is based on public perception and focuses on the reputation of the product in the general perception of consumers in the Contracting Parties from the perspective of consumers, which effectively avoids the lag of the legislation and ensures the accuracy of the perception. On the other hand, China's legislation tends to rely on the identification model of the previous trademark law, which is based on professional knowledge and emphasizes whether the product itself is within the scope of public power or authoritative third parties' constraints.

MPFGI does not show any convergence with the provisions of the RCEP on the composite terminology on GI protection. While the RCEP provides for the date of protection of geographical indications, MPFGI only provides for an annual reporting system for protected geographical indications, but not this system. However, it is worth mentioning that the latter provides for the period for accepting objections to announcements, the implementation of a self-declaration system for special indications, and the information disclosure obligation of the State Intellectual Property Office. This is not only consistent with the information disclosure obligations of the parties in RCEP but also goes further in information disclosure.

4 Analysis of RCEP GI protection rules adaption in China

In the giant international trade agreement represented by RCEP, the concept of rules for the protection of GI is inherited from the concept of the European system for the protection of geographical indications, with the concept of *terroir* as the core, and geographical indications are essentially the natural DNA that nurtures Mother Earth, which cannot be cloned and is difficult to be scientifically interpreted. In other words, GI are the sum of the culture, geography, traditions, heritage, and traditional practices of peoples and nations. [8] As a cultural heritage, the geographical indication is a valuable national treasure and enjoys a high status, which is protected by the state and has the attribute of public power.[9]

Bilateral and regional agreements often lead to the migration of intellectual property standards from developed countries to underdeveloped countries. These transplants are harmful, for example, they tend to overlook the local needs, national interests, technological capabilities, institutional capabilities, and public health conditions of underdeveloped countries; They may also stifle local development, especially when the development paths of the two are inconsistent.Although China's position in the current world intellectual property system is constantly improving[10], from the perspective of per capita income and the overall level of global social development, China is still a developing country. If the standards for intellectual property protection exceed the scope of socio-economic development, it not only cannot have a positive impact on innovation, but may also have adverse effects on public interests and social welfare. Before the implementation of new policies, it is necessary to carefully evaluate the balance between intellectual property protection and social public interests, consider whether these policies are beneficial to the country and meet the development requirements of the country from all aspects, and also consider the relationship between the costs of materials and manpower required and profits. We cannot simply emphasize the interests of one aspect while neglecting the interests of others. It is particularly important not to blindly increase the interests of intellectual property rights holders, but also to take into account public interests and safeguard the overall interests of society.

To some point, China might protect them as intangible cultural heritage at first. The reasons for this are twofold: Legally, China's current intellectual property rights system began with the accession to the Paris Convention for the Protection of Industrial Property in 1985 and has been established through legal transplantation for only more than 30 years now. Due to the lack of modern intellectual property rights awareness, coupled with the long-standing tradition of emphasizing agriculture and suppressing commerce, it is impossible to generate the right to claim the protection of GI. On the other hand, China is a vast country with complex geography, climate, and culture, and the distribution of famous local specialties is extremely dispersed and diverse, involving various industrial fields, without forming a representative geographical indication industry with common interests and a high degree of self-recognition, and lacking the industrial foundation for the establishment of the geographical indication system.

Worse, China is an example of the difficulties faced by countries in the choice of their macro-rule model. Conflicting influences from the US and the EU, but also China's own culture and history, led to the coexistence of trademarks managed by the State Administration for Industries and Commerce (SAIC) and two sui generis models.[11]The historical traditions mentioned above have a direct impact on China's legislation on the protection of geographical indications. However, the domestic geographical indication protection system has obvious redundancy of legal provisions and insufficient protection.

Anyway, the uncertainty of such disputes may shake the confidence of foreign countries in the consistency of China's GI protection, and thus affect the image of China's import and export trade and business environment.

In China, after the new round of institutional reform of the State Council in 2018, although the protection functions of trademarks and GI products were incorporated into the reorganized State Intellectual Property Office, they are managed by the Trademark Office and the Protection of Geographical Indications and Official Signs Division of the Department of Intellectual Property Protection, while the protection functions of GI of agricultural products are still exercised by the Ministry of Agriculture and Rural Affairs alone, and the three management authorities still have problems such as inconsistent administrative status and unsound communication mechanisms. There are still problems such as different administrative statuses and unsound communication mechanisms among the three administrative organs. The fragmented legislation and multiple management have seriously affected the long-term development of GI protection in China and greatly hindered the integrated construction of the protection system.

5 How to improve rules on GI protection in China under RCEP

RCEP is mainly based on TRIPS. The meta-rules established in the TRIPS constitute the overarching level of governance that addresses the outcomes to be achieved by national regulatory frameworks in implementing GI protection. As the TRIPS had to accommodate widely divergent doctrinal views and existing national systems, it provides no more than a broad definition of GI and a minimum scope for protection.[12]

Now, China's legislation still favors technical quality requirements, quality requirements, visibility, and relevance, which is not only contrary to the international legislative trend but also not in line with China's judicial practice regarding the 'relevance' and 'whether confusion can be achieved' determination of preference.

At present, China implements parallel protection for the protection of geographical indications, in addition to a part of geographical indications which are protected with reference to the Trademark Law and are subject to a ten-year protection period, there are also many geographical indications that are protected by other administrative regulations. The latter does not specify the term of protection but only stipulates that the right holder will lose the right of geographical indication if he does not organize the production in accordance with the corresponding standards and management norms, or if he does not use the special sign on the protected geographical indication product within two years. Such a provision will easily lead to disputes in international trade.

Further, it is better to combine both GI protection rules and intangible cultural heritage related rules. According to the definition of UNESCO, intangible cultural heritage refers to various social practices, conceptual expressions, forms of expression, knowledge, skills, and related tools, objects, handicrafts, and cultural venues that are considered as part of their cultural heritage by various communities, groups, and sometimes individuals. Article 2 of the 2011 Intangible Cultural Heritage Law of the People's Republic of China (hereinafter referred to as the 'Intangible Cultural Heritage Law') stipulates that 'Intangible cultural heritage referred to in this law refers to various traditional cultural expressions that have been passed down from generation to generation and are considered as part of their cultural heritage, as well as physical objects and places related to traditional cultural expressions.' Through these provisions, intangible cultural heritage is more of a concept belonging to the cultural system, Its objects mainly include two aspects - intangible forms of traditional culture and the physical objects and places that embody these traditional cultures, with the most crucial being 'culture'. Therefore, the essential connotation of intangible cultural heritage is the experiential knowledge and information created and expressed by humans through the use of wisdom in long-term production and life.

Various symbols reflect human creativity and imagination, and are constantly being created. On this basis, the characteristics of intangible cultural heritage can be summarized as follows: firstly, from the perspective of manifestation, it is reflected as the attribute of immateriality, or 'creative intellectual achievements'. Secondly, from the perspective of its form of existence, it has traditional attributes, has been passed down from generation to generation, has a long history, and is in a dynamic state of change. From the perspective of attribution, although there are a small number of individuals who own such skills, it often contains group and regional characteristics, shared by groups within a certain geographical range, and the owner is uncertain. Finally, from the perspective of value attributes, in addition to possessing the primary cultural value of 'protecting world cultural diversity', its economic and legal 'interest attributes' are also prominent in modern social conditions, which is precisely the reason for the frequent commercial development disputes in the field of intangible cultural heritage. However, although these objects have various names, they can all be covered in two categories: 'intellectual creation achievements' and 'commercial symbols'. The most essential feature of intellectual property is its 'immateriality of the object'. In theory, the two have similarities and can be similarly protected. This is not to prove the intellectual property rights of intangible cultural heritage, but to illustrate the possibility of theoretical connection. After all, the universal and public nature of the term 'culture' in intangible cultural heritage determines that it cannot be used as private property, and it contains too much content and its wording is too grand.

GI and intangible cultural heritage evaluate the same object from different dimensions. Both (intangible cultural heritage and landmarks) focus on ancient creativity and community ownership, rather than new knowledge and private ownership. Due to their inherent similarities, it can be considered to connect the two through procedural and physical similar protection.

Chinese provinces should consider the local conditions and do a good job of legislating in their own provinces, considering the rules on GI protection in China as well as the RCEP, to build free brands and protect their own GI while actively promoting their GI overseas, to let more people feel the colorful charms of China's vast land.

China should expedite the entire protection of GI dispersed in various sectoral laws, establish a library of specific standards for the application of a certain geographical indication, and learn from the legislation of the European Union to adopt a high-standard approach to GI protection in China.

6 Conclusion

RCEP rules on GI protection are trying to take care of the vast number of developing countries among the contracting parties, and the overall legal framework is relatively loose. Regrettably, China's existing relevant rules are still unable to adapt to and converge with this loose legal mechanism. From the acquisition of rights to the maintenance of rights, China's current legal system is a bit confusing and insufficiently enforced. Now that the Law on the Protection of Geographical Indications is about to be promulgated, it is hopeful that China can adapt to RCEP rules on GI protection, improve its legal system, and promote the development of Chinese agricultural products to the world better.

Acknowledgment

Innovation project of Guilin University of Electronic Technology Graduate Education. Supported by 2022 Guilin University of Electronic Technology Graduate Education Innovation Program 'The impact and inspiration of the RCEP geographical indication protection rules on China' (Project Number: 2022YCXS182)

References

1. Bramley, C., Kirsten, J.F, Exploring the economic rationale for protecting geographical indicators in agriculture. Agrekon: Agric. Econom. Res. *Policy Practice Southern Africa*, **46** , 1(2007)
2. Wang, X., Yin Y., The Dilemma and Development of China's Geographical Indication Protection System in the Context of <Civil Code>. *Journal of Guizhou University*, **40**(2022)
3. Yang, G.H. The relationship between RCEP and WTO rules. *International Business Studies*, **5**(2021)
4. Rippon, M.J., Traditional foods, territorial boundaries and the TRIPS Agreement: the case of the Melton Mowbray Pork Pie. J. *World Intellectual Property*, **16**(2013)
5. Delphine, M., LAURENCE, B., JEAN-PIERRE, B., FRANCOIS, C., Are French geographical indications losing their soul? Analyzing recent developments in the governance of the link to the origin in France. World Development, **98**(2017)
6. Ricky, C., Marie M., Protected geographical indications: Institutional roles in food systems governance and rural development. *Geoforum*, **60**(2015)
7. Delphine, M., LAURENCE, B., JEAN-PIERRE, B., FRANÇOIS, C., Are French geographical indications losing their soul? Analyzing recent developments in the governance of the link to the origin in France. *World Development*, **98**(2017)
8. Latha, R. N., Rajendra K., Geographical Indications: A Search for Identity, *Lexis Nexis Butterworths*. (2005)
9. Wang, X.B., Lin, X.Q., A Comparative Study of Geographical Indications Protection in China and the European Union: Taking the Negotiation of the Sino-European Cooperation Agreement on Geographical Indications as a Perspective. *Journal of Xiamen University*, **3**(2012)
10. Jeffrey J Schott, Free Trade Agreements: Boon or Ban of the World Trading System in Schott(ed), Free Trade agreements: US strategies and priorities(institute for international economics, (2004)
11. Xiaobing, W., Kireeva, I., Protection of geographical indications in China: Conflicts, causes and solutions. *The Journal of World Intellectual Property*, **10**(2007)
12. Delphine, M., Estelle, B., The multifaceted role of the state in the protection of geographical indications: A worldwide review, *World Development*, **98**(2017)

Policy Change in the Context of Multiple–Streams Theory–Take China's Centralized Drug Procurement Policy as an Example

Xin Wen*

Beijing Normal University, School of Government, 100875 Beijing, China

Abstract: As an important theory of policy process analysis, multiple–streams theory provides a highly explanatory analytical framework and perspective for exploring the dynamics of policy change. This paper uses multiple–streams theory to analyze the history of china's drug procurement policy changes, found that the china's drug procurement policy has gone through four development stages. Each change in drug procurement policy is due to a particular problem that is highlighted, focal issues enter the policy agenda when problem streams, policy streams and politics streams are highly coupled and the issue window opens, then the policy changes. This paper provides an outlook on the key directions of china's drug procurement policy, in order to provide useful reference for the study of multiple–streams theory and the study of china's drug procurement system.

1 Introduction

Sound drug procurement system is an important task in the reform of the medical and health care system and an important reflection of the ability to improve national governance. Early 21st century, china's pharmaceutical procurement policies entered the era of exploration of centralized procurement, which has gone through three evolutionary processes: state monopoly for purchase and marketing, decentralized purchasing and centralized purchasing, and has significantly reduced the cost of the people's medical. China has issued opinions on promoting the normalization and institutionalization of centralized quantity procurement of drugs, according to a notice released by the General Office of the State Council on February 2021, required all relevant departments to further promote the system of centralized drug procurement, the document became the platform document for centralized volume purchasing. This paper combs the history of china's drug procurement policy changes and analyzes the inner mechanism of drug procurement policy changes by applying multiple–streams theory. In order to enhance the explanatory power of the theory in the study of the dynamics of policy change and to provide a practical basis for further reform of the medical and health care system.

2. Changes in the centralized drug procurement policy

Since the founding of New China, China's drug procurement model has shifted from state monopoly for purchase and marketing to decentralized procurement and centralized procurement, which is a continuous improvement process. The system of unified purchase and sale was implemented before the reform and opening up, and the country comprehensive control over the production, distribution, sale and use of drugs. Due to the system is compatible with the planned economic system and has a weak economic foundation, it differs greatly from the current system and has more influencing factors, so it will not be discussed in this paper. Decentralized purchasing policies occurred from 1984–1999, medical institutions deal directly with drug distributors, thus giving rise to inflated drug prices and massive commercial bribery[1], during this period, governments across the country began to explore a centralized procurement system for addressing market disruptions in decentralized procurement. The evolution of china's centralized drug procurement system is shown below.

2.1 The start-up phase of the centralized drug procurement system (2000-2005)

In the above context, the government summarizes the problems revealed by the decentralized procurement system. National Health Commission of the China jointly with many departments, issued in 2000 "Guidance on the

* Corresponding author: daisybs@163.com

Reform of the Urban Medical and Health System ", "Notice on Strengthen the Management of the Pilot Centralized Bidding and Procurement of Drugs in Medical Institutions", "Notice on Further Improve the Work of Centralized Bidding and Procurement of Drugs in Medical Institutions ", which carried out a nationwide drug centralized bidding and procurement model. In September 2004,national health commission of the china issued "on the further standardization of centralized bidding and procurement of drugs in medical institutions", the document stipulates that medical institutions "sales price shall not be higher than the winning bid price of 15% of the purchase and sale rate", medical organizations are less motivated to purchase due to loss of interest, even, constant opposition from drug companies and distributors has led to frequent deviation from the original policy intent. The document became a milestone document, which played an important guiding role for the main drug procurement mode in the later stage. This phase of centralized drug procurement work to the provincial level, the local government began to clarify the drug collection policy, determine the system, for china's centralized drug procurement system to build a preliminary framework.

2.2 The adjustment phase of centralized drug procurement policy (2006-2014)

The centralized bidding and procurement of drugs has achieved certain results, but the centralized procurement policy of drugs has entered the adjustment period due to the uneven development, differences in procurement policies and intermediary service costs in different provinces. Medical institutions are in a duopoly position in drug distribution, and hospitals have more price advantages and profit margins, in addition, the shortcomings of the "hospitals funding their operations with profits from overpriced drugs "medical compensation mechanism have resulted in the prices of medicines remaining at an inflated level, and high discounts and large kickbacks in the purchasing and selling process. Accordingly, the relevant departments began to rectify the order of purchase and sale, to combat commercial bribery, and in 2009 promulgated the "Opinions on Further Standardizing the Centralized Procurement of Medicines in Medical Institutions", which clearly requires the implementation of province-based online centralized drug procurement system, resumption of power, reduction of the circulation of medicines, expansion of the scope of procurement of medicines, and standardization of the rational use of medicines in medical institutions. Under the active exploration of provinces and cities, the centralized bidding and procurement of drugs has gradually shifted to centralized online procurement, emerging from the drug exchange model and the model of procurement with volume, and shanghai has taken the lead in opening the work of centralized bidding and procurement of medicines.

2.3 Rapid development phase of centralized drug procurement policy (2015-2018)

With the gradual maturity of the drug procurement model and evaluation system, china fully draw on the experience of existing centralized procurement practices. The General Office of the State Council issued the "Guidance on Improving Centralized Drug Procurement in Public Hospitals" in 2015, launching a new round of reform of the centralized drug bidding and procurement system, which proposing "unification of procurement, quantity-price linkage, and classified procurement", proposing hospitals are the subject of tenders and purchases, with categorized purchases and prices based on quantities, clarify that all drugs used in hospitals should be procured through provincial centralized drug procurement platforms, at the same time, the government improved the settlement of purchases and strengthened the comprehensive supervision of the whole process of drug procurement. Moreover, the government encourages inter-provincial joint purchasing methods, such as inter-regional and inter-specialty hospitals. The document marked the formation of a policy framework for centralized bidding and procurement of medicines at the provincial level.

2.4 Stabilization phase of centralized drug procurement policy (2018-present)

In order to break through existing policy barriers and interest fences, to establish a high-quality and efficient healthcare service system and deepen the reform of the medical and health system. National Health Commission of the China jointly with many departments, issued "National–organized centralized procurement of medicines and pilot program "in 2018, proposes that the management of tendering and procurement of medicines be undertaken at the national level, then pilot centralized procurement of medicines started in 11 cities to improve centralized purchasing and market-led drug price formation mechanisms. The General Office of the State Council issued "Opinions on the normalization and institutionalization of centralized procurement of drugs in large quantities "in 2021, the document proposes to institutionalize the centralized procurement of medicines, gradually expand the scope of medicines, pharmaceutical enterprises and medical institutions, and make a comprehensive deployment in improving the procurement rules, supporting policies, etc., and it has become a programmatic document for centralized band purchasing.

3 Multiple–streams Theory

Based on the Garbage Can Model of Cohen, March and Olsen, John W. Kingdon, a famous public policy researcher, established the multiple–streams theory, which first appeared in the book "Agendas, Alternatives, and Public Policies" published by Kingdon in 1984. The theory seeks to explain why certain issue agendas receive the attention of policy makers while others are ignored. The Problem stream of in multi-source theory are various social problems that need to be addressed by the government and these issues are often brought to the attention of government policy makers through a series of indicators reflecting the status and importance of the project, or feedback from focus events and related projects; policy stream policy sources are some specific policy programs or policy proposals put forward by a policy community consisting of government bureaucrats, members of parliament, and scholars; adoption of their policy option depends on the technical feasibility and acceptability of its values. Political Stream include changes in national sentiment, changes in public opinion, ideology, strength of interest groups, etc., national sentiment and the administration have the greatest influence on the policy agenda, policy window in multi–streams theory are opened based on the urgency of the problem and strong political will, such as production safety incidents, political volatility and epidemics. Policy agendas are established in the problem stream and policy stream, and corresponding policy options are provided in the policy stream[2].

Most domestic studies on multiple–streams theory are divided into two categories, one is the review study of multiple–streams theory and the other is the application study of the theory. In terms of theoretical review studies, Jingde Chen suggests that the three dimensions of political source streams should be integrated into one concept, namely the ideology of the ruling party, to better fit the application scenario in China [3].Shuyan Wei and Feng Sun analyze the policy agenda setting under the network society by taking cab reform as an example, and argue that china's policy agenda setting needs to pay attention to the role of new media, network factors in three source streams[4]. Yang Zhijun takes "factor embedding" as the optimization path of the multiple–streams model, and proposes that the three mediating variables of focal events, policy activists and key individuals can enhance the explanatory power of china's policy process[5].

Domestic research on the application of multiple-sources theory. Bicheng Bai using the multiple–streams theory to analyze the dynamics of change in china's housing policy, he argues that the five policy change dynamics can enlighten governments to promote housing policy changes: changes in the problem, increased programmatic feasibility, changes in the political situation, shocks from external events, and the accumulation of positive policy effects[6.] And other scholars use multiple–streams theory to analyze "two separate children"[7], "children of migrant workers to sit NCEE(national college entrance examination) locally "[8] and land system reform [9] etc.

In summary, multiple–streams theory has been widely used to analyze different types of policy changes, and has strong explanatory power for both policy agenda setting and policy changes. In this paper, we use the multiple–streams theory to sort out the history of centralized drug purchasing policy change, and explore what kind of force drives the policy change, in order to predict the future development direction of the policy more accurately.

4 A multiple–streams theory explanation of centralized drug Purchasing Policies

4.1 Problem Stream

First, massive commercial bribery. Whether it is a decentralized procurement model that the public hospitals are the sole purchasers, or a provincial-based bidding and procurement model, it greatly contributes to the problem of cascading transfer of benefits. On the one hand, because public hospitals are the main unit of procurement, they have a dual monopoly of buyer and seller, and medical institutions can interface directly with drug distribution companies and can spontaneously choose which drugs to use and at what price to sell to patients, however, drug distribution companies often subsidize healthcare providers in the form of rebates in order to expand their market size, and this has evolved into a routine operation in drug distribution[10], some medical organizations even regard drug rebates as an important source of income[11]. This directly makes it expensive and difficult for the people's medical care. On the other hand, in order to pursue political achievements, administrative agencies (organizing tenders) have alienated the act of tendering and purchasing into the act of price reduction, and the phenomenon of bad money driving out good money has occurred, and even induced hidden transactions between administrative agencies and tendering enterprises, which has seriously disturbed the market order. As a result, reports and public opinion on drug kickbacks and drug quality were widespread and became a hot social issue at the time, which caused the government to pay great attention and start exploring innovations in drug procurement policies.

Second, the mechanism of the health care system is not sound enough, and it is difficult for the corresponding supporting policies to match it. With the establishment and improvement of the market economy system, the original drug procurement model is unable to meet the people's medical needs. To a certain extent, the decentralized and tendered procurement models have controlled the rise in drug prices, however, after the implementation of the new health care reform, public hospitals above the second level still enjoy the policy of selling drugs at a price difference of no more than 15%, which does not effectively curb the profit-seeking motive of medical institutions in using high–priced drugs. At the same time, local governments blindly invest in and rebuild pharmaceutical enterprises for the sake of political performance, and even interfere in the approval process of foreign pharmaceutical enterprises, which in turn affects the quality of medical resources and the people's demand for medicines, and the role of the bidding and purchasing system in controlling the prices of medicines has been greatly reduced. The Government's efforts to supervise the price and quality of medicines are inadequate; the settlement methods between medical institutions and drug enterprises, and between medical insurance funds and medical institutions need to be improved.

From the different phases of the policy issues mentioned above, it can be observed that they were the most important and specific issues in the political and economic environment of the time and were effectively identified by the government, which in turn shaped the flow of issues in the drug procurement policy.

4.2 Policy Stream

For one thing is the creative policy programs of local governments. Drug rebates caused great social concern in the late 20th century, and as a result, governments across the country began to explore the path to change. Henan Province has taken the lead in making an exploration of a centralized procurement model for medicines. Seven pharmaceutical companies are competitively selected as the fixed-point companies for procurement, then the medical institutions dock with the fixed-point companies for procurement and disclose the relevant information at the service center. This pioneered the shift from a decentralized to a centralized procurement model. During the period of rapid development of the centralized drug procurement system, policymakers in several province broke down the information barriers between regions and took the initiative to build alliances, which provided the prototype for the current drug procurement system, and the shanghai government, in particular, played an important role in the collection work of the "4+7" pilot cities.

And for another, National People's Congress members and scholars in related fields have continued to pay attention to drug prices. Taking china knowledge as an example, a search on keywords such as "centralized purchasing", "volume purchasing" and "government procurement system" reveals that the number of articles published in this field has increased year after year since the beginning of the 21st century. The number of articles published in this field has been increasing year after year since the beginning of the 21st century. In addition, the news media and social organizations have also paid close attention to drug prices and the development of healthcare. When issues come to light, "policy entrepreneurs" respond quickly and offer advice, creating a strong policy stream that drives the policy agenda.

4.3 Political Stream

First, the people's demand for medical services has gradually risen, and dissatisfaction with the current medical system and services is running high. The disorderly expansion of capital and the monopoly of multi-interested parties continue to push up the price of drugs and consumables, resulting in the people's burden of medicine, "difficulties and high expense in medical care " has become a social pain point, which is contrary to the original intent of the policy of standardizing prices. And the network era for the public to provide a more convenient channel to voice, so that focus events spread very fast, dissatisfaction about the high fees charged by medical institutions pervades the country.

Second, standing firm on the people's position is an important guarantee for the continued promotion of reform of the drug procurement system and the most distinctive value orientation of governance. Since 2012, china has adhered to the people–centered development ideology, focusing on solving the people's most realistic interests and improving the quality of medicines and the safety of medicines. At the same time, the government has taken multiple measures to gradually improve the market-led drug price formation mechanism, and has achieved significant results in promoting the supply–side reform of medical services. This reaffirms the importance that the government attaches to improving people's well-being, and lays an important political foundation for the nationwide promotion of the overall path of "state organization, alliance procurement, platform operation".

Third, the government has changed the philosophy of governance in china, with the goal of building a service–oriented government for people. They has optimized its workflow and improved their working methods in administrative approval procedures, the bidding process, procurement management and drug quality supervision, in order to provide high–quality services to enterprises, medical institutions and the people. At the same time, the government has pressed down on the responsibilities of all parties involved in the supply of medical services, to rectified and investigated illegal behaviors, and cleaned up the atmosphere of the medical industry. The shift in governing philosophy has led to an important role for administrative forces in agenda building, providing an opportunity to open the window of opportunity for problems.

The two factors that have the greatest impact on the policy agenda in the political source flow are national sentiment and political attitudes4. Public sentiment changes as a result of the development of a particular issue, while the ability of a particular issue to drive the policy agenda is influenced by the governing philosophy of the rulers, who will move the policy agenda quickly if they are concerned about the event, and vice versa.

4.4 Policy–Window

Policy windows are opportunities for proponents of proposals to bring particular issues to the forefront[2]. They include both political and issue windows. The political window opens in response to shifts in the political situation, and the issue window opens in response to changes in the focal event, which is an important force in the opening of the policy window[12]. Both provide opportunities for policy change. In terms of the development of china's drug procurement policy, the strong reaction of focal events and the change of the top level of the government at each stage triggered the switch of the policy window, realizing the fusion of the window and the source stream, which in turn led to the turnover of the old and the new policies.

In summary, this paper analyzes china's drug procurement policy change through the multiple –sources theory, it can be clearly found that the change of policy is a multi–stage development, the current policy is not directly generated, but in the process of policy change in the source flow of the problem, the source flow of the policy, the source flow of the political around the special issue of the dynamic advancement of the explanation of the logic: the three streams of the convergence of the three streams–policy window opens to the establishment of the policy agenda–the policy change occurs until the emergence of a new political and economic environment of a new special problem, a new round of policy change continues to unfold, as shown in the following figure1.

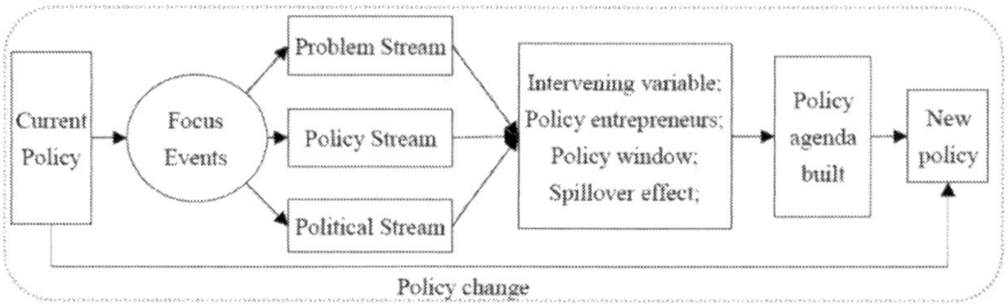

Figure 1. Pathways of policy change

5 Conclusion and discussion

This paper reviews the historical change of drug procurement policy, and then analyzes it by using multiple-streams theory for answering the inner mechanism of drug procurement policy change.

We found that: with the change of history and the evolution of the system, the drug procurement policy has been adapted to the national economic system and the national development strategy, and forming the path of "decentralized procurement–tender procurement-centralized procurement "; facing the opportunities and challenges of the global digital economy, problem stream, policy stream and political stream of each round of change has shown new changes; furthermore, the concept of "whole–process people's democracy "provides more practice fields for drug procurement policy, when "policy entrepreneurs" (the media, the Internet, and NPC deputies) commit to the policy agenda, they not only increase the government's awareness of national sentiment, but also contribute to the formation of a policy agenda for drug procurement.

The current national centralized procurement is gradually institutionalized and a market–led price formation mechanism is gradually established. The accelerated development of population aging has brought great impact and challenges to china's healthcare service system, therefore, the other purpose of this paper is to predicts that the future will focus on solving the problems of healthcare service supply and health insurance fund payment, and promote the industry's high-quality development around value innovation and market innovation.

References

1. LF. Wang, and Z.Ji. Several. issues that need to be addressed in the model of centralized biddin g and procurement of medicines[J].Chinese Healt h Economics,(05):51(2001)
2. Kingdon, JW. Agendas, Alternatives, And Public Policies[M].Longman, Boston,(2011)
3. JD. Chen. "Multiple-streams" Analysis: Unscramb le the New Way of Policy Making[J].Journal of Hubei University of Economics,(03):81-84(2005)
4. SY.Wei and F.Sun. Research on Modernization o f Policy Agenda Setting by Network Society fro m a Perspective of the Multiple-Streams Theory A Case Study of the Taxi Reform[J].Journal of Public Management, **13**(02):1-13(2016)
5. ZJ.Yang. How can the Decision making Model of Policy Process Explain Better the Chinese Ex perience under Ambiguity Conditions? Research on Optimization of Multiple Streams Model Bas ed on the Test of "Stream's Factors and Interme diate Variables"[J].Journal of Public Management, **15**(02):39-51(2018)
6. BC. Bai. Analysis on the Dynamics of Chinese Housing Policy Changes Since Reform and Open ing up:From the Perspective of Multiple Stream s Framework[J].Journal of Public Management,**7**(04):76-85(2010)
7. XM.Long. Interpretation of the "two child alone " policy from the perspective of multiple stream s theory[J].New West,**331**(08):16(2015)
8. N. Zhang, Analysis and Dilemma Cracking of P olicies children of migrant workers sit college e ntrance exams under the Perspective of Multi-so urce Flow Theory[J].Teaching & Administration,**724**(03):36-39(2018)
9. XZ. Zhang, H.Gao, ZZ.Wu. (2023) The logic of land system reform of the spring and autumn an d warring states period based on multiple strea ms theory[J].China Land Science,**37**(04):32-42(20 23)
10. J. Ma, LX.Yan. Trial analysis of the causes and strategies of giving and receiving kickbacks in t he purchase and sale of drugs[J].China Pharmacy,(05):226-227(1996)
11. JS. Xia. Drug purchases and sales in the "kickb ack wind" have a cure? [J].Outlook, (18):24-26(1 997)
12. S. Michaels, N P.Goucher, D. Mccarthy. Policy Windows, Policy Change, and Organizational Le arning: Watersheds in the Evolution of Watershe d Management[J].Environment Management,**38**(0 6):983-992(2006)

Regulating Reverse Payment Agreements in Pharmaceutical Market: A Comprehensive Framework for Antitrust Review

Xupeng Yang*

Ocean University of China, College of Law, 266100 Qingdao, China

Abstract. Reverse payment settlement agreement is a type of arrangement between original and generic drug manufacturers that could potentially violate Antitrust Laws by limiting competition, typically appear in the pharmaceutical market. This paper addresses the impact of these agreements on public interests, and the challenges in conducting antitrust review on these agreements. This paper then proposes a comprehensive framework for regulating reverse payment agreements in the context of China's extensive and complex pharmaceutical market by drawing on the experiences of the United States and the European Union. The framework introduces a two-step antitrust review system for regulating patent settlement agreements, which is suggested by the author to be applied in both patent litigation and daily detection. This paper also advocates establishing an agreement filing system and utilizing the antitrust public interest litigation system in antitrust regulation. This proposed framework provides specific guidance on initiation of antitrust review with respect to multiple backgrounds, in order to protect consumer interests in affordable healthcare, deter anti-competitive behavior, and contribute to the theoretical basis for future legislation.

1. Introduction

Reverse payment settlement agreements, also referred to as "pay-for-delay agreements", typically occur between brand-name drug manufacturers and their generic competitors. Pay-for-delay agreements oblige a party to acknowledge the patent owner's rights and delay their market entry in return for a value transfer (pecuniary or not) from the patentee [1]. Such agreements may violate Antitrust Law, as they can exclude or limit competition in the market.

According to an investigation by the Competition Enforcement Department of the European Commission, generic drugs are the primary source of price competition in the pharmaceutical market, bringing an average 50% reduction in drug prices [2], which substantially diminished the profit of original drug manufacturers but eased the financial burden on consumers. To counteract this impact, original drug manufacturers adapt various strategies to prolong the commercial life of their original drugs. Signing reverse payment agreements with generic drug manufacturers is one of such strategies. Reverse payment agreements are a win-win for businesses, with drug prices remaining high and original drug manufacturers sharing monopoly profits with generic drug manufacturers. However, the cost to manufacturers maintaining monopoly profits is that consumers lose the opportunity to enjoy generic drugs that might be 90% cheaper than the original drug prices. It is estimated that reverse payment agreements are causing loss of $3.5 billion for American consumers every year [3]. Drugs are unique commodities with low price elasticity of demand, and consumers often lack the leverage to reject unreasonably high prices due to their health and life needs. Data shows that in 2021, chemical generic drugs accounted for 83.66% of China's chemical drug market share and 52.80% of the overall pharmaceutical market share [4]. Obviously, China is a significant producer and consumer of generic drugs. Given the vast scale of China's pharmaceutical market, allowing drug manufacturers to maintain unreasonably high drug prices through reverse payments will result in immeasurable damage to public interests.

In December 2021, the Supreme People's Court of P.R.C. (SPC) decided on the patent infringement case between AstraZeneca Ltd and Jiangsu Ausicon Pharmaceutical Co., Ltd. In this case, BMS held a patent effective from August 3, 2005, to March 5, 2021, which was transferred to AstraZeneca on May 23, 2014. Vcare filed a request to invalidate the patent involved in the case in 2011. On January 4, 2012, BMS and Vcare signed a settlement agreement, specifying that Vcare withdraw the request for invalidation of the patent involved in exchange for permission for Vcare and its affiliates to implement the patent involved more than five years before the expiration of the patent rights protection period. Later, Vcare accordingly withdrew the request for patent invalidation, and its affiliate Ausicon implemented the patent involved. Subsequently, AstraZeneca, as the successor patent holder, sued Ausicon, alleging infringement of the patent involved. During the second instance, AstraZeneca

* Corresponding Author: yangxupeng@stu.ouc.edu.cn

applied to withdraw the appeal on the grounds of reaching a settlement agreement. The Supreme People's Court pointed out that the legal review of the withdrawal of the appeal mainly includes: whether the parties have engaged in actions that harm (1) national interests, (2) public interests, or (3) the lawful rights and interests of others [5].

The AstraZeneca case represents the first time the SPC has proactively reviewed a patent settlement agreement in a non-antitrust case to determine whether it may violate the Antitrust Law, marking a watershed in China's judicial practice in regulating reverse payment agreements. This case brings a question to the forefront: In patent litigation, should the court conduct an antitrust review of reverse payment agreements upon receiving a withdrawal application, and to what extent should such review be conducted? Considering reverse payment agreements' secretive and complex nature, relying on the court's proactive review mode has significant drawbacks. Although SPC believes that settlement agreements are suspected violating Antitrust Law and can be transferred to the antitrust enforcement agency, if necessary, the trend towards concealment and diversification of reverse payment agreements is leading to overlooking. This paper focuses on both court's antitrust review of settlement agreements in patent litigation and the regulation on these agreements in the whole market, addresses the challenges in initiating the antitrust review, and proposes a categorized management framework by drawing on the experience of the United States and the European Union. This framework is aimed at providing solutions for China to address the rise of reverse payment agreements and preparing a theoretical basis for future legislative improvement.

2. Analysis of Court's Proactive Antitrust Review in Patent Litigations

2.1 Unavoidable Overlook on Anti-Competition Behaviors

In the AstraZeneca case, the court of first instance, relying on contract law and patent law, determined that Ausicon had the right to implement the patent involved according to the settlement agreement, thereby rejecting all litigation requests from AstraZeneca. This decision reflects that the Nanjing court did not even realize the underlying issues with anti-monopoly law within the settlement agreement [6]. These were only uncovered in the second instance, indicating the incidental nature of the court's review. Moreover, reverse payment agreements often appear lawful and can easily pass as patent settlement agreements through court review. In recent years, such agreements have taken more diverse forms, and the transfer of economic benefits has become more concealed. Many original drug manufacturers no longer directly pay generic drug manufacturers money but instead mask the true nature of their economic benefit transfers through various means, such as providing raw materials [7]. These conducts further reduce the likelihood of detection.

Additionally, under the principle of "no complaint, no action," the scope of the court's investigation can be pretty narrow. When pharmaceutical patent manufacturers and generic drug manufacturers sign reverse payment agreements, they usually include confidentiality clauses. Unless a dispute arises between the parties that lead to court litigation, such an agreement is unlikely to be discovered by the court. Therefore, agreements that have not led to disputes but, in reality, have excluded or limited market competition and violated Antitrust Law will fall into a blind spot in the review process. It is clear that the court's proactive review mode will inevitably overlook anti-competitive behavior. Since reverse payment agreements potentially pose significant harm, a more stringent anti-monopoly review method should be adopted. Furthermore, even if the parties to the agreement have a dispute and resort to the court, the court's maneuverability in reviewing reverse payment agreements remains highly constrained. Even if the court desires to conduct an in-depth review, the "no complaint, no action" principle may make it difficult to take practical action.

2.2 Challenges in Implementing Substantial Review

Another obstacle the court faces is the limited professional expertise, time, and resources needed to review such complex settlement agreements. Suppose the court is required to undertake an anti-monopoly review of such complex agreements independently. In that case, it will consume a substantial amount of time and resources, bringing additional burdens to the already overloaded court system. Additionally, parties may have a motive to conceal reverse payment agreements in their own interest and may selectively submit evidence. They might hide evidence that would lead the court to identify reverse payment agreements or even submit misleading evidence favorable to affirming settlement. They might exaggerate the fairness or necessity of the settlement on the one hand and underestimate the impact of reverse payment agreements on market competition on the other. These practices place the court in a state of information asymmetry when reviewing reverse payment agreements.

Furthermore, the companies' anticompetitive behaviors are becoming less obvious, which contributes to the difficulty. These behaviors include product hops, where a company introduces a slightly reformulated product to replace its current branded drug in the market, thereby thwarting an impending generic entry [8]. With companies often combine strategies to defeat competition that limits their ability to charge monopolistic prices, the court faces difficulties in implementing substantial reviews when examining reverse payment agreements. This not only increases the complexity of the review but may also affect its fairness and accuracy.

2.3 Procedural Barriers in Individual Rights

The effects of reverse payment agreements are extensive and may affect various parties such as generic drug manufacturers, original drug manufacturers, retail manufacturers, and consumers. These effects may extend throughout the entire upstream and downstream of the industry. Parties have different demands in litigation based on different interests. For instance, consumers are more concerned with the price and diversity of drugs, whereas drug manufacturers focus more on developing their own industry. With different interests in mind, litigants will present significant variations in the evidence submitted. These differences will guide the direction and emphasis of the antitrust review, thereby influencing the examination outcome.

Given the diverse litigation interests, courts initiating antitrust reviews on reverse payment agreements in specific lawsuits will face multiple difficulties. Firstly, under the constraint of the principle of "a matter should not be judged twice", which is also referred to as "ne bis in idem" in common law, the subsequent rights to bring a law suit of other related parties will be blocked. The impact of reverse payment agreements goes far beyond the parties involved in the case. Under the principle of "a matter should not be judged twice," the court's review can only focus on the disputes between the parties and judge within the law suit at hand, meaning the court could not reach the external issues. After a reverse payment agreement has been reviewed, it is almost impossible for consumers to bring a law suit claiming damages due to the agreement, as the court will consider the issue to have been already handled and will thus reject to deal with this agreement again. Although litigants can avoid the procedural barrier by submitting different evidence and raising different claims, the court's preexisting conclusion on previous agreements between original drug manufacturers and generic drug manufacturers will indirectly affect subsequent case trials. Changing the review results of reverse payment agreements would essentially undermine the court's judicial authority, a situation that original drug manufacturers, generic drug manufacturers, and consumers would not like to see.

The economic compensation issues related to reverse payment agreements may also become deadlocked due to the court's proactively initiated review. If the court finds a specific reverse payment agreement invalid in particular patent litigation for contradicting public interest, the actual consumers harmed will find it difficult to claim economic compensation, as the active review by the court does not explicitly identify the victimized party. However, if an antitrust review agency discovers and examines the reverse payment agreement, the manufacturers will receive administrative penalties according to Antitrust Law. Meanwhile, consumers can also claim for their rights based on the review results. In this way, the antitrust review agency's examination can simultaneously protect the public interest and individual consumer interests.

3. Comparative Approaches in Antitrust Review Initiation

3.1 United States: FTC and Private Enforcement

Since the Sherman Act, United States Antitrust Law has established a dual enforcement system, dividing the enforcement responsibility between public authorities and private parties, thereby forming two methods: public enforcement and private enforcement. Concerning public enforcement, the US review of reverse payment agreements is primarily overseen by the Federal Trade Commission (FTC). The Federal Trade Commission Act grants the FTC various powers, including investigating businesses or individuals suspected of violating antitrust or consumer protection laws, filing lawsuits in federal court or with administrative judges, and issuing orders. The FTC primarily discovers the existence of reverse payment agreements in two ways: first, by reviewing pharmaceutical manufacturers' agreement records. The Medicare Prescription Drug, Improvement, and Modernization Act of 2003 requires original drug manufacturers and generic drug manufacturers to file agreements relating to the production, sale, and market exclusivity of generic drugs with the FTC and the Department of Justice within 10 days of signing [9]. This provision is believed to have solved the information asymmetry between regulators and regulatees, making reverse payment agreements easier to be detected, thus aiding antitrust review agencies in promptly evaluating agreements and taking further actions [10]. The private enforcement of US Antitrust Law is highly active. Thanks to the establishment of the class-action system, the plaintiff's litigation burden has been reduced, leading to a rapid growth in private enforcement cases. A wide range of victims is included by default in class-action lawsuits, leading to enormous potential compensation amounts, which strongly deter anti-competition businesses.

3.2 European Union: Categorized Management

The review of reverse payment agreements in Europe is primarily the responsibility of the Competition Directorate-General of the European Commission. The categorized management of reverse payment agreements is a highlight of EU's approach, which clarified which types of settlements may warrant further scrutiny. This classification system mainly applies two legal standards: first, whether the agreement restricts the ability of generic drug manufacturers to sell their drugs; second, whether the agreement involves a value transfer from original drug manufacturers to generic drug manufacturers. The applied classification system is as follows in Table 1.

Table 1. Categorization Applied by European Commission

<table>
<tr><th colspan="2" rowspan="2">Categories</th><th colspan="2">Limitation on generic entry</th></tr>
<tr><th>No</th><th>Yes</th></tr>
<tr><td rowspan="2">Value transfer from originator to generic</td><td>No</td><td rowspan="2">Category A</td><td>Category B.I.</td></tr>
<tr><td>Yes</td><td>Category B.II.</td></tr>
</table>

According to the above categorization, agreements that do not restrict generic drug manufacturers from selling their products are classified as Category A, while agreements restricting the entry of generic drugs are classified as Category B. Agreements restricting generic drugs from entering the market are further divided into two categories: B.I. agreements, including those where the original research drug manufacturer does not transfer value to the generic drug manufacturer, and B.II agreements, including those where value is transferred from the original drug manufacturer to generic drug manufacturers.

Generally, Category A settlement agreements do not present Antitrust Law issues, as they allow generic drug manufacturers to bring their products to market immediately. B.I. settlements are similar, but some settlement agreements in this category may be subject to competition law scrutiny, such as settlements reached with patents known not to meet patentability standards, i.e., flawed patents, granted under circumstances where incorrect, misleading, or incomplete information was provided. This situation could still lead to an antitrust review. By contrast, B.II. settlement agreements may be subject to the most rigorous antitrust scrutiny, as they restrict market entry and include value transfer from original research drug manufacturers to non-patented drug manufacturers [11].

Furthermore, within the block exemption system, the EU introduced the concept of "core restrictions" to cover serious restrictive agreements. Legality cannot be assumed without specific scrutiny for these agreements, meaning they cannot be subject to block exemption but must be thoroughly examined according to exemption conditions. If these conditions are met, they may still be lawful. Under the block exemption system, for agreements outside the core restrictions, legality can be recognized without further consideration as long as they do not have the appearance specified in law to simplify the review process as much as possible; for agreements with core restrictions, detailed scrutiny must be conducted regardless of whether they possess the listed appearance features [12]. The EU's block exemption system has dramatically enhanced the efficiency of antitrust review. However, as reverse payment agreements become increasingly complex and concealed, and pharmaceutical manufacturers may take evasive measures against the appearance features listed in the law, the question remains whether the existing block exemption system can still accurately and comprehensively identify potential reverse payment agreements.

3.3 United Kingdom, France, and Germany: Consumer Associations as Viable Entities for Antitrust Public Interest Litigation

Consumer associations as effective organizer for consumers in antitrust claims, have been legally accepted as subjects for antitrust public interest litigation in many jurisdictions. Germany's 1965 "Act against Unfair Competition" granted consumer groups the right to bring collective lawsuits. The 8th Amendment to Germany's "Act of Restrictions on Competition" in 2013 also gave consumer groups the right to sue for inaction, exclusion, and confiscation against monopolistic behavior, forming a consumer group litigation system coordinated with the "Act against Unfair Competition." The UK's 1998 "Competition Act" Section 47B provided a consumer lawsuit request system applicable to antitrust damages and other remedies. According to this provision, consumer groups that meet the neutral, representative of consumer interests, litigation capacity, no trading department, etc., conditions specified by the Secretary of State may sue competition violators in court. France's 1973 "Loi Royer" Article 46 also allows consumer associations to sue in any court for facts that directly or indirectly harm the collective interests of consumers [13].

3.4 Insights from Comparative Perspectives of Regulating Reverse Payment Agreements

The dual enforcement system in the United States, especially active private enforcement, has reference significance for constructing our country's reverse payment agreement antitrust review system. As a developing and populous country with a vast consumer market and a significant producer and user of generic drugs, China is an easy target for foreign innovative pharmaceutical manufacturers' reverse payment agreements. Therefore, strengthening antitrust reviews to protect consumer interests is particularly important. The US and EU experience shows that neither private enforcement nor public enforcement can monopolize the execution of competition law; they must cooperate to achieve the purpose of deterring, investigating, punishing, and compensating the victims of anti-competitive behavior.

Relying solely on public or private enforcement is undesirable. Enforcement of Antitrust Law requires cooperation of public and private enforcement to achieve the best implementation effect. Public agencies' human and material resources are always limited. Private individuals with direct economic interests and consumer associations can mitigate the government's deficiencies in workforce and financial resources, supervise the law enforcement agencies, and help increasing the likelihood of discovering illegal monopolies [14].

A common point in the US and EU regulations on reverse payment agreements is the agreement filing system. Requiring manufacturers to file related agreements with antitrust review agencies when they are reached can effectively improve review efficiency. Moreover, the EU's classification management and exemption system also have reference value. Listing the surface features of agreements and exempting some agreements with a shallow risk of restricting competition also helps achieve maximum efficiency.

4. Institutional Design for Antitrust Review of Reverse Payment Agreements

4.1 Introducing Categorized Management System

In the above analysis, the author found that the court's proactive initiation of an antitrust review of reverse payment agreements can cause many difficulties in patent litigation. The antitrust review agency also needs an effective classification system to improve the efficiency of managing recorded agreements. To avoid overburdening the courts while ensuring the litigation rights of relevant parties when dealing with withdrawal applications, introducing an a classification management system could be a practical choice. By listing the appearance of the agreement, courts, and antitrust review agencies can more easily identify the risk level of reverse payment agreements and thereby make decisions on whether to allow withdrawal or conduct different level of antitrust review on the recorded agreements. However, by simply listing appearances elements the potential anti-competition agreements could not be fully covered. The listed elements may be deliberately avoided in the agreements by pharmaceutical manufacturers, thus causing the whole detection system to be ineffective. The author suggests including a predictive analysis to include those agreements do not have the listed appearance but may lead to increased drug prices or reduced consumer choice in the scope of the review, by which the author constructs a two-step antitrust review initiation system is as follow in table 2.

Table 2. Two-Step Antitrust Review Initiation System

<table>
<tr><th colspan="2">Predictive Market Analysis</th><th colspan="2">Suspicious Appearance Elements</th><th>Outcome</th></tr>
<tr><td rowspan="4">Assume the agreement is executed</td><td rowspan="3">No foreseeable consumer's choice reduction or price increase</td><td colspan="2">No limitation on generic entry</td><td>Free from review</td></tr>
<tr><td rowspan="2">Generic entry delayed</td><td>No value transfer</td><td>Plausible excuses required</td></tr>
<tr><td>Value transfer</td><td rowspan="2">Antitrust review initiated</td></tr>
<tr><td colspan="3">Consumer's choice reduction or price increase foreseen</td></tr>
</table>

In situations where no increase in drug prices or reduction in consumer choice detected and without delaying the entry of generic drug manufacturers into the market, the possibility of reverse payment agreements existing is extremely low. As a result, directly allowing the parties to settle may meet balancing efficiency and risk requirements. In cases where the entry of generic drug manufacturers into the market is delayed, but no value transfer detected, if the parties can provide plausible reasons for the delay in entering the market, they may also satisfy the conditions for free of further review. Otherwise, it should still be transferred to the antitrust review agency for further examination. Including all agreements involving value transfer within the scope of the antitrust review follows the underlying logic that when the patent's validity is weak, the patented drug manufacturer fears that the patent may not meet the granting standards (such as those based on incomplete, misleading, or incorrect information). At this point, if a generic manufacturer files a lawsuit, the court is likely to rule the patent invalid. Therefore, the patented drug manufacturer, fearing the loss of the lawsuit, will tend to enter into a reverse payment agreement [15].

4.2 Establishing Agreement Filing System

In the pharmaceutical market, consumers are usually passive recipients in their interest struggle with original and generic drug manufacturers. Thus, the law must grant more initiative to private parties to fully protect consumer rights. Specifically, the US private enforcement system offers valuable insights. There should be an expansion of the scope of antitrust private enforcement. With the organizational strength of consumer associations, more consumers should be included in class-action lawsuits, providing more individuals with opportunities to assert their rights. Additionally, the agreement filing system adopted by the US and EU has achieved good results, saving the enforcement authorities the cost of discovering reverse payment agreements and reducing their likelihood. China can also adopt the agreement filing system, requiring original research drug manufacturers and generic drug manufacturers to file relevant agreements with the antitrust enforcement department, i.e., the Market Supervision Administration while reviewing the filed agreements and settlement agreements transferred by the courts in patent litigation. Deterrents can be created for manufacturers attempting to conceal reverse payment agreements by linking the related concealment behavior with financing, listing, and the credit system.

4.3 Utilizing the Antitrust Public Interest Litigation System in Reverse Payment Agreements Regulation

Consumer associations can play a significant role in antitrust regulation on reverse payment agreements. Consumer associations are instrumental in monitoring drug manufacturers' anti-competition behavior. Through litigation, lobbying, and public awareness campaigns, consumer associations can on the one hand can supervise antitrust regulators to actively detect and strictly review agreements, on the other hand urge drug manufacturers not to violate competition laws, so that consumers' interests in affordable healthcare are protected.

The newly amended "Antitrust Law" of 2022, in its

https://doi.org/10.1051/shsconf/202317803010

second clause of Article 60, added an antitrust civil public interest litigation system, stating that "if an operator engages in monopolistic conduct that harms the social public interest, People's Procuratorates at or above the district level may file civil public interest lawsuits with the People's Courts according to the law." On August 1, 2022, the Supreme People's Procuratorate issued a notice, requiring the earnest implementation of the amended Antitrust Law, actively and prudently carrying out public interest litigation prosecution work in the field of antitrust, with a focus on areas such as the Internet, public utilities, and pharmaceuticals [16]. In regulating reverse payment agreements, antitrust civil public interest litigation has room to play a role and can be an essential direction for improvement.

Expanding the scope of prosecutorial public interest litigation to the antitrust field helps construct a civil, administrative, and criminal liability accountability system, fully leveraging the advantages and strengths of different departments, which both protects consumer rights and deters pharmaceutical manufacturers through a high amount of compensation [17]. Although Article 60 of the "Antitrust Law" restricts the subjects who can initiate anti-monopoly civil public interest litigation to the People's Procuratorates at or above the district level without specifying the consumer associations' status, this is a regrettable omission. However, according to Article 47 of the "Consumer Rights Protection Law," it is neither suitable nor should it be believed that consumer associations in provinces, autonomous regions, and municipalities cannot be engaged in anti-monopoly civil public interest litigation. Monopoly behaviors are often indistinct and secretive, and the party implementing the monopoly is clearly superior. Therefore, a solid organization to counter monopolistic enterprises is necessary. Consumer associations have been developing in China for over thirty years and have developed and effective a mechanism to protect consumer rights, promptly reporting and stopping monopolistic behavior. Additionally, consumer associations have the ability to organize the individual consumers into groups with shared interest, thus having the strength to counterbalance monopolistic enterprises [18].

5. Conclusion

Currently the precondition for reverse payment agreements to emerge in China's pharmaceutical market is already equipped. It is evident that enhancing regulatory measures for these agreements is imperative. This paper has offered a comprehensive approach to effectively regulate these agreements. By proposing the two-step antitrust review initiation system for patent settlement agreements, advocating for an agreement filing system, and suggesting utilizing antitrust civil public interest litigation, this paper outlines a framework that seeks to balance regulatory efficiency and risk mitigation.

This framework takes inspiration from the experiences of established regulatory frameworks in the United States European Union and European Countries. By adapting these lessons to China's context, government could work towards a more equitable pharmaceutical market where consumer's interest in affordable healthcare is ensured. This framework aligns with the ultimate goal of antitrust regulation, and contributes to the refinement of regulatory strategies in governing reverse payment agreements in China.

References

1. Marco Stief, "No-challenge and Pay-for-delay Agreements in Patent Licence and Settlement Agreements under European Competition Law," Journal of Intellectual Property Law & Practice 18, no. 2 (2023): 135-145.
2. European Commission, Directorate-General for Competition, "Competition enforcement in the pharmaceutical sector (2009-2017): European competition authorities working together for affordable and innovative medicines," Publications Office (2019), https://data.europa.eu/doi/10.2763/218954
3. Federal Trade Commission, "Pay-for-delay: How drug manufacturer pay-offs cost consumers billions: A Federal Trade Commission staff study," (August 5, 2021), https://www.ftc.gov/reports/pay-delay-how-drug-manufacturer-pay-offs-cost-consumers-billions-federal-trade-commission-staff-study
4. China Pharmaceutical Industry Information Center, "China Generic Drug Development Report 2022 Edition," (April 1, 2023)
5. Supreme People's Court, "AstraZeneca Ltd. vs. Jiangsu Osaikang Pharmaceutical Co., Ltd. Patent Infringement Dispute," (2021) Final Civil Judgment No. 388
6. Nanjing Intermediate People's Court of Jiangsu Province, "AstraZeneca Ltd. vs. Jiangsu Osaikang Pharmaceutical Co., Ltd. Patent Infringement Dispute," (2019) Civil Judgment No. 1090
7. K. Lin, M. Lin, "Regulation Research on Reverse Payment Agreements of Pharmaceutical Patents," Strait Law Review, 24(02), 70-76 (2022)
8. M. Marmaro, "Molecule Size Doesn't Matter: The Case for Harmonizing Antitrust Treatment of Pay-for-Delay Agreements," Columbia J. Law Soc. Probl. 54, 169 (2021).
9. U.S. Congress, "Medicare Prescription Drug, Improvement, and Modernization Act of 2003," Public Law 108–173, 117 STAT. 2066, https://www.govinfo.gov/content/pkg/PLAW-108publ173/pdf/PLAW-108publ173.pdf
10. H. Su, "Anti-Monopoly of Drug Patent in U.S. Experience," China Price Supervision and Anti-Monopoly, (03), 20-26 (2017)
11. Competition DG, European Commission, "8th Report on the Monitoring of Patent Settlements," (March 9, 2018), https://ec.europa.eu/competition/sectors/pharmaceuticals/inquiry/patent_settlements_report8_en.pdf

12. Y. Cao, "Regulation of Reverse Payment Agreements in Drug Patent by Anti-Monopoly Law," Hebei University, DOI:10.27103/d.cnki.ghebu.2020.000162 (2020)
13. T. Pan, "Analysis of the Subject Qualification of Consumer Associations in Anti-Monopoly Civil Public Interest Litigation," China Market Supervision Research, (01), 48-51 (2023)
14. J. Wang, "China's Anti-Monopoly Law Should Establish a Dual Enforcement System," Legal Research, (05), 20-25 (2007)
15. L. Yang, "Regulation of Reverse Payment Agreements in Drug Patent under the Perspective of Comparative Law," Xinjiang Social Science, (05), 100-107 (2019)
16. Supreme People's Procuratorate, "Notice on Actively and Steadily Carrying out the Prosecution Work of Public Interest Litigation in the Anti-Monopoly Field"
17. Z. Hu, "Exploration and Improvement Path of Anti-Monopoly Prosecution Civil Public Interest Litigation System," China Price Supervision and Anti-Monopoly, (12), 37-42 (2022)
18. Y. Chen, "Anti-Monopoly Civil Public Interest Litigation: The Road to Relief for Consumers Suffering Monopoly Damage," Modern Law Science, 40(05), 130-144 (2018)

A first look at the advantages of building a new ecosystem of smart cabins

Wenwen Wang ,Gang Deng and Chi Fan*

Qilu Institute of Technology ,3028 Jingshi East Road, Jinan, Shandong,China

Abstract. The construction work of smart civil aviation is an important initiative to promote the high-quality development of the civil aviation industry. The Civil Aviation Administration (CAA) attaches great importance to cabin-oriented Internet applications, and building smart cabins will become an important direction for differentiated development of airlines. This paper takes the Roadmap for the Construction of Intelligent Civil Aviation proposed by CAAC as the starting point, initially analyzes the advantages of building a new ecosystem of intelligent cabin, as well as the possible problems and risks, and puts forward suggestions for airlines.

1 Introduction

Covid-19 caused a sharp decline in global aviation operations over the past three years, but it also pushed forward the digital transformation of civil aviation and the application and development of smart civil aviation.

During the "14th Five-Year Plan" period, in the context of the construction of new digital infrastructure for civil aviation and intelligent civil aviation, the CAAC attaches great importance to the construction of intelligent passenger cabins, and will focus on the gradual transition from the current "LAN" service, which is based on the choices of airlines, to "Internet" service, which is based on the choices of individual passengers, and will include it in the list of services. It will gradually shift the focus from the current "local area network" service, which is mainly chosen by airlines, to the "Internet" service, which is mainly chosen by passengers, and will be included in the Civil Aviation Administration "I do practical things for the public" practical activities. [1]This requires airlines to continuously improve their technological self-innovation capability, strengthen cooperation with carriers, the Internet and other industries, and focus on the construction of smart cabins while accelerating the application of 5G, big data, artificial intelligence and other technologies in the aviation field, in order to enhance cabin protection capability and improve airline revenue by creating a new ecosystem of smart cabins.

The new eco-system of smart cabin is a cabin protection system that provides highly intelligent services to passengers based on the travel characteristics of the crowd in the digital era, with the cabin as the scene and ground-air interconnection, big data, artificial intelligence and other technologies as the driving force during the flight phase of the air route.

On the one hand, the new eco-system of smart cabin can provide passengers with cabin Wi-Fi by means of 5G ATG and other communication technologies to meet passengers' social, shopping, entertainment and business needs; on the other hand, it can continuously explore and strengthen the application of digital intelligence scenarios by means of Internet technologies, so as to realize the transformation from providing a single air Internet service to the overall solution of the new eco-system of integrated cabin. The new ecology of integrated intelligent cabin can be realized gradually through the establishment of "cabin network platform". The "cabin network platform" (hereinafter referred to as the platform) can be understood as an airline company's own platform that integrates the development of "cabin Wi-Fi+Internet+Big Data+Artificial Intelligence", provides diversified Internet contacts, and integrates cabin safety and security, and cabin service guarantee. The following is a summary of the new ecosystem of smart cabins. The following is a preliminary exploration of the advantages of the new eco-system of smart cabin:

2 Improvement of cabin safety and security capabilities

Ensuring cabin safety is a top priority for airlines. Airlines need to continuously explore cabin safety issues through scientific and effective paths. Building a new ecosystem of smart cabins may improve the ability to guarantee cabin safety.

2.1 It is conducive to improving the risk-resistant capacity of airlines' public health.

The normalization of the epidemic has brought about more stringent aviation control requirements, effective

*Corresponding auther: 401768145@qq.com

https://doi.org/10.1051/shsconf/202317803011

prevention and control of public health risks in flight transportation has become more and more important, and airlines need to reduce the spread of diseases in the cabin through the digitalization and intelligent transformation of the cabin in order to improve the risk-resistant capacity of public health in air transportation. Intelligent cabin in the future may appear in the following scenarios: during the epidemic, passengers through the platform to order or cancel meals, reducing the non-essential contact between the flight attendants and passengers; passengers through the platform to understand the real-time restroom queuing situation, reducing the non-essential contact between the passengers and passengers; such as found in the air suspected outbreak of epidemic, the cabin crew can immediately start the relevant prevention and control procedures, to achieve the air-ground seamless connection, to arrange for the transfer of staff and testing, etc.

2.2 It helps to improve the ability of cabin crews to deal with sudden illnesses on board.

Reduced oxygen levels, changes in air pressure, turbulence and psychological or physiological stress during aviation flights can have an impact on human health, so aviation flights may bring about potential health problems or sudden deterioration of certain conditions. Common in-flight emergencies include sudden cardiac arrest, epilepsy, and acute abdominal pain. The cabin crew can transmit passengers' vital signs from the air to the ground at the first time through the "ground-air interconnection + Internet", and realize remote video consultation with ground medical personnel to save passengers with sudden illnesses on board in a timely manner, and help them to turn a critical situation into a safe one.

2.3 It helps to reduce injuries caused by bumps.

During a long-distance flight, the weather conditions of the flight path prepared before the flight will change with time. Cabin crew can browse real-time route weather information through the platform to arrange cabin service plans in advance, reducing injuries caused by improper work plan arrangements; passengers can turn on the bumps warning function on the platform to check whether they are seated and fasten their seat belts in a timely manner before the bumps, reducing injuries caused by lagging information.

Clear air bumps are unpredictable bumps, and preventing them by previewing or receiving information in advance is basically unattainable. At present, the only feasible way to prevent it is to reduce the non-essential movement of on-board personnel. The platform enables efficient communication between cabin crew and passengers, which not only allows cabin crew to provide personalized and precise services to passengers, but also reduces unnecessary movements between cabin crew and passengers, thus reducing the number of major injuries caused by "unnecessary movements".

3 Improving cabin service capacity

Cabin service is an important manifestation of the service level of airlines. Due to the limitations of time, space, facilities and other conditions, the phenomenon of monotonous cabin service and serious homogenization has appeared in various airlines. Building a new ecosystem of smart cabin may redefine the concept of cabin service.

3.1 "Cabin Wi-Fi allows passengers to experience the thrill of surfing the Internet at high altitude, which can effectively enhance the flight experience.

Passengers can realize instant messaging, sending and receiving emails, network meetings, in-flight office, online shopping, online movie watching, games and entertainment in the cabin.Khudhair H. Based on the investigation of the flight process, Y found that in-flight entertainment is one of the main factors affecting customer service experience.[2]The cabin is no longer an information island, and the flight journey is no longer boring. It can be said that "cabin Wi-Fi" has opened a new journey of aviation digitalization.

3.2 The platform brings travelers a new way of personal interactive experience, which is more in line with modern air travel consumption behavior.

Professor Zou Jianjun of Civil Aviation Administration of China (CAAC) mentioned in the ATT forum that air travel consumption behavior is accelerating the transformation of "simplifying the complexity and simplifying the complexity", and the autonomy and intelligence of the whole process of service is the main direction of the transformation. [3]With the support of intelligent technology, the service satisfaction of the whole process of air travel has improved compared with that without technology intervention. Especially in the three links of baggage claim, cabin service and waiting for flights, the most obvious satisfaction improvement is presented, which fully reflects the value and practical significance of the application of intelligent technology in the cabin. The platform allows passengers to choose the service content more independently, so that cabin crew can also provide passengers with more efficient, accurate, rich and personalized services.

3.3 The level of maintenance and protection of cabin service equipment will be significantly improved.

Airlines can establish a structural database for cabin service equipment, retrofit sensors, etc., and deeply integrate the cabin network system with the aircraft maintenance system. Firstly, it is convenient for aircraft maintenance personnel to retrieve complete data of cabin equipment, carry out efficient and accurate maintenance of cabin equipment, and improve the comfort of

https://doi.org/10.1051/shsconf/202317803011

passengers; secondly, it is convenient for cabin crew to clearly understand the current status of cabin equipment, communicate with passengers in a timely manner, and adjust the seats for passengers or make compensation for passengers according to the company's relevant policies. A higher level of cabin service equipment maintenance is conducive to improving the brand image of the airline.

4 Enhancing the economic efficiency of airlines

In recent years, due to the impact of the epidemic, demand in the aviation market has been weak, the number of flights has been reduced, and the debt ratio of airlines has remained high. However, in the long run, the global epidemic situation is improving, the economy and tourism business recovery, civil aviation demand is also an inevitable trend. Airlines surviving the epidemic will usher in a new round of opportunities and challenges, and building smart cabins may become a strategic place for airlines, and smart cabins will become a new revenue growth point.

4.1The provision of "in-cabin Wi-Fi" helps to increase flight occupancy.

On January 13, 2021, CAPSE released the "Domestic Airline Cabin Wi-Fi Research Report", which shows that 23,000 valid samples show that "cabin Wi-Fi" is one of the most popular cabin amenities that attract most passengers to prefer flights, and 98% of the passengers interviewed said that they need cabin Wi-Fi. As shown in Figure 1.

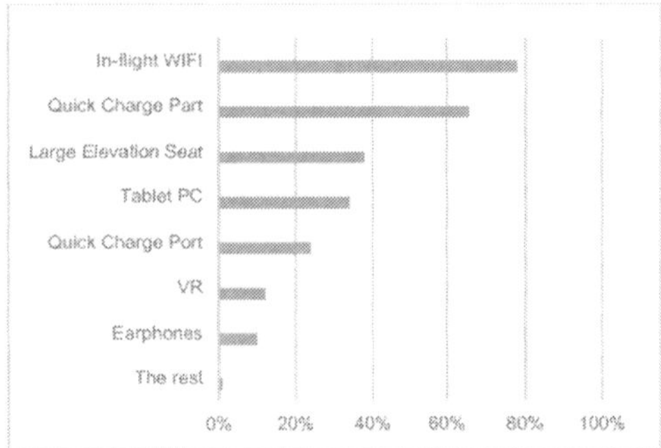

Fig. 1. Attracting passengers to preferred flight cabin facilities and equipment

"Network speed is the key to passengers' experience of Internet access in the air. At present, China Telecom said that after the completion of the 5G ATG air-to-ground cloud network, the peak bandwidth of a single aircraft will be as high as 1Gbps, so that passengers can enjoy high-speed bandwidth Internet experience at a height of 10,000 meters. This will truly solve the urgent need for passengers to access the Internet in the air.

4.2 Turn the platform from a "cost center" to a "profit center".

According to CAPSE research, 63% of travelers are willing to pay for "in-cabin Wi-Fi", which can be in the form of cash, mileage, advertisement time, etc. For frequent flyers, they prefer mileage redemption.While non-members prefer to pay for WiFi, a higher percentage of them choose to redeem their miles. Therefore, in-flight WiFi may be an incentive for travelers to apply for the airline's frequent flyer card. Depending on the amount of mileage redeemed for WiFi benefits, it may be possible to influence member stickiness. [4]Converting travelers' willingness to pay for in-flight Wi-Fi into "real value" may increase the airline's revenue.

Airlines can try to convert travelers' in-flight time into platform consumption time. Airlines can partner with a wider range of service providers to build in-flight transaction platforms using innovative marketing models, big data and artificial intelligence to allow passengers to spend time immersed in flight, and airlines to profit from the experience. According to CAPSE's research on the "Ranking of Passenger Usage of In-Cabin Wi-Fi", as shown in the chart below, the platform can sell products such as: online cinema, destination hotel bookings, tourist attraction tickets, specialty products, online e-books, gaming products, in-flight duty free, airline souvenirs, etc. The platform can also offer differentiated services to customers, such as in-flight duty-free, in-flight duty-free, and in-flight souvenirs.Airlines should apply information technology more widely in this area, working with hotel, rental, insurance and other suppliers to build a more comprehensive distribution system using the resources of related industries.[5]Airlines can use high-level big data programs to analyze consumers' travel preferences to develop digital marketing strategies.[6] Of course, it can also provide customers with differentiated marketing and service methods, such as "blind box" ticket sales, click-through revenue, paid services and so on. In this way, the platform may reshape the new pattern of aviation marketing. As shown in Figure 2.

In addition, according to CAPSE's "China Airline In-Flight WiFi Trend Survey ,young people born after 2000 will become the main force of aviation consumption, and their consumption habits are different from those of the post-80s and post-90s, which is worthwhile for airlines to study and explore. This will be of great help to accurately occupy travelers' minds and increase their stickiness.As shown in Figure 3.

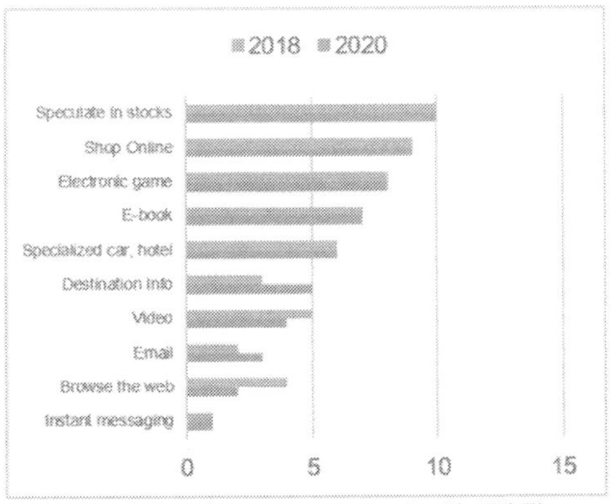

Fig. 2. Ranking of passenger usage of cabin Wi-Fi as a percentage of passenger use

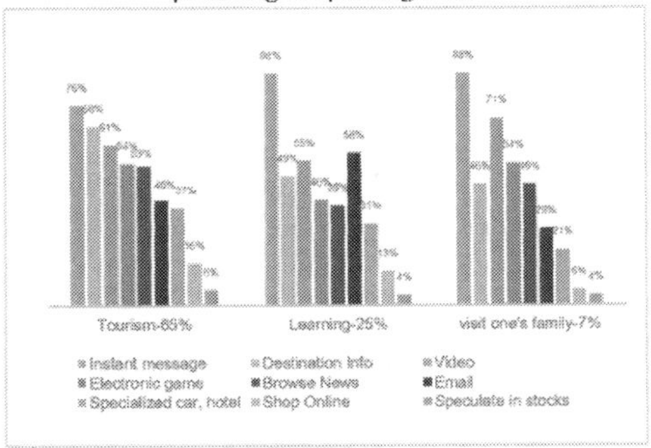

Fig. 3. Differences in the purpose of in-flight WIFI use among post-00s travelers

5 Risks and problems

At present, some domestic airlines have initially realized the cabin intelligent operation, but it is only in the initial stage, the development is not sufficient and there are many problems. Combined with the current operational experience, the following risks and problems may exist in building a new ecosystem of smart cabin:

5.1 Data security risks

Airlines will acquire a large amount of data when building smart cabins. While airlines utilize big data for deep mining and analysis, and process, extract, encapsulate and apply the rich passenger data to their services, they will also face challenges in terms of data security risks. Issues such as data leakage, data rights litigation, and data ownership will all become security risks for airlines. Airlines need to continuously strengthen top-level design, adopt data compliance and protection measures, address data security pain points, and properly handle security incidents and data rights lawsuits.

5.2 Resource constraints

The construction of smart cabin is the main direction of the future high-quality development of airlines, and its development is based on digital transformation, which is a huge and complex systematic project that needs to be supported by a large amount of funds, R&D teams and other resources. In the current situation of erratic epidemics and difficult development of airlines, it is difficult to support the development of digital transformation by relying only on airlines' R&D expenses extracted from their operating revenues. If the construction of smart cabins can be supported by government programs, financial funds or special science and technology funds, then the construction process of smart cabins will be greatly accelerated.

5.3 Not Enough Research

Domestic research results, patents and literature related to smart passenger cabin are relatively few, and there is not much valuable reference data. Most of the literature that can be retrieved domestically is of the overview type, which can only generally understand the current status of the application of domestic smart passenger cabin, and there is little research on the key technologies of specific application scenarios. With the continuous development of domestic smart passenger cabin, a variety of technical problems will be exposed. This requires the relevant fields to strengthen the research on key technologies of smart cabin to cope with the technical difficulties that may arise. If airlines want to gain lasting vitality in the competitive aviation market, they need to improve their independent innovation and R&D capabilities, strengthen cooperation with related enterprises, and do relevant research on building a new ecosystem of smart cabins.

To summarize, building a new ecology of smart cabin has a far-reaching impact on improving the market competitiveness of airlines. Domestic airlines are advised to think deeply about how to improve their risk-resistant ability in the context of the normalization of epidemics, and how to better ensure cabin safety, improve service quality and obtain new revenue growth points in the context of the digital era. Airlines should actively participate in the construction of intelligent civil aviation, do a good job of strategic planning for the operation of the new ecosystem of intelligent cabin, enhance passengers' sense of well-being in the development of civil aviation, and strive to achieve milestones in the early stage of the digital transformation of the industry by 2025, so as to promote the construction of intelligent civil aviation in the 14th Five-Year Plan period of China.

References

1. Zhou Huijun. A brief talk on optimizing the cabin service experience [J] . Civil Aviation Administration, 2020(07) : 42-45.
2. Khudhair H.Y,Jusoh A,Mardani A,etc.Quality Seekers as Moderating Effects betweenService Quality and Customer Satisfaction in Airline Industry[J].International Review ofManagement and Marketing.2019(4):74-79.

3. Shen Qiao. Strengthen cabin safety construction and improve service standardization level [J] . China standardization, 2021(20) : 58-60.
4. Xu Jianghua. Study on the safe man-machine design of aircraft cabin facilities modeling [D] . Wuhan University of Technology, 2020.
5. M.Rebezova,N.Sulima,R.Surinov. Development Trends of Air Passenger Transport Services and Service Distribution Channels[J].Transport and Telecommunication,2015,13(2).
6. Chun Y.H.,Lee S.J.For airline preferences of consumers Big Data Convergence BasedMarketing Strategy[J].Journal of Industrial Convergence,2019,17(3):17~22.

The Concept of New Business, Current Status and Achievements of Research and Development

Yifan Wu*

Harbin University of Commerce, Songbei District, Harbin, Heilongjiang, China

Abstract: The creation of new business is naturally formed by the development of information technology, industrial upgrading, the growth of people's demand, and the replacement of the employment model, and with the support of the government, it will be further developed to realize the replacement of the old and the new and lead to a new generation of employment wave. In this paper, we will analyze the conceptual interpretation, development status and achievements of the new industry to help the new industry to further develop and improve.

1. Introduction

Since its introduction, the new industry has received attention from society and academia, and has made rapid development and achievements in a few years. With the rapid development of Internet technology and artificial intelligence technology, many new forms of employment have emerged, such as takeaway delivery, e-commerce, online car and other emerging occupations, these new occupations will be the business pattern of integration and differentiation again and again, and constantly form a new business form, forming a "new industry". At present, the practitioners of the new industry can't be guaranteed in all aspects, this paper is intended to analyze the development of the new industry through the development of the new industry research to provide suggestions for the development of the new industry.

2. The Concept of New Forms of Employment

The communiqué of the Fifth Plenary Session of the 18th Central Committee of the Chinese People's Political Consultative Conference mentions "promoting employment and entrepreneurship, adhering to the strategy of giving priority to employment, implementing a more proactive employment policy, improving entrepreneurship support policies, strengthening support for flexible employment and new forms of employment, and improving the treatment of skilled workers". For the first time, the concept of "new forms of employment" has been put forward.

New forms of employment are new in that they have new characteristics that are different from traditional forms of employment. New forms of employment can be viewed as occupations that workers engage in during the transition period between two positions, and can be counted as both employed and unemployed at the same time, filling in the income gaps during the period when job seekers are looking for work. The new industry is based on the new information technology revolution and the integration and development of the tertiary industry formed by the creation of a new industrial value chain, but also represents China has entered the old and new kinetic energy conversion of the economic development of the economic development trend of the economic development of the period, which includes in the case of the gradual and full popularization of the Internet technology, many of the offline industry stagnation or slow development, and the new employment forms also take this rapid rise, in such an impact, China's future new changes and trends in employment will become the norm in the coming period, so the new industry also has a sense of the times.[1]

3. Research and Development Status of New Employment Forms

3.1 The current situation of research and development of social insurance for employees in new forms of employment

The social security rights and interests of the employees in new industries can be effectively realized only when they are integrated into the mass groups. 2021 Xi Heng proposed to improve the motivation to participate in the insurance and the awareness of self-protection by popularizing the related to the employees in new industries; to force them to participate in the social insurance according to the law; to lift the restrictions on the social insurance for the movement of the population and the restrictions on the household registration, the purchase of

* author: 1093335822@qq.com

houses, the children's schooling, etc. of the employees in new industries and to standardize the payment of social insurance management by the related departments and the localities. Social insurance management; continuously improve the level of coordination of various types of social insurance.[2]

2022 Wang Lijun proposed to formulate relevant policies and improve laws and regulations for new industry practitioners. It is recommended that workers establish individual accounts that are not linked to their employers and expand the scope of insurance coverage for flexibly employed people; improve the mobility of social insurance and break down the regional barriers for floating populations; set up special subsidies for special new industry workers and appropriately lower the contribution base. [3]

3.2 Current status of research and development of labor rights and interests protection for new industry employment workers

There are certain differences between new employment forms and traditional enterprise mode employment in terms of workplace, work form and labor relations.

Takeaway riders, for example, do not have the awareness and knowledge of rights protection at the individual level, and are weak in defending their rights; there is a lack of clear indicators for determining labor relations at the social level; and there are problems of lax supervision of takeaway enterprises and imperfect employment patterns at the platform level.

Takeaway platforms have adopted various ways to blur or weaken the labor relationship with workers in order to avoid possible labor disputes in the future, as well as to reduce the cost of employment at the same time. In this case, takeaway riders are caught in an awkward situation where there is no social security, no safety guarantee, and no clear labor relationship. This labor relationship makes the rights and interests of takeaway flaggers unprotected, and this labor relationship will seriously hinder the healthy development of the new industry.[4]

2022 Liu Ye put forward three suggestions to improve the labor rights and interests of riders: improve the legal policy system and guide the construction of a third-party consultation mechanism. Enterprises should strengthen industry self-discipline and put an end to the concept of responsibility transfer. Firmly resist the way to avoid risks by treating riders as individual jobs. When signing relevant contracts, they should clearly set out and keep riders informed of the explicit employment rules and employment system, and make the employment conditions and employment rules transparent and clear. The platform of takeaway riders should determine their respective powers and responsibilities according to the principle of reciprocity of rights and obligations, so as to make the rights and responsibilities clear and avoid the phenomenon of shirking responsibilities in the future. Before engaging in the takeaway rider industry, one should take the initiative to learn the knowledge and laws and regulations of the industry one wants to change, and sign a contract with the employing body to do the relevant risk prevention and control; after engaging in the industry, one should regularly keep proof of employment, such as work records or video screenshots; in the case of disputes, one should choose the right way to defend one's rights, establish the awareness of rights defense, and reasonably utilize the existing laws and regulations to protect one's own interests from being infringed upon. [5]

4. Achievements of New Employment Forms

4.1 Promotion of Employment

As early as 2010, China's Internet penetration rate climbed rapidly, greatly promoting employment, with the help of the Internet platform has given rise to a large number of jobs such as takeaway delivery, e-commerce, chauffeur-driven, shared accommodation, and other online careers. The new employment pattern provides more employment choices and employment positions for job seekers, and also provides employment opportunities for disadvantaged groups in employment, and the new employment pattern solves the employment problems of some people to a certain extent.

The new industry has led to a greater fragmentation of work, dividing systematic work into fragments, narrowing the content of work and refining job responsibilities. The way of working has become more platformized, with employees carrying out more work from online network platforms, receiving and submitting work through network platforms, and employees being supervised by the platforms and third parties. Under the new employment model, job opportunities are extremely mobile, and many jobs that do not require the presence of workers can be handed over to non-locals, which, on the one hand, raises the incomes of people in areas with few employment opportunities and remote areas, and on the other hand, prevents outsiders from taking up the employment opportunities of locals in industries such as couriers and take-away workers, which need to be reached quickly and in close proximity to each other, and stabilizes the job market for locals.

A survey by the National Bureau of Statistics shows that the number of flexibly employed people increased nearly threefold from 2020 to 2021.2021 There are about 200 million flexibly employed people, including more than 4 million takeaway riders and more than 1.6 million practitioners in the anchor industry and related industries. Within a month of the resumption of work in the Spring Festival of 2020, the live broadcasting-related industry grew by 166.09% year-on-year, which is more than twice as much as the rate of growth of full-time positions. Baidu library in the first half of 2020 knowledge stores opened more than 40,000, directly driving the employment of nearly 1 million part-time or full-time content creators. [2] Between August 2019 and August 2020, 20.97 million people were engaged in creation, live broadcasting, and e-commerce on the Tiktok platform, most of whom were engaged in the new occupations of Internet marketing is and live salesman, which created and drove the scale of

https://doi.org/10.1051/shsconf/202317803012

employment to a total of 36.17 million jobs on the platform. [6] As of June 2020, the total number of riders earning income on the Meituan platform was 2.95 million, a year-on-year increase of 16.4%. [2] The statistics of the Didi platform as of July 2017, the platform's active practitioners were 21.08 million,[7] and the distribution of the population is shown in Figure 1.

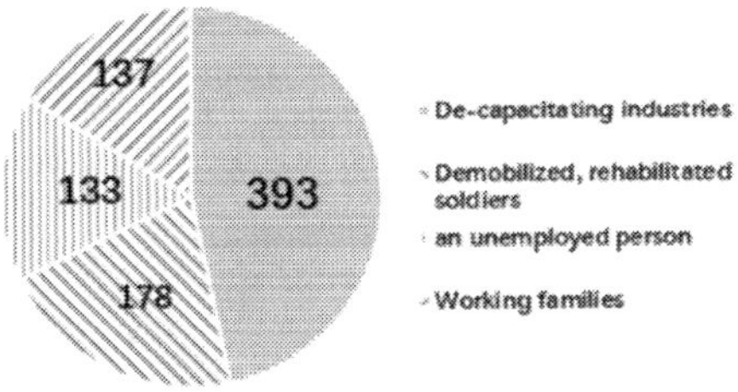

Figure 1 Distribution of New Employment Population in Didi

The working hours of online drivers fully reflect the characteristics of the sharing economy, and the proportion of drivers who are online for less than 2 hours a day is the highest at 50.67%. In 2018, the Meituan takeaway platform provided employment opportunities for "riders" in 781 poor counties nationwide, with a coverage rate as high as 94%, and drove 670,000 people in poor counties to seek employment, accounting for a quarter of the overall number of riders. Didi and Meituan have provided new job opportunities for a large number of laid-off and unemployed people, military personnel who have changed jobs, the disadvantaged groups in employment, and people in the transition period of employment who are awaiting employment. The new industry has made a great contribution to economic stability and social stability.

4.2 Providing a better employment environment for people employed in new industries

Some provinces and cities have carried out skills training, focusing on cultivating diversified employment methods and expanding the scale of employment groups in new industries. With the rapid development of network economy and sharing economy in Guangdong Province, a large number of new professions have emerged, and the number of new professions is gradually increasing. According to the National Bureau of Statistics, the number of new professions in Guangzhou in 2020 will be around 750,000, and in the absence of laws to protect their rights and interests in our country, such a large number of people in employment need to be provided with effective and powerful employment protection. Therefore, Guangzhou is actively promoting the development of new industries while further safe guarding the rights and interests and career development of new industry practitioners, such as Meituan takeaway and other online delivery industry has been developing in recent years, takeaway delivery has become an occupation that does not need to continuously accumulate experience and requires technical content. In order to effectively solve the problems of people engaged in online delivery services, the Labor Union of Lanzhou New District's Party Group Work Department started to operate the "Warm Home" as a station for workers in new forms of employment on September 1, 2022, which provides a place to stay for workers in new forms of employment in Lanzhou New District. It provides a place for workers in new forms of employment in Lanzhou New District to settle down. The station, which is visited about 400 times a day, is a resting station covering an area of about 400 square meters, including restrooms, recreation rooms, study areas, water rooms, catering areas, charging areas, bookstores and other resting areas, and is located in the centralized area of workers' work after scientific site selection and reasonable planning.

5. Labor security of new industry practitioners

New employment form for many years for our country's economic development has made a considerable contribution, especially in the new crown epidemic at the moment, in our country's employment is severe today, plays an immeasurable role. Due to the development of the Internet era of new business also gradually known and growing development, as a new form of employment in the future will also get more attention and support from all walks of life. However, there are still many problems with the new industry in China.

The new industry so far still does not have a complete set of legal constraints on employers, and there is no set of perfect protection to protect the labor rights of employees, most of China's new industry practitioners do not protect the rights and interests of the majority of the new industry practitioners, if these guarantees can not be achieved, most of the new industry practitioners in the transition period after the work of the majority of people will choose to change their jobs. China's social insurance also does not include new industry workers, such as in the pension insurance, they can only choose urban and rural residents pension insurance and commercial insurance, at present China's urban and rural residents pension insurance in the receipt of only a few hundred yuan per month, and the high cost of commercial insurance for most of the new industry practitioners can not afford. In recent years, AI has gradually entered the field of production and life, and for most low-skilled practitioners in the new industry, the career replacement rate is extremely high, in order to ensure that they are not replaced many practitioners will choose to extend the labor time, such as takeaway delivery workers will work late at night, so in recent years, traffic accidents of takeaway delivery workers occur from time to time, but there is no workers' compensation insurance and medical insurance in the event of traffic accidents, they will face a double financial burden of life and medical care. the double financial burden of living and medical care. If these problems cannot be solved, the development of the new industry will be seriously hindered. Therefore, solving the problem of occupational security of the practitioners should be an important issue for the new industry to solve.

At present, the form of social insurance in China belongs to the social insurance and the employer single linked, the social insurance of the employees is mostly paid by the enterprise, but the new industry practitioners do not have the employer directly linked, so the current contribution system is not suitable for the new industry personnel to participate in the insurance. Therefore, the social insurance payment of new industry personnel should be separated from the employer's side, and turn to the direction of individual account contributions, and can set up a special insurance program for new industry practitioners and provide subsidies for practitioners before the introduction of formal policies. In the payment of fees according to the current scale of social insurance contributions can not be fully afforded by individuals, in the absence of employers can be taken by the government, individuals, employment platform tripartite sharing, the three parties in accordance with a certain proportion of a certain share of the share of the insurance. Gradually relax the requirements for participation in the insurance and promote, can be in a single work in a fixed amount of deduction to calculate the cost of participation in the insurance, in addition to liberalization of new industry practitioners in different places to apply for social assistance conditions.[8] The new industry is striving for fair treatment for new industry workers, and resolving claims for occupational injuries, unemployment, medical care and other risks.

6.Conclusion

Current research on social insurance and protection of workers' rights and interests for workers in new industries tends to improve planning at the enterprise, worker and social levels, but there are still deficiencies in overall social security in terms of policies, laws and relevant regulations, and it is necessary to improve relevant legislation, raise individual awareness of their rights and enhance social supervision, so as to provide a better employment environment for workers in new industries.

References

1. Friedman G.,"Workers without Employers: Shado w Corporations and the Rise of the Gig Econom y",Review of Keynesian Economics,2(2),2014,pp. 171-188.
2. XI Heng. Integration and Sharing:The Path to R ealize Social Insurance for New Industry Employ ees[J]. Social Science,2021(06):3-11.DOI:10.13644 /j.cnki.cn31-1112.2021.06.002.
3. Wang Lijun,Tu Yongqian. On the Improvement o f Social Security System for Flexible Employed Persons[J]. Guangdong Social Science,2022(06):2 48-257+283.
4. Lin R,P H D V,Kène H, et al. Employers' views on flexible employment contracts for younger w orkers: Benefits, downsides and societal outlook [J]. Economic and Industrial Democracy,2022,43 (4).
5. LIU Ye,NAN Jiaqi,LI Meng. Reflections on the Protection of Labor Rights and Interests of New Industry Practitioners[J]. Cooperative Economy an d Technology,2022(24):188-189.DOI:10.13665/j.cn ki.hzjjykj.2022.24.028.
6. National Institute of Development and Strategy, Renmin University of China: The Age of Spiritu al Workers: A Research Report on Shakeology P latform for Promoting Employment, Renmin Uni versity of China website, http://nads.ruc.edu.cn/zk cg/ztyjbg/c90764ffcbd641b79e3b9e352abeee61.htm, 2020-09-09.
7. Wang Yan,Zhang Lihua. Research on the new e mployment form of sharing economy in China[J]. Journal of China Institute of Labor Relations,20 19,33(02):49-59+98.
8. Social Security Administration:Social security pro grams,https://www.ssa.gov/policy/docs/progdesc/ssp tw/2018-2019/asia/taiwan.pdf

Institutional quality, infrastructure and economic growth: Evidence from RCEP countries

Chenxue Ren[1,*], lingbo Tan[2]

[1,2]School of Economics and Management, China Jiliang University, Hangzhou, 310000, China

Abstract. Large-scale regional trade agreements have led to a gradual increase in the ease of trade development among countries. The role of geographical distance and other location factors on trade development among countries has gradually diminished, while the institutional conditions of different countries have gradually increased their role on trade development among countries. Given the development process of the RCEP agreement, this study uses panel data from 2011-2021 and explores the mechanism of the institutional quality of RCEP countries affecting China's import and export trade with them based on a gravity model. In addition, this study explores the level of infrastructure in mediating the effect of the process. It is found that the institutional quality of RCEP countries has a substantial impact on China's import and export transactions with them. In addition, infrastructure plays a fully mediating effect in the process. Therefore, bilateral trade can be promoted by upgrading the institutional environment, infrastructure development and optimising the business environment.

1 Introduction

Economic globalization and economic integration have become increasingly prevalent in recent decades, as evidenced in part by a growing number of regional economic cooperation agreements. The Regional Comprehensive Economic Partnership (RCEP), which was officially signed in Vietnam on 15 November 2020, comprises 15 member countries and is the most populous, diverse and dynamic FTA in the world. According to data supplied by the World Bank in 2021, the RCEP region encompasses 3.5 billion people, approximately 47.4% of the world's total population, and has a combined GDP of over US$29 trillion, making it the largest and most influential regional trade free agreement in the world.

With the gradual increase of trade development convenience, the role of geographical distance and other location factors on trade development between countries has gradually diminished, while the role of institutional conditions of different countries on trade development between countries has gradually increased. The institutions that underpin economic growth are endogenous, depending on the collective choices of society[1], and therefore there are differences in institutional arrangements between countries, with different institutional arrangements leading to different market transaction costs. Institutional quality is an important indicator of institutional differences between China and other countries in the world, so studying the effects and mechanisms of institutional quality on China's trade development will be a practical reference for the exploration of China's foreign trade development path.

Most investigations have considered their effects on international trade from a unilateral perspective of institutional quality or infrastructure, with few studies examining their impact on economic growth in the same system, especially for RCEP countries, whose intermediation and regulation mechanisms have not been systematically studied. To address these gaps in the literature, this study uses a panel dataset of 15 countries included in the RCEP agreement from 2011 to 2021 and explores the relationship between institutional quality and import and export trade using a gravity model regression approach. Furthermore, this study sought to investigate the mechanism of institutional quality on import and export trade via infrastructure intermediaries.

2 Research hypothesis

Institutions, according to North, "are the restrictions on human design that define human politics, society, and the economy[2]." Investigate the endogenous linkage between institutional quality and trade development has become an important frontier hotspot of interest in international economics in recent years. A section of scholars has taken the perspective of institutional quality as a notable competitive advantage in international trade, arguing that specialisation and trade in a country does not occur as a result of exogenous endowment differences, but rather from differences in the endogenous institutional quality environment[3]. Another part of the scholars, from the perspective that institutional quality affects the cost of import and export

* Corresponding author: 2012722084@qq.com

trade and trade flows, argue that an increase in institutional quality reduces the transaction costs and information uncertainty risks of trade thus facilitating the growth of trade volumes[4]. In short, a good institutional environment promotes productive activities in a country, eliminates resource allocation distortions and unsustainable consumption, and further lowers the transaction costs of trading with other countries.

Institutional performance theory suggests that institutions are the fundamental factor influencing economic growth. This is because institutional arrangements facilitate economies of scale, reduce information costs, spread risk, etc., and create incentives to change the way transaction costs are incurred[5]. A good institutional environment provides a ' soft environment' for international trade to take place, and as trade facilitation increases, the level of infrastructure development in each country becomes increasingly important in international trade. We propose the following hypothesis in light of the study presented above.

The term "infrastructure" originated as a technical term in railway construction in the early twentieth century and has now been extended to describe the myriad structures that underpin modern society[6] . It is worth noting that trade costs are not only related to distance, but also to the level of infrastructure. Previous research on how infrastructure affects import and export commerce fall into two categories. One is research on the effect of a single type of infrastructure on import and export trade, and this type of research usually selects multiple variables to analyse for a particular type of infrastructure[7]. The second category is the research related to the relationship between infrastructure and export and import trade[8].

The given analysis supports the following hypothesis.

H1: The quality of institutions has a favorable effect on the growth of import and export commerce.

H2: The quality of institutions has a positive contribution to infrastructure development.

H3: Infrastructure plays a mediating role in the impact of institutional quality on export and import trade.

H4: The strength of the infrastructure has a beneficial effect on import and export trade.

3 Methodology and research design

3.1 Empirical model

The baseline model are as follows:

$$\ln Trade_{ijt} = \alpha_0 + \alpha_1 \ln Inst_{jt} + \alpha_2 \ln GDPR_{jt} + \alpha_3 \ln FDI_{jt} + \alpha_4 \ln Open_{jt} + \alpha_5 \ln Dis_{jt} + \alpha_6 \ln Broder_{jt} + year + \varepsilon_i + \mu_{jt} \quad (1)$$

Where i is China, j is RCEP country, and t is time. $\ln Trade_{ijt}$ is the explanatory variable, denoting the natural logarithm of China's import and export trade to country j, $\ln Inst_{jt}$ denoting the level of institutional quality in country j ; $\ln GDPR_{jt}$ denoting the GDP growth rate ; $\ln FDI_{jt}$ denoting foreign direct investment; $\ln Open_{jt}$ measuring trade openness; $\ln Dis_{jt}$ is distance between the capitals of China and those of the RCEP countries; $\ln Broder_{jt}$ is whether RCEP countries borders or not with China, and $year$ is time dummy variable, ε_i for individual effects, and μ_{jt} for the random disturbance term.

The current mechanism test mainly uses the mediation effect model. With reference from Feng et al[9], the proposed stepwise test method is used to test the obvious intermediary effect of infrastructure on institutional quality and import and export trade. Based on the benchmark regression model (1), build an intermediary model:

$$\ln Infra_{jt} = \chi_0 + \chi_1 \ln Inst_{jt} + \chi_2 \ln GDPR_{jt} + \chi_3 \ln FDI_{jt} + \chi_4 \ln Open_{jt} + \chi_5 \ln Dis_{jt} + \chi_6 \ln Broder_{jt} + year + \varepsilon_i + \mu_{jt} \quad (2)$$

$$\ln Trade_{jt} = \kappa_0 + \kappa_1 \ln Inst_{jt} + \kappa_2 \ln Infra_{jt} + \kappa_3 \ln GDPR_{jt} + \kappa_4 \ln FDI_{jt} + \kappa_5 \ln Open_{jt} + \kappa_6 \ln Dis_{jt} + \kappa_7 \ln Broder_{jt} + year + \varepsilon_i + \mu_{jt} \quad (3)$$

Model (1) is the total effect of the improvement of the quality of the RCEP Institutional quality on China's import and export trade. Model (2) represents the influence of the enhancement of the RCEP instituional quality on their nation's infrastructure. Taking the quality of the system and the infrastructure of the intermediary variable as the explanatory variables, model (3) can be constructed to check whether the intermediary variable exists.

3.2 Data descriptive statistics

Table 1 shows pertinent variable descriptive statistics. To test for multicollinearity amongst explanatory variables, we constructed variance inflation factors (VIFs) for the variables of interest. According to the findings, all of the variables had VIFs that were less than 5, with the mean VIF being 1.78[10]. Therefore, there was no multicollinearity problem between the variables.

Table 1. The statistical description of the variables.

Items	Min	Max	Mean	SD	VIF	Mean VIF
lntrade	20.71	26.62	24.37	1.652	-	
lnInfra	0.003	0.740	0.345	0.202	3.020	
lnInst	0.099	0.976	0.468	0.216	2.150	
GDPR	-6.099	8.200	3.498	3.179	1.640	1.780
lnFDI	19.52	25.35	22.60	1.452	1.210	
LnOpen	0.003	0.093	0.033	0.018	1.520	
lndis	10.63	13.86	12.35	0.665	1.570	
Broder	0	1	0.292	0.456	1.340	

4 Empirical results

4.1 Baseline regression

Table 2 characterizes the regression results of the RCEP national institutional quality on China's import and export trade. The baseline regression is displayed in column (1). When compared to non-RCEP countries, those in the RCEP have higher quality institutional environments, which benefits China's exports and imports. The predicted institutional quality coefficient at the 5% level is 0.3301, which is statistically significant. According to this coefficient, China's import and export trade with RCEP nations will rise by 33.01% for every 1 point increase in the institutional quality of those nations. As a result, this investigation supports Hypothesis 1. Table 2 also includes the estimated coefficients of the control variables. lnFDI has a significantly positive coefficient, suggesting that outward FDI accompanies the development of import and export trade. The fact that lnOpen has a negative coefficient suggests that most countries' levels of economic development and trade openness are not aligned. lndis has a negative coefficient, suggesting that distance has a significant impact on import and export trade. gdpr has a significantly. The coefficient on Broder is not statistically significant, suggesting that whether a trading country borders China or not is not sensitive to import and export trade.

4.2 Robustness analysis

This subsection describes a test of the robustness of the results of the benchmark regressions above. First, we check the robustness of the choice of measurement of institutional quality indicators and the institutional quality index was measured using principal component analysis (PCA), before the core model from the previous section was regressed on an individual fixed effects model. The results show in Column (ii) of Table 2. The coefficient on institutional quality is 0.1596, which remains significantly positive at the 1% level and continues to support Hypothesis 1.

Then, the two-stage least squares method is used to analyse the model to test the endogeneity effects. From column (iii), it can be seen that idp<0.1 according to the unidentifiability test, indicating a strong rejection of the original hypothesis of unidentifiability. In terms of the over-identification test, sarganp=0.4532>0.1, indicating that the over-identification test is passed[11]. That is, it shows that the instrumental variables selected for this paper are valid, and then through the results, lnlnst shows a significance at the 0.05 level (p<0.05) and the regression coefficient is 0.4059>0, indicating that the conclusions of this paper are still plausible after the endogeneity correction of the instrumental variables method.

Table 2. Panel quantile regression results.

Variable	Baseline model	Estimated Coefficient	
		Robustness Analysis	
		Use PCA-Inst	IV-2SLS
		variable	Estimation
	(1)	(ii)	(iii)
	lntrade	lntrade	lntrade
lnlnst	0.3301**	0.1596***	0.4059**
	(-0.16)	(-0.06)	(-0.19)
GDPR	-0.0391***	-0.0363***	-0.0331***
	(-0.01)	(-0.01)	(-0.01)
lnFDI	0.1006**	0.1101**	0.0892*
	(-0.05)	(-0.05)	(-0.05)
lnOpen	-1.1151*	-1.2856***	-0.9988***
	(2.58)	(-2.53)	(-2.28)
lndis	-0.1781*	-0.1703	0.3038***
	(-0.1)	(-0.1)	(-0.1)
Broder	0.088	0.0845	0.0861
	(-0.13)	(-0.13)	(-0.1)
Cons	24.2922***	24.1146***	-6.3655
	(-1.78)	(-1.75)	(-3.73)
Country-fixed	Yes	Yes	Yes
Time-fixted	Yes	Yes	Yes
Observations	154	154	126
R^2	0.259	0.28	0.3393
idstat			66.2525
idp			0.0000
widstat			73.8593
sargan			0.5627
Sargan(p)			0.4532

4.3 Mediation effect analysis

This paper refers to Baron & enny's test for intermediation effects[12].The estimation results in Table 3 show model (1), with a coefficient of 0.3301 for lnlnst, indicating the existence and significance of the aggregate effect of the intermediation model. In model (2), the coefficient of lnlnst is 0.3232, which is significant at the 1% level, indicating that an increase in institutional quality significantly contributes to an increase in the level of infrastructure, which further validates hypothesis 2. Further, in model (3), the coefficient of the variable lnlnfra is 1.3191, which is significantly positive at the 1% level, however, the variable lnlnst does not significant. This indicates that the mediating effect in this paper is fully mediated, i.e. it shows that infrastructure can play a fully mediating role in the relationship between the influence of institutional quality on import and export trade, which also proves hypotheses H3 and H4 of this paper. This may be limited by the sample size of this paper, so subsequent studies can further explore it based on expanding the sample size.

Table 3. Estimation of mediation effect.

Model	(1)	(2)	(3)
	lntrade	lnlnfra	lntrade
lnlnst	0.3301**	0.3232***	-0.0962
	(0.16)	(0.06)	(0.16)
lnlnfra			1.3191***
			(0.22)
GDPR	-0.0391***	-0.0179***	-0.0154
	(0.01)	(0.00)	(0.01)
lnFDI	0.1006**	0.0088	0.0890**
	(0.05)	(0.02)	(0.04)
lndis	-0.1781*	-0.2829***	0.1951*

	(0.10)	(0.04)	(0.11)
LnOpen	-1.1151*	1.6495***	-3.2909*
	(2.58)	(0.92)	(2.31)
Broder	0.0880	0.0310	0.0471
	(0.13)	(0.05)	(0.12)
Cons	24.2922***	3.4981***	19.6781***
	(1.78)	(0.64)	(1.75)
Observations	154	154	154
R^2	0.259	0.601	0.425

5 Conclusions and Implications

This study explores the mechanisms by which the institutional quality of RCEP countries affects China's import and export trade, using panel data for the period 2011-2021, and explores the impact of different institutional qualities on trade effects. The intermediary role of infrastructure in import and export trade is then investigated.

Based on the results of the above empirical tests, the following conclusions are drawn from this study. Fisrtly, The RCEP's institutional quality and infrastructure level play an important role in increasing China's import and export trade. In addition, changing the measurement method of comprehensive system quality and the results of 2SLS regression show that infrastructure has an obvious promoting effect on import and export trade. Thirdly, the intermediary mechanism test also demonstrating that institutional quality helps to improve infrastructure level, which is consistent with previous research[13].

In general, the following policy recommendations are given through theoretical and empirical research: Improve the quality of national systems and environment, and reduce trade disputes. Secondly, develop the level of infrastructure and improve trade facilitation. Thirdly, by establishing an international financial cooperation system, RCEP countries can proactively connect with infrastructure construction projects in countries and regions along the route, thereby providing more sufficient funding for project construction.

References

1. Hodgson G.M. (2006) What are institutions?[J]. Journal of economic issues, 40(1): 1-25.
2. North D.C. (1971) Institutional change and economic growth. The journal of Economic History, 31(1), 118-125
3. Nicolini, M. (2011) On the evolution of institutional comparative advantages. Structural Change and Economic Dynamics, 22(2), 162-172.
4. Abreo, C., Bustillo, R., Rodriguez, C. (2021) The role of institutional quality in the international trade of a Latin American country: evidence from Colombian export performance. Journal of Economic Structures, 10(1), 1-21.
5. North D.C. (1993) Institutions and credible commitment. Journal of Institutional and Theoretical Economics, 149, 11–23.
6. Carse A. (2016) Keyword: Infrastructure: How a humble French engineering term shaped the modern world. Infrastructures and social complexity, Routledge. pp.45-57.
7. Fedorenko R.V. (2019) Modern issues of development of the customs and logistics infrastructure of the international North-South transport corridor. Sustainable Growth and Development of Economic Systems. Springer, Cham. pp.63-75.
8. Chang S M, Huang Y Y, Shang K C, et al. Impacts of regional integration and maritime transport on trade: with special reference to RCEP[J]. Maritime Business Review, 2020.
9. Feng Y, Wu H, Jin Y, et al. How does population aging affect carbon emissions?—analysis based on the multiple mediation effect model[J]. Environmental Science and Pollution Research, 2023: 1-16.
10. Akinwande M.O, Dikko H.G, Samson A. (2015) Variance inflation factor: as a condition for the inclusion of suppressor variable (s) in regression analysis. Open journal of statistics, 5(07): 754.
11. Blundell R, Bond S. Initial conditions and moment restrictions in dynamic panel data models[J]. Journal of econometrics, 1998, 87(1): 115-143.
12. Baron R.M, Kenny D.A. (1986) The moderator–mediator variable distinction in social psychological research: Conceptual, strategic, and statistical considerations. Journal of personality and social psychology, 51(6): 1173.
13. Du, X., Zhang, H., Han, Y. (2022) How does new infrastructure investment affect economic growth quality? Empirical evidence from China. Sustainability, 14(6), 3511.

Artificial Intelligence Enabled Double Reduction Policy Path Analysis

Chuanli Wei *, Peng Liu

Harbin University of Commerce, Harbin, 150028, Heilongjiang Province, China

Abstract. The "Double Reduction" policy starts from both inside and outside the school, with a view to combating the excessive academic burden of students and building a favourable educational ecology. However, the implementation of the policy has faced resistance and problems within and outside the school and in the general environment. The application of AI can provide new ideas in improving classroom quality, innovating educational tools, and evaluating and regulating education. At the same time, AI promotes the transformation of the education model and the innovation of education concepts. Artificial intelligence presents both opportunities and challenges for applications in education. This artical examines the obstacles in implementing the double reduction policy and explains the machine learning algorithms. By analysing pertinent education data, it explores AI's role in aiding double reduction policy execution and potential risks.Whilst encouraging their integration, it is important to clarify the instrumental role of AI whilst addressing issues such as educational equity and Digital Gap problems.

1 The reason for integration of artificial intelligence and education

The proliferation of big data and increased computing power in recent years, particularly with the debut of GPT-4 in 2023, has facilitated another booming phase in AI research and application. Through a continuum of advancements, AI's implementation is significantly altering people's livelihoods and productivity at an unparalleled pace and scope. Among them, the education industry is a promising area for AI application scenarios. [1] The double-reduction policy, which is a significant step in the overall reform of China's education development, is also the most far-reaching policy in recent years in the field of education. Some argue that artificial intelligence (AI) technology can enhance policy efficiency, lower resistance to policies, and facilitate education reforms and innovation. However, others contend that AI will impact the education ecosystem. [2] By collecting and summarising the literature, it is found that there are more research results on AI education, but there are relatively few studies on AI-assisted double reduction policies. Therefore, it is worth studying how AI can aid the implementation of the double reduction policy whilst simultaneously minimising associated risks.

2 Dilemmas in the implementation of the double reduction policy

2.1 On-campus dimensional analysis

Traditional homework design is typically uniform in nature, lacking the ability to provide personalised assignments tailored to individual students' knowledge and ability. Moreover, it has increasingly strayed from its nurturing function. When creating homework tasks, excessive emphasis is placed on subject teaching effectiveness, with insufficient attention given to the "learning" that occurs through student-assignment interaction.

The classroom quality may be impacted by the Double-reduction policy, as it results in a longer study time for students at school and decrease the importance of extracurricular tutoring. This could cause a stronger reliance on tutoring within the school, which may lead to a wider gap between schools. The level of teaching is hindered by the personal ability and experience of teachers. Introducing new teaching methods may result in an added workload for teachers, reducing their motivation to reform the current mode of teaching.

The issue of narrowing the role of after-school services is prevalent in practice. Schools can only offer supervised care to complete homework during extended hours without facing increased risks. Additionally, the assessment system for such services is flawed. Insufficient incentive for schools to develop after-school

*1437272939@qq.com

programs can be attributed to the absence of evaluation standards for assessing their level and quality. Insufficient incentive for schools to develop after-school programs can be attributed to the absence of evaluation standards for assessing their level and quality.

2.2 Off-campus Dimensional analysis

Although the outlawing of the subject-based tuition sector by large educational institutions has largely removed it from sight, shadow education persists in both overt and covert forms. New forms of shadow education such as in-residence tutoring and high-end childcare services are increasingly prevalent. Educational institutions operating covertly pose more management challenges, leading families to invest more in education. Unapproved advertisements intermittently intensify anxiety. Short-video platforms have witnessed a spike in "anxiety-trafficking" ads, catering to all levels of education.

Teachers usually take charge, while parents tend to cooperate passively to complete light duties such as correcting homework, preparing school supplies, and ensuring attendance. They play a less active role in participating in their children's educational journey. Sending group messages often becomes repetitive and tedious for parents. The Report on the National Family Education Survey indicates that many parents and even some teachers lack adequate parenting abilities. The study highlights problems, including less frequent communication with their children and high expectations of their offspring, which may leave parents feeling bewildered.

Educational equity is a cornerstone of social equity and a key objective of the double-reduction policy. However, there continues to be a significant gap in the distribution of educational resources across different geographical areas within China. In relation to equity in education, Gardner categorises human intelligence into 6 forms with individuals exhibiting either strong or weak capabilities in each respective area.[3]Typically, traditional education entails a focus on nurturing and valuing mathematical-logical and linguistic intelligences whilst neglecting other areas of intelligence.

3 Basic Theories and Principles of Machine Learning

3.1 Basic Theories and Principles of Machine Learning

Generally, machine learning involves allowing the computer to learn from data in order to acquire certain knowledge or patterns.[4]As a field, machine learning typically refers to a particular problem and the methods used to address it, namely, how to identify patterns from observed data and use the resulting model to predict unknown or unobservable data. In the early years of the engineering industry, machine learning was often referred to as pattern recognition. However, pattern recognition is more commonly applied to specific tasks such as optical character recognition, speech recognition, and face recognition. These tasks are easy for humans to perform, but we do not understand how we do them. Therefore, it is challenging to create a computer programme manually to complete these tasks. A viable approach involves developing an algorithm that enables the computer to acquire understanding from tagged samples and utilise it in accomplishing diverse identification tasks. [5]As machine learning technology is increasingly embraced, pattern recognition is being substituted by the idea of machine learning. It has come to be a comprehensive designation for this sort of difficulties and the techniques employed to solve them.

A tagged feature or label may be conceived as a sample or an illustration. The collection of samples is referred to as a data set or corpus in many fields. Typically, a data set is divided into two categories: training set and test set. The training set comprises training samples that are employed for model training. The test set is composed of test samples that are used to evaluate the model performance. We typically use a dimension vector $D= [x1, x2,...,xD]$ t to represent the vector containing all the characteristics of a mango, also known as a feature vector, where each dimension corresponds to a feature. Assuming the training set $\mathcal{D}$ comprises D samples, where each sample is independent and identically distributed, that is, extracted independently from the same data distribution, they are recorded as:

$$D=\{(x(1),y(1)),(x(2),y(2)),\square(x(N),y(N))\}. \quad (1)$$

Given the training set $\mathcal{D}$, we aim to use a function set $\mathcal{F} = \{f1\ (\boldsymbol{x}),\ f2\ (\boldsymbol{x}),...\}$ to enable the computer to automatically determine an "optimal" function $f*(\boldsymbol{x})$ that approximates the relationship between the eigenvector $\boldsymbol{x}$ and the label y of each sample. By utilising the function $f*(\boldsymbol{x})$, we can predict the label value of a sample $\boldsymbol{x}$. The process of discovering this "optimal" function $f*(\boldsymbol{x})$ typically requires completion through a learning algorithm $\mathcal{A}$ which determines the key to machine learning. In some literature, the learning algorithm is alternatively referred to as the "learner." The process of searching for patterns within the data is typically called "learning." [6] As a result, the learned function can be utilized to predict the sample's quality based on its characteristics during testing as:

$$Acc(f*(\mathbf{x})) = 1\left|\partial'\right|\Sigma(\mathbf{x},y)\in\partial' I(f*(\mathbf{x})=y) \quad (2)$$

3.2 Artificial Intelligence in Education

There are numerous technological applications in the education sector, as demonstrated in Table 1. We compare education and learning outcomes, comparing AI and non-AI learning application data using three mechanisms of mechanical learning. The results are depicted in Figure 1. The evidence demonstrates a significant presence of AI application, warranting further categorisation and meticulous examination.

Table. 1 Main technology used in education

Funtion	Main technology	Technology maturity
Emotion recognition	Natural Language Processing, image recognition	Medium
Intelligent Homework Correction	Optical Character Recognition	High
Student Grade Prediction	Machine Learning	Medium
Courseware Generation	Artificial Intelligence Generated Content	High
Pronunciation Evaluation	Speech Recognition	High
Voice User Interface	Speech Recognition, Natural Language Processing	Low
Adaptive Learning	All of the above	Low

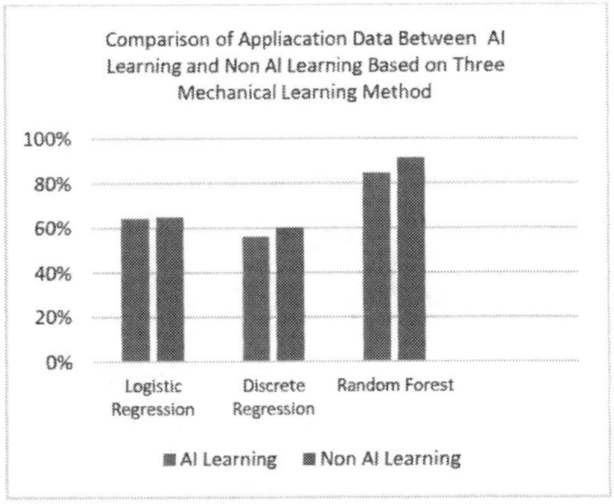

Fig1. Comparison of Application Data Between AI Learning and Non AI Learning Based on Three Mechanical Learning Method

4 Pathways for Artificial Intelligence Technology to Enable Double Reduction Policies

4.1 Analysis from teaching perspective

Using the cognitive diagnostic method of machine learning, and relying on computer vision, we are able to develop and apply algorithms in the field of education. By analyzing the interaction between teachers and students in the classroom, we can diagnose and evaluate their performance. This enables us to make timely adjustments to teaching methods as well as provide individual students with reminders, resulting in the efficient development of classroom teaching activities. Teachers can utilize the data analysis code automatically generated by ChatGPT to examine teaching data. Relying on the smart classroom to collect students' learning data for generating learning reports, teachers can use relevant reports and data to focus on explaining common issues, thereby enhancing classroom efficiency. By utilising AR technology, abstract concepts are conveyed through digital modelling, which is more effective in aiding students' memory. Based on a deep neural network for learning situation analysis, this study analyses the actual needs, ability levels, and cognitive tendencies of students. Characteristics of the students are comprehensively assessed and the advantages of interest are explored.

The use of augmented and virtual reality technology as a supplement to formal teaching not only engages students in a unique way and enhances their ability to learn visual information, but also decreases the risks of extending school hours. ChatGPT can offer guidance for extracurricular activities such as programming, reading, writing, and more, enriching the range of after-school services.

4.2 Analysis from student perspective

Using the "Smart Homework" system, homework assigned by teachers is personalised and structured hierarchically by utilising the big data of students' learning conditions. This method guarantees the accurate fulfilment of teaching requirements whilst avoiding redundant assignments that could demotivate students from learning. Relying on the "learning community" to overcome the limitations of time and space in providing homework explanations, this system uses natural language processing (NLP) technology to categorise and store incorrect questions. Subsequently, it presents the related questions to students to verify their comprehension. Developing the "Error Book" tool involves utilizing big data from homework evaluations, the forgetting curve, NLP technology, and "intelligent questions" to compile students' incorrect answers for targeted practice.

As AI is continuously being applied to educational scenarios, students can receive knowledge education at home without specific spatial constraints. Artificial intelligence can be integrated into educational activities in various ways as needed by students, teachers, and parents. This results in a transformation of the original subject-role relationship, with all three individuals assuming the roles of learners, knowledge transmitters, and supervisors, as per the specific learning requirements.

4.3 Analysis from family perspective

It is imperative to supervise the course syllabus and employ semantic analysis techniques to assess audio and video recordings in order to thwart the proliferation of unlawful or unsuitable educational resources. Regulations concerning illegal advertisements for tutoring require extracurricular tutorial institutions, whether online or offline, to advertise on major platforms using text, voice, and images. To aid supervision, ASR/NLU technology must be employed to identify and analyze reviews. Any distress caused by unauthorized advertisements should result in their expeditious removal and the pursuit of responsibility. The external training institutions must install cameras and an intelligent AI analysis system to capture faces and perform analysis during the entry and exit of students.

The integration of AI can mitigate family conflicts resulting from the shift towards home-based formal education. Furthermore, ChatGPT's capacity to produce simulations of family interactions permits training in psychology and pedagogy to enhance relationships within the family. The intelligent correction system, which fuses image recognition and natural language processing, may assist in alleviating the pressure on teachers and parents whilst reviewing homework. The incorporation of artificial intelligence and a pupil's family allows parents to engage more actively in their offspring's education, amplifying their comprehension of advancement and facilitating productive guidance.

The development of open courses in "smart classrooms" aims to facilitate learning for faculty members across different locations while complementing each other's skillsets. AI has the potential to replace costly one-on-one private tutoring, accommodating students who cannot afford this aspect of their education, and eventually bridging this gap. AI tutoring programmes, such as Khanmigo at Khan Academy, are based on OpenAI's ChatGPT technology. [7]The practical application of Multiple Intelligences theory is constrained by certain conditions. However, the emergence of new technologies liberates educators from rudimentary, repetitive tasks, like using ChatGPT. This tool can enabling a more diverse and effective approach to pedagogy.

5 Conclusion

The emergence of AI products highlights constraints in the initial evaluation system. The development of a more accessible means of learning and consolidating knowledge has rendered aptitude for AI, such as creative thinking, emotional intelligence, leadership, etcetera, highly significant, and fostering these capabilities is of critical importance in implementing a more dynamic training approach and enhancing the adaptability of the educational evaluation process. The Double Reduction policy presents an opportunity for transformation and is a significant step in education reform. Advances in new-generation information technology, including cloud computing, artificial intelligence, big data, virtual reality, augmented reality, and 5G, have the potential to enhance the effectiveness of policy implementation. The establishment of a smart learning environment, which includes process recording, precise analysis, intelligent counselling and immersive learning, will comprehensively enhance the integration of information technology and education. This will reduce the burden on all educational participants and increase efficiency. Artificial intelligence technology presents a two-fold scenario, and one must not overlook the possible hazards whilst utilizing such innovative advancements prudently. No changes needed as the text lacks context and already adheres to the principles.

Therefore, there are feasibility and problems in the expansion of the scale of case disclosure and collection in the field of education, information and data security, horizontal and vertical data comparison and even cross regional and national comparison. With the further evolution and development of machine learning and big data analysis, it will certainly be more applied and developed in the field of education in the future. We can expect it to play a greater role in the field of education and promote the further development of education.[8]

References

1. S.F. Ahmad, M.K. Rahmat, M.S. Mubarik, M.M. Alam, S.I. Hyder, Artificial intelligence and its role in education. Sustainability 13, 12902 (2021)
2. H.Y. Zhou, Y.Y. Li, Impact of ChatGPT on education ecology and countermeasures. J. Xinjiang Normal Univ. 44, 102 (2023)
3. H. Gardner, The theory of multiple intelligences. Ann. Dyslexia 37, 19 (1987)
4. J. Baum, C. Laroque, B. Oeser, A. Skoogh, M. Subramaniyan, Applications of big data analytics and related technologies in maintenance—literature-based research. Machines 6, 54 (2018)
5. D. Saidulu, R. Sasikala, Machine learning and statistical approaches for Big Data: issues, challenges and research directions. Int. J. Appl. Eng. Res. 12, 11691 (2017)
6. V.K. Shukla, A. Verma, Enhancing LMS experience through AIML base and retrieval base chatbot using R language. In 2019 International Conference on Automation, Computational and Technology Management (ICACTM), (2019)
7. S. Guevara, Khan academy: Learning made personal. Comput. Libr. 39, 6 (2019)
8. W.P. Lee, J.Y. Huang, H.H. Chang, K.T. Lee, C.T. Lai, Predicting drug side effects using data analytics and the integration of multiple data sources. IEEE Access 5, 20449 (2017)

Problems in the development of new agricultural management entities and Suggestions on fiscal and tax policies

Jiahao Liu*

Harbin University of Commerce, School of Finance and Public Administration, 1995, Harbin,China

Abstract. Promoting the development of new agricultural management entities is an important measure to realize agricultural industrialization and an important way to realize agricultural and rural modernization. This paper analyzes the existing problems of the new agricultural business entities, aiming at the problems that the effect of its financial support policy is not obvious, the degree of land transfer can not meet the large-scale operation, the financing difficulties of the new agricultural business entities and the imperfect insurance system, and the relevant tax policies are still not perfect. It is proposed to increase the intensity of financial support, improve the form of financial subsidies, strengthen financial subsidies and social security, promote land transfer with double incentives, strengthen the support mode of financial credit, improve the system of agricultural insurance, and optimize the current agricultural tax policy recommendations. It is hoped that it can play a certain role in promoting the development of new agricultural management entities.

1 introduction

The new agricultural management subject is an important subject to promote the comprehensive promotion of rural revitalization and accelerate the modernization of agriculture and rural areas. The report of the 20th National Congress of the Communist Party of China clearly stated : " Consolidate and improve the basic rural management system, develop a new type of rural collective economy, develop new agricultural management entities and socialized services, and develop moderate-scale agricultural operations." As a large agricultural country, under the guidance of the national strong agricultural policy, new business entities such as agricultural leading enterprises, agricultural professional cooperatives, family farms, and professional large households have developed rapidly. However, at present, some areas in China still have high yield but low efficiency, and the development of agricultural scale and intensification is lagging behind. As an important macro-control tool of the government, fiscal and taxation policies play a vital role in promoting the development of new agricultural business entities. Therefore, it is very important to formulate corresponding fiscal and taxation policies.

2.Definition and classification of the concept of new agricultural business entities

2.1 Definition of the concept of new agricultural business entities

The new agricultural management entity is the product of the reform and development of China's rural management system. Before 2012, new agricultural business entities were used in some articles on theoretical research and policy research on agriculture, rural areas and farmers. In 2012, the concept of new agricultural business entities was first mentioned in official documents. Researchers in the theoretical circle and the government also put forward different definitions of new agricultural business entities according to their respective understandings, but at present, the state has not yet defined them in a unified and standardized way. This paper defines it as : the new agricultural management entity refers to the professional farmers who are educated, skilled and able to operate on the basis of improving the household contract responsibility system, and the agricultural management organization with large-scale operation, high degree of intensification and market competitiveness. It mainly includes large professional households, family farms, farmers' professional cooperatives, agricultural leading enterprises and other operational agricultural socialized service organizations.[1]

*Corresponding author:396490630@qq.com

2.2 Classification of new agricultural business entities

This paper divides the new agricultural business entities into three categories, namely family farms, farmers' professional cooperatives and agricultural leading enterprises. The term "family farm" originated from European and American countries. It refers to the large-scale farmers in developed countries such as Europe and the United States. They use family members as the main labor force, do not employ workers or only employ a small number of workers in busy farming seasons, engage in large-scale, intensive and commercial production and operation of agriculture, and use agricultural income as the source of family income.[2]Farmers' professional cooperatives are mutual economic organizations which are based on the rural household contract management, the production operators of similar agricultural products or the providers and users of similar agricultural production and management services, voluntary joint and democratic management.[3]Agricultural leading enterprises are agriculture-related enterprises with modern enterprise system in the industrialization of agriculture. They form a mechanism of interest linkage with farmers through contracts or shares, and rely on agricultural and sideline products production bases to establish large-scale, stronger production technology, strong radiation and leading role, and strong economic strength. It can provide direct or brief services to the production of agricultural products and enterprises producing agricultural products.

3 Problems in the development of new agricultural management entities

3.1 The effect of financial support policy is not obvious

3.1.1 Financial support is not enough and the standard is too high

Although the government has invested a lot of financial support for agricultural development, most of the funds it has invested are used for the construction of agricultural infrastructure, the protection of cultivated land fertility, the purchase of agricultural machinery and retail agricultural subsidies. Therefore, the funds used to support new agricultural business entities are very few, resulting in insufficient support for them. Many of the government's financial support policies for new agricultural business entities have set fixed deadlines and must meet the prescribed conditions, and the policy support standards for small and medium-sized new agricultural business entities with greater capital demand are higher,[4] They are difficult to meet the requirements for obtaining financial support, resulting in a smaller coverage of policy support.

3.1.2 The form of financial subsidies is not reasonable enough

At present, government's subsidies for new agricultural business entities are mainly direct subsidies, and the subsidy method is relatively simple. It is a long-term process to promote the development of new agricultural business entities. Most of the relevant subsidy policies have short-term problems, often only temporary implementation, so it is difficult to promote the sustainable development of new agricultural business entities. In addition, there are still some problems in some financial subsidy policies. There are great differences in the development of new agricultural business entities in different regions, so it is very important to implement subsidy policies accurately. However, because the government 's understanding of the development of new agricultural business entities is not deep enough, it has not been able to know its actual needs. In view of financial subsidies, it often fails to consider the micro needs of the main body, thus affecting the implementation effect of financial subsidies, which is not conducive to the development of new agricultural business entities.

3.2 The degree of land transfer can not meet the scale operation.

3.2.1 Farmers' uncertainty of land transfer income

Agricultural scale operation is the basis of new agricultural management entities. Whether it is family farms, farmers' professional cooperatives or agricultural leading enterprises, large-scale and large-scale land is needed. Although China's land transfer has achieved certain results, there is still a big gap with the complete formation of modern agricultural management. The ratio of youth labor outflow is quite different from the ratio of land transfer, indicating that there are still many migrant workers' land not transferred out. This is because the current social security system for farmers is not perfect and popular enough, some financial subsidy policies are not strong enough, and the pertinence is not strong. Land is the only guarantee for farmers' life. They are unwilling to face the uncertainty of future income and expect to obtain stable income. Therefore, the land transfer rate is low and the development of new agricultural management entities is restricted.

3.2.2 The aging of farmers in the county has affected the transfer of land

With the transfer of a large number of young labor force to cities, the proportion of the elderly population is increasing year by year, especially in rural areas. There are many elderly people in China's rural areas who cannot farm their own land. Because most rural elderly people do not have a stable income from pension insurance, they are not willing to transfer their land. The difficulty of land circulation leads to the obstruction of intensive and large-scale operation of new agricultural management entities, and also makes it impossible for many large

https://doi.org/10.1051/shsconf/202317803015

households or enterprises willing to form new agricultural management entities to continue.

3.3 The financing of new agricultural business entities is difficult and the insurance system is imperfection

3.3.1 Credit financing difficulties

The development of new agricultural business entities require a lot of financial support. Agriculture is an industry with large investment scale, high risk, long term and low profit.[5] At present, commercial banks have high financing costs for agricultural credit products, complex evaluation indicators and short-term loans. For most of the start-up small and medium-sized new agricultural business entities, their economic strength is weak, and there is no corresponding collateral, so it is difficult to obtain bank credit financing. In addition, the richness of credit products for new agricultural entities is not enough. Most of them are still approved according to the original traditional agricultural loan model, resulting in the loan amount and interest rate cannot be adapted to the development of new agricultural business entities. [6]

3.3.2 The relevant agricultural insurance system is imperfection

Agricultural management faces many risks. In the process of operation, new agricultural management entities cannot avoid natural risks. Especially in such a large-scale operation, when faced with natural risk disasters, the losses caused are huge. At this time, relevant insurance support is needed as the last line of defense for new agricultural business entities. Compared with Western agricultural developed countries, China's agricultural insurance has many problems. Especially for the insurance system of new agricultural business entities. Insurance companies are less willing to develop agricultural insurance, lack the motivation of agricultural insurance with high operating risk and low income, and the types of related insurance are limited, and the insurance amount of most types of insurance is high. Therefore, the imperfection of the insurance system also indirectly affects the development of new agricultural business entities.

3.4 Relevant tax policies are still not perfect.

3.4.1 The content of tax policy is not comprehensive

The existing tax policies only stipulate some preferential tax policies for agricultural leading enterprises and farmers' professional cooperatives. However, family farms account for a large proportion in the development of agricultural industrialization, and the lack of corresponding tax incentive policies will inhibit the development of new agricultural management entities. In addition, the tax policy for agricultural leading enterprises and farmers' professional cooperatives only stays on direct tax incentives. Direct tax incentives only promote the development of new agricultural business entities in the initial period through the reduction and exemption of tax amount, and have no incentive effect on their subsequent development and progress, so they cannot continuously promote the development of new agricultural business entities.

3.4.2 The content of tax policy lacks pertinence

At present, the current tax policies on new agricultural business entities have certain inclusiveness, and are still in the stage of imitation and exploration. Without investigation and understanding of the current situation of new agricultural management entities in the region, the preferential tax policies introduced are not targeted and efficient, and the incentive effect on the development of new agricultural management entities is not good.[7]

4 Suggestions on fiscal and taxation policies to promote the development of new agricultural business entities

4.1 Increase the intensity of financial support, improve the form of financial subsidies

4.1.1 Strengthen financial support and enrich agricultural investment models

At present, most of the agriculture-related funds are inclusive, and the promotion effect on agricultural development is not obvious. Therefore, the government should reasonably formulate the allocation plan of financial support for agriculture according to different development projects, and at the same time increase the proportion of capital allocation for new agricultural management entities, strictly distinguish it from other agriculture-related funds, and make special funds available for special use, so as to fundamentally solve the problem of insufficient financial support for agriculture. In addition, the income of county finance is mostly derived from the transfer payment of the higher government, so it is inevitable to make the best use of it and set certain regulations on the object of obtaining financial support. For the small and medium-sized new agricultural management entities that have just started, they are difficult to actually obtain financial support. Therefore, the government needs to appropriately reduce the adjustment policy standards and appropriately extend the funding period for small and medium-sized new agricultural operators to strengthen their development foundation. At the same time, considering the limited government revenue, we can also guide and encourage agricultural leading enterprises, social capital and other enterprises or individuals with certain economic strength to set up special support funds, which can not only reduce the financial burden but also support the development of new agricultural business entities.

4.1.2 Improve the form of financial subsidies

Direct subsidies have a certain effect on the early incentive of new agricultural business entities, but it is often necessary to establish a long-term mechanism for the development of new agricultural business entities. Therefore, the government should innovate a variety of forms of subsidies, such as indirect subsidies, tax incentives, awards instead of subsidies, and give different subsidy policies at different stages of its development. For example, in the early stage, it focuses on giving direct subsidies, and in the middle and late stages, it encourages its development with indirect subsidies and awards instead of subsidies. In addition, the government should deeply understand the current development of new agricultural business entities, know their actual needs, clarify the focus of current subsidies, carry out precise policies, implement corresponding subsidy policies for entities at different stages and different development situations, give full play to the efficiency of subsidy policies, and fundamentally promote the development of new agricultural business entities.

4.2 Strengthen financial subsidies and social security, and promote land transfer with double incentives

4.2.1 Increase the intensity of land transfer financial subsidies

It is an inevitable trend for rural young and middle-aged labor force to flow to cities. To improve their willingness to transfer land, it is necessary to increase the intensity of land transfer subsidies so that they can truly enjoy the preferential treatment of policies. Through the policy of "collective transfer, subsidy increase", we can promote multi-agent collaborative participation in the implementation of the policy, and allow villagers with the ability to call for spontaneous organization of joint transfer, which not only increases the efficiency of land transfer, but also adds some additional benefits. The problems faced by the elderly may be more complicated, and the dependence of the elderly on land may be higher than that of the young and middle-aged groups, because most of the rural elderly have not paid pension insurance and can only rely on land to maintain their lives. Therefore, it is necessary to increase their land transfer subsidies. The government can divide the amount of subsidies according to the age group. The higher the age, the more the subsidies. The annual increase in the amount of subsidies can promote them to actively transfer their land.

4.2.2 Improve the social security system of farmers

Although there are a large number of young and middle-aged labor force transfer to the city every year, but its social security is still not perfect phenomenon, which indirectly led to its unwilling to transfer their land, land as income security. To this end, the government can give it a certain off-site resettlement subsidies, special preferential pension, medical, unemployment insurance and other policies, so that it can take root in the city and strengthen its income security, so as to improve its land transfer rate. For the elderly in rural areas, the government can increase its efforts to promote the rural special pension insurance policy, relax the age limit for farmers to pay insurance, subsidize or appropriately reduce the farmers who pay pension insurance, and encourage them to be insured, thereby increasing the land transfer rate.

4.3 Strengthen the support model of financial credit and improve the system of agricultural insurance

4.3.1 Strengthen financial credit financing support

The government can cooperate with financial institutions to solve the agricultural difficulties of new agricultural management entities. The central bank should flexibly use monetary policy tools such as re-loaning, rediscounting, and targeting to increase credit supply and reduce financing costs.[8]Agricultural Bank of China, Rural Commercial Bank, Postal Savings Bank of China and other commercial banks to serve agriculture, rural areas and farmers should also cooperate with the central bank to increase the ratio of agricultural credit, enrich the form of credit products, standardize the mode of loan approval, and simplify business processes. At the same time, the government should cooperate with commercial banks and financing guarantee institutions to innovate loan mortgage methods and increase the types of mortgaged goods such as large-scale agricultural machinery and equipment, land management rights, etc.[9] to escort the credit delivery of commercial banks, and also facilitate the financing of small and medium-sized new agricultural business entities.

4.3.2 Improve the relevant agricultural insurance system

At present, the main reason why insurance companies are not enthusiastic about developing agricultural insurance is that agricultural risks are high and force majeure factors are in the majority. Therefore, the government should increase the insurance company's development of agricultural insurance from the aspect of policy incentives. It can give financial subsidies and preferential tax policies to insurance companies related to agricultural insurance,[10]encourage them to broaden the scope of agricultural insurance, increase investment in agricultural insurance, enrich and improve relevant insurance types, promote the new agricultural business entities should be fully insured, strengthen their ability to resist risks.

4.4 Optimize the current agricultural tax policy

4.4.1Broaden the scope of tax incentives policy

The government should improve and formulate preferential tax policies to promote the development of

https://doi.org/10.1051/shsconf/202317803015

new agricultural business entities, increase tax relief for agricultural leading enterprises and farmers' professional cooperatives, supplement and formulate tax policies for relevant family farms, and vigorously promote indirect tax incentives, such as the enterprise income tax reserve system and tax credits, etc., so that tax incentives run through the production and operation of new agricultural business entities, and effectively play the role of preferential tax policies in promoting new agricultural business entities.

4.4.2 Flexible configuration of preferential tax policies

The government should fully investigate the current development and actual needs of new agricultural business entities, and flexibly formulate and implement corresponding tax preferential policies according to different entities in different regions. Tax reduction and exemption policies, tax rebate policies, tax threshold, tax rate, deferred tax and other tax expenditure forms can be freely combined and implemented according to the current situation to meet the development needs of new agricultural business entities. At the same time, to do a good job in the supervision and management of personalized tax policies, the government should establish a special tax preferential supervision mechanism and supervision department for new agricultural business entities, standardize and supervise the use of tax preferential funds, so that personalized tax preferential funds can be truly implemented.

5.Conclusion

The development of new agricultural management entities has become an important basis for promoting agricultural modernization. Therefore, the government should deeply understand the development status of the current new agricultural business entities, clarify the shortcomings of their development, and formulate and improve the corresponding fiscal and taxation policies in combination with the local actual development situation, so as to promote the high-quality development of the new agricultural business entities.

Reference

1. Wang Jingli.Analysis of the reform and development of China 's agricultural management system in the past 70 years [J].Financial theory and teaching, 2023(01): 98-101. https://doi:10.13298/j.cnki.ftat.2023.01.015.
2. Robert A.Hoppe,David E.Banker.Structrue and Finances Of U.S.Farms[J].Family Farm Report,2010,(10):55-58.https://xueshu.baidu.com/usercenter/paper/show?paperid=9e9f2d4467be7f61c032563759409cf0&site=xueshu_se
3. Guo Jingwen,Xu Mengdi.Research on the innovation of pesticide supply chain model based on agricultural cooperatives[J].China Storage and Transportation, 2023(03):189-190. https://doi:10.16301/j.cnki.cn12-1204/f.2023.03.030.
4. Ying Haifen. Research on the Dilemma and Countermeasures of the Industrialization Development of New Agricultural Management Entities [J].Agricultural Economy,2019(03):62-64.https://doi:10.3969/j.issn.1001-6139.2019.03.024.
5. Paul.Samuelson.(2014)Economics.In:Beijing CommercialPress.Beijing.pp.https://www.cp.com.cn/book/978-7-100-08406-2_51.html
6. Wu Chenghao.Research on financial policy of cultivating new agricultural business entities in Henan Province[J].Financial theory and practice,2019(08):100-105.https://kns.cnki.net/kcms/detail/41.1078.f.20190730.1759.028.html
7. Wei Zibai. Research on fiscal and taxation policy support for cultivating new agricultural business entities in the new era [J].Agricultural economy, 2021 (05):105-107.https://doi:10.3969/j.issn.1001-6139.2021.05.039
8. Tan Wei. Practice and Reflection on Financial Support for the Development of New Agricultural Business Entities - Based on the Investigation and Analysis of Heilongjiang Province [J].Financial Development Review, 2018 (05) : 66-71. https://doi:10.19895 / j.cnki.fdr.2018.05.009.
9. Lian Xiaoyue, Changpingping. Research on fiscal and tax policy support for the cultivation of agricultural industrialization consortium in the new era [J].Agricultural economy, 2021 (01) : 97-99. https://doi:: 10.3969 / j.issn.1001-6139.2021.034.
10. Zhu Runxi, Zhao Fushun. Research on fiscal and taxation policies to support the development of new agricultural business entities [J].Journal of Inner Mongolia University of Finance and Economics, 2018,16 (02):1-4. https://doi:10.13895 / j.cnki.jimufe.2018.02.001.

Optimization of strategic closed-loop management mechanism for large-scale power grid enterprises adapting to digital transformation

Xiaofeng Zhang*, Xiaobin Lin, Jing Li

Institute of Corporate Strategy, State Grid Energy Research Institute, 102209 Beijing, China

Abstract. When facing increasingly complex and fast-changing strategic situation, powerful strategic management system is crucial for large-scale power grid enterprises to ensure scientific decision-making, coordinated action and efficient management. This paper analyzes the new strategic management situation of large-scale power grid enterprises under the background of the integrated development of digital revolution and energy revolution. Based on the new situaiton, this paper carries out the demand analysis of the optimization of strategic management mechanism for power grid enterprises, and puts forward the core content of the strategic closed-loop management mechanism optimization. The mechanism optimization will ensure the forward-looking of power grid enterprises' strategy formulation under the background of digital transformation. Additionally, the mechanism optimization will promote accurate identification of changes, scientific response and effective prevention of strategic risks.

1 Introduction

Power grid is important infrastructure, which closely related to the national economy and people's livelihood. Power gird has the characteristics of long industrial chain, strong driving force and wide influence. These characteristics requires the State Grid company must accurately and sensitively perceive the dynamic changes of the external environment, and respond to them in a timely and effective manner.

In recent years, the external environment of large power grid enterprises becomes variability, uncertainty, complexity and fuzzy increasingly. In terms of scientific innovation, the international competition around the key core technologies of energy and power is becoming increasingly fierce, so that the "subversive innovation" and "cross-border operation" in the field of energy and power has gradually become norm[1]. In terms of institutional reform, a new round of power reform and state-owned enterprise reform has continued to advance in depth, which profoundly change the industry ecology and development pattern, bring more market competition, while strengthen the demonstration effect and chain reaction of reform increasingly. In terms of industry transformation, a new round of energy and digital revolution go hand in hand, which jointly drive the transformation and upgrading of the power grid to a smarter and more efficient energy Internet[2]. In terms of customer service, with the transformation and upgrading of energy consumption concepts and patterns, diversified, personalized and interactive demands are put forward for power services. In terms of international business, events, such as China-Us trade frictions and the COVID-19 epidemic, have stirred up changes in the global situation, which result in uncertainties affecting international business increasing significantly. In general, these complex and profound changes in the external environment continue to refresh the external environmental situation of large power grid enterprises. Only by keenly perceiving the complex and profound changes in the external environment and dynamically optimizing and improving the strategic closed-loop management system, power grid enterprises can make correct strategic choices, avoid strategic risks and master the initiative of development. This requires the use of big data and quantitative analysis technology to improve the scientific and intelligent level of strategic situation research and judgment, which promote the transformation and upgrading of strategic management from relying on experience to data.

From the international experience, world-class enterprises generally attach importance to strategic environment perception and intelligence analysis. They establish their own unique analysis tools, such as BP's enterprise brain, SAP and IBM's business intelligence[3]. However, most of these tools are core business secrets, which can't be obtained. Meanwhile it is difficult to adapt to China's large power grid enterprises. Therefore, by means of independent innovation, it is of great theoretical and practical significance to develop a set of strategic closed-loop management mechanism, which can fully perceive the dynamic changes of environment, fit its own reality and operate efficiently.

* Corresponding author: sgerizxf@163.com

2 Demand analysis for optimization of strategic closed-loop management mechanism of large power grid enterprises

Strategic management is a series of dynamic management activities for enterprises to control the whole process of strategy around the realization of strategic objectives, which generally covers strategic analysis, strategic revision, strategic decomposition, strategy implementation, strategy evaluation, etc. It is a closed-loop work[4]. In recent years, a new round of industrial technological revolution has deepened, and the internal and external situation of large power grid enterprises has undergone profound changes. They encountered multiple challenges such as overlapping difficulties, concentrated contradictions, and arduous tasks, which result in many major issues that need to be continuously studied, and new requirements for accelerating the innovation strategy closed-loop management mechanism.

Firstly, it is required to strengthen strategic propaganda and change the thinking philosophy. For enterprises, the first of digital transformation is a revolution in thinking, which requires continuous expansion of ideas and horizons, a deep understanding of Internet concepts such as big data, platform and ecological. Then they need to integrated new thinking philosophy in-deathly with business philosophy, business operation, innovation process, and business model. By emphasizing process innovation, rapid iteration, and "agile response", strategic management needs quickly adjust and respond to environmental changes and situation requirements.

Secondly, it is required to strengthen strategic training and improve team capacity. To promote implementation of the power grid's digital transformation strategy under new situation, it is necessary to continuously strengthen strategic management and research work, improve business capabilities, strengthen style construction, and strengthen self-restraint. Through strengthening internal training and work exchanges, the whole team of strategic management and research should be promoted to exchange learning and workout boost.

Thirdly, it is required to be customer-focus and highlight grassroots innovation. Under the new situation, power grid enterprises need to pay more attention to customer demand and market changes in business development. And then need to give full play to the role of grassroots in strategic management. By building a platform for communication and unified interaction between the upper and lower levels, the penetration of strategic management need to be enhanced. And the grassroots is promoted to strengthen tracking research and feed-backs on the strategic environment.

Fourthly, it requires integrate internal and external resources and greater open sharing. Under the new situation, the development concept of power grid enterprises emphasizes openness, innovation, symbiosis and win-win. By enhancing the openness and inclusiveness of strategic management, integrating and utilizing internal and external expert resources of power grid enterprises, deepening the research on strategic issues, it is required to accelerate the formation of a professional think tank system with efficient coordination and internal and external collaboration, which provide support for the innovation of strategic management mechanism and serving the strategic decision-making of enterprises.

3 Research on optimization framework of strategic closed-loop management mechanism of large power grid enterprises

In general, the promotion of strategic management mechanism innovation of large-scale power grid enterprises under the background of digital transformation mainly focuses on the four elements which are "enterprise mission, industry trend, external environment and internal capability". On the one hand, around the overall goal of promoting the implementation of the power grid enterprise strategy, power grid enterprises should fulfill the development mission of building energy Internet, ensuring national energy security, and serving the people's better life. And power grid enterprises should conform to the industry trend of the integrated development of energy revolution and digital revolution. On the other hand, power grid enterprises should adapt to the rapidly complex changing external environment, enhance the ability to perceive the external environment, scientifically judge the company's strategic situation, and provide strong support for corporate strategic decision-making. Meanwhile, it is necessary to make great efforts to improve the quality of the strategic management team, use new management theories, methods and tools, and continuously improve the strategic management and analysis capabilities. The "four elements" constitute the boundary of the enterprises' strategic management work, and are also the starting point and foothold that strategic management must adhere to for a long time. On this basis, combined with the closed-loop management of the whole process of strategy, power grid enterprises explore to promote the innovation of strategic management mechanism based on digital transformation.

3.1. Strategic advocacy training mechanism

The new generation of information and communication technology provides new technical means for large-scale power grid enterprises to consolidate the strategic publicity and training mechanism, which is conducive to promoting employees to better understand and master the essence of strategy, further condense consensus, and promote the implementation of corporate strategy. Firstly, power grid enterprises need to explore the mechanism for formulating and revising "The Outline of Enterprise Development Strategy". Facing the implementation of digital strategy, enterprises formulate the text of the enterprise development strategy and revise it on a rolling basis every year, interpret the development strategy system such as the mission purpose, strategic objectives, strategic positioning, strategic path, and corporate culture in detail. Then enterprises distribute it to front-line employees, and continue to further promote the interpretation and publicity of the strategic concept and connotation. Secondly, enterprises should form a normalized strategic publicity mechanism. Through new media and online platforms, they innovate to organize news and publicity topics, and publicize the strategy.rely on various training courses such as "cloud classroom".

3.2. Strategic collaborative research mechanism

Under the background of digital transformation, large power grid enterprises face a higher degree of complexity, ambiguity and uncertainty in the external environment, It is necessary to strengthen the working mechanism of strategic collaborative research, organize and coordinate relevant core business departments, second-level units and relevant think tank institutions, and continue to deepen collaborative research on strategic issues. The power grid enterprises need to innovate a rapid response mechanism to major issues. They should carry out major strategic research in a forward-looking manner, pay attention to the refinement of research results, and provide strong support for corporate strategic decision-making. For example, focusing on hot issues such as enterprise digital transformation, customer privacy data protection, and network infrastructure utilization, then can sort out key influencing factors, put forward solutions and suggestions, and provide support for promoting the implementation of the strategy.

3.3. Strategic implementation promotion mechanism

To adapt to the trend of digital transformation, large power grid enterprises need to actively solidify the mechanism of strategic deployment implementation, use quantitative tools to improve the closed-loop management of corporate strategic goals, promote the implementation of strategic goals, and evaluate the implementation. (1)Mechanism for indexing strategic objectives. Through the analysis and decomposition of the connotation of the strategic goals of large-scale power grid enterprises, advanced strategic management tools are used to form a complete set of index systems to characterize the strategic objectives of enterprises. (2)Stage goal clarity mechanism. Near-medium-term goals (annual, three-year, and five-year goals) are made as clear and explicit as possible, and are further linked with annual target indicators such as peer benchmarking, performance evaluation, and comprehensive planning.

3.4. Strategic management communication mechanism

Under the background of digital transformation, the development concept of power grid enterprises emphasizes openness and sharing, and the business sinks further to the grassroots level, which requires higher openness, inclusiveness and adaptability of enterprise strategic management. To ensure the high-quality implementation of the strategy, it is necessary to coordinate and link up the whole enterprise, and promote the innovation of strategic management communication mechanism at various levels. (1)Internal strategic implementation issue exchange mechanism. Under the new situation, the role of grassroots in the strategic management of large-scale power grid enterprises is more prominent. It is necessary to build a strategic research, management and discussion mechanism that runs through "group headquarters - provincial companies - city and county level companies". And focusing on the problems existing in the process of strategy implementation, headquarters "decentralization, management and service", and the development of supervision business and non-supervision business classification, etc., it is necessary to invite enterprise managers and professional and technical personnel at all levels to deeply exchange the implementation of strategy and related business development, analyze the external environment, find out the existing problems and causes, and discuss coping ideas and measures. (2) External expert strategic situation discussion Mechanism. Adhering to the concept of opening the door to construction and innovation, enterprises fully absorb the views and suggestions of external industry experts and research institutions, organize seminar on the strategic situation with inviting a number of academicians, experts and senior scholars. In the seminar, enterprises discuss with experts on major issues such as how to promote China's energy transformation, accelerate the construction of the energy Internet, and discuss construction ideas and prospects. At the same time, enterprises establish a tracking mechanism for important views of external experts, and timely track the analysis comments of external experts, research institutions, news media on the development strategies of large power grid enterprises.

3.5. Strategic environment awareness mechanism

Under the background of digital transformation, the external environment of large-scale power grid enterprises is becoming increasingly complex, and the source of influence and causal chain are becoming more and more ambiguous, which requires accurate analysis and comprehensive research and judgment. At the same time, as a regulated natural monopoly industry and public utility enterprise, the power grid has high social attention. Meanwhile, the power grid enterprises engaged in the power grid, finance, industry, international, and strategic emerging businesses, result in facing different industry policies, upstream and downstream of the industrial chain, market competition. Based on the above factors, the power grid enterprises urgently need to innovate the strategic environment perception mechanism, classify and deal with the impact of various environments, and give differentiated and precise countermeasures. (1)Strategic environment dynamic monitoring and analysis network and monthly strategic environment dynamic reporting mechanism. The power grid is the central link connecting energy production and consumption, with wide coverage, complex organizational system, and many branches, and the grassroots units of various regions and business sectors provide the most direct observation channels for identifying environmental changes. In order to further strengthen the comprehensive analysis of the strategic environment and timely grasp the important situations and dynamics such as the macro situation, policy trends, and market changes in various regions (industries), it is necessary to explore the construction a mechanism that relies on second-level units to submit strategic environmental information on a monthly basis to enhance the ability to respond to environmental changes. The specific content should include major policy adjustments, reform progress, market competition situation, major technological breakthroughs, emergencies in the region and industry. (2)Strategic environmental analysis evaluation model innovation. Based on the latest theories and practical cases of strategic environment analysis, new technologies and new tools are used to construct an external environment analysis framework from both macro and industry environment[5]. The macro environment is based on the PEST environmental analysis framework and combined with the characteristics of power grid enterprises, which mainly refers to the external overall environment faced by the enterprises, including policy environment, economic environment, social environment, scientific and technological environment, international environment and other dimensions. The industry environment is expanded based on Porter's five forces model, covering energy situation, industrial chain situation, market competition and other dimensions. On this basis, the secondary dimension, tertiary index and specific subdivision index are further subdivided and clarified around each dimension, and a key indicator system of the strategic environment of large-scale power grid enterprises is formed, which provides support for enhancing the ability of enterprises to perceive the external environment and judge the strategic situation. (3)Innovation of strategic environment analysis mechanism based on big data. At present, more and more innovative enterprises have begun to cultivate big data genes, which are applied to all aspects of production and operation such as production, operation management and scientific research, so as to tap the potential value of data and promote the application of big data in enterprises. The strategic environment of enterprises implies external environmental factors that may have a significant impact on the survival and development of the enterprise. And the strategic environment plays a vital role in the key resources and competitiveness of the enterprise. Traditional strategic environment analysis is difficult to fully grasp enterprise strategic intelligence in the new situation of rapid increase in data volume, large volume, multiple types, and high real-time in today's era. Relying solely on human judgment often loses valuable sources of information and increases corporate strategic risks. Under the background of digital transformation, the power grid enterprises need to be guided by strategic environment analysis and strategic research needs, analyze the application scenarios of big data technology in strategic management. Meanwhile, the enterprises need to explore and develop strategic management methods and tools based on big data technology. And they should combine big data search and mining tools such as web crawlers, to realize the "online capture" and "intelligent analysis" functions of dynamic information of the external strategic environment. Finally, they need to form output display materials such as fluctuations of key environmental indicators, hot information word clouds, and visual data charts. Which are used to identify important developments that require attention.

4 Conclusions

This paper analyzes the new situation faced by the strategic management of large power grid enterprises under the background of the integrated development of digital revolution and energy revolution. Then we analyze the needs of strategic management mechanism optimization of power grid enterprises from four aspects, which are strengthening strategic publicity, improving research capabilities, highlighting grassroots innovation, and emphasizing openness and sharing. Then we put forward the overall framework for optimizing the strategic closed-loop management mechanism around four elements of "enterprise mission, industry trend, external environment, internal capability". This paper also propose that the core contents of the innovation of strategic closed-loop management mechanism for large-scale power grid enterprises under the background of digital transformation, which include strategic publicity and training mechanism, strategic collaborative research mechanism, strategy implementation promotion mechanism, strategic management communication mechanism and strategic environment perception mechanism. Optimizing relevant mechanisms can help ensure the forward-looking and advanced strategy formulation of power grid enterprises under the background of digital transformation, and promote accurate identification, scientific response and effective prevention of strategic risks.

References

1. J. Wang, K. Meng, J. Cao, Energy internet information technology research overview. Computer research and development, **52**, 18 (2015)
2. Z. Dong, J. Zhao, F. Wen. From smart grid to energy internet: Basic concept and research framework. Automation of Electric Power Systems, **38**, 15 (2014)
3. R. Shara, D. Delen, E. Turban. *Business Intelligence: a management perspective on data analytics*.(China Machine Press, Beijing, 2018)
4. H. Michael, I. Duane, H. Robert. *Strategic Management: Concepts and Cases (13th edition)*. (China Renmin University Press, Beijing, 2021)
5. Y. Yi, Y. Liu, Y. Li. Analysis of enterprise strategic environment in dynamic environment. Journal of Xi 'an Jiaotong University: Social Science Edition, **23**, 6 (2003)

Research on the Path of Improving Enterprise Performance through Employee Stock Ownership Plan in Technological Enterprises

Song Zhou[1,a], Nanyun Xiao[2,*]

[1]Department of Management,Harbin Finance University,Harbin Heilongjiang 150030,China
[2]Department of Accounting,Harbin Finance University,Harbin Heilongjiang 150030,China

Abstract. In the period of rapid development of the knowledge economy, human capital plays a very prominent role in technology-based enterprises. Technological enterprises have characteristics such as high R&D investment, intensive technological talent, and significant impact of human capital on their competitiveness. Employee stock ownership plans can stimulate innovation vitality and improve innovation efficiency. The research on employee stock ownership in technology-based enterprises is still a relatively new field, and there are still some challenges. Employee stock ownership in technology-based enterprises is an incentive mechanism that can effectively improve corporate performance. This article elaborates on the characteristics of employee stock ownership plans, analyzes the process of their impact on corporate performance, and proposes countermeasures and suggestions. Employee shareholding in technology-based enterprises is not a panacea for all problems, and its successful implementation requires specific design and adjustment based on the company's own situation. In addition, the implementation of employee stock ownership plans may also face some challenges, such as equity allocation, liquidity management, etc., which require rational allocation of resources and handling of interest relationships.

1 Introduction

Employee stock ownership plan（ESOP）refers to an incentive equity arrangement in which employees within a company enjoy share dividends by acquiring a portion of the company's equity[1]. It was developed in the 1960s in the United States. After more than 30 years of policy changes since the 1980s, China introduced the "Interim Measures" in 2012 and officially issued the "Guiding Opinions on the Pilot Implementation of Employee Stock Ownership Plan by Listed Companies" (hereinafter referred to as the "Guiding Opinions") in 2014. China's employee stock ownership system has officially entered a new stage of deepening standardization and comprehensive improvement. Technological enterprises, as important economic entities leading innovation and technological iteration, are key to promoting national industrial transformation and upgrading, as well as high-quality economic development. As a knowledge and technology intensive enterprise, human capital can be regarded as one of its most important resource endowments. In addition to various incentive policies at the national level, most technological enterprises attach great importance to exploring the value of internal employee human capital. The employee stock ownership plan has become an important institutional arrangement for promoting the value of human capital in enterprises due to its advantage of achieving benefit sharing.

2 Characteristics of employee stock ownership plans in Technological enterprises

2.1 High proportion and stable development

As of the end of 2022, a total of 1100 listed companies in China have announced employee stock ownership plans, of which 986 are technological enterprises, accounting for 89.6%. The number of companies that announced their employee stock ownership plans in 2022 reached a six year high.

2.2 Long duration and short lock-in period

The average lock-in period of employee stock ownership plans in technological enterprises is 19 months. The majority of employee stock ownership plans with a lock-in period of 12 months account for 67% of the total number of plans, while those with a lock-in period of 36 months or more account for 23%. The average duration of the plan is about 40 months, with a maximum of 150 months.

* Corresponding author: 724259289@qq.com, a qiuyefengliang1@126.com

2.3Wide employee coverage and low proportion

The employee stock ownership plan has a lot of discretion in scheme design. Compared to equity incentive plans, it covers a wider range of employees, has looser participation conditions, and has more flexible funding sources and management methods[2]. However, the proportion of participants in employee stock ownership plans to the total number of employees in the enterprise is not high overall, with an average participation rate of 12% and a median of only 6%, and has shown a downward trend in recent years.

2.4Diversified sources of stocks and abundant sources of funds

According to the employee stock ownership plan announced by technological enterprises, the main sources of stocks include: secondary market purchases, block trading, non-public offerings, shareholder gifts, repurchase of stocks, targeted transfers, targeted repurchases, negotiated transfers, and other methods permitted by laws and regulations. The employee stock ownership plan of technological enterprises is also very diverse in terms of funding sources. In addition to employee compensation, it also includes self raised funds, financing loans, interest free loans, performance bonuses, rewards for net profit extraction, special funds for the company's stock ownership plan, and other methods permitted by laws and regulations.

2.5Low performance indicators and reduced risk

The performance restrictions of employee stock ownership plans are relatively limited, and the unlocking restrictions are mainly limited to time[3]. The vast majority of plans can be traded according to the plan agreement after the lock-in period ends. Given the flexibility of employee stock ownership plans, when faced with force majeure factors such as internal and external environmental changes, enterprises have a high degree of autonomy to adjust the plan, which can reduce losses, reduce risks, and protect the rights and interests of participants to a certain extent.

3 The process of the impact of employee stock ownership plans on corporate performance in technology enterprises

3.1 Stimulate employee motivation and motivation

A stock ownership plan is a way to motivate employees to invest in their work and pursue excellence. Employee stock ownership plans can motivate employees to invest more energy and time in their work, as they directly benefit from the development and profits of the company. When employees hold equity, they are closely linked to the interests of the company, sharing its growth and profitability. This direct interest correlation will stimulate employees' motivation for innovation, improve production efficiency and quality. Employees will be more proactive in thinking about how to improve workflow, enhance the quality and competitiveness of products or services, thereby promoting the performance improvement of the enterprise. Stock ownership plans allow employees to connect their interests with the long-term development of the company, enhancing their sense of identity and loyalty[4].The reasons why employee stock ownership plans can improve the quality and competitiveness of products or services are as follows:

3.1.1 Benefit sharing

After employees hold equity, they will directly share in the company's profits and growth. This benefit sharing mechanism motivates employees to be more proactive in the development of the enterprise, and they will pay more attention to the quality of products or services to obtain higher profits and returns.

3.1.2Long term perspective

Employee shareholding means that they bear certain risks in the success and failure of the enterprise, so they usually have a longer term perspective. They will pay more attention to the long-term development of the enterprise and strive to improve the quality of products or services, thereby enhancing the competitiveness of the enterprise.

3.1.3Innovation motivation

The employee stock ownership plan has stimulated employees' innovation motivation. Employees have direct economic interests and are more inclined to think about how to improve work processes, improve the quality and competitiveness of products or services. The shareholding plan encourages employees to propose new ideas and innovative solutions, thereby driving continuous innovation in the enterprise.

3.1.4Sense of responsibility and commitment

The employee stock ownership plan closely connects the interests of employees and the company, making them more responsible and committed. Employees are willing to make more efforts for the enterprise and actively participate in team cooperation to improve the quality and competitiveness of products or services.

3.2Enhance employee loyalty and stability

The employee stock ownership plan enhances employees' loyalty and sense of belonging to the company. Employees are more inclined to stay in the company for a long time and grow together with the company, as they receive returns on the company's operating performance through shareholding. The stability and loyalty of employees help establish a stable team, reduce recruitment and training costs, and maintain continuous business operations[5]. This stability and loyalty are crucial for the long-term development of the enterprise.The reasons why employee stock ownership plans enhance employees' loyalty and sense of belonging to the company are as follows:

3.2.1 Interest correlation

After employees hold equity, they will directly benefit from the development and profits of the enterprise. This interest correlation makes employees more concerned about the long-term development of the enterprise and closely linked to the interests of the enterprise. They are willing to make more efforts for the enterprise, promote its success, and thus enhance their loyalty to the enterprise.

3.2.2Identity

The employee stock ownership plan turns employees into partners and shareholders of the enterprise. They are no longer just employees, but a part closely related to the interests of the enterprise. This identity makes employees more proud and belonging to the company, willing to fight for the honor and reputation of the company.

3.2.3Participation in decision-making

Stock ownership plans usually grant employees certain decision-making or participation rights. Employees can participate in the discussion and formulation of important decisions, express their opinions and suggestions. This sense of participation enhances employees' sense of belonging to the enterprise, and they will be more engaged and responsible in the development of the enterprise.

3.2.4Long term stability

Stock ownership plans encourage employees to establish long-term stable relationships with the company. Employees holding equity means they take risks in the success and failure of the enterprise, so they are usually more willing to stay in the enterprise for long-term development. This long-term stability enhances employees' loyalty and sense of belonging to the company.

3.3Promote cooperation and sharing among employees

The employee stock ownership plan promotes cooperation and sharing among employees. Holding equity allows employees to recognize their mutual interests and establish a sense of common goals and teamwork. They are more willing to share knowledge and skills, work together to solve problems, and jointly promote innovation and development of the enterprise[6]. Collaboration and teamwork among employees can help improve work efficiency, strengthen internal communication and coordination, and thus improve the performance of the enterprise.

3.3.1 Incentive mechanism

The employee stock ownership plan is an incentive mechanism that can motivate employees to actively share knowledge, experience, and skills. The number of shares held by employees is related to their positions and contributions in the enterprise, so employees will learn and collaborate with each other to improve the performance level of the entire team.

3.3.2Collaborative innovation

Employee stock ownership plans can also encourage collaborative innovation among employees, creating more innovative ideas and solutions. Employees can share their innovative ideas and experiences with other team members, thereby generating more new products or services and improving the competitiveness of the enterprise.

3.4Improving corporate governance and transparency

Employee stock ownership plans can help improve the governance level and transparency of enterprises. Employees, as shareholders, participate in the decision-making process of the enterprise and have direct influence on the development direction and strategic decision-making of the enterprise[7]. This participation can promote corporate governance practices, increase the transparency of decision-making, and improve the quality and efficiency of decision-making. Excellent corporate governance structure and processes contribute to the establishment of a stable enterprise management system and provide guarantees for the long-term development of the enterprise.

3.4.1 Democratic decision-making

Employee stock ownership plans can encourage enterprises to implement democratic decision-making. As shareholders, employees have the right to participate in the process of enterprise management and decision-making. Enterprises can rely on employees' opinions and suggestions to form consensus and make more democratic and scientific decisions. This democratization of decision-making not only enhances the level of corporate governance, but also enhances employees' sense of participation and belonging to the company.

3.4.2 Information Disclosure and Transparency

The Employee Stock Ownership Plan requires companies to engage in more proactive and extensive information disclosure to enhance transparency. Enterprises need to disclose relevant information such as their financial status, business operations, and strategic planning to their employees and shareholders. This kind of information disclosure encourages the management of enterprises to pay more attention to the transparency of information, improve their sense of responsibility and professional level.

4 Countermeasures and suggestions

4.1The government encourages enterprises to implement employee stock ownership plans in multiple aspects

Firstly, the government can provide tax incentives for employee stock ownership plans through tax policies to encourage companies and employees to participate.
Secondly, the government can establish specialized funds to provide financial support to enterprises and help them implement employee stock ownership plans.

Thirdly, the government can improve relevant laws and regulations, clarify regulatory provisions and operational procedures for employee stock ownership plans, and provide legal protection for enterprises.

Fourthly, the government can cooperate with relevant institutions to provide training and guidance on employee stock ownership plans for enterprises, helping them understand the advantages, operational methods, and management skills of employee stock ownership plans.

Fifthly, the government can increase the awareness and understanding of employee stock ownership plans among enterprises and employees through promotional activities, and promote joint participation and support between enterprises and employees.Through the encouragement and support of the above measures, more technological enterprises can be promoted to implement employee stock ownership plans, thereby enhancing the sharing of benefits between enterprises and employees, and improving enterprise performance and competitiveness.

4.2Enterprises actively motivate employees

Firstly, encourage enterprises to support employee stock ownership plans through internal accumulation or self owned funds. Enterprises can establish specific savings funds or profit retention plans, and use a certain proportion of profits to purchase employee shares.

Secondly, appropriately expand the scale. Ensure that the scale of the employee stock ownership plan is appropriately expanded to allow more employees to participate. Enterprises can establish clear participation standards based on factors such as employee contributions and service years, and encourage more employees to participate in shareholding plans. The third is a high level of participation. Encourage employees to actively participate and increase their shareholding ratio. By setting up a mechanism for purchasing shares in stages, corresponding discounts can be given based on employee participation time and performance.

Thirdly,companies can provide more incentive measures, such as equity incentives or dividend mechanisms, to encourage employees to hold and increase shares. The fourth is to provide information transparency. Ensure that employees have a clear understanding of the company's operational and financial situation. Establish effective communication channels, regularly disclose the company's business situation, financial reports, and progress of shareholding plans to employees, and enhance employees' confidence and participation in shareholding plans. The fifth is incentive and training plans. Provide incentive and training plans to help employees understand the incentive mechanism and long-term benefits of employee stock ownership plans.

4.3Improving the Employee Stock Ownership System in Enterprises

To improve the employee stock ownership system, enterprises can take the following measures:

Firstly, Set clear goals and policies: Enterprises need to clarify the goals and policies of the employee shareholding system, including employee shareholding ratio, distribution method, equity incentive plan, etc. These goals and policies should be closely aligned with the long-term strategy and development plan of the enterprise, ensuring that the employee stock ownership system can effectively promote the development of the enterprise and employee motivation.

Secondly, Fair and reasonable incentive mechanism: Establish a fair and reasonable employee shareholding incentive mechanism, considering factors such as employee contributions, positions, and performance to determine the equity ratio and distribution method. The incentive mechanism should have operability and sustainability, which can not only motivate employees' enthusiasm, but also meet the interests and long-term development needs of the enterprise.

Thirdly, Transparent and open information disclosure: Strengthen information disclosure to ensure the transparency of the employee shareholding system. Timely disclose important information such as the financial status, business operations, and development plans of the company to employees, enabling them to understand and participate in the decision-making process of the company. At the same time, encourage communication and exchange with employees to answer their questions and concerns about the employee shareholding system.

Fourthly, Sound governance mechanism: Establish a sound corporate governance mechanism, including organizational forms such as shareholder meetings and board of directors. Ensure that the implementation of the employee stock ownership system complies with the requirements of laws, regulations, and the company's articles of association, and follows the principles of fairness, fairness, and transparency. Strengthen supervision and constraints on management to prevent abuse of power or conflicts of interest.

5 Conclusions

The employee stock ownership system encourages employees to be more actively engaged in work, improving production efficiency and innovation ability. At the same time, employees becoming one of the shareholders of the enterprise will provide more effective supervision and management, promoting the healthy development of the enterprise. The employee stock ownership system promotes the construction of corporate culture and forms the concept of "common development and common honor". Employees will pay more attention to the cultural atmosphere and values of the company, actively participate in the construction and promotion of corporate culture. The win-win significance of the employee stock ownership system is not only reflected in the development and profitability of the enterprise, but also in improving employee income, corporate governance level, corporate culture, and other aspects, building a close interest relationship between the enterprise and employees, and achieving the common development of the enterprise, employees, and society.Summarizing the research results, this article proposes practical suggestions for the employee stock ownership incentive mechanism in technological enterprises. Based on the research findings, provide reference for enterprise decision-makers and guide technological enterprises on how to reasonably design and implement employee stock ownership plans to improve corporate performance.

Project source

2018 Heilongjiang Provincial Universities Basic Scientific Research Business Expenses Scientific Research Project, No.2018-KYYWF-014; 2023 Heilongjiang Provincial Universities Basic Scientific Research Business Expenses Scientific Research Project, No.2023-KYYWF-E009

References

1. Dai Cheng. Research on Employee Stock Ownership in State owned Science and Technology Enterprises - Taking Employee Stock Ownership in a State owned Science and Technology Enterprise as an Example [J]. Enterprise Reform and Management, 2021 (17): 67-68
2. Qin Yuanjian, Yang Jie. Employee Stock Ownership and Enterprise Innovation in Listed Companies [J]. Monthly Journal of Finance and Accounting, 2020 (16): 96-104
3. Zhang Jinfa. Brief Discussion on Employee Stock Ownership of Agricultural Research Institutes and Technology Enterprises [J]. Xinjiang Agricultural Reclamation Technology, 2017,40 (08): 64-66
4. Xie Linfang. Exploration and Practice of Employee Stock Ownership in State Owned Enterprises: Taking Sichuan Furong Technology Co., Ltd. as an Example [J]. Fujian Metallurgy, 2021,50 (04): 56-58
5. Le P L, Nguyen H T T. EFFECTS OF EMPLOYEE STOCK OWNERSHIP PLANS ON FIRM PERFORMANCE–EVIDENCE FROM LISTED COMMERCIAL BANKS OF VIETNAM[J]. 2023.
6. Mwenda B, Ngollo M I. Do Employee Stock Ownership Plans Affect the Quest for Firm Profitability? Evidence from Tanzania[J]. Jurnal Manajemen Teori dan Terapan, 2023, 16(1).
7. Kruse D, Blasi J, Weltmann D, et al. Do employee share owners face too much financial risk?[J]. ILR Review, 2022, 75(3): 716-740.

The Relationship between Innovation and Strategy

Zhuolun Liu*

University of St Andrews, St Andrews KY16 9AJ, United Kingdom

Abstract. The study emphasizes the significance of design strategy as a means to accomplish the innovation process, wherein companies utilize design thinking to develop innovation strategies encompassing various dimensions such as business, organizational, marketing, and technological innovation. Additionally, the study discusses disruptive innovation strategies and their impact on industries and markets. The study concludes that while innovation can result from Strategy, the absence of Strategy and innovation can lead to business failure. Timely adjustment of strategies and alignment with focus areas and available resources lay the foundation for successful innovation.

In summary, this research highlights the causal relationship between Strategy and innovation, emphasizing the importance of design strategy and disruptive innovation in driving organizational competitiveness and market success. Integrating innovation into Strategy and embracing design thinking as a problem-solving approach is crucial for companies to thrive in today's dynamic business environment.

1 Introduction

The design of strategic approaches plays a crucial role in fostering innovation within organizations. Strategic approaches provide a roadmap for organizations to identify and capitalize on opportunities for innovation. By setting clear objectives, allocating resources, and aligning efforts towards a common vision, organizations can create an environment conducive to innovation.

Disruptive innovation strategies also play a significant role in the relationship between Strategy and innovation. Disruptive innovation refers to the introduction of new products, services, or business models that disrupt existing markets and create new ones. Organizations that adopt disruptive innovation strategies actively seek out opportunities to challenge the status quo and create breakthrough innovations[1]. These strategies often involve a reevaluation of traditional strategic frameworks and a willingness to take risks.

However, it is important to acknowledge that innovation can also occur without a deliberate strategic framework. There are instances where serendipitous discoveries or spontaneous creative ideas lead to significant innovations. These "accidental" innovations may not have been a part of a formal strategic plan but can still drive organizational growth and competitive advantage[2]. This suggests that innovation can emerge organically within organizations, even in the absence of a predefined strategic framework.

In conclusion, the relationship between Strategy and innovation is multifaceted. While strategic approaches provide a foundation for organizations to foster innovation, disruptive innovation strategies and accidental innovations demonstrate that innovation can also occur independently of a formal strategic plan. Organizations that recognize the importance of both proactive strategic planning and cultivating an environment that encourages creativity and risk-taking are more likely to sustain their competitive advantages in today's rapidly evolving market environment.

2 Innovation and Strategy

First, we consider innovation based on design strategy. More and more companies use design strategy as a critical driver of their innovation practices to gain a competitive advantage [7]. So, no matter how the industry changes, it is possible to be more competitive in design strategy. Design strategy is not design but Strategy, i.e., how to use the concept of design and design innovation to develop an innovation strategy to achieve growth and transformation of the company; this involves not only business innovation but also organizational, marketing, and technological innovation. At the same time, Strategy itself is a dynamic balancing and iterative process, and innovation needs to be integrated with the company's situation and strategic development to form a good strategy.

To accomplish design strategy, one must first have design thinking. Many scholars have published articles on the nature of design thinking and how to implement the design thinking process. Owen [1] defines design thinking as solving real problems for specific groups of people by

*lil.zhuolun@gmail.com

creating meaningful ideas and thoughts. Design thinking can help people from all professional backgrounds to solve problems through innovative solutions. Steve Jobs, the founder of Apple, once made a classic statement that people think design is all about appearance and how a product looks, but that is only part of it. Design is how it works. This statement speaks to the essence of design thinking.

Secondly, in extreme markets, an extreme strategy emerges, the disruptive innovation strategy. Disruptive innovation is a fierce strategic tool that, while a powerful means of developing or broadening markets, can destroy existing market relationships while exposing itself to significant risk [8]. The theory was developed by the famous innovation scholar Clayton Christensen, who, in his article, refers to disruptive innovation leading to the replacement of companies, products, and even entire industries. Moreover, it implies that the balance of something will be upset [4].

Understanding the reasons for increasing the speed of disruption helps predict the outcome but does not change how disruption is approached. A fast disruption process is not fundamentally different from any other disruption process. They follow the exact causal relationship and, in principle, do not require different responses [3].

Design strategy is an essential strategic way of accomplishing the innovation process. According to the study of Mintzberg's strategy theory, one of the characteristics of design strategy is innovation [6]. To maintain design as a core competency, companies must use effective product design strategies to maximize the various design resources to respond to changes in the broader environment. So in the Strategy, companies can use design thinking for innovation purposes. Herbert A. Simon introduced the concept of design thinking, an innovative human-centered approach to solving complex problems that use the designer's understanding and approach to match technical feasibility, business strategy, and user needs, which translates into customer value and market opportunities. As a way of thinking, it is generally regarded as being able to understand the context in which problems arise, to generate insight and solutions, and to analyze and identify the most appropriate solutions rationally [5].

The best example of this is Apple, one of the leading companies in the field of innovation, which is closely related to its adoption of design thinking. Steve Jobs' flexible use of design thinking reflects his vision for Apple products, such as meeting the masses' needs and desires and building simple and user-friendly products. Hence in modern Apple products, we can see the above vision. While other competitors at this stage focus on the features and performance of their products, Apple focuses on the overall user experience. The iMac, for example, is known for its quiet, fast boot-up, and high-quality display. At the same time, Apple's platform strategy has dramatically increased customer loyalty.

Apple's history in innovation provides us with a clear idea of design thinking as a way of thinking that can firmly bind Strategy and innovation together, so here we can conclude that organizations can innovate and become more competitive through Strategy, i.e., innovation can be a result of Strategy. Design and innovation can turn a company's failures into market successes and lead the way in a highly competitive market. Design thinking helps Apple to innovate while putting the consumer at the heart of the innovation process. On the other hand, innovation with design thinking as a strategy certainly helps to build a successful business.

Companies can innovate through Strategy and innovate because using an innovative strategy, which is as numerous as a weed, is undoubtedly an asset for a company's success. However, due to its difficulty and high threshold limitations, this innovation strategy requires substantial financial and technological power and a certain amount of luck. A representative example of this kind of disruptive innovation is Tesla. Tesla's disruptive nature is reflected in its industry disruption, with its battery technology, software, disruptive experience, and self-driving capabilities being solid competencies. Tesla's disruptive nature as a start-up is also reflected in its cross-border partnerships, as there are clear internet traces in Tesla. Its use of Internet thinking is undoubtedly a dimensionality reduction strike for the traditional car manufacturing industry. This strategic model brings a premium for the product and strategic success in the commercial sector.

Therefore, the study concludes that a disruptive innovation strategy's success leads to a company's success and its ability to become a leader in a new industry. Using innovative strategies can take a company up a notch in its nature. However, this model of innovation is not recommended because there is only one Tesla and only one Apple in the world, and disruptive innovation strategies exist only for extreme innovation phenomena occurring in extreme companies in extreme markets.

While using Strategy to innovate will not necessarily make a business successful, not using Strategy to innovate will not make a business successful, and not innovating will also make a successful business obsolete. The bankruptcy of Kodak is an excellent example of this. Once the world's largest manufacturer and supplier of imaging products and related services, Kodak was synonymous with cameras and images, employing nearly 150,000 people worldwide at its peak and operating in over 150 countries worldwide (Jonathan & Hongyi, 2017). However, Kodak, the 'king of the film' in the 20th century, has been tragically eliminated by the rapid development of digital technology in recent years.

In January 2012, the company, which had been in business for over 100 years, was declared bankrupt. Why did Kodak go bankrupt quickly just a few years after entering the new century? The root cause of Kodak's bankruptcy was its refusal to innovate. Kodak has long been content with its market share and monopoly position in traditional film products and needs to refocus the company's business strategy promptly; This has led to a situation where the company's resources would be allocated to its traditional business units as a priority, resulting in a slow transition to its emerging markets. When digital cameras appeared, Kodak made no effort to keep up with the trends but instead assumed an arrogant posture towards market demand and refused to innovate with technology. It was inevitable that they would be eliminated from the market.

The case of Kodak serves as a sobering reminder of the importance of innovation in today's fast-paced and ever-changing business landscape. It highlights the need for organizations to continuously adapt their strategies and embrace innovation to stay relevant and ahead of the competition. Companies that fail to recognize the significance of innovation risk losing their competitive edge and, ultimately, their market share.

Innovation and Strategy are not mutually exclusive but rather interdependent concepts. While innovation can be the result of a well-thought-out and implemented Strategy, it can also be the driving force behind the formulation of Strategy. Organizations must recognize the symbiotic relationship between the two and leverage it to their advantage.

Successful companies like Apple and Tesla have demonstrated the power of combining Strategy with innovation. Apple's adoption of design thinking as a way of thinking has allowed them to create user-centric products that revolutionized the tech industry. By focusing on the overall user experience rather than just product features, Apple has been able to cultivate customer loyalty and maintain a competitive edge. Similarly, Tesla's disruptive innovation strategy, with its emphasis on battery technology and self-driving capabilities, has positioned the company as a leader in the electric car market.

However, it is important to note that while disruptive innovation strategies can yield remarkable results, they are not suitable for every company or industry. The success of Tesla's disruptive approach does not guarantee the same outcome for other organizations. Disruptive innovation requires significant financial and technological resources, as well as a certain amount of luck. Companies should carefully assess their capabilities and market dynamics before embarking on such strategies.

On the other hand, organizations that fail to innovate are at risk of becoming obsolete. The pace of technological advancement and changing customer preferences demand constant adaptation and innovation. Companies that refuse to embrace innovation and rely solely on their existing market share are vulnerable to disruption from competitors who are willing to push boundaries and meet evolving customer needs.

In conclusion, Strategy and innovation are intricately tied to each other. Innovation can result from Strategy, and Strategy can be enhanced through innovation. Organizations must recognize the importance of innovation in maintaining and enhancing their competitiveness. Design thinking and disruptive innovation strategies are powerful tools that can drive innovation and business success. However, companies must carefully evaluate their capabilities and market dynamics before embarking on disruptive strategies. Failure to innovate can have dire consequences, as illustrated by the downfall of Kodak. To stay ahead in today's fast-paced business environment, organizations must prioritize innovation and continuously adapt their strategies to align with market demands.

3 The Importance of Continuous Learning and Adaptation

This study further delves into the different aspects of the relationship between Strategy and innovation. One aspect to consider is the role of strategic intent in driving innovation. Strategic intent refers to the long-term vision and goals of an organization. When Strategy is aligned with a clear intent for innovation, it can provide a framework for guiding and prioritizing innovative efforts. By integrating innovation into the core strategic objectives, organizations can foster a culture of innovation and encourage employees to think creatively and explore new ideas.

Another dimension to explore is the influence of external factors on the relationship between Strategy and innovation. In today's dynamic market environment, organizations are exposed to various external forces such as technological advancements, changing consumer preferences, and competitive pressures. These external factors can shape the strategic choices and innovation priorities of organizations. Effective strategic planning involves continuously monitoring and analyzing these external factors to identify emerging opportunities and threats that may impact innovation efforts. By understanding the external landscape, organizations can proactively adapt their strategies to capitalize on opportunities for innovation.

Furthermore, it is crucial to examine the extent to which innovation drives strategy or vice versa. While Strategy often provides the overarching direction for an organization's innovation efforts, innovation itself can also shape and influence Strategy. Breakthrough innovations can disrupt existing markets and necessitate strategic shifts. Organizations that are responsive to the potential impact of innovative technologies or business models can adjust their strategies to leverage these disruptions and gain a competitive edge.

Ultimately, the relationship between Strategy and innovation is complex. Organizations must strike a balance between planned strategic approaches and the ability to embrace creative and spontaneous innovations. By fostering a strategic mindset while also encouraging a culture of experimentation and idea generation, organizations can position themselves for sustained success in a rapidly evolving market environment.

Another aspect to consider in the relationship between Strategy and innovation is the role of experimentation and risk-taking. Innovation often involves venturing into uncharted territory and trying new approaches that may carry certain risks. Organizations that are willing to take calculated risks and encourage experimentation are more likely to foster a culture of innovation.

Taking risks and experimenting with new ideas can lead to breakthrough innovations and transformative outcomes. By encouraging employees to think outside the box, challenge conventional wisdom, and explore unconventional solutions, organizations can uncover new possibilities and drive innovation. This requires creating an environment where failure is seen as a learning opportunity, where employees are encouraged to learn from their mistakes and iterate on their ideas.

One company that exemplifies the importance of experimentation and risk-taking in driving innovation is Amazon. Known for its "fail fast" culture, Amazon encourages its employees to experiment with new ideas and initiatives, knowing that not all will succeed. This culture of experimentation has allowed Amazon to continually explore new business models, develop innovative products and services, and stay at the forefront of technological advancements. Through bold experimentation and calculated risks, Amazon has transformed from an online bookstore into a global powerhouse in e-commerce, cloud computing, and other industries.

In addition to experimentation and risk-taking, adaptability is essential for continuous learning and adaptation in the face of changing circumstances. Organizations must be willing to embrace change, pivot when necessary, and adjust their strategies based on market feedback and evolving customer needs. This requires a willingness to let go of outdated practices, challenge the status quo, and adapt to new realities.

Flexibility and adaptability are especially critical in today's rapidly evolving digital landscape. Technology advancements, shifting consumer expectations, and global disruptions can quickly render traditional business models obsolete. Organizations that are agile and nimble in adapting to these changes can position themselves for long-term success.

Another aspect to consider is the importance of nurturing a diverse and inclusive workforce. Studies have shown that diverse teams are more creative and innovative, as they bring a wide range of perspectives, experiences, and ideas to the table. By fostering a culture that values diversity and inclusion, organizations can unlock the full potential of their employees and foster an environment that encourages collaboration, creativity, and innovative thinking.

Furthermore, organizations should not underestimate the power of collaboration and knowledge-sharing. By fostering an environment where employees are encouraged to collaborate, share ideas, and work together towards common goals, organizations can harness the collective intelligence of their workforce. This can lead to the generation of innovative ideas and the development of groundbreaking solutions.

In conclusion, continuous learning and adaptation require organizations to embrace experimentation, take risks, be adaptable, foster diversity and inclusion, and promote collaboration and knowledge-sharing. By creating an environment that encourages these behaviors, organizations can drive innovation, stay competitive in the market, and position themselves for long-term success. The relationship between Strategy and innovation is dynamic, and organizations must continuously assess and adjust their strategies to align with evolving market trends and customer needs. By embracing continuous learning and adaptation, organizations can future-proof themselves and seize new opportunities for growth and innovation.

4 Conclusion

Innovation and strategy are intricately linked, with each influencing and shaping the other. While innovation can result from strategy, it is equally essential for organizations to adopt a strategic approach to guide and drive their innovation efforts.

Design thinking, disruptive innovation strategies, and continuous improvement are powerful tools that organizations can use to achieve innovation and gain a competitive advantage. However, it is essential for companies to carefully assess their capabilities, resources, and market dynamics before embarking on disruptive strategies.

Effective strategic planning ensures that innovation aligns with the organization's goals and objectives. It guides resource allocation, provides a clear vision for innovation, and ensures that innovation efforts contribute to the company's long-term success.

Successful companies like Apple, Tesla, and Amazon have demonstrated the power of combining strategy and innovation. By leveraging design thinking, disruptive innovation, and continuous improvement, these organizations have achieved remarkable success and transformed their respective industries.

In today's rapidly evolving market environment, organizations must recognize the critical role of innovation in sustaining their competitive advantages. Embracing a strategic approach to innovation can position companies for long-term success and ensure their relevance in an ever-changing business landscape.

Reference

1. Charles Owen (2007). Design Research Quarterly: Design Thinking: Notes on its Nature and Use. Vol.2, No.1, pp.16-27
2. Chris Carter, Stewart R Clegg, Martin Kornberger (2008). A Very Short, Fairly Interesting and Reasonably Cheap Book About Studying Strategy. 1st edition, London: SAGE Publications Ltd. ISBN 978-1-4129-4786
3. Clayton M. Christensen, Michael Raynor, & Rory McDonald (2015). What Is Disruptive Innovation? Brighton: Harvard Business Review.
4. Clayton M. Christensen, Rory McDonald, Elizabeth J. Altman, Jonathan E. Palmer (2018). Journal of Management Studies: Disruptive Innovation: An Intellectual History and Directions for Future Research. Volume.55, Issue.7, pp.1043 – 1078 https://doi.org/10.1111/joms.12349
5. Herbert A. Simon (1979) The American Economic Review: Rational Decision Making in Business Organizations. Vol. 69, No. 4, pp. 493-513 https://www.jstor.org/stable/1808698
6. Jonathan C. Ho, Hongyi Chen (2018). Review of Policy Research: Managing the Disruptive and Sustaining the Disrupted: The Case of Kodak and Fujifilm in the Face of Digital Disruption. Volume35, Issue3, pp 352-371 https://doi.org/10.1111/ropr.12278
7. Marvin Berkowitz (1987). Journal of Product Innovation Management: Product shape as a design innovation strategy, Volume 4, Issue 4, pp.274–283. https://doi.org/10.1016/0737- 6782(87)90031-2
8. Yu Dan and Hang Chang Chieh (2008). A thoughtful review of disruptive innovation theory, PICMET '08 - 2008 Portland International Conference on Management of Engineering & Technology, pp. 402-414, DOI: 10.1109/PICMET.2008.4599648.

Research on Evaluation of Tourism Live Marketing Effect

Hao Dong[1], Yanjun Liu[1], linfeng Dong[2*]

[1]Griffith University, Master of International Tourism and Hospitality Management Gold Coast, Queensland, Australia
[2]Professor, Postdoctoral fellow, Hainan University, Haikou, Hainan Province, China

Abstract:With the development of Internet and tourism industry, live tourism marketing has gradually become an important way of tourism marketing. Among them, tourism live marketing belongs to one of the big categories of live marketing. This paper first determines the evaluation index through the analysis of the influencing factors of live marketing, and further determines the final evaluation index through the use of Delphi method and network research. Then the weights of the evaluation index system are determined by AHP. Finally, the rationality of this evaluation index system is verified through the calculation and comparison of specific cases and evaluation index system.

1. Introduction

Based on the background of the Internet era, tourism live streaming has become popular. Live tourism marketing refers to a marketing model that diversifying tourist destinations or tourism products, presenting them in an all-round way by planning live broadcast of different live broadcast themes, or conveying personal experience and emotion after the host experiences the tourism projects or tourism products. Compared with live TV and network broadcast, such methods have lower cost, stronger sense of presence, and stronger interaction with users. Let the traditional "one to one" mode into "one to many" service mode, so as to improve service efficiency. And the definition of network broadcast marketing is to use the network broadcast platform to achieve remote in-depth communication and interaction with users, so as to improve the visibility, reputation and social recognition of the enterprise, and achieve the goal of directly or indirectly promoting the company(Wang, 2017)[1].

At present, research on tourism live broadcasting mainly focuses on theoretical research on the theory of tourism live broadcasting, research on the interaction and presence of tourism live broadcasting on tourism consumers' willingness to participate and marketing effects as well as part of the research from the perspective of live broadcast marketing. For example; Shalini et al. (2020) started from tourists' consumption views, and proposed that quality, price, social status, preference and information value determine their attitudes towards online travel platforms[2]. DeVos J (2020) points out that people's travel needs cannot be met due to the COVID-19 pandemic, and walking, cycling and online travel have become important ways for people to maintain mental happiness [3]. Ji et al. (2020) conducted research on social media platforms such as Twitter and YouTube, and proposed a new computer system providing data support for decision-making and marketing[4].

Based on the analysis of the influencing factors of tourism live broadcast, this article uses research to initially determine the rating index of tourism live broadcast marketing effect. On this basis, taking tourists as the second research object, determine the final evaluation index, and calculate the weight of each index based on the AHP analytic hierarchy process, and establish the evaluation system.

2. Analysis of Influencing Factors of Tourism Live Marketing

Tourism live broadcast marketing can be divided into two categories: (1) "Live broadcast + e-commerce". (2) "Live broadcast + content"(Chen et al., 2021)[5]. The article based on the analysis of the factors influencing the marketing effectiveness of tourism live streaming, preliminary evaluation indicators for the marketing effectiveness of tourism live streaming have been selected. There are a total of 15, including: spread index, user preference, recommendation index, brand awareness, brand trust, interactivity, professional, fan conversion rate, impact index, live streaming frequency, number of fans, live room page views, number of online viewers, live broadcast duration, single popularity peak.

3. Screening of evaluation indicators

In the previous part of this article, 15 evaluation indicators for the marketing effect of tourism live streaming have been summarized. In order to conduct an in-depth study on the evaluation of the marketing effect of tourism live broadcast, this paper uses the expert interview method to further test and modify the

*Corresponding author: dlfphd@163.com

evaluation indicators. Based on this, the author has interviewed 30 experts. According to the valuable opinions put forward by experts, the evaluation indicators of tourism live broadcast marketing effects are screened and supplemented. Later, through online research, it was found that some indicators needed to be deleted. Respondents generally believe that its indicators are "not important". Therefore, a secondary censorship is performed on the indicators. Determine the final tourism live broadcast marketing effect evaluation index including 9 items. There are spread index, recommendation index, brand awareness, brand trust, interactivity, fan conversion rate, live room page views, number of online viewers, sales index.

4. Construction of evaluation index system for tourism live broadcast marketing effect

(1) Selection of research samples and design of questionnaire

1.Selection of research samples

This survey takes tourists from Qilou Old Street, Nanmen Snack Street, and Baishamen in Haikou City, Hainan Province, China, as the research objects, and distributes as well as fills in online questionnaires. The three places will have nearly 15,000 tourists from March 5 to 10, 2023. So the total sample size of this questionnaire is 15,000. With a confidence level of 90%, a maximum error of 5% is allowed. According to the statistical simple random sampling calculation formula $n_1=(Z^2*S^2)/d^2$, $n_2=(n_1*N)/n_1+N$ (Note: n_1 is the initial sample size; Z is the statistic under the confidence level; S is the overall Standard deviation, take 0.5; d is the maximum allowable error; n_2 is the minimum effective sample size; N is the target total). Calculate the minimum effective sample size for a total sample size of 15,000. The minimum effective sample size at the beginning of the calculation is 264. During the investigation process, a total of 300 questionnaires were issued, and 300 were returned, with a return rate of 100%. After removing the invalid questionnaires, there were 285 valid questionnaires, and the effective rate of the questionnaires was 95%. If the number of valid questionnaires recovered in this survey meets the above-mentioned minimum valid sample size requirements, the survey data are valid and reliable.

2.Questionnaire design

First, the use of the AHP method requires the construction of a judgment matrix. Therefore, in this study, the DELPH method is used to construct the judgment matrix. Secondly, the judgment matrix is not directly given in the questionnaire, but the Likert scale transformed into the attitude scale method is used. According to the content of the research, the attitude scale in the questionnaire needs to be converted into an importance scale. The five options of the scale are very unimportant, unimportant, general, important, and very important. Finally, explain each indicator in detail to avoid the influence of the respondents on the weight judgment due to the ambiguity of the meaning of the indicators, thereby affecting the accuracy of the research results.

(2)Trust level analysis

The method of testing reliability in this paper is to use cronbach α, When Cronbach $\alpha<0.3$, $0.3\leq$Cronbach $\alpha<0.4$, $0.4 \leq$ Cronbach $\alpha\leq 0.5$, $0.5\leq$Cronbach≤0.7, $0.7\leq$Cronbach$\alpha<0.9$, $0.9 \leq$Cronbachα, respectively corresponding to not credible, barely believable, believable, very believable (most common), very believable (less common), vital believable.

Through the analysis of the data, the corresponding cronbach α value is obtained, and the α reliability coefficient of each index is greater than 0.6, which proves that the consistency between each index and the corresponding topic is good, and the reliability of the scale is good , the recovery data is relatively reliable.

(3)The weight determination and analysis of the index system.

When using the AHP method to construct the judgment matrix, the author draws on the 1-9 scale method to reduce the 9-level scale to 5-level scale, so as to reduce the data errors caused by the subjective will of the respondents when filling out the questionnaire. Sum the values of the questionnaire survey matrix and take the arithmetic average to obtain the judgment matrix. Follow the steps above. Using SPSS to perform matrix operations, the results of AHP hierarchical analysis are shown in the table 1.

Table 1. Analyze the data results table.

Index	Eigenvalues	Weights	Largest eigenvalue λmax
Spread index	1.067	11.842%	
Recommendation index	1.059	11.747%	
Fan conversion rate	1.001	11.103%	
Interactivity	0.994	11.025%	
Number of online viewers	1.056	11.717%	9.064
Live room page views	1.023	11.352%	

Sales	0.940	10.436%
Brand awareness	0.940	10.322%
Brand trust	0.942	10.456%

In Analytic Hierarchy Process. If CI=0, it means that the judgment matrix has complete consistency. Its calculation formula is CI=(λmax-n)/(n-1)， n=9， According to the table 1, the maximum eigenvalue λ max=9.064，CI=0.008. Calculate the CR value according to the CI value and RI value to check whether the judgment matrix meets the consistency test. The formula is CR=CI/RI.

When CR ≤ 0.1, it indicates that the judgment matrix has consistency. When n=9, RI=1.46, calculated CR=0.05<0.1. Therefore, it is proven that the consistency test of the judgment matrix and the calculated weights have consistency.

Determination of indicator weight value:spread index, recommendation index, brand awareness, brand trust, interactivity, fan conversion rate, live room page views, number of online viewers, sales index. These nine items account for 11.84%, 11.75%, 11.10%, 11.72%, 11.03%, 11.35%, 10.44%, 10.32%, 10.46% respectively.

5. Establishment of evaluation index system for tourism live broadcast marketing effect

(1)Index value representation

The interactivity of travel anchors is represented by the number of likes. The spread index, recommendation index, fan conversion rate, number of online viewers , live room page views, and sales data are all from Xindou(https://xd.newrank.cn/home/search).Brand trust and brand awareness are represented by the "Alexa ranking" and "web page PR value" in the comprehensive tourism website ranking list (http://top.chinaz.com/list.aspx) of the website ranking network. Among them, the higher the PR value, the higher the brand awareness of the website. Alexa ranking refers to the world ranking of the website, the higher the ranking, the stronger the authority of the website, and the user's higher trust in them.

(2) Normalization of indicators

Since the magnitude and dimension of each indicator above are different, it is necessary to normalize the indicators. The specific method is $X_i=(X_i-X_{min})/(X_{max}-X_{min})$When $X_i=X_{max}$ or X_{min}, take the decimal part expressed in scientific notation, n=the power of 10.

(3)Establish system

The weighted summation method is used to establish the evaluation system. Multiply the normalized value of the index with the corresponding index weight(Y), and finally sum to obtain the final marketing effect M.The formula is M=\sum^{n}_{i=1} (X_i ·Y_i) +(-(X2·Y2))Among them, n=(1,3,4,5,6,7,8,9).

6. Verification of the evaluation system for tourism live broadcast marketing effects

Author uses Douyin as the live broadcast platform and selects the Douyin accounts officially certified by Ctrip and Fliggy as examples. Using the system established above, the theoretical value of the tourism live broadcast marketing effect of the two websites was calculated (a total of 10 live broadcasts). Among them, table 2 is the Ctrip Index normalized data table.

Table 2. Ctrip index normalized data table.

	X_i	X_i'
Spread index	66	0.00026
Recommendation index	6.7	0.00003
Fan conversion rate	24%	0.24000
Interactivity	3616	0.01446
Number of online viewers	24	0.00010
Live room page views	9700	0.03880

Sales	250000	2.50000
Brand awareness	7	0.00003
Brand trust	922	0.00369

M_X=0.00026×11.84%+0.00003×11.75%+0.24000×11.10%+0.01446×11.03%+0.00010×11.72%+0.03880×11.35+2.50000×10.44%+0.00003×10.32%-0.00369×10.46%=0.29

And Fliggy M_F=0.17

Conclusion: The marketing effects of the two tourism live broadcasts are W_X>W_F. That is, the marketing effect of Ctrip's travel live broadcasts is greater than that of Fliggy's travel live broadcasts.

A total of 400 valid questionnaires were filled in and returned through the survey questionnaire, 266 valid web-based questionnaires and 134 valid paper-based questionnaires. Through the research proves that the evaluation index system established above is reasonable and feasible.

7. Conclusion

This paper takes tourism live broadcast as the research object, determines the evaluation indicators through literature review and questionnaire survey, collects data through secondary research, analyzes the data to obtain the weight of each index, and establishes the evaluation index system of tourism live broadcast marketing effect. Finally, the evaluation index system is verified, and the verification proves that the index system is correct.

References

1. Wang Qiong. (2017). Research on the New Tourism Marketing Model of "Tourism+Live Broadcasting" *Journal of Hebei Tourism Vocational College* (02). https://wenku.baidu.com/view/c1d5c06091c69ec3d5bbfd0a79563c1ec5dad79c.html?_wkts_=1691443528508
2. Shalini Talwar, Amandeep Dhir, Puneet Kaur &Matti Mäntymäki. (2020).Why do people purchase from online travel agencies (OTAs)? A consumption values perspective. *International Journal of Hospitality Management*, (88). https://www.researchgate.net/publication/341283298_Why_do_people_purchase_from_online_travel_agencies_OTAs_A_consumption_values_perspective
3. Jonas De Vos. (2020). The effect of COVID-19 and subsequent social distancing on travel behavior. *Transportation Research Interdisciplinary Perspectives*, (05), 100121. https://doi.org/10.1016/j.trip.2020.100121
4. Chen Zhen, Shu Siqi &Xing Xiaoli. (2021). Exploring the Path of Live Tourism Industry Assisting Rural Revitalization Based on the Background of Hainan Free Trade Port. *Modern Economic Information* (24), 158-160. https://wenku.baidu.com/view/6bc5d12c17791711cc7931b765ce050877327540.html?_wkts_=1691443084030&bdQuery=%E5%9F%BA%E4%BA%8E%E6%B5%B7%E5%8D%97%E8%87%AA%E7%94%B1%E8%B4%B8%E6%98%93%E6%B8%AF%E8%83%8C%E6%99%AF%E4%B8%8B%E7%9B%B4%E6%92%AD%E6%97%85%E6%B8%B8%E4%BA%A7%E4%B8%9A%E5%8A%A9%E5%8A%9B%E4%B9%A1%E6%9D%91%E6%8C%AF%E5%85%B4%E8%B7%AF%E5%BE%84%E6%8E%A2%E7%A9%B6
5. Ji Han, Dongmyung Park &Hannah Forbes. (2020). A computational approach for using social networking platforms to support creative idea generation, *Procedia CIRP*, (91), 382-387. https://doi.org/10.1016/j.procir.2020.02.190

Research on Local Characteristic Industry Construction in Characteristic Towns under Public-Private Partnership Development Mode

Shyng, Jyh-Hang *

Department of Landscape Architecture, School of Life Sciences, Zhaoqing University, Zhaoqing, Guangdong, China

Abstract. In recent years, there's been a global trend of emphasizing the role of local characteristics in steering economic growth. To foster the advancement of local characteristic industries in characteristic towns, the public-private partnership (PPP) model is being adopted. This strategy leverages social resources and governmental incentives to drive both urban and rural progress, thus becoming a silver bullet to bolster the growth of characteristic towns. However, achieving "town-industry integration" and "sustainable development" necessitates shared consensus and understanding among all stakeholders. Given the intricate nature of the PPP mechanism, a comprehensive theoretical framework tailored to Chinese local characteristics is essential. This study scrutinizes the integration of the PPP model into characteristic industries within small Chinese towns. It follows a framework encompassing theoretical characteristics, feasibility analysis, existing problems, countermeasures, and recommendations, as well as future trends. The study employs both inductive reasoning and comparison. The research findings serve to establish a theoretical consensus for the PPP model's integration with characteristic industries in these towns, benefiting all involved parties.

1. Introduction

Governed by the state and facilitated by the business entities in the market, the construction of characteristic industries in characteristic towns under the public-private partnership (PPP) mode centres on characteristic industries and emerging industries. During practical implementation, it is imperative to underscore the pertinent theories applicable to these unique industries. Owing to the broad scope of PPP, this study aims to delve into the development of such industries within specific towns and the specific operation mechanism of PPP. It emphatically elaborates on the innovation of the new operation mode of PPP in China and probes into the strategic feasibility, application necessity, existing challenges and drawbacks that must be resolved in the accelerated construction of characteristic towns and villages.

Specifically, the paper follows the framework of theoretical characteristics, strategic feasibility analysis, application necessity, existing problems, countermeasures and suggestions, and future development trends, to investigate the introduction of PPP mode into the development of characteristic industries in small towns with Chinese characteristics.

2.Theoretical Characteristics of Introducing PPP Mode into Local Characteristic Industries

The development of characteristic towns can be sub-categorized into "industry-oriented," "community-oriented," and "tourism-oriented." Hence, we can summarize the theoretical framework of the application of the PPP mode in characteristic industry market promotion to outline the essential theoretical characteristics of introducing the PPP mode into characteristic industries of characteristic towns, thus explaining the key to the successful application of the mode.

Targeting the protruding problems in the development of characteristic industries in characteristic town projects, such as backward policies and regulations, inadequate talent provision, difficult approval of enterprise financing, constrained risk management ability, and poor later business performance, the promotion of PPP mode in the construction of characteristic industries in characteristic towns can fully mobilize the enthusiasm of social enterprises for investment and development, reduce local financial pressure, and improve operational efficiency.

* Corresponding author: hidosing@outlook.com

2.1. Theory and feasibility of introducing PPP mode into characteristic industries.

Constructing characteristic industries in characteristic towns under the PPP mode is a "partnership between government and social capital." It is a mutually beneficial cooperation mode authorized by construction projects and related services, in which the government and the public and private sectors provide various infrastructure, public utilities, natural resources and environmental benefits, etc., relying on the policy advantages provided by the government [12].

This operational mode of public project investment can adopt a variety of forms, including Build-Operate-Transfer (BOT), Build-Own-Operate-Transfer (BOOT), Transfer-Operate-Transfer (TOT) and other ways of combination. Among them, the public sector can help alleviate the social burden brought by public finances, complete project construction, and fulfil government services, while the private sector can share certain risks through financing and obtain a reasonable return [3].

2.2. Advantages of introducing PPP mode into characteristic industries

The introduction of PPP mode into characteristic industries can:

-(1) improve the business financial level of public organizations and staff in the general private sector.

-(2) reduce the economic and financial pressure imposed on governments at all levels.

-(3) encourage innovation and a more flexible way in certain aspects of risk management within the private sector than in the public sector.

-(4) contribute to the sharing of risks.

-(5) boost the overall reform of the socialist market system construction.

-(6) expand the investment field of social capital.

2.3. Application necessity of introducing PPP mode in characteristic industries

Presently, the PPP operation mode is progressively emerging as one of the top two choices for financial guarantees employed by local administrations to establish diverse characteristic towns. The advancement and creation of infrastructures within these towns have notably propelled the growth of the PPP operational model in China, which stimulates the internal vitality of investment in characteristic towns across the country.

-(1) The introduction of the PPP operation mode can further reduce the financial pressure faced by the grassroots government in investment. Since the early investment and construction management in the development of characteristic towns has a high barrier and the investment and construction turnover of capital often takes longer, the essential constraint for the planning and construction of characteristic towns lies in insufficient total financial funds.

-(2) In addition, it is helpful to further reduce the project capital risk brought by the small-scale, scattered construction of characteristic towns.

-(3) Moreover, it can effectively improve the overall construction service quality and efficiency, as well as the operational efficiency of small towns with Chinese characteristics. The market-oriented construction of characteristic town incubators and its close combination with industrial park PPP projects and other innovative modes will become an inevitable trend [11] [15].

2.4. Operation mode of PPP projects in characteristic industries

The selection of PPP operation mode is mainly determined by factors such as PPP project type, financing demand, reconstruction and expansion demand, charging and pricing mechanism, return of investment, the basic framework of risk allocation, and disposal upon expiration.

In China, the naming of operation mode is based on "the control state of ownership/ right to use of public assets." Another international parallel naming convention is based on "the functions transferred by the government to social capital [8]."

PPP concession generally refers to the financing, operation and management and other management work related to projects in small towns with Chinese characteristics under the PPP mode, including products with incidental attributes intrinsically linked to project construction. The relationship network of the concessionaire's development environment is as follows (Fig 1) [13] [14]:

3. Feasibility of PPP Projects in Characteristic Industries of Characteristic Towns

3.1.Obtaining policy support from the central and local governments.

The overall development ideology of characteristic towns should reflect the concepts of "innovation, openness, sharing, environmental protection, and ecological coordination." A platform for the tourism industry should integrate elements including local industries, tourism, cultural activities, and community function [2].

The central department shall offer financial support to various policies. Provincial and municipal governments in areas such as Zhejiang, Guangzhou, and Fujian can generally apply the PPP cooperation mode in a few pilot regions to develop and construct characteristic industries in characteristic towns, directing targeted policy support to the local area.

3.2 Meeting the capital needs of participants by attracting broader social capital.

Given the lack of local government financial support in the early development stage, the construction of characteristic towns needs developers injecting a lot of idle funds. The financing operation mode of PPP projects, due to its stable economic return, can attract a broad range of social capital to intervene. It is impossible, however, to rely merely on government finance for the construction funds of characteristic towns; social forces in social capital should be leveraged extensively. Improving the return on investment is key to attracting social capital. PPP operation financing mode is considered an effective solution to this problem [2].

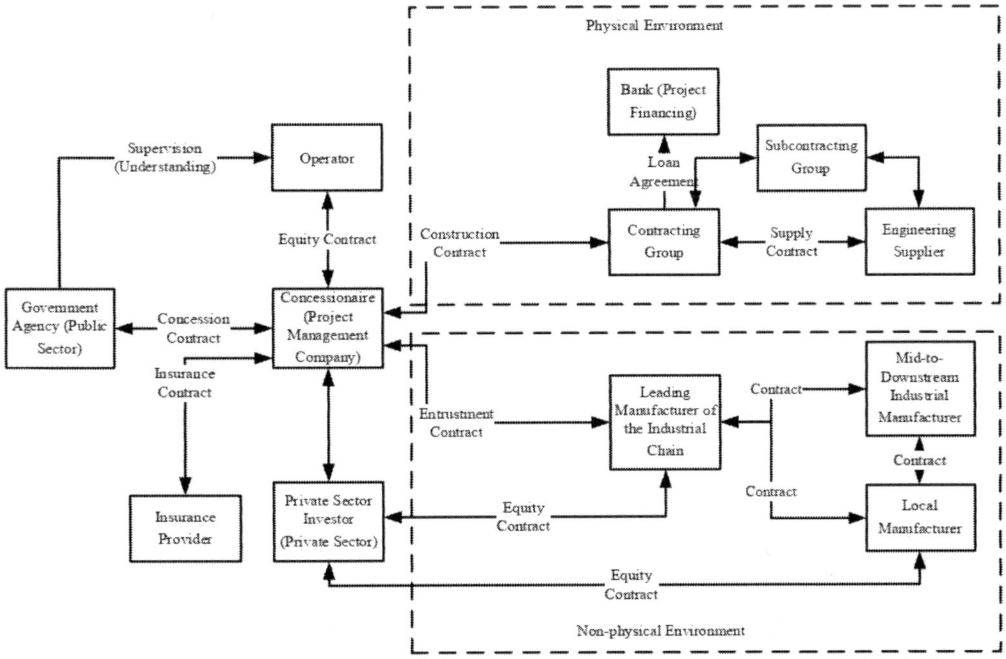

Fig 1: Network Relationship of Concessionaires' Development Environment for Characteristic Industry Growth under the PPP Mode

3.3 Introducing new profit model of PPP projects in local characteristic industries.

The specific profit model is the key to supporting the operation of PPP projects in characteristic industries. The classification of PPP projects based on the profit model of concessionaires is listed as follows (Table 1):

Table 1 Profit Model, Contents, and Facilities of PPP Projects in Local Characteristic Industries

Profit Model	Franchise Business Contents	Facilities
(1) User-pays mode	Usually adopted for infrastructure projects with high operationality and high economic benefit and provide services directly to end users.	Municipal water supply, city pipeline gas, toll roads, etc.
(2) Government-pays mode	Usually adopted for terminal infrastructure projects that do not directly provide services to end users.	Municipal sewage treatment plants, waste incineration power plants, or non-profitable infrastructure projects such as municipal road construction and river regulation.
(3) VGF mode (Viability Gap Funding/Subsidy VGF)	Insufficient part of users' payments by social capital will be subsidized by the government with financial subsidies, equity investment, preferential loans, financing guarantees, and other preferential policies.	Hospitals, schools, cultural and sports venues, social security housing, network-based municipal public projects with reduced price policies or insufficient demand, toll roads with insufficient traffic flow, etc.

3.4 Favourable operating conditions and practical foundation

Although both the adoption of the PPP operation mode in some characteristic towns in China and the path toward future industrialized development are still under exploration and research, we can sum up the experience of other public infrastructure projects and draw insights from the successful experience of foreign PPP projects and solve the application problems based the actual situation of specific domestic projects [2].

4. Problems Existing in PPP Projects of Characteristic Industries in Characteristic Towns

The application of the Chinese-style PPP operation mode is a Chinese-style investment and financing operation practice drawing lessons from the PPP mode within the comprehensive urban planning policy for Chinese characteristic towns. Nevertheless, PPP projects entail substantial investment, long-term concession contracts, and intricate technologies, potentially giving rise to numerous implementation risks that can lead to project setbacks [4].

Risks associated with characteristic industrial projects in the PPP mode encompass the following aspects: the first aspect involves inadequate policy and regulation development, along with imperfect systems. The second aspect encompasses a dearth of skilled personnel, insufficient project management, and limited financial resources. The third aspect entails inefficient bank loans and approvals, leading to uncertain risks. The fourth aspect is relevant to weak risk management capabilities, outdated methods, and one-sided thinking. Lastly, unsatisfactory long-term operational outcomes for characteristic towns also matter [7].

5. Countermeasures and Suggestions for Construction of Characteristic Towns and Characteristic Industries under the PPP Mode

5.1 Suggestions

We should optimize the approval process and regulate the implementation of policies; innovate financing modes to attract social capital to participate in PPP financing in characteristic towns; enhance core competitiveness to attract highly skilled personnel; and evaluate market demand based on local conditions [5].

5.2 Sustainable development countermeasures

The sustainable development countermeasures include the design and implementation of a PPP cooperation mode that can realize the stable development of the local economy, the rise of social living standards, and the improvement of the industrial development and investment environment. The government should provide guidance in the investment and operation of funding projects and the investment and financing cooperation in urban regional economic and social development.

The specific countermeasures include establishing an information disclosure platform, improving the system of laws and regulations, and supplementing the government's decision-making and supervision mechanism [5].

6. Development Trend of Construction of Characteristic Towns and Characteristic Industries under the PPP Mode

In a future society, characteristic towns will promote the reform and innovation of people's lifestyles; an enterprise-led, city-supported, and villager-involved innovative development style will be established. Advantageous private enterprises will be ushered to establish a green product circle that favours a healthy life. The development will be based on expanding investment methods. Hence, the PPP model will offer more cutting-edge technologies for characteristic town development, enhanced management practices, and decreased investment risks [10].

6.1 Propulsive effect of the PPP mode in the development of characteristic towns

-(1) As an important innovation in infrastructure construction and service provision, the PPP mode helps improve the infrastructure construction level and operation ability of characteristic towns.

-(2) PPP mode emphasizes market-oriented operation, which can absorb extensive social resources to invest in the construction of characteristic towns. Marketization is also a major direction for the government to promote the development of characteristic towns.

-(3) The PPP model can alleviate the fiscal expenditure burden. Although characteristic small towns can draw substantial investments, securing such investments is challenging. Moreover, the prolonged operation and maintenance require extensive financing. The PPP mode can pool a large amount of social capital and realize intellectual investment, providing a whole-life-cycle investment scheme design for PPP projects in small towns with Chinese characteristics.

-(4) The PPP mode can encourage intensive and large-scale supply of social funds through supply chain operations. The important feature of PPP projects in characteristic towns is that it contains many sub-projects with relatively complicated boundaries. Through the PPP mode, a large amount of social funds can be integrated into the early planning and construction of small towns, which can resolve the difficulties in the scattered design and operation of small towns with Chinese characteristics [9].

6.2 The direction and trend of characteristic towns under the PPP mode

The future direction and trend of characteristic towns under the PPP mode are:

-(1) clarify the development direction of characteristic industries [6];

-(2) broaden financing channels to meet capital needs;

-(3) realize the organic combination of project and industry.

-(4) create a new characteristic of integrating the PPP mode and intelligence for the construction of small towns with Chinese characteristics.

PPP + BIM intelligent technology can enhance the advancement of various intelligent technologies in project management. Anchored in the project's entire lifecycle, it yields digitized outcomes for decision-making and error avoidance. Furthermore, it carries the capability of reintegration to cater to diverse stakeholders, thereby enhancing the conventional approach and methodology of project management.

-(5) reasonably spread the risks of PPP projects in characteristic towns.

7. Conclusion

Currently, the PPP operational approach is essential for crafting distinctive Chinese towns. Examining its necessity and significance, adeptly employing the PPP model assists in easing financial pressures on local governments, mitigating capital risks tied to construction, and, to some extent, enhancing construction quality and efficiency [1].

Acknowledgment

This paper is one of the phased achievements of the project "Intelligent Management Practice of Promoting Local Characteristic Industry Construction under PPP Mode in Taiwan Province" (FW202208) supported by Zhaoqing University Scientific Research Fund.

References

1. Lai, Y., & Wu, S. (2019). Review of studies on risk early warning of public-private partnership projects in character town. IOP Conference Series. Earth and Environmental Science, 330 (2).
2. Vinod N. Sambrani. (2014). PPP from Asia and African Perspective towards Infrastructure Development: A Case Study of Greenfield Bangalore International Airport, India. Procedia-Social and Behavioural Sciences.
3. Liu Q.D., Cheng P., Xu J. (2019) Research and analysis on the application of PPP mode in the construction of characteristic towns [J]. Engineering Technology Research, 4 (17): 217-218.
4. Liu Q.J. (2017) Discussion on promoting PPP financing mode in characteristic towns [J]. Journal of the Party School of Leshan Municipal Committee of C.P.C, (6): 5.
5. Zhu Y.Y. (2019) Analysis of profit model of construction and operation of characteristic towns under PPP mode [J]. Engineering Economy, 29 (04): 72-76.
6. Xu X. (2019) Discussion on the planning and design of characteristic towns under PPP mode [J]. Engineering Technology Research, 4 (03): 204 +206.
7. Gu Y.Y. (2019) Introduction of PPP mode into the construction of characteristic towns in Jinan [J]. China Business, 000 (003): 49-50.
8. Shyng J.H. (2023) Public infrastructure idleness and public-private partnership (PPP) activation mechanism practice [M], Beijing: Xinhua Publishing House: 10-15
9. Fei J.X., Liu H.P. (2019) Research on characteristic town + PPP mode [J]. Enterprise Reform and Management, 14: 21-212.
10. Nie D.J., Li W., Guo S. (2018) Thoughts and suggestions on boosting the development of characteristic towns under PPP mode [J]. China State Finance, 18:39-40.
11. Guo C., Yan W. (2021) PPP Financing Mode in Characteristic Town Construction—A Case Study of Shenqi Town in Henan Province [J]. Journal of Shangqiu Normal University, 37 (06): 41-44.
12. Gürlebeck, K., & Kiesel, K. (2019). The PPP development model: a tool for improving PPP project success. Journal of Public Procurement, 19(3), 277-303.
13. Woldeamanuel, M.G., Brilman, J., & Zevenbergen, J. A. (2018). The potential of PPP development model to manage infrastructure in the water sector of Ethiopia. Journal of Water Supply: Research and Technology-AQUA, 67(5), 412-422.
14. Jayawardena, S., Ranasinghe, M., & Amaratunga, D. (2018). Understanding the PPP development model in the context of affordable housing in Sri Lanka. Built Environment Project and Asset Management, 8(5), 559-569.
15. Wandeler, C., & Zaugg, E. (2019). PPP development model for urban infrastructure: A case study in Switzerland. Engineering, Construction and Architectural Management, 26(11), 2348-2363.

 https://doi.org/10.1051/shsconf/202317803021

Research on transmission model for technology rewards to enhance enterprises innovation ability: Based on the case of power enterprises.

Jianfei Lu *, Dan Wang, Xin Li, Wei Zhu, and Yan Chang

State Grid Energy Research Institute Co., Ltd, 102209 Beijing, China

Abstract. This study refines the formation of key indicators of science and technology reward evaluation, determines the key evaluation indicators affecting the development of electric power science and technology innovation and the key contents of electric power science and technology development according to the characteristics and features of the electric power industry and the energy Internet enterprises; taking the key indicators as a breakthrough, this study constructs the role transfer model between the influence of the evaluation indicators of scientific and technological rewards on the development of science and technology innovation in the electric power industry and carries out empirical analyses to verify and optimize the influence transfer model.

1 Introduction

Scientific and technological awards have an important incentive effect on individual scientific and technological workers, awarding organizations and the scientific community as a whole. A scientific research project winning a scientific and technological award will promote the enhancement of the enterprise's innovation capacity in several aspects: technology innovation, economy, policy innovation and team cooperation; the enhancement of the enterprise's innovation capacity will further increase the innovation motivation of scientific researchers, increase the innovation resources, and promote a new round of innovation activities. Even if scientific research projects fail to win scientific and technological awards, they can slowly improve the innovation capacity of enterprises in the process of transforming the results; however, the promotion effect is weak because of fewer rewards and the impact of the incentive function[5].

2 Incentive function model of scientific and technological rewards

From a psychological perspective, Scientific and technological rewards can be viewed as psychological incentives that give researchers a sense of satisfaction and contentment[3]. Motivation theory holds that motivation is "a psychological process that constantly stimulates the engine", and that The higher the level of motivation, the greater the effort and satisfaction in achieving the goal, the higher the work efficiency, including Maslow's Hierarchy of Needs Theory, Alder's ERG Theory, Herzberg's Two-Factor Theory, Fromm's Expectation[1] Theory of Work Motivation, and Adams's Equity Theory, among others. The theory of incentives holds that, since human needs are diverse, If we can take appropriate incentives to satisfy human needs and targets, human needs can turn into good work performance. Thus, exploring the incentive mechanism of scientific and technological awards for scientific and technological innovation will help to rationalize the operation of scientific and technological awards and further mobilize the enthusiasm of scientific research workers. With reference to the "Diamond Model", this paper focuses on four factors influencing technology innovation: technology prowess, financial support, policies, and group work[2].

2.1 Diamond Model for Technology Rewards

The national science and technology awards will motivate the award-winning enterprises to strengthen the motivation of technology innovation and stimulate the technology innovation intention [4]. Besides, from the perspective of innovation motivation, the element parts needed for enterprise innovation conclude technology, talents, data and capital, shown as Fig. 1.

* Corresponding author: feixiaruc@163.com

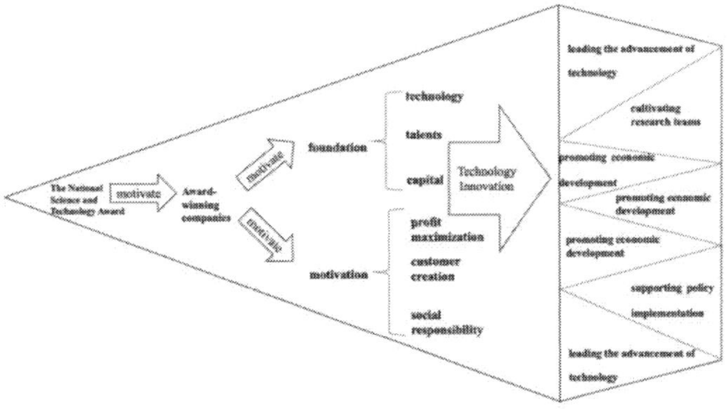

Fig. 1. Diamond Model for Rewards

First, science awards are the society's priority for scientific discovery and recognition of scientific research labour achievements, the national technological awards refer to high-end scientific research results in China. Therefore, after obtaining the recognition of the national technological award, the enterprise will make a leading part in the transformation and development of the industry, bring disruptive innovation to the entire industry, and then stimulate the wave of scientific and technological innovation. Taking power enterprises for instance, technological achievements of winning the National Science and Technology Prize will realise a technological breakthrough or a series of technological breakthroughs, which usually leads to local industrial upgrading in the field of power generation and electricity consumption, or promotes the energy transformation and decarbonization[7]. The realisation of a particular core technology identified by the national technological prize can have a revolutionary impact on an industry or technological field, change the original development mode, bring non-gradual leapfrog development, and finally achieve scientific and technological progress.

Secondly, scientific and technological awards can further help enterprises to support the scientific research work of their research teams, while at the same time recognising the scientific contribution of existing projects. Enterprise science and technology personnel to obtain scientific and technological awards, on the one hand, can enhance the value of self-realisation, to obtain legitimate wealth, so as to continue to shine in the field of scientific research, to expand the influence of the original scientific research, to win more scientific research funds, to help individuals to promote a higher level, and then to achieve some individual development chances. Besides, it can enhance the visibility and academic authority of researchers, thus obtaining more academic resources to help them incubate more innovative talents and provide sufficient reserves for scientific and technological innovation. In addition, for scientific and technological staff who have not received scientific and technological awards in the same enterprise, scientific and technological awards are one of the incentive factors and goals with good attractiveness, which can stimulate their enthusiasm for work to a certain extent and promote the overall improvement of scientific and technological strength[6]. On the whole, from the perspective of enterprises, the issuance of science and technology awards will enable the unit to invest more financial help and personnel support to the award-winning projects to help incubate more outputs; On the other hand, it will help the unit to improve its visibility in the country and recruit more high-quality research teams, thereby improving the construction of scientific research teams.

2.2 Evidence from power enterprises

For the power industry, on the one hand, the national science and technology awards have a chain effect, and winning an award enhances the recognition and influence of the project, which is conducive to continuing to receive awards at other levels; The award highlights the positive side of the allocation of innovative resources, promotes the emergence of relevant knowledge achievements, and guides the public to pay more attention to the protection of intellectual property rights, and further develops orderly market competition; the award-winning project directly improves the development and continuity of the organisation, and expands its influence; it indirectly promotes the extension of relevant disciplines to universities and research institutes[8]. On the other hand, at the team level, the national science and technology award "planted plane trees and attracted golden phoenixes", relying on science and technology awards to cultivate a group of national leading talents; Guide all sectors of society to actively build a group of innovative teams, and cultivate more high-quality technical and skilled personnel, skilled craftsmen, and craftsmen from major countries; It is also an effective guide to gather global brains; Meanwhile, It also has a positive impact on the increased stability of the project team and the level of scientific and technical output of the team.

Finally, the most direct rewarding dividend that science and technology awards bring to enterprises is more funding, which can lead to more economic benefits through the transformation of results, as science and technology awards increase the visibility of the project and the likelihood of obtaining more project opportunities and co-operation opportunities. As far as

Taking power industry for instance, National Technological Award is positive to promote the innovation and promotion of scientific research achievements, which has a very good impact on the industrial development of the power industry, as well as achieving independent innovation ,independent production and batch production, promoting the whole power industry innovation level, help the electric power industry and related industrial structure adjustment and upgrading, promote the better development of the electric power equipment manufacturing industry, drive the electric power industry upstream and downstream related industries to develop together. It has injected a stronger impetus to the rapid development of the economy and society, showing extremely strong economic benefits[9].

On the other hand, from the perspective of motivation to innovate, a firm's motivation to innovate consists of three main assumptions:"maximum profit", "customer creation" and "social responsibility". Taking the power industry for example, the program honoured with awards are more in line with the strategic needs of the country, such projects are generally centred on the relationship between the people's livelihood and national security of the major technical issues for scientific research, scientific research results speech directly support the implementation of the national strategy, a strong guarantee of the country's energy security[11]. Besides, power projects honoured with awards still play a dominant role in fostering new industries, gradually opening up the upstream and downstream of the industrial chain, driving equipment manufacturing to the international market, and playing a significant part in the implementation of the "One Belt, One Road", safeguarding China's energy safety, high-quality development, scientific and technological innovation, breaking neck problems, and revitalisation of rural areas, and other major development strategies related to the lifeblood of the country's national security and the national economy.

3 Guiding functional model of scientific and technological rewards

The orienting role of technological awards is to guide scientific researchers to carry out innovative activities oriented to national needs and to solve hot, difficult and key problems related to economic construction, social development and national security. The state formulates development plans (such as the Outline of the Fourteenth Five-Year Plan for the National Economic and Social Development of the People's Republic of China and the Outline of the National Medium- and Long-Term Scientific and Technological Development Plan (2021-2035)), which specify the key development areas, main objectives and deployment of major projects during the current planning period;;; on this basis, the main bodies of innovation in various industries and sectors formulate development plans in accordance with their own development status[12] On this basis, innovation subjects in various industries and sectors will formulate development plans according to their own development conditions, and specify the development goals and key tasks of their own industries and units in the current planning period; scientific researchers will carry out corresponding innovation activities in accordance with the development plans of the State, industries and innovation subjects, and in the light of their own actual conditions, including the declaration of new scientific research projects, and the reasonable selection of scientific research projects for the transformation of achievements[14]. After the end of the current planning period, the state, industry and innovation main body will formulate the development plan for the next planning period according to the current development situation at home and abroad, taking into account the fulfillment of the development goals.

The guiding role of science and technology awards is essentially a regulating role. After the introduction of the national development plan, the reward orientation and selection criteria of science and technology awards should be consistent with the key areas and main objectives of the national development plan. In the process of innovation subjects formulating their own development plans according to the national development plan and industry development plan, and in the process of scientific researchers carrying out innovative activities according to the development plans of the country, industry and innovation subjects, science and technology awards play a positive regulating role.

Scientific and technological awards should be closely integrated with national development plans. Commodity prices are determined by the supply and demand of commodities, when supply exceeds demand, the demand for commodities is large and the supply is insufficient, the price of commodities will rise; when there is an oversupply of commodities, the demand for commodities is small and the supply is too much, the price of commodities will fall. The supply and demand of commodities is determined by the stage of socio-economic development, and the demand for commodities is different at different stages of socio-economic development, which leads to fluctuations in commodity prices. Similarly, at different stages of socio-economic development, there are differences in the level of demand for and supply of different industries and technologies, and industries and technological fields with high demand but insufficient supply will become key areas[10]. By analysing the current level of national economic and scientific and technological development, the national development plan specifies the key industries and technological fields for the coming period, leading the direction of national economic development and technology innovation. In terms of scientific and technological awards, the Regulations on National Science and Technology Awards clearly stipulate that the positioning of awards and selection criteria for scientific and technological awards should be closely integrated with the major strategic needs of the country and the medium- and long-term scientific and technological development plan, so as to guide the gathering of innovative resources and innovative talents in key areas and to promote the rapid development of China's economy and society.For example, new awards can be set up to increase the number of awards in key industries and key technology fields, highlighting the outstanding contribution of scientific and technological innovation achievements to socio-economic development, and so on, so as to guide the advantageous scientific and technological resources to gather in key scientific and technological fields and strategic emerging industries.;For non-key industries and backward technology fields, the negative effects can be suppressed by reducing the number of awards, so as to realize a good interaction between S&T innovation and social and economic development. Therefore, the positioning of S&T awards is in line with the national development plan.

Scientific and technological awards can promote the development plans of innovative subjects and the innovative activities of researchers. Reputation theory holds that organizations have the need to pursue a good reputation, strive for social recognition, obtain resources, opportunities and support, and then accomplish value creation. Maslow's Hierarchy of Needs Theory holds that human needs are divided into five types from low to high like a ladder, i.e. physiological needs, safety needs, social needs, respect needs and self-actualization needs. Only after satisfying the low-level needs can we pursue the high-level needs[13]. Therefore, as high-level talents, researchers will have a strong need to pursue self-actualization on the basis of satisfying physiological needs, safety needs, social needs and respect needs. As science and technology awards are in line with national development plans and industry development plans, science and technology awards are an institutional arrangement for recognizing the innovative achievements of scientific researchers, which can satisfy both the pursuit of organizational reputation by innovation subjects and the need for self-realization by scientific researchers. Therefore, whether it is the implementation of national and industrial development planning or the pursuit of organizational reputation, the innovation subject should actively formulate development planning, clarify the strategic objectives and work deployment in the coming period, so that its national and industrial development objectives are consistent with the positioning of scientific and technological rewards; whether it is to strive for scientific research funding, improve scientific research, or declare scientific and technological rewards, and pursue the realization of value, the scientific researchers should actively adjust the innovation Direction, in the key areas determined by the national development plan, in accordance with the industry development plan and the scientific and technological innovation tasks determined by the development plan of the main body of innovation, actively declare scientific research projects, rationally select the results of the transformation projects, and strive to achieve significant social and economic benefits, and to meet the selection criteria for scientific and technological awards.

4 Conclusions

In summary, the reward positioning of science and technology awards is consistent with the national development plan, and scientific and technological innovation activities in key areas are more in line with the needs of social and economic development at this stage, and it is easier to achieve significant economic or social benefits, and meet the selection criteria of science and technology awards. Therefore, the guiding role of science and technology rewards is reflected in the fact that it can promote the formulation of the development plan of innovation subjects, promote the adjustment of the innovation direction of scientific researchers, and thus promote the gathering of superior scientific and technological resources and scientific and technological talents in key technical fields and industrial directions.

Acknowledgments

This work was supported by the science and technology project of State Grid Corporation of China, "Research on promotion path, power transitive model and empirical research of Energy Internet firms' innovation capability" (1400-202157233A-0-0-00).

References

1. Haidong C, Zineng K E. Comparison Between Chinese and American Non-governmental Science and Technology Awards[J]. Science & Technology Review, 2007.
2. Wan-Qiang X U, Xiao-Gang X, Yue L I. A Study on the Will of Science Researchers to Apply for Science and Technology Awards Managed by Non-Government Bodies[J]. China Soft Science, 2010.
3. Yang Z. Research on Evaluation Mechanism of National Science and Technology Awards[J]. Evaluation & Management, 2007.
4. Tao H, Fang L. An Exploration on the Common Features of the Winners of the State Science and Technology Awards[J]. Studies in Dialectics of Nature, 2011.
5. Yu X, Zhao Y Q, Shi C S. System dynamics modelling for promoting technological innovation via science and technology awards of national defence[J]. Transactions of Beijing Institute of Technology, 2010, 30(12):1505-1504.
6. Xian-Fei M, Zheng-Feng L I, Technology and Society, et al. Research on national science and technology awards evaluation process from the perspective of actor network theory[J]. Studies in Science of Science, 2016.
7. Liao S. Research on the Development of Non-governmental Science and Technology Awards of China: Environment, Problems and Countermeasures[J]. International Journal of

Economics, Finance and Management Sciences, 2015, 4(1):1-9.

8. Ting X. A Comparative Study of Chinese and International Science and Technology Awards: An Empirical Analysis Based on Chinese and International Science and Technology Awards[J]. Global Science, Technology and Economy Outlook, 2019.
9. Chuang-Mu Z, Tian-Jin C, Lei W U, et al. Food Science Research Progress Based on State Science and Technology Awards from 2000 to 2016[J]. Food and Nutrition in China, 2017.
10. Liao S, Zhang Y, Wei J. Research on the Development of Non-governmental Science and Technology Awards of China: Environment, Problems and Countermeasures[J]. International Journal of Economics, Finance and Management Sciences, 2015.
11. Shanghai Municipality Science and Technique Awards Management Office Bao Yu. The Science and Technology Awards Information System[J].Electrical Automation, 2005.
12. Zhi-Cong S, Su-Qin T, Hui Y. A Survey on the Non-government Awards of Science and Technology in China[J].Journal of Dialectics of Nature, 2009.
13. Bent F .From Nobel Prize to Project Management: Getting Risks Right[J].Project Management Journal, 2006, 37(3):5-15.
14. Cañedo Andalia,Rubén.The Centre for the Research and Rehabilitation of Hereditary Ataxias Received the National Prize of the Cuban Science Academy[J].Arkivoc, 2002, 2002(5):92-101.

Supply chain integration, circulation industry development and unified large market---From the perspective of Internet popularization

Xiangyi Meng*

Shandong Agricultural University
No. 67 Daizong Street, Taiqian Street, Taishan District, Tai 'an City, Shandong Province, China, 271018

Abstract: The construction of a unified big market is the inevitable requirement of China's continuous deepening of reform, which plays an important role in eliminating the urban-rural gap and promoting the domestic and international double cycle. The key to building a large, unified national market is to promote high-standard connectivity of market facilities, continue to deepen and popularize through the Internet, combine supply chain integration with the development of the circulation industry, reduce institutional transaction costs and explicit costs such as raw materials, labor and logistics, break local protection, curb market segmentation, and update infrastructure. In this way, market transaction costs can be reduced, scale advantages can be leveraged and resource allocation can be optimized.

1. China's unified large market

On March 25, 2022, the Central Committee of the Communist Party of China and The State Council issued the Opinions on Accelerating the Construction of a Large and Unified National Market, which emphasized the need to continue to promote the efficient and smooth expansion of the domestic market and the need to promote high-standard connectivity of market facilities, including the construction of a modern circulation network, improving market information exchange channels and promoting the optimization and upgrading of trading platforms. The central idea of a unified Grand market is that in a regional market, goods, services or elements should not be divided according to their source, and that subjects within the market have the same rights, responsibilities and obligations[1].

1.1 Building a unified large market is an inherent requirement for socialist common prosperity

Common prosperity is the goal of the CPC, the essential requirement of socialist society, and an important feature of Chinese path to modernization.For a long time, China's economic development has been unbalanced between the eastern and western development, unbalanced urban-rural development, and unbalanced resident income. For example, in 2022, the per capita disposable income of rural residents reached 20133 yuan, a real increase of 4.2%, and the per capita income ratio of urban and rural residents was 2.45;In 2022, the per capita GDP of 7 provinces and cities along the eastern coast is over 100000 yuan, the per capita GDP of 8 provinces in the central region is between 70000 and 100000 yuan, and the per capita GDP of 9 provinces and cities in the western region is less than 70000 yuan. The building of a large unified national market will benefit more low-income people, stimulate development momentum in rural areas, further remove barriers to the free flow of goods, and firmly promote common prosperity.

1.2 Building a unified large market is an important basis for the "double cycle"

After the novel coronavirus epidemic, the world economy is in a bit of a slump, the tide of anti-globalization is rising, and the driving effect of China's net exports on the economy is slowing down, and the domestic market needs to play a bigger role in economic growth. The national unified big market can expand domestic demand and effectively increase the consumption potential of the people, which is the top priority of the new development pattern.Local protection, market segmentation, oligopoly, and urban-rural disparities are issues that arise from both local governments and enterprises, as well as imperfect market mechanisms. If these problems are allowed to exist, a unified market cannot be established, and the "internal circulation" cannot be smooth, thus laying the foundation for the "dual circulation" development pattern.

1.3 The popularization of the Internet is a new advantage of the unified large market

According to the 51st Statistical Report on the Development of the Internet in China, as of December

*Corresponding author: Xiangyi Meng, E-mail:2060524097@qq.com

2022, the number of Internet users in China was 1.067 billion, and the Internet penetration rate reached 75.6%, an increase of 2.6 percentage points year-on-year, especially in rural areas, the Internet penetration rate increased to 61.9%, further narrowing the gap with urban and rural areas. Data from the National Bureau of Statistics show that from 2012 to 2021, the scale of China's digital economy grew from 11 trillion yuan to more than 45 trillion yuan, the proportion of digital economy in GDP increased from 21.6% to 39.8%, and the popularization of Internet applications accelerated, deepening the application in the circulation industry, supply chain integration and other fields. The rapid development of emerging technologies such as 5G, intelligent manufacturing, and artificial intelligence has injected new vitality into technological innovation and high-quality economic development in all walks of life.

2. Supply chain integration is the micro basis of the national unified large market

Supply chain integration refers to the process of a company working with its supply chain partners to manage the supply process within and between its organizations, so as to more effectively integrate raw materials, semi-finished products, information and financing for production.

First, supply chain integration can reasonably allocate supply and demand resources. Supply chain integration can calculate the optimal cost and benefit plan for enterprises, and realize the value improvement of the whole supply chain through the collaborative integration of upstream and downstream industries. Take a manufacturing enterprise as an example, through close cooperation with suppliers, it organizes raw material assembly according to order scheduling, and allocates raw materials directly and accurately to the production line to maximize storage pressure, and in the same production line, supply chain integration can reach more than twice the output in the past[2].

Second, supply chain integration and optimization of resource allocation. The survey shows that supply chain integration through changing production plans, optimizing procurement processes, diversifying procurement, warehousing, order, inventory management and transportation management, Achieve improvements in cost, quality, diversity and service levels[3]. Taking Jingdong Group as an example, as of December 31, 2022, Jingdong Logistics operated more than 1,500 warehouses, adding more than 700 warehouses in the past three years, and the total area of the warehouse network, including the cloud warehouse area managed by Jingdong Logistics, has exceeded 30 million square meters. The number of employees of Jingdong system has increased from 260000 at the beginning of 2020 to more than 550000 at the end of 2022, and 290000 new employees have been added. The huge supply system makes Jingdong achieve "same-day" and "next-day" in major cities, significantly improving the competitiveness of Jingdong Mall[4].

Third, the popularization of the Internet strengthens the integration of supply chains. The Internet has a key role to play in advancing the goal of supply chain integration, and access to the Internet can play an important role throughout the manufacturing lifecycle, including services such as design analysis and simulation, suppliers, manufacturers, maintenance, and quality testing. Moreover, the Internet is capable of redefining back-end operations, from product design and development, procurement, production, inventory, distribution, after-sales service support, and even marketing.Through the Internet, these processes are not only able to connect, but more importantly, they can work together to form an efficient and flexible innovation system, achieve unified procurement channels, establish centralized warehousing, unified distribution systems, implement supplier management, etc., continuously optimize business processes, achieve close collaboration between upstream and downstream enterprises, and become one of the core competitiveness of enterprises.

By adopting an e-commerce approach to supply chain integration, businesses can realize significant returns through increased efficiency, improved asset utilization, faster time to market, reduced total order fulfillment time, enhanced customer service and responsiveness, penetration of new markets, improved return on assets, and ultimately increased shareholder value.

Fourth, unified large market and supply chain integration promote each other. The formation of a unified large market can form diversified consumer demand, which forces the supply chain to carry out product style reform, customized production, and shorten the innovation time. With the development of digital technology, especially the continuous deepening and popularization of the Internet, the members of the supply chain can optimize the allocation of resources and improve the effective utilization of resources through efficient information transmission. Taking BYD Automobile as an example, in the transformation of individual consumer demand in the market, it makes use of its own vertical supply chain system, and through independent research and development of core components such as batteries, motors, electric controls, etc., its products cover from A-class cars to D-class cars, and from MPV (multi-Purpose Vehicles). utility vehicle) to sport utility vehicle (SUV)[5-6].

3. The development of the circulation industry is the inevitable requirement of the national unified large market

Generally speaking, the circulation industry refers to the intermediate link from production to consumption, usually referring to the wholesale industry, retail, catering and logistics industry four major circulation branches. In 2022, the total amount of social logistics in China reached 347.6 trillion yuan, an increase of 3.4 percent year-on-year based on comparable prices. The total cost of social logistics was 17.8 trillion yuan, up 4.4 percent year on year[7]. The ratio of total social logistics expenses to GDP was 14.7%, 0.1 percentage points higher than the previous year.

First, the circulation industry promotes economic transformation and upgrading. The circulation industry uses modern technology to improve efficiency, such as improving storage technology, optimizing resource flow, and reducing costs. In the process of China's economic development, social logistics has been constantly upgrading the value chain to promote economic transformation and upgrading. The total social logistics increased from 197.8 trillion yuan in 2013 to 347.6 trillion yuan in 2022, a total increase of 75.73%, and the total social logistics cost increased from 10.2 trillion yuan in 2014 to 17.8 in 2022, a total increase of 74.51%[8].

Second, the popularization of the Internet is a new trend in the development of circulation industry. The popularization and deepening of the Internet has brought about fundamental changes in the logistics industry. The market scale of e-commerce, mobile payment, online ride-hailing, and mobile takeout has continued to expand. In areas with higher concentration of the logistics industry and resource endowment, the efficiency of Internet popularization has become more obvious. At the same time, under the background of the deepening popularization of the Internet, it has promoted the rapid development of logistics distribution and information dissemination. Data from the National Bureau of Statistics show that in 2022, online retail sales in China reached 13.79 trillion yuan, an increase of 4.0% over the previous year on a comparable basis. Online retail sales of physical goods reached 11.96 trillion yuan, an increase of 6.2% over the previous year on a comparable basis, accounting for 27.2% of total retail sales of consumer goods[9].

Third, the circulation industry narrowed the gap between urban and rural areas. The continuous development of the circulation industry can accelerate the docking of production and consumption, promote the sustainability of consumption, promote the formation of rural industrial clusters and the innovation of agricultural industrial structure, and connect the logistics network between urban and rural areas more closely with the logistics hubs between various regions. On the one hand, the expansion of large supermarkets and convenience store chains has driven consumer demand in rural areas, and on the other hand, it has also provided channels for agricultural products to urban areas. By 2022, China will increase the storage capacity of agricultural products by 10.8 million tons, build 1500 county-level logistics and delivery distribution centers, 7600 township express delivery and mail processing stations, and 95% of administrative villages will achieve direct express delivery[10].

4. Policy suggestions

The construction of a unified big market is a global, gradual, and leapfrog process, which is a process of quantitative change leading to qualitative change[11]. It can take five key breakthroughs: internet popularization, government support and guidance, promoting technological innovation, supply chain finance, and rural revitalization, and lead the construction of a unified big market nationwide through a point to area approach.

First, vigorously promote the construction of digital platforms, give full play to the positive role of Internet popularization, and effectively improve the scale effect of supply chain integration and the synergy effect of circulation industry development. Build an efficient circulation system, reduce barriers to the circulation of factors between urban and rural areas and between regions, and take differentiated measures according to local conditions. In 2022, the proportion of online retail sales in the east, central, west and northeast regions in the country will be 83.92%, 8.87%, 5.73% and 1.48%, respectively, and there is still considerable room for Internet popularization.

Second, increase government public budget expenditure and promote the construction of supporting systems and facilities such as transportation and industrial standards[12]. The spread of the Internet has also improved the efficiency of government services. According to the United Nations e-Government survey report, China's e-government service Index ranking has risen from 78th in the world in 2012 to 9th at present. Relying on the policy dividend of building a national unified large market, break the urban-rural division and local protection of the development of the circulation industry, and promote the digitalization, intelligence and modernization of the circulation industry to improve the economic quality.

Third, adhere to technological innovation and improve the efficiency of technology landing. Establish a supply chain science and innovation center to provide supply chain solutions to more enterprises in a market-oriented way. We will support enterprises to increase investment in innovation and research and development for their own supply chain technologies, protect intellectual property rights, and expand spillover effects. We will jointly tackle key problems with enterprises, universities and research institutes, and promote cooperation in major technology R&D among research institutions and upstream and downstream of the industrial chain.

Fourth, do a good job in supply chain finance. Supply chain finance is a financing activity provided by financial institutions with banks as the main body for core enterprises, logistics enterprises and upstream and downstream supporting enterprises, and is an integrated solution of supply chain capital flow management. Taking the opportunity of building a unified large market, explore and improve the online and digital level of supply chain financing settlement, increase support for core enterprises and logistics enterprises, and improve the financing efficiency of small, medium-sized and micro enterprises' accounts receivable.

Fifth, strengthen the construction of rural logistics infrastructure and narrow the consumption gap between urban and rural areas[13]. Further strengthen the construction of logistics and communication infrastructure such as roads, transportation, warehousing, mobile base stations, optical cables, etc., increase support for rural agricultural farmers, promote the digital transformation of rural circulation industry, improve agricultural production efficiency, increase farmers' income, and expand the consumption demand of rural residents with the modernization, digitization, and intelligence of rural circulation industry[14]. Create a good cycle between rural revitalization and the construction of a unified market.

Reference

1. Lloyd P J. What is a Single Market?: An Application to the Case of ASEAN[J]. ASEAN Economic Bulletin, 2005, 22(3): 251-265. https://search.ebscohost.com/login.aspx?direct=true&db=buh&AN=34932975&site=ehost-live
2. Awad D. Supply chain integration: definition and challenges[J]. Management and Technology (IJIMT, 2010, 1(1). https://www.zhangqiaokeyan.com/academic-conference-foreign_meeting-270528_thesis/0705016727553.html
3. Smart A. e-Business and supply chain integration[J].Journal of Enterprise Information Management, 2013, 21(3):227-246. DOI:10.1108/17410390810866619.
4. Ciorciari J D. Chiang Mai Initiative Multilateralization: International Politics and Institution-Building in Asia[J].Asian Survey, 2013, 51(5):926-952. DOI:10.1525/as.2011.51.5.926.
5. Tiwari S. Supply chain integration and Industry 4.0: a systematic literature review[J]. Benchmarking: An International Journal, 2021, 28(3): 990-1030. DOI:10.1108/BIJ-08-2020-0428.
6. Swatman P M C, Swatman P A. Europe's Single Unified Market - 1992 - A Spur To The Development Of Electronic Data Interchange?[J]. Adjunct proceedings of european symposium on ambient intelligence, 2000. DOI:10.1145/1031419.1031437.
7. Nguyen N T, Nguyen D T, Tran T Q,et al. Assessing the change of water quality and quantity in the upper basin of Thac Ba reservoir under the impacts of future land-use scenarios[J]. Modeling Earth Systems and Environment, 2023, 9(3):3709-3720. DOI:10.1007/s40808-023-01732-3.
8. Hanaoka S, Matsumoto K, Kitazawa T, et al. Cost of illness of the dementia in Japan[J].The Journal of Japan Society for Health Care Management, 2016, 17(1):8-13. DOI:10.11191/jhm.17.8.
9. Polo M, Pino F, Delgado J C A, et al. Assessment of a NaIL detector performance for radiation monitoring applications[J]. The European Physical Journal Special Topics, 2023, 232(10):1477-1486. DOI:10.1140/epjs/s11734-023-00896-4.
10. Wang Y, Liu J, Wang J,et al. Exploring the effect of city size on carbon emissions: Evidence from 259 prefecture-level cities in China[J].Environmental Science and Pollution Research, 2023, 30(36):86165-86177. DOI:10.1007/s11356-023-28564-z.
11. Trandafirescu B C. The process for the construction of a uniform framework of contracts in the European Union[C]. Perspective ale dreptului afacerilor în mileniul al treilea, ASE Bucure ş ti, 2013. DOI:10.13140/2.1.3590.3368.
12. Wu X, Liang H. Government procurement issues and supporting role of standardization in China[C]//International Conference on Information Management. IEEE, 2017:262-265. DOI:10.1109/INFOMAN.2017.7950388.
13. Shuanjun W, Guizhen S, Yaqing L. Analysis of Rural Informationizion Strategy in China[J]. Chinese Agricultural Science Bulletin, 2009, 4:177-178. DOI:10.2307/1005094.
14. Kelum W A, Bandara S. Methodology for Selecting Suitable Eco-friendly Traces for Road Railway Planning and Development that Minimizes Impacts due to Natural Disasters[J]. Proceedings of International Forestry & Environment Symposium, 2014, 18. DOI:10.31357/fesympo.v18i0.1991.

Research on The Construction Strategy of Heilongjiang Digital Agriculture Pilot Area

Yingying Sun*

School of Economics and Management,Suihua University , Suihua152061, China

Abstract. Digital agriculture is the organic integration of digital information technology and agricultural products. Using big data technology and artificial intelligence technology, it can effectively improve the output, efficiency and economic development of agricultural products. At present, there have been some important progress in the observation of digital agriculture and remote sensing research of agricultural problems. However, the framework and technical system of digital agriculture have not yet been formed, and digital agricultural products in most regions have not been widely used. This paper will focus on the current development situation and development strategy of Heilongjiang province are analyzed in detail, and according to the actual situation of Heilongjiang province, its development direction is discussed, and finally gives the construction of the province 's digital agriculture first demonstration area.

1.INTRODUCTION

The conflict between population growth and resource scarcity is intensifying, especially in agriculture, which is in the middle of population and resources and has the closest relationship with the two. With the rapid development of modern digital technology, 'digital agriculture' has emerged in China's rapidly developing digital economy. Applying digital technology to agriculture is one of the driving forces for making decisive progress in rural revitalization and basically realizing agricultural and rural modernization. Digital agriculture is the organic integration of digital information technology and agricultural products[1]. Using big data technology and artificial intelligence technology, it can effectively improve the output, efficiency and economic development of agricultural products. In the 1950s, there was a concept of digital farm abroad, and computer technology was also used in it. In addition, by using the new standard networking technology, a new stage of three digital agricultural informatization is realized[2]. The development of digital agriculture in China is lagging behind. In the early stage, it mainly relied on government policy and financial support. By 2017, the development of digital agriculture has entered the digital stage[3,4,5]. Although there have been some important progress in China 's agricultural problems, China 's digital technology and framework have not yet been formed, and most regions have not yet been fully realized. Therefore, this article will focus on the advantages and disadvantages of Heilongjiang Province, and its development strategy is prospected. According to the reality of Heilongjiang province, this paper mainly discusses four problems : what to develop, what to develop leading industry, how to develop, expand and develop market. Finally, this paper makes a preliminary discussion on the construction of the first demonstration area of digital agriculture in the whole province, so as to improve the economic benefits and output level of agriculture in the whole province, the effectiveness of work and the consumption of labor force[6].

2.DEVELOPMENT STATUS OF DIGITAL AGRICULTURE IN HEILONGJIANG PROVINCE

2.1 The hardware configuration status of agricultural resources in Heilongjiang Province

Heilongjiang Province is located in the eastern part of Eurasia, the west coast of the Pacific Ocean, and northeast China[7,8]. Northeast China is one of the four black soil areas, which plays an important role in protecting regional ecological environment, food security, promoting food security and promoting sustainable development of China. However, with the high level of use in China for many years, the black soil production in China is decreasing year by year[9,10]. The content of soil organic matter decreased year by year. The soil surface becomes shallow and hard. The coordination of soil moisture, fertilizer, air and heat is obviously weakened. Continuous plough layer soil showed acidic tendency. At present, based on the urgent need of black soil degradation, restoration, utilization and protection in China, the protection and management of black soil must be accelerated.

* Corresponding author:sunyingying2019@126.com

2.2 The current situation of digital technology platform construction in Heilongjiang

The agricultural information platform of Heilongjiang Province has been initially used in all aspects of agricultural production, and has explored high-level agricultural Internet of Things platform, agricultural production trusteeship service platform, and agricultural machinery dispatching command platform in agricultural production services. China has built the first set of agricultural machinery production scheduling command system, using GPS and GPRS dual mode, to meet the monitoring requirements of various field operations. BeidahuangSanjiangQixing Company has built the first demonstration project of farmland Internet of Things in China, which adopts three modes of automatic collection, artificial intelligence input and continuous upgrading of management software in production. Under the support of digital technology, the agricultural scientific research institutions with Heilongjiang Academy of Agricultural Sciences as the core have 11 national field observation experimental stations, 12 Internet of Things monitoring platforms, 8 professional data management platforms, and 1 comprehensive meteorological station. It has accumulated 42 years of soil fertility, 50 years of soil sample data, 21 years of image data of staple crop area, growth and disaster in the province, and 6 years of monitoring and verification data of cultivated land fallow rotation and high standard farmland construction pilot in the province. These accumulations provide technical and data support for digital agriculture.

2.3. The current situation of digital technology infrastructure construction in Heilongjiang province

On the basis of digital technology, the existing agricultural production base and agricultural production base in Heilongjiang Province have been basically completed. During the 'Tenth Five-Year Plan' period of the agricultural reclamation area, 230 million yuan was invested in informatization, and the farms of Baoquanling, Hongxinglong, Suihua, Jiusan, Beian and Qiqihar were built. The expansion and expansion of the optical fiber and switching system of the production team were completed. A total of 12,000 fiber optic cables and more than 460,000 switches were built. And it has built 13 wide area networks and 21 local area networks, opened the 'Beidahuang Information Port Station', with tens of thousands of computers and more than 20,000 broadband users. At present, the power of agricultural machinery in this area reaches 433,000 kilowatts, and the mechanical power of per capita arable land is 2 kilowatts. At present, there are 26,775 large, medium and small tractors, 4,166 100 tractors, 31 agricultural machinery 31 tractors, and the comprehensive mechanization rate is 92 %. A variety of field operations have realized aircraft operations, which is the leading level in the country.

3.ANALYSIS OF THE PROBLEMS EXISTING IN THE DEVELOPMENT OF DIGITAL AGRICULTURE IN HEILONGJIANG PROVINCE

At present, the basic conditions of digital agriculture in Heilongjiang Province have been significantly improved, and the application of digital technologies such as remote sensing, big data and Internet of Things in agriculture has gradually deepened. With the increasing abundance of data resources in agricultural production, operation and management, the gradual deepening of digital industrialization, new industries and new formats are emerging. However, the construction of agricultural demonstration areas in Heilongjiang Province is still lagging behind on the whole, and there are many problems.

3.1 Lack of sufficient digital technology platform

At present, although there is a considerable scale of agricultural big data in Heilongjiang Province, due to the high data barriers among government departments, scientific research units, enterprises and the three parties, and the lagging development of relevant laws and regulations, the lack of public platforms and sharing channels and other factors, agricultural data are mostly scattered in different departments and fields, resulting in the information island formed by the division of data departments, which makes the agricultural big data resources in the era of big data unable to be shared.

3.2 Lack of high-level talents and high technology

In the digital farms developed in Heilongjiang Province, there are a large number of ' small data ' and ' limited samples ', which cannot be accurately managed. Although China 's use of big data is not inferior to the international, but in the basic technology can not be in line with international standards[11]. A large number of data storage, data cleaning, data analysis and mining technologies, due to the lack of research on its process combing, index monitoring, problem diagnosis and evaluation, have led to inefficient data-driven decision-making.

3.3 Insufficient infrastructure construction

At present, the digital technology infrastructure in Heilongjiang Province is still in its infancy, such as efficient and accurate agricultural machinery, Beidou foundation enhanced base station, high-precision operation of agricultural machinery and precise navigation system. And the lack of agricultural

https://doi.org/10.1051/shsconf/202317803023

production process management system, including paddy field precision finishing and operation system, sowing monitoring system, pest control system. Data processing, the operating system has problems such as inefficiency.

4. FEASIBILITY ANALYSIS OF DIGITAL AGRICULTURE PILOT ZONE CONSTRUCTION IN HEILONGJIANG PROVINCE

4.1 The site selection level of digital agriculture pilot area test base

Digital agriculture is a comprehensive utilization of modern science and technology, with huge investment, huge risk, but also requires a better experimental and development environment[12,13]. Select areas with good scientific research environment, good farmland construction, and certain agricultural information construction to carry out national rural digital demonstration pilot work. Carry out scientific research on digital agriculture at all levels, apply existing scientific and technological achievements and theoretical results to production practice in time, and improve and accelerate research and development. Heilongjiang has a vast territory and can be used as a typical pilot unit to promote the digital development of Heilongjiang Province through pilot projects and build the requirements of ' digital first demonstration area.

4.2 economic aspects

The connection function of the Internet and the effective configuration of the big data platform can improve the efficiency of agricultural production, bring about the effect of increasing marginal returns, and lead to the exponential growth of output[14]. The construction of digital agriculture pilot area is conducive to giving full play to the economic effect of digital technology in agricultural production, accelerating the construction of modern agricultural industrial system, production system and management system, making up for the shortcomings of rural development, and empowering rural revitalization and development. The establishment of digital agriculture pilot demonstration area has low economic requirements and less economic burden on Heilongjiang Province.

4.3 Technology aspects

Digital technology has expanded the business scope of traditional industries, promoted the interconnection, intelligent control, precise operation and scientific development of various production factors, and promoted the steady growth of agricultural products in China. Digital technology provides enterprises with information, intelligent and intelligent management means, which makes the traditional means of production more intensive, from extensive to digital, intelligent and intensive[15]. The ecological transformation of Heilongjiang can reduce the uncertainty of agriculture, reduce the consumption of resources and reduce the burden on the environment. For remote sensing technology, GIS, satellite positioning system and other high-end technology has not yet reached effective integration, can not achieve the intelligent control of information systems, precise operation and scientific comprehensive needs.

5. HEILONGJIANG PROVINCE DIGITAL AGRICULTURE PILOT AREA CONSTRUCTION STRATEGY FORMULATION

5.1 Development Planning Idea of Digital Agriculture Pilot Area in Heilongjiang Province

The establishment of ' digital agriculture first demonstration area ', in order to develop modern agriculture and cultivate intelligent agriculture as the goal, vigorously support the training and introduction of scientific and technological personnel, promote the deep combination of ' agriculture ' and ' digitization ', in order to improve agricultural labor productivity, resource utilization and cultivated land output as the main purpose, to build a precise input, personalized service as the characteristics of agricultural production technology system, service system and market system, to achieve agricultural intelligent perception, intelligent control, independent operation, smooth digital agriculture development new model.

The construction of digital agriculture demonstration area is a comprehensive project. First of all, under the guidance of the development of digital agriculture in Heilongjiang Province, on the basis of the digital agricultural technology system, the key technology of great significance is selected as the breakthrough point[16]. On the basis of unified technical specifications, we should concentrate our efforts and work together to jointly develop the core technology of digital agriculture, establish the digital agricultural technology system of Heilongjiang Province, and provide support for the development of national economy. Secondly, to build a digital pilot zone of " increasing the income of workers and increasing the efficiency of enterprises, " we should highlight the characteristics and advantages of " digital agriculture first zone, " and emphasize the policy of " based on reality, highlighting key points, overall coordination, standardization, improving institutions, sharing, training, promotion, promotion, demonstration and stability. " Third, we should base on high-level development, attach importance to the introduction and absorption of foreign advanced technology, selectively absorb foreign scientific and technological achievements, and achieve ' information leap '.

5.2 Heilongjiang province digital agriculture pilot area policy support strategy

In August 2019, Heilongjiang Provincial Department of Agriculture, Heilongjiang Provincial Agricultural Investment Group and Heilongjiang Branch of China Construction Bank signed the " Strategic Cooperation Agreement for Heilongjiang Digital Agricultural Comprehensive Service System " in Harbin. The purpose of the agreement is to further promote the deep integration of Internet, big data, artificial intelligence and other technologies with the agricultural field, promote the digital, networked and intelligent transformation of Heilongjiang agriculture, and lay a solid foundation for the new format of digital agriculture in Heilongjiang Province, the establishment of a national digital agriculture demonstration area and the northern portal of the digital Silk Road.

At present, Heilongjiang Province has fully implemented the service system of digital rural integration, but in the support of digital technology and the construction of infrastructure, it is necessary to strengthen the popularization of data network, strengthen the training of infrastructure and talents, strengthen financial investment, strengthen the management of agricultural technology and finance, and make full use of the support function of policy to agricultural scientific research. Support commercial banks to carry out agricultural science and technology projects, promote the transformation of technology and scientific and technological achievements, and actively explore new financing channels.

5.3 Infrastructure and digital technology platform planning

Consolidate and improve rural broadband projects, continue to deepen the reduction of fees and popularize telecommunications services, continue to expand the coverage of 5 G and optical fiber networks in rural areas, radiate network services to households, and greatly increase rural Internet penetration ; By grasping the application terminal, the focus of rural informatization was shifted to the field, the agricultural Internet was developed, and the promotion of 5 G in agriculture was actively promoted, which laid a material foundation for the development of digital agriculture in China.

Based on the agricultural data resource directory and data sharing and exchange system, the data model, algorithm model and scheme model are refined to provide meaningful data assets, services, model services and knowledge experience for agricultural data[17]. The data processing of the agricultural industry is carried out by means of API, and it is integrated with the actual situation of the agricultural industry. Big data and artificial intelligence technology are used for data processing, and intelligent planning and design are carried out for agricultural production, management, management, service and other links, so as to establish a unified, reliable, practical, technologically advanced and safe digital technology platform.

6. CONCLUSION

Digital agriculture is the frontier field of agricultural development in the 21 st century. Heilongjiang Province has the basis for the development of digital agriculture, but it is still in its infancy in terms of technology and personnel training. This is far from the needs of digital agriculture development. Heilongjiang Province must formulate a development strategy in line with the development of digital agriculture according to the actual situation of Heilongjiang. In the specific implementation, it should be carried out as a whole, step by step, from point to surface, step by step ; avoid repeated waste and mistakes. In the early stage, technical specifications should be unified, with digital platforms, infrastructure, etc.as the main content. On this basis, the relevant application mode of digital agriculture is established, the key technologies of pilot digital agriculture are discussed in depth, and the integration and promotion are carried out, so as to form the first demonstration area of Heilongjiang digital agriculture.

Acknowledgments:

2021 Research Start-up Fund of Suihua University (SQ21006);

Key topics of the 14th Five-Year Plan for Education Science of Heilongjiang Province in 2022 (GJB1422388).

REFERENCES

1. Yuxia Song. Solving practical problems in the development of digital agriculture[J]. Financial Time. 2021(09).
2. Weijie Wang, Zilai Wang. 'Digital Engine ' : A New Picture of Yuan Liwono[J]. ChangDe Daily. 2021.(09).
3. Yong Sun. Fingertip management field, agricultural planting with new technology[J]. Shanxi Science and Technology Journal. 2021.(09).
4. Yanan Wei. Promoting the development of digital agriculture needs to strengthen the system concept[J]. Chinese information community. 2021.(08).
5. Regional industrial structure and agglomeration economies: An analysis of productivity in three manufacturing industries. Joshua Drucker;;Edward Feser.Regional Science and Urban Economics,2011
6. Xuecheng Lv. Digital Agriculture, Creating the Future[J]. Oriental urban and rural newspaper. 2021(07).

7. MinYu. Countermeasures for the Development of Digital Agriculture[J]. Agricultural Technology and Equipment. 2021(09).
8. Agricultural view : how does German digital agriculture do[N]. Agricultural Materials Guide. 2021.09.17(D03)
9. Liang Hu,Yongbo Liu,Lingshan Qing,Jiangyun Tang. Analysis of the construction and development mode of digital village[J]. Sichuan Agricultural Science and Technology.2021(09):68-71
10. Ecology and valuation: Big changes needed. Bryan G. Norton;;Douglas Noonan.Ecological Economics,2007
11. SenYu. Digital agriculture and three-dimensional planting, agriculture to play a new pattern[J]. Hunan Agriculture.2021(09):55.
12. Xiao Niu. Promote the digitization of agriculture and rural areas with ' Hema Fresh '[J]. Modern Communicative.2021(16).
13. Yanan Wei. Promoting the development of digital agriculture needs to strengthen the system concept[J]. Chinese information community.2021(04).
14. Lan Huang. On the improvement of farmers ' information literacy under the background of digital agriculture and rural construction[J]. Modern Rural Science and Technology. 2021(09):1-2.
15. Chengdu promotes digital agriculture and rural development to help rural revitalization[J]. Sichuan agriculture and agricultural machinery.2021(04):11.
16. Lu Wang. The transformation of agriculture from intelligence to digitization[N]. Agricultural Materials Guide.2021-08-13(D03).
17. Huaxi Li, Leontief, 1941,The Economic Structure of the United States,Harvard University Press,America.

AUTHOR INDEX